Advances in Polar Ecology

Volume 1

Series editor
Dieter Piepenburg
University of Kiel, Institute for Ecosystem Research, Kiel, Germany

Advances in Polar Ecology

In recent years, the polar regions have received increased scientific and public interest. Both the Arctic and Antarctic have been recognized as key regions in the regulation of the global climate and polar ecosystems have been identified to be particularly susceptible to the ongoing environmental changes. Consequently, the international efforts in polar research have been enhanced considerably, and a wealth of new findings is being produced at a growing rate by the international community of polar researchers. The aim of the book series Advances in Polar Ecology is to foster the progress in the scientific knowledge about the polar and sub-polar regions of both hemispheres by contributing to the fast and wide-ranging dissemination of novel scientific information gained from recent studies of sea, freshwater and land biota among polar researchers, environmental managers and policy makers. Advances in Polar Ecology's broad ecology-oriented scope encompasses environmental and biological research of both recent and past ecosystems. The Series offers an outlet to publish contributions (monographs, edited works, conference proceedings, etc.) addressing scientific topics that need more comprehensive and in-depth analysis than the length of typical journal articles allow for. These include, but are not limited to, thorough accounts of the current status of active research areas of wide importance that outline promising perspectives for future research. An international editorial board oversees the rigorous peer review that all contributions will be subjected to.

More information about this series at http://www.springer.com/series/10290

Volker Siegel
Editor

Biology and Ecology of Antarctic Krill

Springer

Editor
Volker Siegel
Thuenen Institute of Sea Fisheries
Hamburg, Germany

ISSN 2468-5712 ISSN 2468-5720 (electronic)
Advances in Polar Ecology
ISBN 978-3-319-29277-9 ISBN 978-3-319-29279-3 (eBook)
DOI 10.1007/978-3-319-29279-3

Library of Congress Control Number: 2016945250

Printed on acid-free paper

This Springer imprint is published by Springer Nature
The registered company is Springer International Publishing AG Switzerland

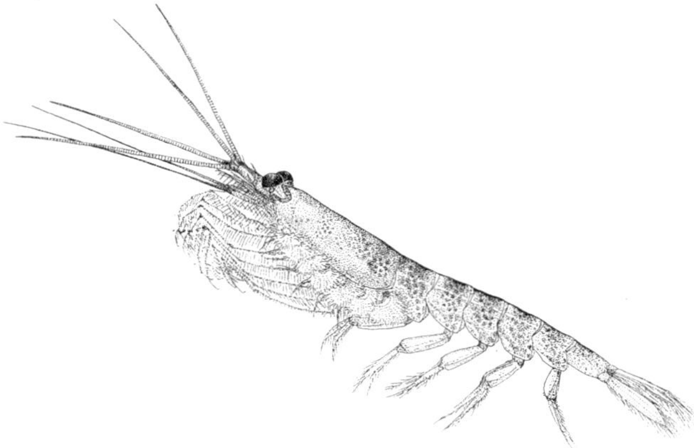

mobilis in mobile

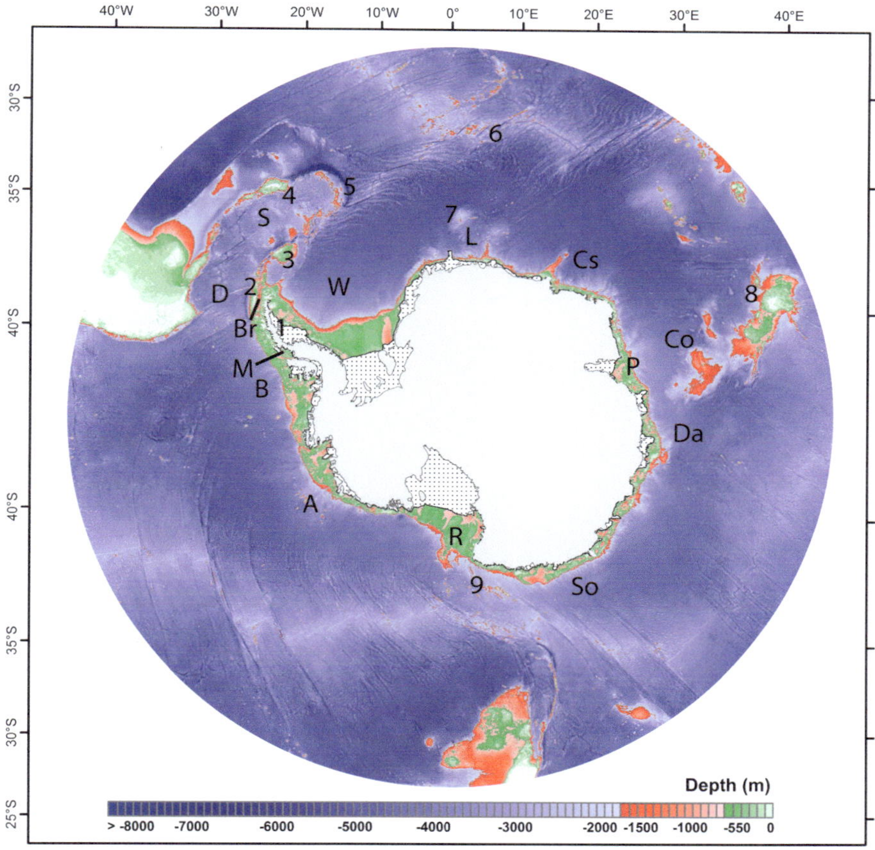

Map of the Antarctic Ocean with geographical names used throughout the chapters of the current book. Names are referring to references in the biological literature of Antarctic krill; however, some names may not be officially approved by the International Hydrographic Organization (IHO) or any other organization. A Amundsen Sea, B Bellingshausen Sea, Br Bransfield Strait, Co Cooperation Sea (60°–85°E), Cs Cosmonaut Sea (30°–50°E), Da Davis Sea (82°–96°E), D Drake Passage, L Lazarev Sea (0°–14°E), M Marguerite Bay, P Prydz Bay, R Ross Sea, S Scotia Sea, So Somov Sea (150°–170°E), W Weddell Sea; 1 Antarctic Peninsula, 2 South Shetland Islands, 3 South Orkney Islands, 4 South Georgia, 5 South Sandwich Islands, 6 Bouvet Island, 7 Maud Rise, 8 Kerguelen, 9 Balleny Islands

Preface

More than half a century ago, James Marr published his famous work *Natural History and Geography of the Antarctic Krill* (*Euphausia superba*) (Marr 1962). Since the start of Antarctic whaling, whalers knew that *E. superba* was the staple food of baleen whales. Thus, they searched for reddish surface waters, which indicated huge krill swarms in those days. In the wake of the whalers, scientific work on krill started. For half a century, research remained motivated by the importance of whales and whaling. The *Discovery* Expeditions were financed by the levies of the whaling industry operating out of South Georgia and other British bases. Marr's treatise is mainly the compilation of the results of three decades of the *Discovery* Programme.

Research on krill did not die with the whales. After the depletion of the Antarctic stocks of whales, new interest in krill arose in the 1970s. At that time, the report of the Club of Rome and other warnings had directed public attention to the sea as a source of protein for nourishing an ever-growing human population. Simultaneously, the new Law of the Sea expelled foreign distant-water fishing fleets from the exclusive economic zones (EEZ). Those fleets looked for new fishing grounds and target species. The Southern Ocean was one of the few open-access regions, and krill fishing was considered the answer to global protein shortage. Estimates of krill production were based on assumptions of the total consumption of krill by the former whale stocks and by rather inadequate figures of annual primary production. In the light of these figures, industry and politics became inspired by the simplistic idea of fisheries potentially replacing the whales as the main consumers of krill. Experimental fishing and processing at sea demonstrated the feasibility of successful exploitation. A new phase of krill research culminated in the international BIOMASS Programme with its multi-ship surveys FIBEX and SIBEX in the 1980s. Scientists from many countries contributed to the vast amount of information on the biology of *E. superba*, its distribution and abundance and its fluctuations. The dreams of a multi-million ton exploitation of Antarctic krill faded away due to economical and technical reasons. Furthermore, BIOMASS came up with much lower figures of krill abundance than previously estimated. Finally, new thinking in terms of the conservation of the Antarctic marine ecosystem called for precautionary management of krill exploitation. At the end of this second phase of krill research and four decades after James Marr's *opus magnum,* Inigo Everson edited a comprehensive summary of the knowledge of *E. superba* and related euphausiid species (Everson 2000).

Towards the end of the twentieth century, public interest in Antarctic krill was out of the focus of marine biological research, but then "Global Change" provided a new impetus: ecologists became increasingly aware of the variations in ice cover and their effects on krill recruitment and distribution. In view of the central role of krill in the complex marine ecosystem, krill research gained international attention again. CCAMLR and SCAR developed international cooperative projects for studies at sea, supplemented by laboratory work and modelling. Large-scale surveys indicated a substantial overall decline in krill biomass in various parts of the Southern Ocean over the past three decades, particularly in the region of the Antarctic Peninsula.

The present book reflects the progress made over recent years in all lines of krill research. For 40 years, its editor Volker Siegel has devoted his scientific work to studies of the population dynamics of Antarctic krill. The list of contributors to the book is impressive and represents the worldwide community of krill researchers.

Krill in a changing environment is the "leitmotiv" of the book in all its 11 chapters, particularly in Chaps. 2 and 3 on distribution, biomass and population dynamics, where fieldwork and computer modelling are intimately linked. The highly complex genetics of Antarctic krill (Chap. 7) are related to the extraordinary population size and the expanse of the circum-Antarctic distribution. Extensive recent studies have rejected former assumptions on the genetic subdivisions of *E. superba*.

Chapter 1 provides a general introduction to the present day's knowledge of Antarctic krill by the editor himself. Studies in ecophysiology, including metabolism, have made great progress over the past couple of decades (Chap. 4). They cover all environmental stages and focus on seasonal and ontogenetic variability in response to varying environmental factors. They consider temperature tolerance and adaptation at the cell, tissue, organism and population levels. Chapter 6 stresses inter alia the vital importance of prolonged ice cover for survival of the late larval stages during the first winter. New optical, acoustic and electronic tools have provided spectacular insights into the behaviour of krill in ice and swarming in open water (Chap. 8).

E. superba is one of the largest and most abundant pelagic Crustacea and the species with the highest total biomass of 10^7–10^8 tons, next to *Homo sapiens*. Linked to that is its unusually versatile feeding, ranging from nano-filtering to raking of ice-algae and raptorial grasping of zooplankton and ice fauna. Thirty years after the ground-breaking work by Kils on the filter basket of krill, Chap. 5 deals with this very intricate multifunctional feeding apparatus. Krill is the staple food for a great variety of predators like fish, squid, penguins and flying birds as well as whales and seals (Chap. 9), which largely depend on the availability of high concentrations of krill. These food web studies are of high relevance to the discussions on the growing krill exploitation and on ecosystem conservation in times of shifting ice regimes (Chap. 11).

James Marr was intrigued by the complex reproduction and early life history of krill. Chapter 6 describes the great progress made in the understanding of the synchronisation of maturation cycles with the seasonality of the light regime, ice cover and availability of adequate food. For the first time, the very scattered information on parasites and diseases in krill has been compiled and analysed in Chap. 10. Obviously, krill is host to a great number of ciliates, gregarinids and helminths, and it suffers from bacterial infections and infestations by yeasts.

The concluding chapter deals with the harvesting of krill by modern fisheries and its potentials and ecological risks. Up to now, fisheries are far below the sustainably allowable levels of exploitation, and their impacts are locally restricted. But the economic interest is increasing in krill as a source of valuable biochemical materials. Much of the information on krill ecology gathered in this book will prove relevant to the development of future fishery management with regard to growing exploitation at times of deterioration of the environmental conditions for the krill population with the rise in global warming.

Springer Publishers are congratulated for inviting the editor of the journal *Polar Biology*, Dieter Piepenburg, to create a series of books related to key issues of polar ecology. Undoubtedly, the book on Antarctic krill is a good start to the series. Thanks to its authors and the great efforts of the editor, it became an excellent state-of-the-art review of research in one of the most fascinating and ecologically important creatures on Earth.

Kiel, Bremen, Germany Gotthilf Hempel

Acknowledgements

The initial inspiration to undertake this krill related volume of the Springer book series "Advances in Polar Ecology" was provided by Prof. G. Hempel who also kindly agreed to take on responsibility to prepare the preface of this specific volume. I thank the series editor Dieter Piepenburg for offering me the great opportunity to contribute to this exciting and significant project. This volume is a result of a collective effort by many people. Thanks are extended to all authors of individual chapters who immediately agreed to take over responsibility for this challenging and time-consuming task and were putting so much effort into the venture which greatly encouraged me to complete this project. I am extremely grateful to authors as well as reviewers who were uniformly generous and patient with all aspects of manuscript preparation, review and publication. All reviewers are sincerely thanked for their timely efforts and very constructive reviews in assuring that the quality of the chapter manuscripts was maintained. Once again I thank the series editor Dieter Piepenburg for his efficient handling of the independent review process. No volume of this nature would be completed in such a short time without the efficiency and professionalism of an excellent publisher team ensuring the completion of the project. I would particularly like to thank Margaret Deignan publishing editor from Springer Publishers (Dordrecht, The Netherlands) for her efforts and readiness to help and solve all problems, may they have been trivial or substantial. I also wish to acknowledge our librarian Heike Müller (Thünen Institut Hamburg) who has enthusiastically sought publications from obscure sources and I am grateful to her for her perseverance. Last but not least, it is a pleasure to acknowledge the great technical assistance and patience of my colleague Susanne Schöling who has been involved in most aspects of this work, both practical and secretarial.

Hamburg, December 2015 Volker Siegel

Contents

Contributors

Angus Atkinson Plymouth Marine Laboratory, Plymouth, UK

Bruce E. Deagle Australian Antarctic Division, Kingston, TAS, Australia

Sophie Fielding British Antarctic Survey, Natural Environment Research Council, Cambridge, UK

Jacqueline Foster Kirkwood, QLD, Australia

Jaime Gómez-Gutiérrez Instituto Politécnico Nacional, Centro Interdisciplinario de Ciencias Marinas, La Paz, Mexico

Gotthilf Hempel Kiel, Bremen, Germany

Simeon L. Hill British Antarctic Survey, Natural Environment Research Council, Cambridge, UK

Simon N. Jarman CIBIO-InBIO, Centro de Investigação em Biodiversidade e Recursos Genéticos, Universidade do Porto, Vairão, Portugal

Australian Antarctic Division, Kingston, TAS, Australia

So Kawaguchi Australian Antarctic Division, Kingston, TAS, Australia

Antarctic Climate and Ecosystems Cooperative Research Centre, Hobart, TAS, Australia

Bettina Meyer Alfred Wegener Institute, Helmholtz Centre for Polar and Marine Research, Section Polar Biological Oceanography, Bremerhaven, Germany

Institute for Chemistry and Biology of the Marine Environment, Carl von Ossietzky University of Oldenburg, Oldenburg, Germany

José Raúl Morales-Ávila Instituto Politécnico Nacional, Centro Interdisciplinario de Ciencias Marinas, La Paz, Mexico

Stephen Nicol Institute for Marine and Antarctic Studies, University of Tasmania, Hobart, TAS, Australia

Christian S. Reiss Antarctic Ecosystem Research Division, Southwest Fisheries Science Center, NOAA Fisheries, La Jolla, CA, USA

Katrin Schmidt Sir Alister Hardy Foundation for Ocean Science, Plymouth, UK

Volker Siegel Thuenen Institute of Sea Fisheries, Hamburg, Germany

Geraint A. Tarling British Antarctic Survey, Natural Environment Research Council, Cambridge, UK

Mathias Teschke Institute for Chemistry and Biology of the Marine Environment, Carl von Ossietzky University of Oldenburg, Oldenburg, Germany

Philip N. Trathan British Antarctic Survey, Natural Environment Research Council, Cambridge, UK

Jonathan L. Watkins British Antarctic Survey, Natural Environment Research Council, Cambridge, UK

List of Referees

Ainley, David G.	H.T. Harvey & Associates, San Jose CA, USA
Buchholz, Friedrich	Alfred Wegener Institute, Bremerhaven, Germany
Clarke, Andrew C.	British Antarctic Survey, Cambridge, UK
Daly, Kendra L.	University of South Florida, St. Petersburg FL, USA
Endo, Yoshinari	Tohoku University, Sendai, Japan
Kaartvedt, Stein	University of Oslo, Oslo, Norway
Krafft, Bjørn A.	Institute of Marine Research, Bergen, Norway
Landers, Stephen C.	Troy University, Troy AL, USA
Pakhomov, Evgeny A.	University of British Columbia, Vancouver, Canada
Reid, Keith	Commission for Conservation of Antarctic Marine Living Resources, Hobart TAS, Australia
Ritz, David A.	University of Tasmania, Hobart, Australia
Ross, Robin M.	University of California, Santa Barbara CA, USA
Saborowski, Reinhard	Alfred Wegener Institute, Bremerhaven, Germany
Small, Hamish J.	Virginia Institute of Marine Science, Gloucester Point VA, USA
Song, Sun	Institute of Oceanology, Qingdao, China
Spiridonov, Vassily A.	P.P. Shirshov Institute of Oceanology, Russian Academy of Sciences, Moscow, Russia
Zane, Lorenzo	University of Padova, Padova, Italy

Chapter 1
Introducing Antarctic Krill *Euphausia superba* Dana, 1850

Volker Siegel

Abstract The first description of Antarctic krill (*Euphausia superba*) dates back to 1850. After a period of taxonomic work and the description of several synonyms, detailed studies on the morphology, internal anatomy and general biology of this species commenced with the 'Discovery' investigations beginning in the 1920s. During the BIOMASS project (Biological Investigations Of Marine Antarctic Systems and Stocks) in the 1980s, studies expanded on the biology and ecology of Antarctic krill, including all aspects of zoogeography and zooplankton community structure. The developing krill fishery and the establishment of CCAMLR (Convention for the Conservation of Antarctic Marine Living Resources) resulted in the pressing need for more quantitative data, which required the use of standardized methods of surveying, sampling, and measuring krill. Despite krill becoming one of the most well-known marine species, many questions remain unanswered and require – in addition to well-established methods – different or even new approaches in future krill research.

Keywords Taxonomy • Morphology • Zoogeography • Sampling • Length-weight relationship

1.1 Morphology and Taxonomy

The Antarctic krill, *Euphausia superba* Dana, 1850 (Fig. 1.1), is one of the largest of 86 exclusively marine species of Euphausiacea including the most recently described *Thysanopoda minyops* Brinton, 1987 and the reinstated *Stylocheiron armatum* Colosi, 1917 (Baker et al. 1990). Together with Decapoda the Euphausiacea belong to the Superorder Eucarida crustaceans and are divided in two families, the Bentheuphausiidae representing a single species and the Euphausiidae. Whilst Bentheuphausiidae lack photophores, they have eight well developed thoracic legs. In males, the endopodites of the first and second pairs of pleopods are not modified as sexual organs, the petasmata. In *Euphausia superba*,

V. Siegel (✉)
Thuenen Institute of Sea Fisheries, Palmaille 9, 22767 Hamburg, Germany
e-mail: volker.siegel@thuenen.de; siegel.v@web.de

V. Siegel (ed.), *Biology and Ecology of Antarctic Krill*, Advances in Polar Ecology,
DOI 10.1007/978-3-319-29279-3_1

Fig. 1.1 Antarctic krill, *Euphausia superba* Dana, 1850

like all other Euphausiidae species, it possesses a number of well-developed photophores and the eighth thoracic leg is extremely minute. The petasma is present on the first pleopod of mature males and can be used in euphausiid systematics (Mauchline and Fisher 1969; Guglielmo and Constanzo 1978). The present knowledge of euphausiid photophores (Peterson 1968, for a review see Herring and Locket 1978) is predominantly based on their morphology, whilst no unequivocal information exists on their assumed role in intraspecific communication (Fregin and Wiese 2002). Experiments demonstrated the complex communication function of the photophores (Fregin and Wiese 2002) indicating that the emission of flashes of light plays a role in individual recognition between conspecifics, in mate recognition, and in keeping swarms together. The hypothesis that photophores are arranged ventrally and point away from the light to reduce the shadow against the lighter water surface, "countershading", could not be verified in experimental studies, as the intensity of light emission surpasses the level of intensity necessary for creating a shading effect (Fregin 2011).

The term "krill" was derived from the Norwegian word for whale food, and originally included other organisms such as pteropods and medusa. William Scoresby Jr, sea captain and naturalist, wrote in 1820 about squillae and shrimps or insects he observed in whale stomachs which are undoubtedly called krill in later reports (Mauchline and Fisher 1969). In later scientific publications the term "krill" was primarily used for euphausiids in general and more recently for *E. superba* specifically. For the remainder of this volume the terms 'krill' and 'Antarctic krill' are used interchangeably, and both refer exclusively to the species *E. superba*. The scientific name *Euphausia* is derived from the luminescence produced by the light organs (from Greek eu-+ phausis "a lighting", see phaino "light").

Euphausia superba was originally described by James Dwight Dana in 1850. He was an American geologist, mineralogist, and naturalist at Yale College. He gained field experience during the 1838–1842 United States Exploring Expeditions

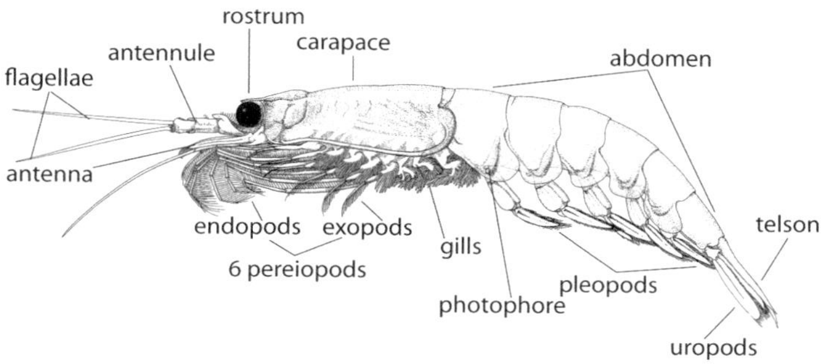

Fig. 1.2 General external morphology of the Antarctic krill (total body length around 60 mm)

commanded by Charles Wilkes. The sample Dana used for the description of *E. superba* was obtained in the Antarctic off Van Diemen's Land at 66°05′S latitude and 157°E longitude. *E. superba* was one of the first euphausiid species scientifically described. Dana (1850, 1852) listed the species under Crustacea, Mysidea and he first described the morphology as follows:

> *Carapax with a very short and acute beak. Sixth segment of abdomen hardly longer than fifth. First joint of inner antennae produced at apex and obtuse. Basal lamina of outer antennae very slightly shorter than base. Last joint of feet very slender, much shorter then penult. Last abdominal segment slightly shorter than adjoining caudal lamina. Branchiae furnished with setiform branchlets; the posterior pair involuted at base, and having branches radiating around somewhat wheel-like, branches curving, and furnished on outer (posterior or inferior side) with long setiform branchlets, subradiate or spiral in position.*

The general morphology of krill is displayed in Fig. 1.2. Since Dana's description is rather short compared to modern taxonomic species descriptions, more recent identification keys – such as the ones published by Mauchline and Fisher (1969) or Baker et al. (1990) – highlight several characteristics of *E. superba* as distinguishing it from other Euphausiacea:

- Ten photophores of complex nature are located one in each eye stalk, one pair at the bases of the second and seventh pair of thoracic legs, respectively, and a single one lying between each of the first four pairs of pleopods
- No mid-dorsal spines on abdominal segments
- Thoracic legs nearly uniform in structure
- 7th and 8th pairs of thoracic legs quite rudimentary
- Compound globular eye well developed
- Antennular peduncle lappet extends to almost half of the length of the 2nd segment
- 6th abdominal segment as long as 5th

Like in all other Euphausiidae, the endopodites of the first two pairs of pleopods in the males are transformed into copulatory organs. The first pair develops lobes provided with protrusions, which together constitute the petasma, the morphology

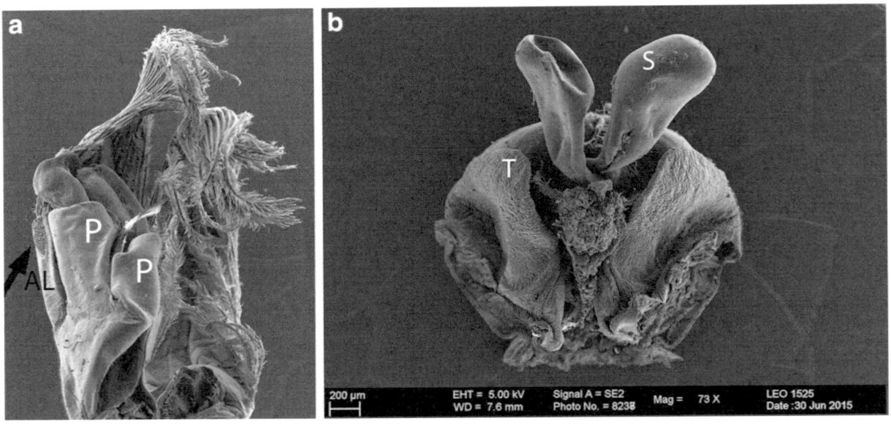

Fig. 1.3 Scanning electron microscope (SEM) photo of Antarctic krill. (**a**) Endopodite of the first pleopod adult male petasma. (**b**) Adult female thelycum with spermatophores attached

of which is species-specific and was described in detail by Guglielmo and Costanzo (1978). The first who used the morphological development of petasma for maturation staging was Bargmann (1945) who proposed separate staging systems for external and internal genitalia. This was later used by Makarov and Denys (1981) for the classification of sequential maturity stages (Fig. 1.3a).

In females, the lateral oviducts arise from the ovaries and open by gonopores under the coxae of the sixth pair of thoracopods. On the sternite of the sixth thoracic somite and near the arthrodial membrane, the integument folds to form a tri-lobed structure, the thelycum (Fig. 1.3b). The morphology of the thelyca is species-specific. In the process of mating, the male attaches spermatophores to the thelycum of the female. The appearance of secondary sexual characters (petasma and thelycum) marks the onset of the juvenile phase.

Another interesting morphological feature is present at the endopodites of the pleopods of male and female krill. Bargmann (1937) described the endopodites as a large blade-like setigerous lobe with a smaller rod- or finger-like protrusion, the appendix interna or on the petasma it is termed the auxiliary lobe. This appendix interna carries an assemblage of small, uniform hooks near the tip, the so-called cincinnuli. Despite her clear description, Bargmann (1937) did not provide a detailed drawing of the cincinnuli, nor did she discuss their potential utilisation. Nicol and Nicol (1983) provided the first scanning electron microscope (SEM) photos for several Euphausiacea species. These authors hypothesize that arrangement in a zipper-like manner and the juxtaposition of the patches of hooks suggests an adhesive function which would allow the endopodites to maintain their close opposition and possibly enhancing synchronized swimming strokes of the pleopods. Indeed, when preparing photos for Fig. 1.4 it became evident that the patches of cincinnuli can be found on the appendix interna of all five pairs of male and female pleopods, supporting the general idea to enable synchronized movement of

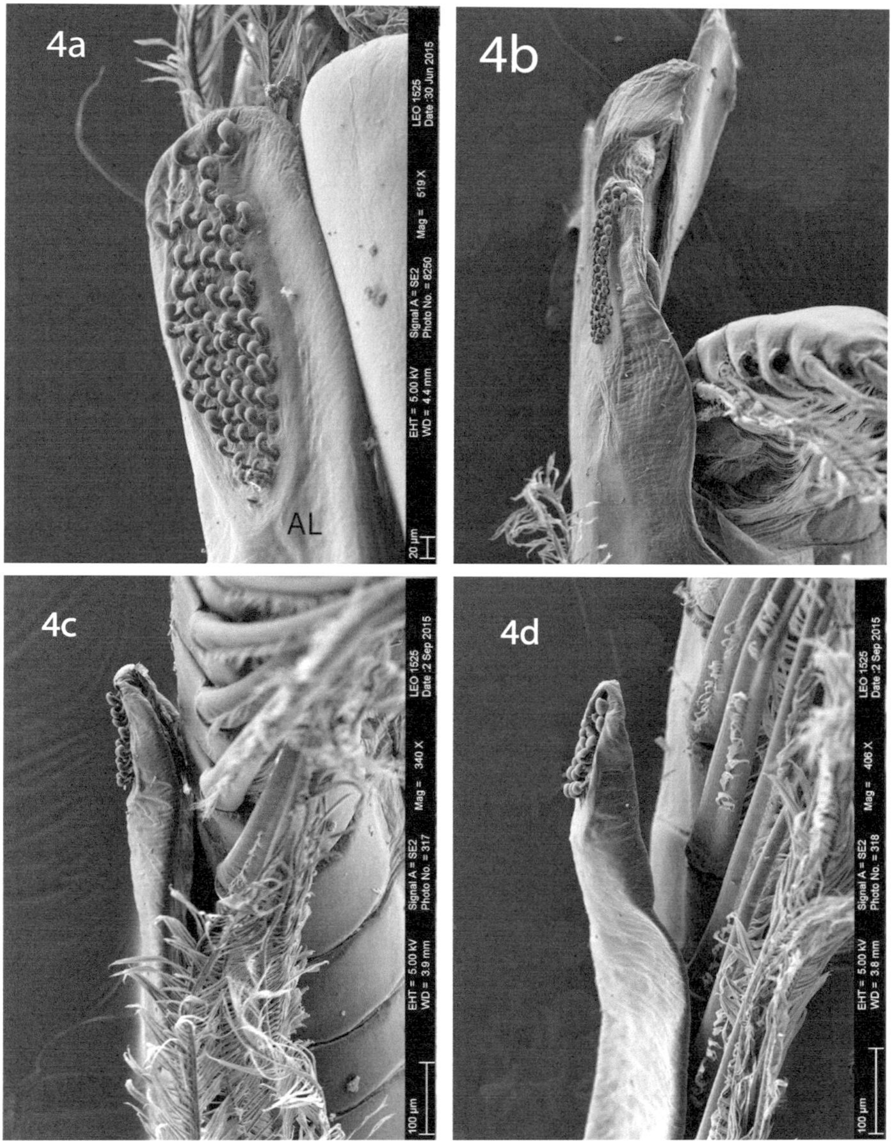

Fig. 1.4 Scanning electron microscope (SEM) view of the auxiliary lobe/appendix interna (AL) which carry an assemblage of small, uniform hooks near the tip, the "cincinnuli" (**a**) male endopodites 1st pleopod, (**b**) male 2nd pleopod, (**c**) male 3rd pleopod, (**d**) male 4th pleopod, (**e**) male 5th pleopod, (**f**) female endopodites 1st pleopod

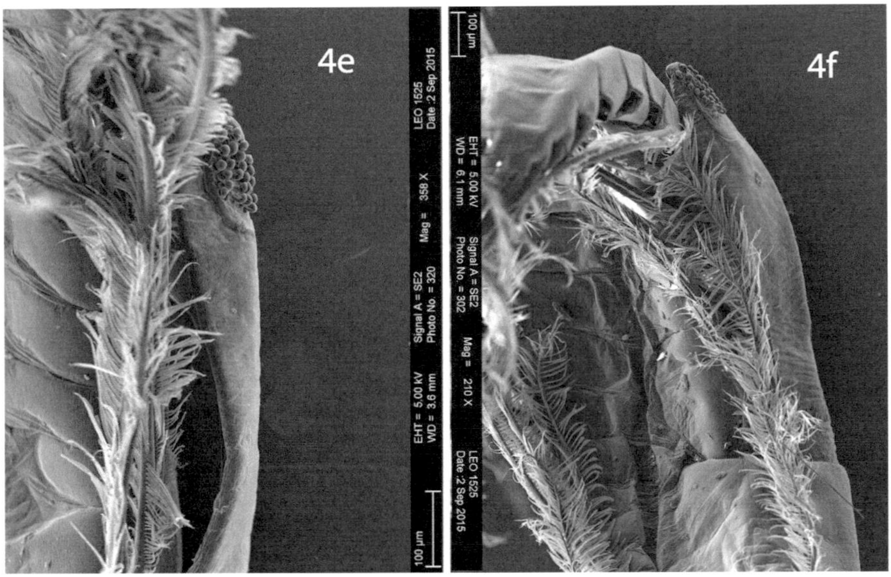

Fig. 1.4 (continued)

the pairs of pleopods and enhancing their swimming propulsion. Details on swimming behaviour will be discussed in Chap. 8 by Tarling and Fielding (2016).

Several other descriptions of the Antarctic krill were published under different species names; however, these were often confused with larval or juvenile stages of the same species and thus have to be regarded as synonymized names: *Euphausia murrayi* G.O.Sars, 1883; *E. antarctica* G.O.Sars, 1883; *E. glacialis* Hodgenson, 1902; *E. australis* Hodgenson, 1902. Detailed description of the morphology and internal anatomy of euphausiids in general have been given by Bargmann (1937), Mauchline and Fisher (1969), and Spiridonov and Casanova (2010). Identification keys for the different species worldwide were provided by Mauchline and Fisher (1969) and Baker et al. (1990).

1.2 Biological Observations

Zoogeographically, *E. superba* is classified as an endemic Antarctic species. It inhabits shelf areas as well as shelf-slopes and deep-ocean basin regions south of the Polar Front. *E. superba* is one of five to six (true) polar stenothermic species that occur mainly or exclusively in Antarctic waters: the neritic ice krill *E. crystallorophias*, the mesopelagic species *E. frigida* and *E. triacantha*, as well as the very common *Thysanoessa macrura* and *T. vicina* in the Antarctic Polar Front (APF) region. The latitudinal succession of the *Euphausia* species has been described by John (1936). This south polar endemism contrasts to those species occurring in the northern polar regions, such as *T. inermis*, *T. raschii* and *Meganyctiphanes norvegica* (Einarsson 1945; Mauchline and Fisher 1969). Since the Arctic system is intimately linked to the North Atlantic system, eurythermal species of southern origin populate – sometimes only periodically – boreo-Arctic waters (Hempel 1985; Siegel 2000).

The Antarctic ecosystem is much older than the Arctic, characterized by a long evolutionary history. Molecular-clock estimates for divergence times of Antarctic and sub-Antarctic species of *Euphausia* of ~20 million years ago were fairly close to the time of formation of the Antarctic Polar Front (Lagabrielle et al. 2009), consistent with their vicariant speciation (Jarman et al. 2000, see also Jarman and Deagle (2016) Chap. 7 this volume). The austral species of the genus *Euphausia* are usually grouped into two lineages, probably reflecting peculiarities of colonization of the Southern Ocean. John (1936) distinguished between the group *E. lucens*, *E. vallentini*, *E. frigida*, *E. superba* and *E. crystallorophias*on the one hand and *E. hanseni*, *E. spinifera*, *E. longirostris*, and *E. triacantha* on the other hand. Whilst comparative studies on the morphology of preanal spine by Lomakina (1972) and Spiridonov (1992) corresponded well with this grouping, results also indicated a somewhat separate position of *E. superba* and *E. crystallorophias* in the genus *Euphausia*, which might be related to the paleontological age (at least Pliocene) of this species (Spiridonov 1996).

Antarctic krill have survived for million years in much warmer and colder conditions than today (Spiridonov 1996). The great temperature oscillations over the last 20 million years produced a periodic shrinkage and expansion of suitable krill habitat. In cold eras, when ice closed off Antarctica's shelves, the only shallow water would have been near just a few isolated islands—very different to today (Spiridonov 1996). However, it is no longer believed that most krill concentrate in shallow coastal waters but most krill occur in oceanic waters (see Chap. 2, Siegel and Watkins 2016). Especially in high-latitude shelf areas Antarctic krill is often replaced by its neritic congenial ice krill *E. crystallorophias*. In contrast to the shrinking shelf habitat, the basic ACC circulation is bathymetrically constrained (Orsi et al. 1995) and appears to have persisted over a geological time scale.

The Antarctic Ocean south of the Polar Front is a region of high physical and biological variability (Hempel 1985). In this region krill, salps, and herbivorous copepods are the main contributors to total zooplankton stock in relation to both abundance and biomass (Voronina 1998). *Euphausia superba* is not just a very abundant species, it is a key species of the entire Antarctic ecosystem and food web (see Chap. 9, Trathan and Hill 2016), and is also a part of a wider zooplankton community. In principle, three major epipelagic zooplankton communities have been identified south of the Polar Front, a northern oceanic community with *Salpa thompsoni* and *Themisto gaudichaudii* as dominant components, a main oceanic community with copepods, *E. superba* and *Thysanoessa macrura* dominating, and a neritic community with *E. crystallorophias* and the Antarctic silverfish *Pleuragramma antarctica* but also *E. superba* as indicator species (Boysen-Ennen and Piatkowski 1988; Siegel and Piatkowski 1990; Hosie 1994). Despite increasing knowledge of zooplankton community structure and their distribution patterns, the grazing impact of key grazers such as krill, salps and copepods are still poorly documented. The important part of krill feeding biology will be discussed in some detail in Chap. 5 of this volume (Schmidt and Atkinson 2016).

1.3 Measurements

Only few specimens are necessary to provide data for species identification and the general description of the species distribution, while quantitative population ecology studies require a large number of samples and measurements. The body sizes of a representative number of specimens in samples are measured to accumulate length frequency/density data for population growth rates, age at sexual maturity, longevity, mortality and recruitment of a species (Chap. 3, Reiss 2016). Older individuals are also larger, and size differences affect maturity and fecundity (Chap. 6, Kawaguchi 2016). Therefore, it is essential that information is gathered by means of standardized methods to derive basic data on body size and stages of sexual maturity.

In his BIOMASS handbook, Mauchline (1980) presented a variety of alternative length measurements for krill. Some of these are quite tedious and require a low-power stereo microscope (e.g. "reference measurement"). For practical reasons, the Antarctic krill science community has limited itself to two methods, the "standard 1 measurement" (S1) and the "Discovery measurement" (AT). S1 is defined by Mauchline (1980) as:

> *the lateral or dorsal distance between the anterior tip of the rostrum and the posterior end of the uropods, excluding their terminal setae*

The 'Discovery' method (AT) was not mentioned in Mauchline's handbook, but is one of the most widely used measurements for total length of Antarctic krill. It was applied throughout the "Discovery investigations" and currently by many research institutions. Everson (2000) reviewed the method and total length (AT) is defined as:

> *the total length from the **A**nterior margin of the eye to the tip of the **T**elson, excluding the terminal spine.* (Figure 1.5)

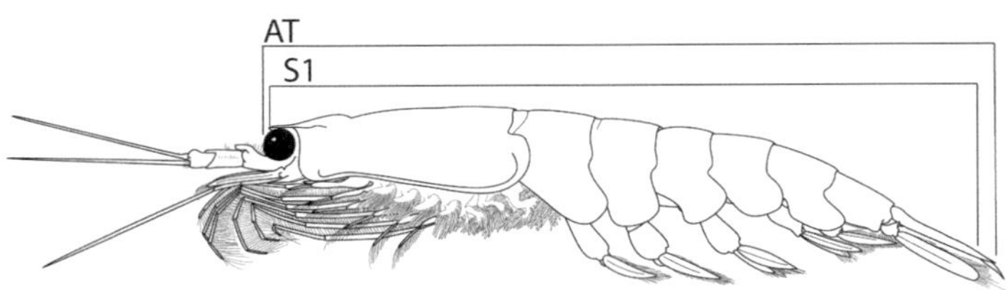

Fig. 1.5 Common types of routine measurements for Antarctic krill, AT total length 'Discovery' method, S1 standard 1 length according to Mauchline (1980)

The question of length measurements was reviewed by CCAMLR to establish standard methods for the scientific observer programme on commercial krill vessels. The current Scientific Observers Manual (CCAMLR 2011) requires measurements according to the 'Discovery' standard of total length (AT) to the nearest millimetre below. This was thought to be most appropriate to be undertaken at sea on commercial vessels.

The use of different methods is sometimes confusing and may create bias, although the difference between the two methods seems to be relatively small. However, the deviation can be 1–2 mm, depending on the actual size of the specimens. To facilitate direct comparisons among data sets, Siegel (1982) provided linear regressions to convert the length data:

$$AT = 0.133 + 1.03 * S1$$
$$S1 = 0.024 + 0.967 * AT$$

An additional finding in the study of Siegel (1982) and a similar analysis of Miller (1983) showed that carapace length is less suitable as a standard method to estimate krill length. Differences occur in carapace length between males and females, and the relationship between AT or S1 and carapace length becomes non-linear with increasing numbers of adult gravid maturity stages in the samples.

For more generalized studies of sexual maturation of individual krill and in the framework of investigations of population dynamics, it was necessary to develop a simple classification of external morphological indications of sexual maturity (Makarov and Denys 1981). Such stepwise systems sometimes cause difficulties in interpretation, especially for females. Spent females lose the attached spermatophores after their first moult, and the ovary is reduced in size which then is often confused with the pre-spawning female stage without spermatophores and an ovary in its early stage of development. This problem is partly due to re-maturation (Makarov 1975) or re-juvenation (Poleck and Denys 1982) of the long-living krill and the regression of external female and male sexual characteristics as a regular process in natural populations (see also discussion by Siegel 2012). Due to these caveats, Cuzin-Roudy and Amsler (1991) categorized maturity stages according to the state of the gonads (see also Chap. 6 of this volume, Kawaguchi 2016). The classification of the internal developmental stages is more accurate in its description of the maturation process but not always in phase with the external characteristics with the later often lagging behind the internal development. However, the study of the gonad development is far too complex and time-consuming to be used as a routine measure for stock assessment work. The classification scheme of Makarov and Deny allows for relatively fast categorization of the status of the spawning stock. Although the full classification scheme is published previously in the former BIOMASS Handbook but might not be easily accessible and only a shortened version of the classification scheme was presented in Everson (2000), I document the full sequence of developmental stages here.

Maturity stages are divided into three general classes according to their secondary sexual characteristics given as Roman numerals: I – juveniles, II – subadults, and III – adults, capital letter A refers to males, B to females:

I – Juveniles – secondary sexual characteristics (thelycum and petasma) are not visible
IIA1 – subadult male – petasma is a single undivided lobe
IIA2 – subadult male – petasma is two-lobed, no wing present
IIA3 – subadult male – petasma is two-lobed, wing present
IIIA – adult male – petasma fully developed, no fully developed spermatophores present in ejaculatory ducti
IIIB – adult male – spermatophores present, and are easily ejected by pressing on the ducti
IIB – subadult female – developing thelycum present, colour is feeble or absent
IIIA – adult female – thelycum fully developed, red in colour, no spermatophores present, body not swollen
IIIB – adult female – spermatophores attached, body not swollen
IIIC – gravid female – ovary fills whole thoracic space and extends into first and second abdominal segment, body not noticeably swollen
IIID – gravid female – thorax and first and second abdominal segment swollen by enlarged ovary
IIIE – spent female – eggs have been laid, body cavity empty with a large hollow space and small ovary located in the middle of this space, spermatophores present or absent

Another important variable in biological studies is body weight. Biomass and production estimates from net sampling and acoustic surveys are expressed as wet weight and are derived from quantitative length frequency/density data. Physiological studies often convert length into dry weight or ash-free dry weight. Whilst a direct precise measurement of weight of sorted plankton specimens is often difficult to obtain aboard vessels, length measurements can easily and rapidly be carried out. An assessment of length-weight relationships has been carried out by Morris et al. (1988), who investigated the importance of maturity stages in improving the precision of calculating weight from length data. However, Morris et al. (1988) did not consider the seasonal aspect in variation of krill body weight. Therefore, Siegel (1992) summarized information on length-weight relationships from all seasons with additional information on the maturity stage of krill. Table 1.1 does not list all available data from the published literature, but gives examples for different months of the year and for different maturity stages dominating the stock at that time. More data can be found in Siegel (1992), including length-weight relationships based on different methods of length measurement (e.g. standard length 1).

Morris et al. (1988) noted that "surprisingly, the simple division of krill into male and female categories is of little practical use in improving the precision of

Table 1.1 Length-weight relationships for krill *Euphausia superba* for different seasons and dominating maturity stages; coefficients apply to the formula $W = a*AT^b$ with *W* weight in mg, *AT* total length in mm, *WW* wet weight, *DW* dry weight, *AFD* ash-free dry weight, *M* males, *F* females, maturity stage information according to classification of Makarov and Denys (1981); data are cited from Siegel (1992), Morris et al. (1988), Jazdzewski et al. (1978), Sahrhage (1978), Kils (1979), and Retamal and Quintana (1982)

Month	Sex	Type of weight	Regression coefficient a	Regression coefficient b	Dominant adult maturity stage
October	All	WW	0.00236	3.251	Resting
December	All	WW	0.00086	3.551	Gravid
January	All	WW	0.00158	3.40	Pre-spawning/gravid
January	All	WW	0.00205	3.325	Gravid-spent
February	All	WW	0.00083	3.561	Gravid
February	All	WW	0.00165	3.380	Spent
Feb/ March	All	WW	0.00385	3.20	Gravid-spent
March	All	WW	0.00193	3.325	Spent
April	All	WW	0.0018	3.343	Spent
June	All	WW	0.00353	3.151	Resting
October	M	WW	0.00236	3.251	3A resting
	F	WW	0.00242	3.247	3A resting
November	M	WW	0.00315	3.207	3A resting
	F	WW	0.00430	3.102	3A resting
December	M	WW	0.00083	3.561	3B gravid
	F	WW	0.00115	3.457	3C, D gravid
January	M	WW	0.00156	3.403	3A, B gravid
	F	WW	0.00282	3.234	3D, E gravid-spent
Jan/ February	M	WW	0.00423	3.170	3B prespawning
	F	WW	0.00573	3.080	3B prespawning
February	M	WW	0.00111	3.507	3A spent
	F	WW	0.00211	3.302	3E, A spent
Feb/ March	M	WW	0.00613	3.0766	Gravid
	F	WW	0.00289	3.270	Gravid-spent
	F	WW	0.01088	2.9077	Only non-gravid F
	F	WW	0.00975	2.9809	Only gravid F
	F	WW	0.03548	2.590	Only spent F
June	M	WW	0.00328	3.176	Resting
	F	WW	0.00441	3.084	Resting
October	M	DW	0.00060	3.030	Resting
	F	DW	0.00080	2.971	Resting
November	M	DW	0.00076	3.071	Resting
	F	DW	0.00105	2.965	Resting
December	M	DW	0.00019	3.435	3B gravid
	F	DW	0.00025	3.357	3C, D gravid
January	M	DW	0.00036	3.277	3A, B gravid-spent
	F	DW	0.00075	3.066	3D, E gravid-spent

(continued)

Table 1.1 (continued)

Month	Sex	Type of weight	Regression coefficient a	Regression coefficient b	Dominant adult maturity stage
February	M	DW	0.00009	3.694	3A spent
	F	DW	0.00031	3.306	3E spent
Jan/ February	all	DW	0.00007	3.760	Pre-spawning/gravid
Feb/ March	all	DW	0.00010	3.799	Gravid
Feb/ March	M	DW	0.00238	2.93	Gravid
	F	DW	0.00024	3.55	Gravid-spent
	F	DW	0.00139	3.0737	Only non-gravid
	F	DW	0.00199	3.0438	Only gravid
	F	DW	0.00106	3.150	Only spent
October	M	AFD	0.0005	3.022	3A resting
	F	AFD	0.0008	2.906	3A resting
December	M	AFD	0.00011	3.527	3B gravid
	F	AFD	0.00014	3.460	3C, D gravid
January	M	AFD	0.00029	3.277	3A, B gravid-spent
	F	AFD	0.00066	3.041	3D, E gravid-spent
February	M	AFD	0.00007	3.720	3A spent
	F	AFD	0.00026	3.205	3E spent

any prediction of weight". However, Morris et al. (1988) only analysed samples taken over a short period during the spawning season (end of February to beginning of March 1985), when most adult krill were still in the gravid maturity stage. The result they obtained was to be expected, because for the spawning season, covariance analyses of length-weight relationships indicated no differences between male and female krill (Siegel 1989).

However, the seasonal change in krill weight clearly showed that the results may differ by 15 % when using length-weight relationships for krill with different dominant maturity stages. Siegel (1989) already noted that length-weight relationships during the post-spawning and winter seasons are significantly different from those obtained during the spawning season, with minimum weight during winter. Moreover, a difference in the weight of both sexes has also been noted by Nemoto et al. (1981) and Retamal and Quintana (1982), with males being heavier than females at similar sizes. However, this difference in weight between males and females is statistically significant only during the post-spawning period, when females have shed their eggs (Siegel 1989). After the ovaries had recovered, and during the winter resting stage, there was no significant deviation in weight (length-weight relationship) between males and females.

1.4 Sampling

Most of what we know historically about the at-sea distribution, abundance and ecology of all species of Euphausiacea comes from studies using scientific nets (Atkinson et al. 2004). Increasing quantitative data sets have been gathered by acoustic sampling from ships to study krill distribution, biomass and krill behaviour since the 1980s (Macaulay 2000), see also Chap. 8 of this volume (Tarling and Fielding 2016). There are, however, limitations on the types of data these two techniques can provide. Nets offer the highest accuracy in species identification, but suffer from avoidance (Everson and Bone 1986; Everson 2000) escapement, damage and integration (Watkins 2000) and biases of animal size depending on swimming speeds and net mesh sizes (Skjoldal et al. 2013). In order to obtain a sufficient number of animals, large volumes of water containing dispersed (or aggregated) animals are integrated into one net sample. Large plankton nets such as the RMT partly obviate these caveats for krill and produce some meaningful data sets, at least for certain size ranges of krill (Siegel 2007). Size, mobility and variable behaviour of krill are a challenge for sampling, either with nets or acoustics (Watkins 2000, see also Chap. 8, Tarling and Fielding 2016). However, in contrast, hull-mounted acoustic techniques offer a method for rapidly covering larger space scales of the distribution of krill than nets. Despite rapid developments in acoustic technology and analytical models (Macaulay 2000; Demer and Conti 2005), this method suffers from a challenge of species identity (Horne 2000), ambiguity in the relationship between acoustic backscatter and animal number or biomass. Another caveat includes near-surface and seabed dead-zones (Ona and Mitson 1996; Atkinson et al. 2012).

Many aspects of krill biology and ecology are based on a small number of observations or locally restricted time series. Thus, hypotheses and models had been developed to explain or generalize observations. These are necessarily based on incomplete knowledge and often cannot describe the complexity of the krill life cycle with all different aspects of the various habitats krill are living in. Atkinson et al. (2012) listed four under-sampled components of krill habitat. The first is the underside of sea ice, which is probably important year-round. The second is the 0–10 m layer of the open ocean/sea-ice zone, which is 'invisible' to acoustics. The third undersampled habitat is the seafloor, although many records exist on krill using this habitat. However, these observations have always been assumed as exceptional or anecdotal. The fourth vertical habitat is the water column below 200 m depth. There are indications that krill could be found here in unknown quantities, and more detailed discussion on these points can be found in Chap. 2 of this volume (Siegel and Watkins 2016). The issue of undersampled krill has fundamental implications beyond determining krill stock biomass, but there is differentiation between life stages among different components of the habitat. This may lead to undersampling particular life-history stages, which has a direct impact on the estimation of recruitment strength (Atkinson et al. 2012).

The use of nets and acoustics, and their limitations and prospects, have been reviewed numerous times in earlier publications, and the reader is referred to the detailed discussion in these publications (see Everson 2000; Nicol and Brierley 2010; Atkinson et al. 2012). The future study of krill in their different environments requires that krill researchers adopt a variety of new technologies and apply them in conjunction with traditional approaches (Nicol and Brierley 2010).

It can be foreseen that much future data collection from the Antarctic krill will take place with autonomous instrumentation, including fixed or stationary moorings, buoys and rigs, and mobile platforms, such as autonomous underwater vehicles (AUVs) and gliders (Guihen et al. 2014). First steps in improving current acoustic methods relate to reducing vessel noise, which enhances signal-to-noise ratio and thus extends depth range where krill targets can be distinguished from background noise. This will also enable the use of multiple-frequency techniques for species identification at greater depths. Most importantly, this improvement also facilitates quality data from krill distributed in deep water and towards the bottom (Godø et al. 2014). In terms of advanced instrumentation, multibeam echosounders have the potential to supersede vertical echosounding; multibeam echosounders and sonars are also a useful tool for swarm studies (Cox et al. 2010).

Acoustic Doppler Current Profilers (ADCPs) have been in use for measuring current velocities (Roe and Griffiths 1993) and examining distribution and abundance of pelagic organisms, including krill, both from ships (Brierley et al. 1998) and from moorings (Cisewski et al. 2010). Moorings equipped with ADCPs can extend the time window of observations of the relative abundance of krill and are capable of collecting extended time series of bioacoustical data at remote locations on diel and seasonal vertical migration, even during the hostile winter period.

Remotely Operated Vehicles (ROVs) may be other tools needed for deep-water studies on krill. Moorings and towed vehicles instrumented with upward looking sensors offer important tools to address the challenges of investigating the near sea surface and the under-ice habitat of krill (La et al. 2015). Other techniques that still require further development before they may become of future interest in krill research are the ocean acoustic waveguide propagation (OAWRS) and the LIDAR (Light Detection and Ranging) techniques (Carrera et al. 2006). Further discussion on this subject can be found in Chap. 8 (Tarling and Fielding 2016).

Indirect sampling by making use of krill predators has a history in CCAMLRs ecosystem monitoring programme. This approach may "widen the net" (Reid et al. 2004) and allow year-round repetitive sampling. Considerable information has now been collected on krill availability and demographics from sampling the diet of krill-feeding organisms. However, the method relies on the availability of krill to various predators with different requirements and is not a substitute for directly obtaining physical samples of krill populations (Nicol and Brierley 2010, see also Chap. 9, Trathan and Hill 2016).

Scientific studies and monitoring of the krill resource include measurements across a variety of disciplines. Research vessel capacity for Antarctic research is rather limited. Modern fishing vessels are normally better equipped for trawling than research vessels. In addition, they are often equipped with multibeam echosounders and sonars or may even possess an instrument drop keel for the transducers to minimize the impact of bubble noise. Krill fishing vessels could offer a good opportunity to collect additional krill data related to resource management, since they operate almost all year-round and could increase the spatial and timely coverage of krill investigations across the fishing grounds (Godø et al. 2014).

As Nicol and Brierley (2010) clearly stated, it is unlikely that any single technique or instrument will be capable of accurately and precisely measuring krill distribution and abundance, whilst at the same time provide the necessary range of information on the biology of a population. Some of the problems are common to most pelagic organisms, but in several instances Antarctic krill is also unique. The species shows great plasticity in its distribution and behaviour between summer and winter, as well as between lower and high latitude areas. Seasonal horizontal migration into offshore areas for spawning (Siegel 1988; Lascara et al. 1999), seasonal differences in vertical migration (Taki et al. 2005), seasonal change from filter-feeding in open water to grazing under the ice all require different approaches, methods and techniques. Traditional techniques have allowed for a good first detailed insight into krill distribution, abundance, life cycle and population dynamics. With development of the krill fishery, there is a huge requirement for more and better data for management to address the challenges underlying the sustainable harvest of krill through CCAMLR's ecosystem-based management approach (see also Chap. 11, Nicol and Foster 2016). However, climate change will present challenges to management on top of the rational use of krill through the trophic 'cascade' effect that will impact many krill-dependent species, as well as the entire Antarctic ecosystem. Many new pressing questions on the biology and ecology of krill, its resilience, flexibility and survivorship will have to be answered, all of which will require different and often new techniques.

Nevertheless, few species of pelagic invertebrates have received as much research attention as Antarctic krill. This is because this species plays a key role in the food web of the entire Antarctic Ocean ecosystem (Croxall et al. 1999) and is the focus of a valuable fishery (Nicol et al. 2011). At the time when Miller and Hampton (1989) produced their first review on the biology of krill, a little more than 900 publications were published on this species since 1975. When Everson (2000) presented his review, with a strong emphasis on the biology, ecology and fisheries of various Euphausiacea species, another 750 publications on Antarctic krill had to be considered. Between the years 2000 and 2015, the Web of Science lists more than 1,630 new publications, which include Antarctic krill in their title or abstract. This volume aims to assist students, researchers and managers and facilitate their work to identify key references and sources of data and to support essential and comprehensive information for all aspects of the biology and life cycle of Antarctic krill.

Acknowledgement Special thanks go to Renate Walter (Biozentrum Grindel, Zoologisches Institut Universität Hamburg) for the preparation of the SEM photographs.

References

Atkinson A, Siegel V, Pakhomov E, Rothery P (2004) Long-term decline in krill stock and increase in salps within the Southern Ocean. Nature 432:100–103

Atkinson A, Nicol S, Kawaguchi S, Pakhomov E, Quetin L, Ross R, Hill S, Reiss C, Siegel V, Tarling G (2012) Fitting *Euphausia superba* into Southern Ocean food-web models: a review of data sources and their limitations. CCAMLR Sci 19:1–26

Baker A, Boden BP, Brinton E (1990) A practical guide to the euphausiids of the world. Natural History Museum Publications, London, p 96

Bargmann HE (1937) The reproductive system of *Euphausia superba*. Discov Rep 14:325–350

Bargmann HE (1945) The development and live history of adolescent and adult krill *Euphausia superba*. Discov Rep 23:103–176

Boysen-Ennen E, Piatkowski U (1988) Meso- and macroplankton communities in the Weddell Sea. Polar Biol 9:17–35

Brierley AS, Brandon MA, Watkins JL (1998) An assessment of the utility of an Acoustic Doppler Current Profiler for biomass estimation. Deep-Sea Res I 45:1555–1573

Carrera P, Churnside JH, Boyra G, Marques V, Scalabrin C, Uriarte A (2006) Comparison of airborne LIDAR with echosounders: a case study in the coastal Atlantic waters of Southern Europe. ICES J Mar Sci 63:1736–1750

CCAMLR (2011) Scientific Observers Manual – 2011. Observation guidelines and reference material. CCAMLR, Hobart, p 71

Cisewski B, Strass VH, Rhein M, Krägefsky S (2010) Seasonal variation of diel vertical migration of zooplankton from ADCP backscatter time series data in the Lazarev Sea. Antarct Deep-Sea Res I 57:78–94

Cox MJ, Warren JD, Demer DA, Cutter GR, Brierley AS (2010) Three-dimensional observations of swarms of Antarctic krill (*Euphausia superba*) made using a multi-beam echosounder. Deep-Sea Res II 57:508–518

Croxall JP, Reid K, Prince PA (1999) Diet, provisioning and productivity responses of marine predators to differences in the availability of Antarctic krill. Mar Ecol Prog Ser 177:115–131

Cuzin-Roudy J, Amsler MO (1991) Ovarian development and sexual maturity staging in Antarctic krill, *Euphausia superba* Dana (Euphausiacea). J Crust Biol 11:236–249

Demer DA, Conti SG (2005) New target-strength model indicates more krill in the Southern Ocean. ICES J Mar Sci 62:25–32

Einarsson H (1945) Euphausiacea 1. North Atlantic species. Dana Rep 27:1–185

Everson I (2000) Krill biology, ecology and fisheries, Fish and aquatic resources series. Blackwell Science, Oxford, p 372

Everson I, Bone DG (1986) Effectiveness of the RMT8 system for sampling krill (*Euphausia superba*) swarms. Pol Biol 6:83–90

Fregin T (2011) Untersuchungen zu Form und Funktion der Leuchtorgane des Nordischen Krill *Meganyctiphanes norvegica* (M. Sars, 1857) (Crustacea: Euphausiacea). PhD Thesis, University Hamburg, Germany, p 289

Fregin T, Wiese K (2002) The photophores of *Meganyctiphanes norvegica* (M. Sars) (Euphausiacea): mode of operation. Helgol Mar Res 56:112–124

Godø OR, Reiss C, Siegel V, Watkins JL (2014) Commercial fishing vessel as research vessels in the Antarctic – requirements and solutions exemplified with a new vessel. CCAMLR Sci 21:11–17

Guglielmo L, Constanzo G (1978) Diagnostic value of the thelycum in euphausiids, II. Oceanic species. Genus *Euphausia* Dana, 1852. Arch Oceanogr Limnol 19:143–155

Guihen D, Fielding S, Murphy E, Heywood K, Griffiths G (2014) An assessment of the use of ocean gliders to undertake acoustic measurements of zooplankton: the distribution and density of Antarctic krill in the Weddell Sea. Limnol Oceanogr Methods 12:373–389

Hempel G (1985) On the biology of polar seas, particularly the Southern Ocean. In: Gray JS, Christiansen ME (eds) Marine biology of polar regions and effects of stress on marine organisms. Wiley, Oxford, pp 3–33
Herring PJ, Locket NA (1978) The luminescence and photophores of euphausiid crustaceans. J Zool 186:431–462
Horne JK (2000) Acoustic approaches to remote species identification: a review. Fish Oceanogr 9:356–371
Hosie GW (1994) The macrozooplankton communities in the Prydz Bay region, Antarctica. In: El-Sayed SZ (ed) Southern Ocean ecology. The BIOMASS perspective. Cambridge University Press, Cambridge, pp 93–123
Jarman SN, Deagle BE (2016) Genetics of Antarctic krill. In: Siegel V (ed) The biology and ecology of Antarctic krill, *Euphausia superba* Dana, 1850. Springer, Cham, pp 247–278
Jarman SN, Elliott NG, Nicol S, McMinn A (2000) Molecular phylogenetics of circumglobal *Euphausia* species (Euphausiacea: Crustacea). Can J Fish Aquat Sci 57(S3):51–58
Jazdzewski K, Dzik J, Porebski J, Rakusa-Suszczewski S, Witek Z, Wolnomiejski N (1978) Biological and populational studies on krill near South Shetland Islands, Scotia Sea and South Georgia in the summer 1976. Polskie Archiwum Hydrobiologii 25:607–631
John DD (1936) The southern species of the genus Euphausia. Discov Rep 14:193–324
Kawaguchi S (2016) Reproduction and larval development in Antarctic krill (*Euphausia superba*). In: Siegel V (ed) The biology and ecology of Antarctic krill, *Euphausia superba* Dana, 1850. Springer, Cham, pp 225–246
Kils U (1979) Schwimmverhalten, Schwimmleistung und Energiebilanz des antarktischen Krills, *Euphausia superba*. Ber Inst Meereskd Kiel 65:1–58
La HS, Ha HK, Kang CY, Wahlin AK, Shin HC (2015) Acoustic backscatter observations with implications for seasonal and vertical migrations of zooplankton and nekton in the Amundsen shelf (Antarctica). Estuar Coast Shelf Sci 152:124–133
Lagabrielle Y, Goddéris Y, Donnadieu Y, Malavieille J, Suarez M (2009) The tectonic history of Drake Passage and its possible impacts on global climate. Earth Planet Sci Lett 279:197–211
Lascara CM, Hofmann EE, Ross RM, Quetin LB (1999) Seasonal variability in the distribution of Antarctic krill, *Euphausia superba*, west of the Antarctic Peninsula. Deep-Sea Res 46:951–984
Lomakina NB (1972) Phylogenetic relationship in the species of the genus Euphausia of the southern Ocean. Zool Zhurn 51:812–820 (in Russian)
Macaulay M (2000) Sampling krill – acoustic estimation of krill abundance. In: Everson I (ed) Krill biology, ecology and fisheries. Blackwell Science, Oxford, pp 20–32
Makarov RR (1975) A study of the second maturation of euphausiid (Eucarida, Euphausiacea) females (in Russian). Zool Zhurn 54:670–681
Makarov RR, Denys CJ (1981) Stages of sexual maturity of *Euphausia superba* Dana. BIOMASS Handb 11:13
Mauchline J (1980) Measurement of body length of *Euphausia superba* Dana. BIOMASS Handb 4:9
Mauchline J, Fisher LR (1969) The biology of Euphausiids. Adv Mar Biol 7:1–454
Miller DGM (1983) Variation in body length measurement of *Euphausia superba* Dana. Pol Biol 2:17–20
Miller DGM, Hampton I (1989) Biology and ecology of the Antarctic krill (*Euphausia superba* Dana): a review. BIOMASS Sci Ser 9:1–166
Morris DJ, Watkins JL, Ricketts C, Buchholz F, Priddle J (1988) An assessment of the merits of length and weight measurements of Antarctic krill *Euphausia superba*. Brit Antarct Surv Bull 79:27–50
Nemoto T, Doi T, Nasu K (1981) Biological characteristics of krill caught in the Southern Ocean. In: El-Sayed S (ed) Biological investigations of marine Antarctic systems and stocks, vol II, BIOMASS. SCAR, Cambridge, pp 47–63
Nicol S, Brierley AS (2010) Through a glass less darkly – new approaches for studying the distribution, abundance and biology of Euphausiids. Deep Sea Res II 57:496–507

Nicol S, Foster J (2016) The fishery for Antarctic krill – its current status and management regime. In: Siegel V (ed) The biology and ecology of Antarctic krill, *Euphausia superba* Dana, 1850. Springer, Cham, pp 387–422
Nicol S, Nicol D (1983) A unique adhesive structure on the pleopods of euphausiids. Crustaceana 44:163–168
Nicol S, Foster J, Kawaguchi S (2011) The fishery for Antarctic krill – recent developments. Fish and Fishery 13:30–40
Ona E, Mitson RB (1996) Acoustic sampling and signal processing near the seabed: the deadzone revisited. ICES J Mar Sci 53:677–690
Orsi AH, Whitworth T III, Nowlin WD Jr (1995) On the meridional extent and fronts of the Antarctic Circumpolar Current. Deep-Sea Res 42:641–673
Peterson G (1968) Studies on photophores in the Euphausiacea. Sarsia 36:1–39
Poleck TP, Denys CJ (1982) Effect of temperature on the moulting, growth and maturation of the Antarctic krill *Euphausia superba* (Crustacea: Euphausiacea) under laboratory conditions. Mar Biol 70:255–265
Reid K, Jessopp MJ, Barrett MS, Kawaguchi S, Siegel V, Goebel ME (2004) Widening the net: spatio-temporal variability in the krill population structure across the Scotia Sea. Deep-Sea Res II 51:1275–1287
Reiss CS (2016) Age, growth, mortality, and recruitment of Antarctic Krill, *Euphausia superba*. In: Siegel V (ed) The biology and ecology of Antarctic krill, *Euphausia superba* Dana, 1850. Springer, Cham, pp 101–144
Retamal MA, Quintana R (1982) Basic biological studies relating to the population dynamics of krill *Euphausia superba* Dana 1850. Cer Sci INACH 28:175–190
Roe HSJ, Griffiths G (1993) Biological information from an Acoustic Doppler Current Profiler. Mar Biol 115:339–346
Sahrhage D (1978) Zur Längen-Gewichts-Beziehung beim antarktischen Krill (*Euphausia superba*). Meeresforschung 26:47–49
Schmidt K, Atkinson A (2016) Feeding and food processing in Antarctic krill (*Euphausia superba* Dana). In: Siegel V (ed) The biology and ecology of Antarctic krill, *Euphausia superba* Dana, 1850. Springer, Cham, pp 175–224
Siegel V (1982) Relationship of various length measurements of *Euphausia superba* Dana. Meeresforschung 29:114–117
Siegel V (1988) A concept of seasonal variation of krill (*Euphausia superba*) distribution and abundance west of the Antarctic Peninsula. In: Sahrhage D (ed) Antarctic Ocean and resources variability. Springer, Berlin/Heidelberg, pp 219–230
Siegel V (1989) Winter and spring distribution and status of the krill stock in Antarctic Peninsula waters. Arch FischWiss 39:45–72
Siegel V (1992) Review of length-weight relationships for Antarctic krill. SC-CAMLR Sel Sci Pap 9:145–155
Siegel V (2000) Krill (Euphausiacea) demography and variability in abundance and distribution. Can J Fish Aquat Sci 57(Sup 3):151–167
Siegel V (2007) Size selectivity of the RMT8 plankton net and a commercial trawl for Antarctic krill. SC-CAMLR-WG-EMM-07/28, pp 1–8
Siegel V (2012) Krill stocks in high latitudes of the Antarctic Lazarev Sea: seasonal and interannual variation in distribution, abundance and demography. Polar Biol 35:1151–1177
Siegel V, Piatkowski U (1990) Variability in the macrozooplankton community off the Antarctic Peninsula. Polar Biol 10:373–386
Siegel V, Watkins JL (2016) Distribution, biomass and demography of Antarctic krill, *Euphausia superba*. In: Siegel V (ed) Biology and ecology of Antarctic krill. Springer, Cham, pp 21–100
Skjoldal HR, Wiebe PH, Postel L, Knutsen T, Kaartvedt S, Sameoto D (2013) Intercomparison of zooplankton (net) sampling systems: results from the ICES/GLOBEC sea-going workshop. Prog Oceanogr 108:1–42

Spiridonov VA (1992) The preanal plate and spine of Euphausiacea (Crustacea Eucarida), their variability and homology (with particular reference to the austral species of the genus Euphausia DANA). Arthropodan Selcta 1:3–15

Spiridonov VA (1996) A scenario of the Late-Pleistocene – holocene changes in the distributional range of Antarctic krill (*Euphausia superba*). Mar Ecol 17:519–541

Spiridonov VA, Casanova B (2010) Order Euphausiacea Dana. Crustaceana 61:5–82

Taki K, Hayashi T, Naganobu M (2005) Characteristics of seasonal variation in diurnal vertical migration and aggregation of Antarctic krill (*Euphausia superba*) in the Scotia Sea, using Japanese fishery data. CCAMLR Sci 12:163–172

Tarling GA, Fielding S (2016) Swarming and behaviour in Antarctic krill. In: Siegel V (ed) The biology and ecology of Antarctic krill, *Euphausia superba* Dana, 1850. Springer, Cham, pp 279–320

Trathan PN, Hill SL (2016) The importance of krill predation in the Southern Ocean. In: Siegel V (ed) The biology and ecology of Antarctic krill, *Euphausia superba* Dana, 1850. Springer, Cham, pp 321–350

Voronina NM (1998) Comparative abundance and distribution of major filter-feeders in the Antarctic pelagic zone. J Mar Syst 17:375–390

Watkins J (2000) Sampling krill. In: Everson I (ed) Krill biology, ecology and fisheries. Blackwell Science, Oxford, pp 8–39

Chapter 2
Distribution, Biomass and Demography of Antarctic Krill, *Euphausia superba*

Volker Siegel and Jonathan L. Watkins

Abstract There are increasing concerns over potential long-term changes in krill distribution and abundance, determining whether such changes are related to climate change and/or harvesting and how they might be distinguished from multi-scale variability. With nearly a century of observations on krill distribution, the general patterns of occurrence were determined over 50 years ago, however the last decade has seen an important consolidation of details of seasonal and inter-annual distributions, although some potentially important krill habitats such as the Bellingshausen, Amundsen and Ross Seas are still poorly sampled. Moreover the unexplored perennial multi-year pack-ice regions still await survey by remote under-ice samplers. Krill consistently occur in the upper 200 m of the ocean but various recent observations show that a substantial proportion of the population may be found below this pelagic zone. To further understand variability and change there has been a focus on the importance of historic data with the compilation of comprehensive databases, re-analysis of large scale synoptic surveys and the generation of smaller scale time series that now span several decades. In particular the compilation of 90 years of net samples has provided a key resource both in showing changes in large-scale abundance over the last 50 years and together with acoustic data in generating modern estimates of overall krill abundance and biomass. Meanwhile, the increasing availability of ocean circulation models together with remotely sensed data, has enabled the exploration of large-scale teleconnections at both the regional (Scotia Sea) and circum-polar scale and at the seasonal to decadal time scales.

Keywords Vertical distribution • Production • Larvae • Seasonal variability • Time series

---

V. Siegel (✉)
Thuenen Institute of Sea Fisheries, Palmaille 9, 22767 Hamburg, Germany
e-mail: volker.siegel@thuenen.de; siegel.v@web.de

J.L. Watkins
British Antarctic Survey, Natural Environment Research Council, High Cross, Madingley Road, CB3 OET Cambridge, UK
e-mail: jlwa@bas.ac.uk

V. Siegel (ed.), *Biology and Ecology of Antarctic Krill*, Advances in Polar Ecology,
DOI 10.1007/978-3-319-29279-3_2

2.1 General

The knowledge of the distribution and abundance of Antarctic krill is a key element in understanding the ecology of the species and also vital in the sustainable management of the krill fishery in an ecosystem that has been exploited for the last two centuries. At the circumpolar scale the distribution can only be determined by considering data from many years, collected by many nations and then collated and analysed jointly. In addition the accessibility of krill to surveys has been made more difficult by presence of sea-ice and so it is perhaps not surprising that even after nearly a century of scientific study there are still areas that have not been sampled extensively. The early estimates of abundance came from net samples and some visual observations of surface swarms. The difficulty associated with the extreme patchiness of krill, linked to the large area of potential occurrence, led to attempts to estimate krill biomass from independent methods such as estimating predator consumption or amounts of primary production. Later with the development of acoustic techniques it has been possible to have much greater sampling coverage (both vertically and horizontally) but the acoustic techniques require methods to identify what proportion of the acoustic signal can be attributed to krill (target identification) and also to scale the signal to generate krill density (Target Strength). Net and acoustic estimates were frequently considered to give very different results but more recently there has been more of a convergence in the combined use of the estimates derived by different methods.

In this chapter we consider initially the large scale distribution, abundance and production estimates, particularly focussing on the compilations of data that are now available through international databases such as KRILLBASE, before going on to discuss the different conditions in the sectors of the Southern Ocean relating to the Atlantic, Indian and Pacific oceans. Finally we focus on the processes occurring at the smaller scales and particularly at the meso-scale (10–100s km) where annually repeated surveys now provide much information on the variability and change occurring within the Southern Ocean.

2.2 Circumantarctic Distribution, Global Biomass and Production

2.2.1 Horizontal Distribution

The habitat of Antarctic krill, *Euphausia superba*, is actually a fraction of the approximately 32 million km^2 area of the entire Southern Ocean south of the Polar Front. According to Mackintosh (1973) Antarctic krill are spread over some 19×10^6 km^2 around the Antarctic Ocean. Compilations of acoustic surveys prior to 2000 estimated the size of the krill habitat as 8.4×10^6 km^2 (Nicol et al. 2000a), while Siegel (2005), taking account of post 2000 acoustic surveys increased the estimate to 12.5×10^6 km^2. However it wasn't until the publication by Atkinson et al. (2009) which takes account of the majority of net samples collected over the last 90 years, that the values stated by Mackintosh of a habitat area of 19×10^6 km^2 around the Antarctic Ocean have been confirmed.

Marr (1962) used the 'Discovery' net samples for the first broad view of the overall krill distribution. The northern limit of krill distribution is south of the Polar Front. This northern limit reaches a considerably lower latitude in the Atlantic under the influence of the Scotia Arc and Weddell Gyre than, for example in the Indian Ocean sector. Marr (1962) also concluded that the main krill concentrations are confined to the Weddell-Scotia Sea and that east of 30°E is a poor krill habitat.

Mackintosh (1973) tried to combine net data with the distribution data of pelagic whaling – including the sightings of krill surface swarms – to better delineate and measure the area inhabited by krill and to indicate where higher concentrations might be expected. The general distribution of post-larval krill in summer is well summarized in his figure 14 (and reproduced here in Fig. 2.1). For comparison Fig. 2.2 displays the mean krill distribution across the entire Southern Ocean generated using all available historic krill density data of the KRILLBASE data base (see Atkinson et al. 2004, 2008) from net sampling surveys including additional data sets mentioned in the figure legend. In comparison to Marr's work, Mackintosh (1973) more clearly defined the northern limits and furthermore indicated five to six stocks around the continent; the Weddell-Scotia stock, the Enderby stock, the Kerguelen-Gaussberg stock, two smaller stocks in the northern Ross Sea and the Bellingshausen stock. However, Mackintosh (1973) highlighted that these areas of higher krill density should not be regarded as isolated populations whereas Latogursky (1979) speculated that the three main gyre systems around the continent might be regarded as independent populations.

Between the northern belt of krill occurrence in the eastern Atlantic and the southern Weddell Gyre branch, Mackintosh (1973) and Marr (1962) observed a gap of low krill abundance (Fig. 2.1). They concluded that the scarcity of krill here (65–68°S and 30°W–20°E) might be explained as a kind of backwater between the stocks of the Scotia Sea and the southern Lazarev Sea. Similar observations from net sampling surveys were made by Boysen-Ennen and Piatkowski (1988) and Siegel (2012) and are apparent in Fig. 2.2.

In the Indian Ocean sector Marr (1962), Mackintosh (1973) and Fig. 2.2 show that krill in the 90–150°E region are confined to a narrow belt close to the coast. Oceanographic data indicate a southward extension of warmer water (Bindoff et al. 2000) and Lanin (1993) showed that the region between 20°E and 120°E is the only region where the 2 °C isotherm crosses 60°S and the 1.5 °C isotherm is recorded in its southernmost position which may explain the constraining of the krill habitat to coastal waters (Nicol et al. 2000b). Consequently, krill presence may be high in the Kerguelen-Gausberg stock east of 115°E, but the overall biomass is not great because of its restricted latitudinal distribution (Mackintosh 1973). The Australian BROKE survey in 1996 showed similar results (Pauly et al. 2000; Nicol et al. 2000b) and confirmed that krill are confined to a narrower belt near the coast than anywhere else around the continent.

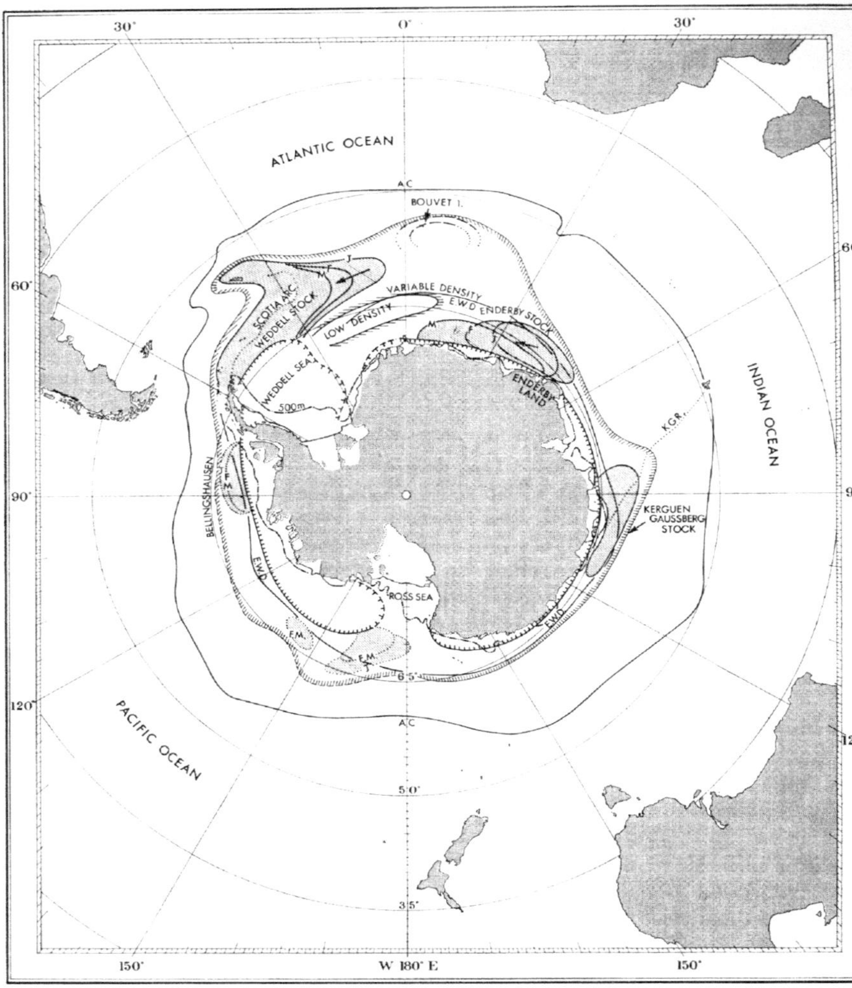

Fig. 2.1 Circumpolar distributions of postlarval Antarctic krill taken from Mackintosh (1973) Discovery Report 36, with permission from the National Oceanography Centre, European Way, Southampton

In the Pacific sector Mackintosh (1973) observed that in spring (October–December) in the East Pacific most net hauls caught no krill in contrast to the Atlantic and Indian Ocean sectors where krill were caught in low numbers at this time of the year. He concluded that the krill habitat in the Pacific is almost entirely covered by ice until late December. However, the very scarce information from the Pacific led to the conclusion by Marr (1962) that the Pacific is relatively poor in krill and that the stock is located almost entirely south of 65°S. Very few additional data have been collected since then in the region 150°E–80°W in the Amundsen/Bellingshausen Sea (Murray et al. 1995; Siegel and Harm 1996). Studies of large-

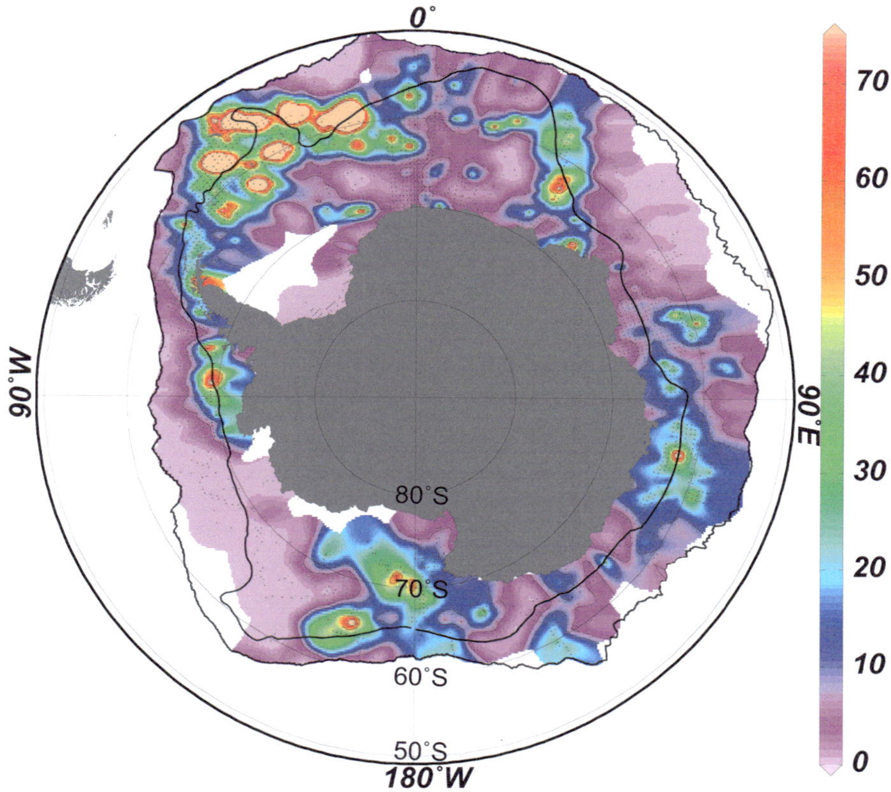

Fig. 2.2 Circumpolar distribution of Antarctic krill, average densities (number per m^2) for 0.5°lat by 1°long from net sampling surveys, data origin KRILLBASE (with permission of data holders, Atkinson et al. 2004, 2008), additional data sources are from publications by Fevolden (1979, 1980), Boysen-Ennen and Piatkowski (1988), CPR data (Hosie et al. 2003), Sala et al. (2002), Siegel (2005, 2012), Siegel et al. (2013), Nicol et al. 2000b; Kawaguchi et al. (2010a)

scale coastal polynyas revealed that the ice krill *Euphausia crystallorophias* dominates these regions (La et al. 2015). Italian acoustic and net surveys in the eastern Ross Sea (Sala et al. 2002; Azzali et al. 2006) indicated that the northern and northwestern areas of the Ross Sea are dominated by *E. superba*, but ice krill replaces Antarctic krill in the High Antarctic Zone south of 74°S (Azzali and Kalinowski 2000; Sala et al. 2002; Azzali et al. 2006). Very similar observations are reported from the south-eastern Weddell Sea, where Antarctic krill not only diminish but almost completely disappear from the region south of 74°S and being replaced by ice krill (Fevolden 1979; Siegel 1982; Boysen-Ennen and Piatkowski 1988).

Even after over 90 years of work on Antarctic krill in the Southern Ocean there are still areas that have not been sampled. The perennial multi-year pack-ice regions of the western Weddell Sea as well as the eastern Ross Sea are still blank spots on the map, and sampling of these regions will have to wait for the development of remote sampling systems that can operate under the permanent ice.

Few studies have demonstrated a clear and consistent relationship between simple environmental factors and krill. For temperature, salinity, oxygen concentration, nutrient level, dissolved organic matter, seawater stability, frontal systems and latitude: no single environmental factor shows a predictable relationship with krill density (Witek et al. 1981; Weber and El-Sayed 1985; Weber et al. 1986). Hydrographic features are known to affect the distribution of krill at broad range of scales (Nicol et al. 2000c). On the large scale (100–1000s km) Strutton et al. (2000) found that lower phytoplankton biomass and productivity coincided with extension of warmer oceanic water and a shift from krill/diatom to salp/picoplankton communities. Although Siegel et al. (2013) found a general correlation between the abundance of krill and chl-a in the Antarctic Peninsula region, this relationship was weak. Satellite-based estimates of ocean colour (e.g. SeaWiFS) may provide useful proxies for measuring and mapping phytoplankton concentrations, however, this information relates to ocean surface conditions and does not provide information on the quality of food that krill prefer. The offshore distribution of adult krill along the Antarctic Peninsula was delineated by the warm (~1 °C), low salinity (33.8) water at 30 m, suggesting that most krill were present shoreward of the Southern Boundary of Antarctic Circumpolar Current Front (SBF).

On the meso-scale krill may interact with the physical structure and circulation (Simard and Lavoie 1999; Allen et al. 2001). The only fine-scale (1–10s km) relationship between krill and the environment occurs between krill density and bathymetry (Trathan et al. 2003) with the highest densities of krill occurring in the shelf-break zone as demonstrated by the consistent concentration of the krill fishery into these areas in the Scotia Sea and Antarctic Peninsula. In addition in the Antarctic Peninsula region the presence of Circumpolar Deep Water (CDW) interacts with the shelf break and where deep canyons onto the shelf exist the UCDW water transports krill onto the shelf (Ashijan et al. 2004; Lawson et al. 2004). The UCDW provides the deep troughs and canyons with nutrients leading to increased total phytoplankton and diatom biomass (Kavanaugh et al. 2015). By modifying their behaviour (diel vertical migration, swarming, and continuous swimming) krill distribution may result in retention and accumulation in such local areas. Areas like the Bransfield Strait in particular, with the steep bathymetry, deep basins, and inflow of water from the ACC in the west (Savidge and Amft 2009; Jiang et al. 2013), and the Weddell shelf in the east of the Peninsula (von Gyldenfeldt et al. 2002; Thompson et al. 2009) create a unique coastal basin habitat. The bathymetrically complex habitats together with the cyclonic circulation and eddies over the basins (Zhou et al. 2002; 2006), may be an important mechanism for retaining krill within the Bransfield Strait (Allen et al. 2001; Hofmann and Murphy 2004).

While simple environmental correlates are not obvious, the recognition of potential krill habitat using multi-variate techniques has been undertaken. A recent analysis by Cuzin-Roudy et al. (2014) predicts suitable krill habitat using multivariate regression modelling techniques (Boosted Regression Trees) to relate the observed presence-absence of krill with a suite of biologically relevant local environmental variables. Their model of habitat suitability explained 63 % of variance, and has been used to infer the presence of *E. superba* in regions where

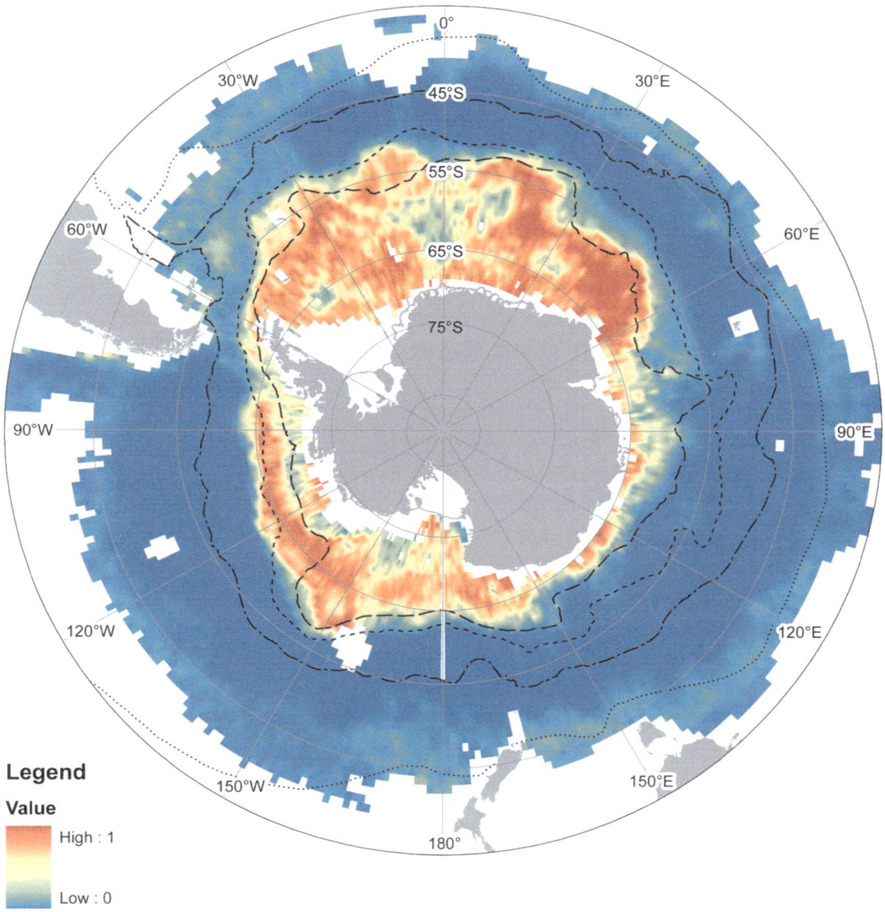

Fig. 2.3 *Euphausia superba* modelled habitat suitability using presence-absence data and environmental variables; from Cuzin-Roudy et al. (2014) (with permission)

sampling was scarce. Their map (Fig. 2.3) shows a high probability of presence almost everywhere south of the Polar Front, and a low probability beyond it. When compared with Fig. 2.2 it is notable that several areas where krill are not seen in the current distribution would appear to be areas of suitable habitat. In particular, it predicts that the regions between 90°W and 150°W, the Amundsen Sea region, and 40°E to 60°E are suitable krill habitat and therefore the infrequent detection of krill may be due to the scarcity of the sampling. In this model *E. superba* was predicted to preferentially inhabit regions with cold temperature at depth (200 m) between −1.5 and +1.5 °C, high oxygenated near surface water, shallow to intermediate mixed layer depths between 30 and 50 m, productive waters with high silicates, nitrates and chlorophyll which are usually met in the marginal ice zone of the retreating sea-ice in spring and near the coast in summer (Cuzin-Roudy et al. 2014).

Cuzin-Roudy et al. (2014) used presence/absence krill data and Fig. 2.3 represents the probability of presence of the species based on the environment. Inside this generally suitable habitat, krill may show even higher preferences for some areas depending on other factors such as currents, gyres and sea-ice which would add further explanations why krill density and biomass in the overall suitable habitat are not evenly distributed but show clear centres of aggregation in the Southern Ocean.

Biomass calculations by Atkinson et al. (2009) indicated that 70 % of the entire circumpolar population is concentrated within the Southwest Atlantic (0–90°W), but that this area only accounts for 25 % of the overall available habitat (see also below under biomass). One possible explanation for the high krill density in the region might be that the distribution range of krill along the Antarctic Peninsula and the western Scotia Sea are narrowed down by the Drake Passage and krill aggregations may be condensed by the channelling effect. A complementary possible explanation may be the complex bathymetry of the Atlantic sector which creates a greater accumulation of concentrations with longer retention times. Although krill occur in low densities over wide areas, it is the size of the area which constitutes the high biomass, notably the oceanic areas between 30°W and 30°E. Only 13 % of the total krill stock lives over shelf/shelf break areas. Overall 87 % of the stock is found over deep oceanic water (Atkinson et al. 2008). The oceanic distribution provides a refuge from most land-based predators. Here krill aggregations are more dynamic and less predictable in time and space than in near-shore eddies. Migration and advection control, concentration and retention in productive sectors as well as reduced mortality/predation loss represent a blend of top-down and bottom-up control and could explain most of the unusual asymmetrical distribution of krill (Atkinson et al. 2008).

2.2.2 *Vertical Distribution*

Although high abundance regions are well known for Antarctic krill (see above, Constable and Nicol 2002; Nicol 2003), the small-scale (<1 km) spatial distribution of krill is quite variable and difficult to predict. Krill are typically distributed in aggregations or swarms which can show differing degrees of concentration and dispersion over diel cycles. Swarming is likely a response to predation but may also offer energetic advantages in comparison to dispersed krill (Ritz 2000). They also may also show a diel vertical migration which may be strong or absent (Hamner and Hamner 2000; Siegel 2005), and may even use the bottom as a habitat (Kawaguchi et al. 1986; Gutt and Siegel 1994; Schmidt et al. 2011). Aggregation behaviour as well as vertical migration will be one of the main focus topics of Chap. 8 (Tarling and Fielding 2016). Here the focus is on the general pattern of vertical distribution.

During routine krill surveys standard net tows are carried out in the upper 200 m water column (or slightly less). Vertical extent of krill occurrence has been reported from the summer period and occurs generally within the upper 150 m water column

(Miller and Hampton 1989; Demer and Hewitt 1995; Lascara et al. 1999), while in more northern latitudes like South Georgia krill may occur deeper in the water column, although still above 200 m (Marr 1962). Acoustic observations also tend to support the concept of a major portion of the population occurring in the upper 250 m of the water column, thus during the CCAMLR 2000 survey acoustic sampling covered the upper 500 m of the water column but no significant biomass of krill was found below 150 m (Demer 2004). The depth distribution of krill in the vicinity of 13 day-time stations indicates that krill tended to be concentrated near the top of the thermocline. It is suggested that this was due to lateral current shear in the boundary region creating favourable conditions for feeding and aggregation (Hampton 1985).

Similar results were reported by Jarvis et al. (2010) from the large-scale BROKE-WEST survey to the Indian Ocean sector in austral summer 2006. The majority of acoustic krill detections were in the top 100 m of the water column centered around 50 m depth. Most of the krill population appeared to reside between the bottom of the surface mixed layer and the winter thermocline, and often in high concentrations around the summer thermocline. For the Antarctic Peninsula region the minimum depth at which *E. superba* biomass was detected acoustically in the water column during summer was usually at depths shallower than 50 m. However, in winter krill were absent throughout most of the area (Siegel 1989) and occurred in large but less dense aggregations deeper than 100 m (Lascara et al. 1999). The total depth range of biomass steadily declined from the most common range of several meters down to a relatively rare range of around 200 m (Lawson et al. 2004). Seasonal net sampling surveys in the Lazarev Sea also provided evidence that the summer surface krill concentrations move deeper in the water column in autumn. RMT (Rectangular Midwater Trawl) net tows confirmed that krill concentrated in water layers deeper than 200 during daylight hours and only migrated into the 0–200 m depth stratum during night (Siegel 2012).

However, there are an increasing number of observations that krill may regularly be found in deeper water layers. For instance, Marin et al. (1991) caught gravid krill specimens in depth stratified net samples to at least 1000 m depth in Drake Passage over deep water, while benthopelagic krill in summer have also been observed by ROVs at depths down to 480 m where krill were swarming close to the bottom (Gutt and Siegel 1994). Clarke and Tyler (2008) presented photographs of krill, possibly gravid females, from abyssal depths down to 3500 m. Recently Kawaguchi et al. (2011) provided video footage of an entire sequence of the mating behaviour of Antarctic krill near the sea floor at depths between 400 and 700 m. A detailed summary of observations is given by Schmidt et al. (2011). Such direct observations of krill across a wide range of depths are also complemented by data from predator feeding studies which indicate that some predators have been taking krill well below the traditionally accepted upper 150 m. Taki et al. (2005) found evidence for a seasonal pattern in the krill fishery in the Scotia Sea to fish deeper during winter and shallower during summer, which they related to changes in the mean depth and amplitude of the species vertical migration.

Schmidt et al. (2011) argue that when calculating the total krill abundance over the sampled water column, the low krill densities at depth would translate into relatively high total numbers, because the sampled habitat volume below 200 m was twice (shelf) or even nine times (ocean) as large as the epipelagic layer. Therefore, Schmidt et al. (2011) estimated that up to 20.3 % of the sampled oceanic krill population, if extrapolated down to 2000 m depth, and 8.5 % of the shelf population may be found below 200 m. In contrast, Atkinson et al. (2009), on the basis of the KRILLBASE deep-towed nets, concluded that 97.3 % of the total krill population were located in the upper 200 m in summer. While from survey data collected along the Antarctic Peninsula in the very good krill year 1982, Siegel (1985a) reported krill presence in net samples between 400 and 600 m, however, their density rapidly decreased with depth and >95 % of total krill abundance occurred in the upper 200 m in summer.

In general results suggest that even though krill densities are low in deeper waters below 200 m, it is common to encounter small numbers. If these amounts of deep krill are really important at the population size level, then further studies that take into account seasonal, latitudinal, shelf/oceanic changes in the proportion occurring at depth will be important both for understanding the biology as well as in stock assessment procedures.

2.2.3 *Biomass*

The extreme variability in the density of krill caused by its swarming behaviour can cause problems in estimating biomass and production. Early attempts were made to estimate krill biomass directly from net samples, or indirectly from estimates of predator consumption, or primary and secondary production assuming phytoplankton/krill conversion ratios. Gulland (1970) assumed that 50 % of the total zooplankton biomass might be attributed to krill, which would lead to a biomass of 750 million tonnes (Mt). Several authors tried to estimate krill biomass from annual consumption by whales and came up with figures between 38 and 270 Mt.

From plankton net samples Marr (1962) provided two estimates of total krill biomass: 44 and 521 Mt. The first biomass value was based on Marr's original visual observation estimate of 2.5 g m^{-2} while the latter figure was derived by Marr on a recalculation using potential consumption of krill by whales and a revised estimate of 29 g m^{-2}. These broad limits were extended further by Makarov and Shevtsov (1972), who quoted a range of 953–1350 Mt of biomass. Voronina (1998) used net sample densities from three main zones based on commercial fishing zones of regular and rare concentrations as well as dispersed krill occurrence. Data were grouped according to these zones, means were calculated and the total biomass of krill was found to be 272 Mt. Clearly, these levels of uncertainty in estimates made it impossible for quite some time to establish a reliable management of the krill resources. Further information on these early estimates can be found in Everson (1977) and Miller and Hampton (1989), who presented biomass estimates ranging from 14 to 7000 Mt. More recent estimates have been based on standardized net or multi-frequency acoustic sampling surveys and will be discussed below in more detail.

Over the past two decades, hydro-acoustic surveys have become a routine technique for estimating Antarctic krill biomass and now most of the circumpolar regions have at least been surveyed once. Nicol et al. (2000a) estimated the overall krill biomass from all available acoustic meso- and large-scale data sets around the continent and derived biomass estimates between 60 and 155 Mt covering a krill habitat of 8.4 million km^2. The advantage of the study is the circum-polar coverage of the combined data sets based on a single method. Even so there are several caveats as not all acoustic data had been analysed by the same analytical procedures or to the same level of coverage. Out of all the areas surveyed the Southwest Atlantic provides the most comprehensive set of results.

Siegel (2005) updated the estimate of Nicol et al. (2000a) including new acoustic biomass density data from surveys such as the CCAMLR 2000 Survey (Hewitt et al. 2004) and the Italian surveys in the Ross Sea (Azzali and Kalinowski 2000). The mean instantaneous biomass ranged from 67 to 297 Mt for an area covering 12.5 million km^2, depending on whether low or high survey results from different years were included in the estimate. The high variability around a mean of approximately 170 Mt does not seem to be unrealistic, given that it is unlikely that extremely favourable or poor conditions will prevail around the entire continent at the same time.

Atkinson et al. (2009) used an extensive database (KRILLBASE) of published net haul density and length frequency data from the beginning of the 'Discovery' era to 2004. After a series of selection criteria were applied (spring–summer–autumn period using untargeted sampling, individuals >19 mm long), the extracted net sampling data were standardized and krill density was averaged over the period 1926–2004, during which krill abundance has fluctuated substantially. The overall mean estimate of krill biomass was 379 Mt. This standardized net-based estimate falls well within the range of acoustic estimates of 60–420 Mt (Nicol et al. 2000a; Siegel 2005). While it has been problematic to relate net catches and acoustic estimates at the scale of the individual net haul or aggregation (Watkins 2000; Everson 2000) at the larger scale of the regional populations there is now considerable convergence in the biomass estimates derived from net sampling and from acoustic survey (Atkinson et al. 2009). The high biomass reflects a massive distributional range spanning over 19 million km^2. Obviously high-latitude species such as *E. superba* all produce only one generation per year and additionally have a non-growing season of up to 8 months. It takes the species several years of a multiyear life cycle to build up this high biomass (Siegel 2000).

The calculations indicated that 58–71 % of krill are located within the Southwest Atlantic, a phenomenon that has been reported repeatedly (Siegel 2000; Nicol et al. 2000a; Constable et al. 2003; Atkinson et al. 2004). The biomass studies made aspects obvious that were already discussed by Marr (1962) and Mackintosh (1973): although krill show a circumpolar distribution, the distribution is not only asymmetrical on a circumpolar scale, but krill also show marked regional differences in biomass densities and lower densities seem to be real in the Indian Ocean (see also the discussion about differences in swarm density and inter-aggregation spacing between the Atlantic and Indian Ocean sector (Miller et al. 1993) with larger aggregations in the Atlantic, swarms closer to the surface and smaller inter-swarm distances than in the Indian Ocean).

Despite a widespread distribution, it is well documented that krill aggregate in hotspots (Huntley and Niiler 1995; Constable and Nicol 2002). Such areas likely provide both good food availability for krill and shelter from offshore currents heading towards less productive areas (Hazen et al. 2013). On the other hand current fisheries management considers only large scale distribution and abundance information. Identifying smaller scale krill hotspots as a mid- to end-trophic level prey resource is significant as these meso-scale hotspots may warrant protection to ensure key ecosystem functions and resilience and may be important for the conservation of krill-dependent predators (Santora et al. 2011).

2.2.4 Production

For the understanding of the quantitative role of a species in the transfer of organic material from primary production to higher trophic levels, it is essential to interpret and compare information about krill production. Instantaneous and local krill biomass estimates have often been criticized as being too low to sustain the large predator populations in the Southern Ocean. However, such calculations only consider one element of what is needed to assess the availability of krill to the predators, it does not take into account that it is not the overall or local biomass that is important for the predator, but it is the krill availability determined through factors such as production and flux into the predator foraging ranges that are also of vital importance.

Production is generally defined as the total amount of energy fixed or metabolized by an individual or an animal population. Since this is a dynamic function, it has hardly ever been measured directly. Consequently, production has usually been estimated by three indirect methods (Allen 1971):

1. Methods which estimate production from the number, length and weight of individuals of a given population at the beginning and end of a set period of time (e.g. a year);
2. Methods based on estimates of the amount and energy content of food consumed by a population, considering metabolic losses as well as the efficiency of food assimilation of the individual animal;
3. Methods based on estimates of the amount and proportion of a population consumed by predators

Production has rarely been estimated for Antarctic krill, and early estimates for total production ranged from 75 to 1350 Mt per year (Miller and Hampton 1989). Murphy et al. (2004a) argued that production is at least four times larger than the estimated krill consumption by the main krill predators in order that the krill stock may be sustained against the predation. This hypothesis was first put forward by Yamanaka (1983). However, the factor was based on a number of tentative assumptions (predation rates, predator nutrition turnover rates, modest krill stock size of 10^8 tons). Yamanaka (1983) himself carefully stated that due to the uncertainty in using the various parameters of unknown accuracy, his computations aimed to set up the model and find out what parameters are important rather than to obtain the quantitative result itself.

Several attempts had been made to estimate the ratio of production to biomass (P/B). Allen (1971) was the first to calculate a ratio of P/B = 1.8–2.3 and stated the lower end of the range being more likely. The P/B estimations of Allen (1971) were carried out for various growth and mortality models, but demonstrated that the applied short longevity and high mortality rates are fundamentally important and lead to an overestimate of the P/B ratio. Based on the potential reproduction of an average female krill and a mortality rate of Z = 5.5, Yamanaka (1983) suggested the production-biomass ratio to be least 1.19 between the parent at spawning and 1 year of age. Miller et al. (1985) obtained different values ranging from P/B = 1.19–2.77. They used acoustic measurements of krill standing stock and concurrent primary productivity measurements together with data on krill energy requirements to assess the productivity of the species. Again it is important to note that these results are based on growth rates for a short krill life cycle of 2 years. Tseitlin (1989) argued for an even higher P/B ratio of 3.6; in his opinion krill production should be similar to copepods. However, Aseev et al. (1988) doubted the similarity of krill and copepods and estimated a P/B ratio of 1.03 for the krill population in the Indian Ocean based on the age structure derived from composite length frequency data (cited in Menshutkin 1993).

Siegel (1986, 2000) applied the approach by Winberg (1971) which defines production as the sum of growth increments (weight increase of all individuals during a 1 year period) of all post-larval age groups. This reflects the annual production of a monocyclic species, i.e. the population is represented by one age group when the reproductive period is of limited duration. It includes the annual elimination (mortality) which is defined as the difference between original and final abundance multiplied by the arithmetic mean of the starting and final individual weight of the respective age group. The production of a species with a multi-year life-cycle is obtained by summing up the results for the single age groups (age based catch curve). Krill production was estimated as the relative relationship between production and biomass and ranged from P/B = 0.83–1.1 (mean 0.95) for different years. The yearly biomass production is highest in age group 1+ and 2+ and can account for 64–67 % of the total annual stock production (Siegel 1986).

Voronina (1998) applied the P/B ratios by Allen (1971) and biomass densities for the year 1984 (17–29 g m^{-2}) and estimated an annual krill production for the entire Antarctic of 215 Mt. Using a mean P/B ratio of 0.95, assuming a krill distribution area of 18 million km^2, and using the biomass density data from the FIBEX expedition in 1981, Siegel (1986) estimated a total annual krill production of 220 Mt.

Rather than attempting to calculate net production (as described above), which reflects both growth and mortality operating together, Atkinson et al. (2009) have separated the two processes by calculating gross production. This is the theoretical increase in mass of the krill population over the growing season due to growth, egg production and lipid build-up, and in the absence of mortality. They used three different growth models to calculating gross production. The first two methods are based on the instantaneous growth rate (IGR) method. The first model was used to provide a circumpolar-scale prediction of daily growth rates in mm per day based on ambient food, temperature and krill length (Atkinson et al. 2006). The second model by Kawaguchi et al. (2006) includes a longer period of the growth season, but does not account for environmental variation. The third method was based on field-derived composite monthly length-frequency data from KRILLBASE as an index of growth. All of these methods provide a series of increasing lengths of krill throughout the growing season. These were converted to mass increase over time. Results from all three methods were in broad agreement and converged on values of total annual production of 342–536 Mt y^{-1}. Atkinson et al. (2009) highlight that the main result is that gross annual production slightly exceeds the biomass of 379 Mt estimated as a long-term annual mean for the entire population. This turnover rate is lower than that of Antarctic copepods but it is still substantial for a polar species that grows to a size of 6 cm. Current best estimates of total predator consumption range from 128 to 470 Mt y^{-1} (Mori and Butterworth 2006), which is in the lower range of the above gross production estimates. Considering gross production therefore resolves the often-reported discrepancy between estimates of instantaneous krill biomass and predation loss.

2.2.5 *Krill and Sea-Ice*

Seasonal sea-ice plays an integral role in the ecology of krill. Sea-ice coverage around Antarctica varies seasonally between 4×10^6 km^2 in summer and 20×10^6 km^2 in winter (Gloersen et al. 1992). Although the physical ice conditions did not appear to directly affect krill, observations on the seasonal distribution and behaviour of krill were interpreted to be a function of the need for food supply and predator avoidance (Daly and Macaulay 1991).

First basic geographical descriptions of the austral winter pack-ice zone in the Bellingshausen Sea were given by Cook (1900) on the drift of the 'Belgica' in 1898/1899 south of 69°S. From this cruise we also gain the first information on the occurrence of ice algae and large quantities of krill feeding on diatoms under the ice flows as well as the occurrence of krill in crabeater seal stomachs (Hansen 1908).

Shipboard observations provided anecdotal information of krill clinging to the undersurface of overturning ice floes. Highest densities (tens to hundreds m^{-1} cruise track) were recorded by Marschall (1988) when the vessel broke through pressure ridges. ROV surveys indicated average krill densities of 0.1–21 krill m^{-2} under smooth-bottomed ice and 12–800 m^{-2} under rugged ice. Field observations by SCUBA divers revealed mean densities of 2–24.6 krill larvae m^{-2} in early and late winter sampling periods (Frazer et al. 1997). Krill preferred two types of ice habitat (a) smooth-bottomed flat sheets and (b) rugged ice caused by piled up floes of pressure ridges or by melting, in contrast ice floes underlain with a billowy layer of unconsolidated ice platelets were usually avoided by krill (Spiridonov et al. 1985; Garrison et al. 1986; O'Brien 1987; Marschall 1988; Bergström et al. 1990). Highest densities were recorded in caverns and crevices of pressure ridges and the branched melting cavities during November (Marschall 1988). The feeding behaviour of krill under the ice in winter described by Garrison et al. (1986), O'Brien (1987), Stretch et al. (1988), and Marschall (1988) will be discussed further in the chapters on physiology and on food and feeding (see Chaps. 4 and 5, in this volume, Meyer and Teschke 2016; Schmidt and Atkinson 2016).

Although Marschall (1988) argued that the ROV surveys revealed that krill occurred almost exclusively in close association with the ice, net catches and acoustic measurements as well as SCUBA dives confirmed that krill were also present in the water column beneath the ice. O'Brien (1987) reported on small swarms swimming freely in the water column with densities between 10 and 20 krill m^{-3}, while densities ranged from 1 to 100 krill m^{-2} when feeding on the undersurface of the ice. Most of the krill were of juvenile/subadult stages with a mean length of approximately 36 mm.

At least in the upper 100 m depth layer of the marginal ice zone of the northern Weddell Sea, Daly and Macaulay (1988) recorded krill aggregations or loosely defined layers by acoustics and estimated krill densities between 1 and 68 g m^{-2}. Theoretically, these aggregations could have also been *E. crystallorophias* according to O'Brien (1987) who observed more ice krill in the water column under the ice than Antarctic krill. However, O'Brien's study area was located on the high latitude shelf close to the slope which is the preferred habitat of *E. crystallorophias*, whereas Daly and Macaulay (1988) worked in an area over deep sea ocean far away from any continental shelf where *E. crystallorophias* does not occur. Their net samples revealed mostly krill of 9–25 mm length in spring, which are late larvae and juveniles of the age-group 0.

Brierley and Watkins (2000) acoustically detected midwater krill aggregations under pack-ice in the Scotia, Weddell and Bellingshausen Seas. They could not observe significant differences between ice-covered or open waters in terms of the horizontal and vertical extent of swarms or inter-swarm distance. However, krill swarms in the Weddell (summer) and Scotia Seas (spring) were generally shallower under ice than in open water, whereas in the Bellingshausen Sea (spring) they occurred deeper in the water column. Sprong and Schalk (1992) acoustically sampled along a north-south transect running into the ice between November to January in the Weddell Sea; during this period krill abundance remained relatively constant under the ice although the location of peak abundance shifted north with time.

Although discrete spot-measurements have detected krill under ice, there is a paucity of quantitative information on the meso-scale distribution and abundance of krill beneath ice. From an early spring (October) survey to the northern Weddell Sea, Siegel et al. (1990) calculated from RMT net sampling mean krill densities between 1 and 27 g m^{-2} in the closed pack-ice zone. Krill length frequency data from the consolidated pack-ice area covered only 18–34 mm small juvenile krill (modal length 23 mm). Adult krill of 20–58 mm size dominated in the ice free area to the north or sometimes they were also detected in the marginal ice zone in the depth layer 60–200 m beneath the 0–60 m depth stratum inhabited by juveniles. Otherwise, krill length frequency distributions obtained by divers from the ice undersurface corresponded well with those samples taken at the same station by RMT nets in the water column (Bergström et al. 1990). On the role of sea-ice for the overwintering survival and the recruitment of krill see Chaps. 3 and 4 of this volume.

Using an autonomous underwater vehicle (AUV) equipped with an upward looking echosounder, Brierley et al. (2002b) surveyed krill under pack-ice in summer. Krill were concentrated within a 13 km wide band of the marginal ice zone south of the ice edge. Within this band, mean density under ice was 61.6 g m^{-2} and fivefold greater than that of open water. However, the acoustic dead-zone extended approximately 1 m beneath the ice-water interface also prevented acoustic data sampling there. This would indicate that the estimated density under sea-ice is probably conservative because of the association of juvenile krill in particular which are mainly associated with crevices within the ice. The AUV survey demonstrated that krill abundance declined rapidly after 13 km from the ice edge and moving from the dynamic and heterogeneous environment of the marginal ice zone into the area of 100 % sea-ice concentration.

To study the immediate surface and under-ice habitat (0–2 m) Flores et al. (2012a) used a specially designed Surface and Under Ice Trawl (SUIT). During the LAKRIS surveys in the Lazarev Sea they sampled the sea-ice water column interface in summer, autumn and winter and krill densities were compared with RMT net haul data taken from 0 to 200 m depth at the same stations. The study provided evidence that Antarctic krill in high latitudes are associated with the under-ice habitat year-round and especially in winter krill not only use the marginal sea-ice zone but spread hundreds of kilometres into the ice covered area. Surface densities in the 0–2 m depth stratum were generally higher under sea-ice than in open water with highest surface densities during winter time. The water column beneath the surface layer (0–200 m) had lower krill densities in winter. This was due to a few exceptionally high 0–2 m SUIT catches indicating – at least in winter – an even higher variability in krill densities of the under-ice habitat than in the water column beneath the ice. The numerical differences were not related to differences in stock composition (small versus large krill), because no significant difference was apparent when comparing surface SUIT and epipelagic RMT krill length frequency data.

2.3 Distribution and Abundance of Larvae

2.3.1 Developmental Ascent and Vertical Distribution

From the extensive "Discovery" collection Fraser (1936) concluded that the eggs and nauplii are found in deep rather than shallow water. Marr (1962) presented the hypothesis on the developmental ascent of larval stages with eggs sinking from the spawning surface layer to greater depth where nauplii hatch in depths as deep as 2000 m. The larvae pass through two stages of nauplii and one of metanauplius as they ascend in the water column. Feeding does not occur before calyptopis I stage reaches the surface layer. Mackintosh (1972) noted that the "Discovery" net tows down to 1000 m may not even be quantitative, because there was evidence that eggs sink beyond the reach of the deepest nets. Experimental work confirmed that eggs are sinking continuously and hatching occurs after about 6 days, however, in some cases the estimated sinking rate was relatively low so that hatching would generally occur in 850 m depth (Quetin and Ross 1984). According to George and Strömberg (1985) experimentally determined sinking rates are extremely variable even within a single brood. So it is not surprising that other experimental studies on sinking rates and incubation time concluded a hatching depth of 1250–1850 m (Marschall 1983) which is much closer to at sea findings.

All these studies assume a spawning depth of 50–200 m. However, more recently an entire sequence of the mating behaviour of Antarctic krill in the wild was captured on underwater video and provided evidence that mating can take place at depths of 400–700 m (Kawaguchi et al. 2011). While no spawning was observed in this sequence krill are assumed to spawn shortly after mating and so it is quite possible that spawning could be occurring deeper than 200 m. Furthermore, more stratified sampling programs down to 2000 m depth confirmed Marr's (1962) conclusions of a developmental ascent. Hempel and Hempel (1986) reported on net samples in the Scotia Sea as deep as almost 3000 m over water depths of more than 4000 m. Marin et al. (1991) sampled Drake Passage and the deep basins of Bransfield Strait. In both studies eggs and nauplii abundance peaked between 1000 and 2000 m and even below 2000 m eggs and nauplii were present. The abundance of eggs was two orders of magnitude higher in the deep layer compared to the 500–1000 m layer. Also the majority of eggs were at a multicellular stage which corresponds to an age of just 1 day after spawning but were already more than 1000 m deep. Nauplii and metanauplii were observed in lower numbers in 200–1000 m depth, where calyptopis stage 1 dominated. In the top layer only calyptopis larvae were found (Hempel 1985b; Hempel and Hempel 1986). Similar results were reported by Makarov (1982) who recorded eggs and nauplii in the Scotia Sea mostly over areas where the water depth was between 2000 and 3000 m.

Makarov sampled from 0 to 500 m (sometimes 1000 m) depth by vertically stratified net tows. In late spring and summer eggs and nauplii mostly occurred deeper than 500 m with catches an order of magnitude greater between 500 and 1000 m. In addition virtually all metanauplii were recorded in that deeper layer.

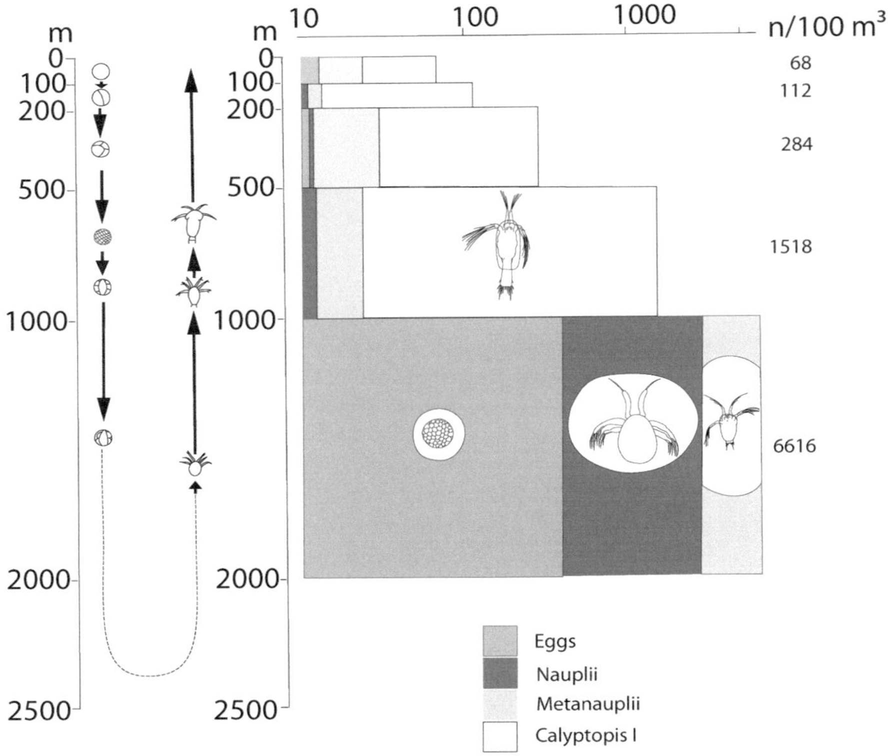

Fig. 2.4 Vertical distribution of early krill eggs and larval stages in different depth layers, larval stages as % within each layer bar. On the *left* side: Marr's (1962) hypothesis of the developmental ascent; from Hempel and Hempel (1986) (with permission)

Early in the season, between December and February, 40–80 % of the calyptopis 1 larvae tended to occur between 200 and 500 m deep. Later in the season in February–May 40–60 % were concentrated in 100–200 m and sometimes up to 80 % in the upper 25 m surface layer. Although in February–March furcilia 1 and 2 seemed to prefer the 200–500 m depth zone, in April–May almost all furcilia 1–6 had moved into the upper 25 m layer (Makarov 1982). Besides the ontogenetic developmental ascent and the seasonal vertical migration of the different larval stages, calyptopis and furcilia performed a diel vertical migration mostly within the 0–100 m layer occurring deeper during the night. This leads to the conclusion that krill spawning is not necessarily confined to shelf areas as suggested by Marr (1962) who considered that eggs and early larvae were carried in the bottom water away from the continental shelf to undergo the development in the oceanic deep water. Figure 2.4 shows an up-dated graph on the quantitative vertical distribution/developmental ascent given by Hempel and Hempel (1986).

Along the shelf break and at offshore stations of the south-western Antarctic Peninsula, Ashjian et al. (2008) found that larval krill often occur in or just above the pycnocline during autumn as well as winter (although not specified these larvae were probably mostly furcilia). During autumn larval krill would then be most abundant in the leftover winter water and in the Antarctic Surface Water which confirms Makarov's (1982) findings of larvae moving to shallower water depth during autumn. Ashjian et al. (2008) used the Video Plankton Recorder between 20 and 250 m depth and reported on significantly reduced krill larval abundance during winter which might be explained by the observation that high densities of larvae occur on the under-ice surface in winter when sea-ice is present (Garrison et al. 1986; Kottmeier and Sullivan 1987). Frazer et al. (1997) reported that ice-associated larval krill abundances were significantly greater during late winter than during early winter. Neuston sledge sampling in winter showed that even in open water krill furcilia larvae obviously preferred the upper 1 m water layer (Siegel 1989, Krakatitsa et al. 1993 cited in Pakhomov 2000).

2.3.2 Horizontal Distribution and Abundance

Although some local studies on the deep distribution of early life stages have been carried out since the time of the 'Discovery' expeditions, studies with a wider spatial coverage on krill eggs and nauplii stages are still missing. Several large scale surveys (e.g. GAMLR, FIBEX, CCAMLR 2000, BROKE, BROKE-WEST, CPR) have at least added valuable data sets on the distribution and abundance of surface larvae. These large-scale data including several national meso-scale surveys have been used to produce an up-dated map of the average distribution of krill larvae around the continent (Table 2.1, Fig. 2.5a, b).

Voronina (1974) found a correlation between the distribution of early calyptopis larvae and the depth of the upper boundary of dense bottom water. She concluded that the successful spawning of krill is limited to those areas where the dense bottom water is not deeper than 1800 m. Although this 1800 m boundary does encompass many of the localities in which early larval stages were found, many positive records fall outside this limit and so it probably does not fully describe the northern extent of successful spawning.

Within the total habitat of krill, the early larvae tend to be restricted to zones where low sea surface temperatures extend to greater depths if not to the sea floor, often characterized by the disappearance of the thermocline in the East Wind Drift or enclosed basins such as the Bransfield Strait (Mackintosh 1972). Marr (1962) demonstrated that surface larvae first occur in the western Weddell drift and later in the eastern and in the more southerly parts. Surface larvae usually reach their maximum quantities in March and April (for more details on the timing of spawning see Chap. 6, Kawaguchi 2016). The western Scotia Sea is thought to represent the most productive spawning area of the circum-Antarctic krill populations (Marr 1962). Figure 2.4a, b are composite maps produced from multi-year observations which indicate where hotspots of krill larvae may generally been found. Fortunately, the FIBEX and CCAMLR 2000 large-scale surveys

Table 2.1 Examples of mean krill larvae densities from large-scale surveys, Cal calyptopis, Fur furcilia stages, Numbers for areas are references to CCAMLR Statistical Subareas/Divisions

Year/survey	CCAMLR Area	Longitude	Mean Cal	Density Fur	($N\ m^{-2}$) total	References
1981 FIBEX	SW Atlantic 48.1–48.3	30°W–70°W	18,602	435	19,308	Siegel et al. (2004)
2000 CCAMLR	Scotia Sea East 48.3–48.4	20°W–70°W	2	0.2	2	Siegel et al. (2004)
2000 CCAMLR	Scotia Sea West 48.1–48.4	36°W–70°W	1842	203	2044	Siegel et al. (2004)
2011 Polarstern and USAMLR	Antarctic Peninsula 48.1	50°W–70°W	910		910	Siegel et al. (2013)
1979 Polarsirkel	Weddell Sea East 48.5	20°W–30°W			324	Fevolden (1980)
1979 Polarsirkel	Weddell Sea South 48.5	(South of 74°S)			8	Fevolden (1980)
1980 Polarsirkel	Weddell Sea East 48.5	10°W–30°W			367	Hempel and Hempel (1982)
2004 LAKRIS	Lazarev Sea 48.6	6°W–3°E			358	Siegel (2012)
1981 VNIRO	Lazarev Sea 48.6	5°W–15°E			1010	Makarov et al. (1985)
2006 BROKE West	Cosmonaut and Cooperation Sea 58.4.2	30°E–80°E	17	0.3	17	Kawaguchi et al. (2010a)
1984 SIBEX	Cosmonaut Sea 58.4.2	50°E–65°E	8		8	Miller (1986)
various	Prydz Bay 58.4.2.	60°E–80°E			22	Pakhomov (2000)
1985 SIBEX	Cooperation Sea/Prydz Bay 58.4.1. and 4.2	58°E–93°E	97		97	Hosie (1991)
various CPR	East Antarctica 58.4.1. and 4.2	70°E–140°E	121	45	166	McLeod et al. (2010)
1996 BROKE	East Antarctica 58.4.1.	80°E–150°E	205	0.5	206	Nicol et al. (2000b)
1982, 1984	Somov Sea, Ross Sea 88.1	150°E–170°E			>1000	Timonin (1987)

included krill larvae as an integral part of their research objectives. This gives us a synoptic view of large-scale larval distribution and density.

The hypothesis that the Scotia Sea is a major reproduction centre for krill was in principle confirmed during the international FIBEX expedition in 1981 (Rakusa-Suszczewski 1984) and the CCAMLR 2000 Survey (Siegel et al. 2004). These two surveys showed highest concentration of krill larvae in the western part of the

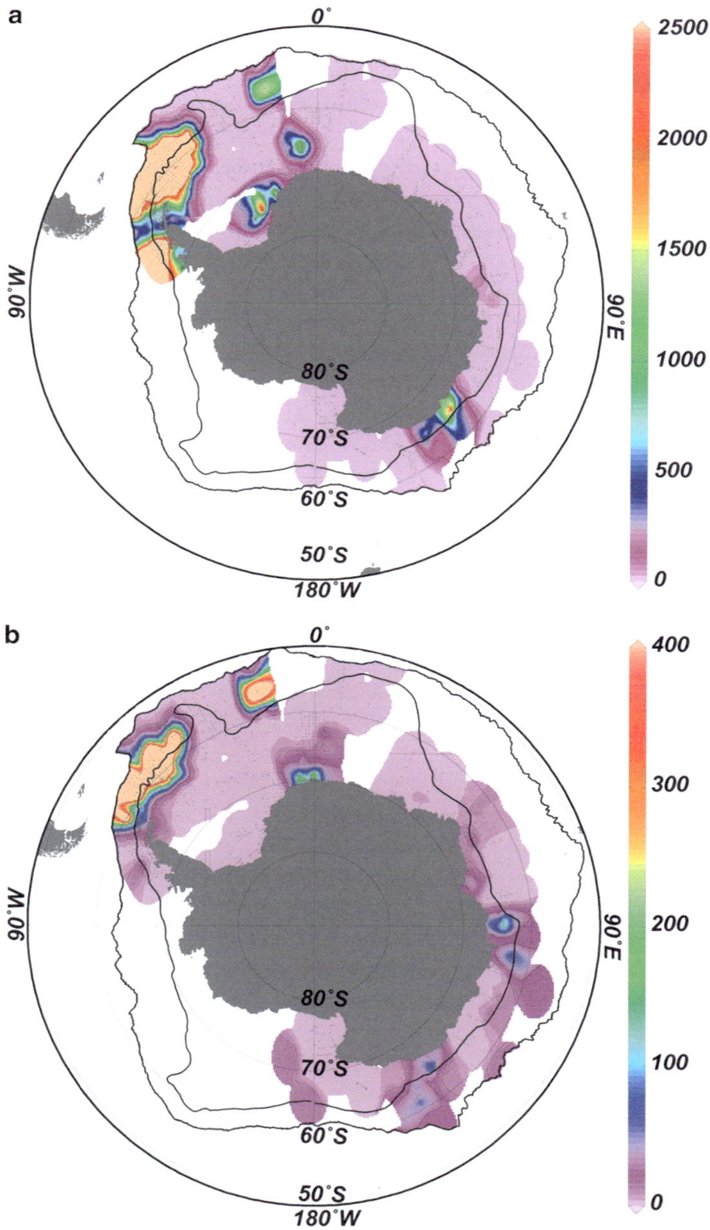

Fig. 2.5 Global distribution of Antarctic krill larvae, densities are averaged for geographical bins of 0.5° latitude and 1° longitude given in numbers m^{-2}; (**a**) calyptopis and (**b**) furcilia; note different scales for calyptopis and furcilia densities; data are from Hempel and Hempel (1978, 1982, 1986), Fevolden (1979, 1980), Siegel (1982), FIBEX by Rakusa-Suszczewski (1984), Miller (1986), CPR data by Hosie et al. (2003), Menshenina 1992, BROKE by Nicol et al. (2000b), CCAMLR 2000 Survey by Siegel et al. (2004), Siegel (2005, 2012), BROKE-WEST by Kawaguchi et al. (2010a), Siegel et al. (2013)

Atlantic sector (Fig. 2.5a, b). In 2000 there were a few scattered stations with records of calyptopis larvae in the South Sandwich Islands region, but there was a sharp boundary around 36°W where larval distribution rapidly petered out. The FIBEX season was very special with a mean abundance of >19,000 calyptopis m^{-2} and a maximum of 1.8×10^6 C larvae m^{-2}; densities were about three orders of magnitude higher than in other years. Similar distribution patterns were observed by Hempel and Hempel (1978) who surveyed the Scotia Sea between 60°W and 14°E, but found very few larvae east of 36°E until mid-May despite high catches in the central Scotia Sea.

It is clear that krill larval abundance in the western Atlantic sector is subject to marked annual fluctuations. Another example of a large scale analysis is given by Brinton et al. (1986) who counted average larval densities of 35 m^{-2} in the Scotia Sea during summer 1984. This is an example of a relatively poor year for larval krill. The CCAMLR 2000 survey yielded a mean density of larvae in the western Scotia Sea of 2044 m^{-2}, but only scattered aggregations of larvae were recorded in mid-summer east of 36°W (2 m^{-2}) (Siegel et al. 2004). Larvae were concentrated along both sides of the Weddell Front (WF) and occurred in somewhat lower abundance up to the approximate position of the SACCF (Southern Antarctic Circumpolar Current Front), while gravid females remained south of the SACCF (Siegel et al. 2004). Previous work on euphausiid larvae from the 'Melville' expedition (Brinton 1985) has highlighted the importance of this frontal region for reproduction in Antarctic krill. However, *E. superba* calyptopis stage 1 larvae were widespread but had a more southerly distribution while progressively older calyptopes and furcilia stages were found further north and east in surface waters (Ward et al. 2004).

In the South Atlantic the distribution patterns of larval stages showed a continuous band of larvae from the central Scotia Sea, along the northern slope of the South Orkney Islands and Elephant Island to the western side of the Antarctic Peninsula. According to data from Siegel (1989 his Fig. 8) and the LTER program (Ross et al. 1996) this band extended upstream as far as Adelaide Island west of the Antarctic Peninsula at the entrance of the Bellingshausen Sea.

High concentrations of krill larvae have always been found along the Antarctic Peninsula from Adelaide Island in the south to the eastern Bransfield Strait in the north (Fraser 1936; Marr 1962; Hempel et al. 1979a; Kittel and Jazdzewski 1982; Mujica and Asencio 1983; Witek and Kittel 1985; Siegel 1989; Ashjian et al. 2008; Siegel et al. 2013). In some years calyptopis and furcilia larvae were most abundant on the shelf and in the continental slope areas (Makarov et al. 1990a) with few in the oceanic zone; in 1976 highest concentrations exceeded 2000 larvae m^{-2} mostly in the eastern Bransfield Strait and around Elephant Island whereas the south-western side of the Peninsula showed less dense concentrations. In other years larval distribution patterns showed the opposite with high mean densities of 3590 larvae m^{-2} and >8800 m^{-2} in 2001 encountered in the slope area (Ashjian et al. 2008; Pakhomov et al. 2004) or mean densities of 3885 larvae m^{-2} in 2011 in the southern oceanic zone vs 15 m^{-2} in the northern Peninsula region (Siegel et al. 2013). Those high densities of larval krill observed in Marguerite Bay are obviously carried by advective transport of the Circumpolar Deep Water (CDW) onto the shelf and via the deep canyons (Wiebe et al. 2004, 2011; Ashjian et al. 2008). Although spawning seems to occur in the shelf regions of the Peninsula, it can be inferred from historic records of early larval stages in the oceanic waters of the Scotia Sea (Hempel and Hempel 1978; Rakusa-Suszczewski 1984; Hempel 1985a; Brinton et al. 1986; Siegel et al. 2004) that the inner shelf zone and its function as a successful spawning habitat is less well supported.

South Georgia is well-known for its high concentration of adult krill (Marr 1962; Brierley et al. 1997; Watkins et al. 1999). However, it is questionable whether it is an important or successful spawning ground for Antarctic krill (Ward et al. 1995; Atkinson et al. 2001). Catches of early larvae were small to moderate in some years, e.g. 1978 and 1979 (Hempel 1981; Witek et al. 1980) and 2001 and 2003 (Tarling et al. 2007) but close to zero in most years (Mackintosh 1972; Hempel and Hempel 1978; Witek 1979; Ward et al. 1990; Siegel et al. 2004). Since Antarctic krill off South Georgia are living close to the northern distributional limit several options exist to explain the recruitment failure, such as the inability of individuals to attain sexual maturity or whether larvae hatch and develop successfully (Ward et al. 1990; Peck et al. 2004); unfavourable environmental conditions or high levels of predation could also preclude larvae from recruiting into the juvenile/adult stage (Murphy et al. 1998; Murphy and Reid 2001).

Since gravid females are often present in the South Georgia area, Tarling et al. (2007) used adult abundance data in a semi-empirical model, to determine the rate of larval development and assumed a mortality rate to predict larval concentrations in the area. According to their model results, they concluded that spawning takes place and the descent-ascent cycle should be completed successfully in off-shelf regions. Tarling et al. (2007) discussed some of the most likely potential causes for the absence of larvae in the area which may be predation by chaetognaths, myctophids and *Themisto gaudichaudii*, or advective export, although over the time-scale of weeks advective forces in the South Georgia region are unlikely to displace early larvae very far from where they were spawned. Since many South Georgian larvae do not make it beyond the early developmental stages there must be further factors that prevent the successful development of local larvae. *E. superba* is a stenothermal, cold-water species and Atkinson et al. (2006) note that South Georgia summer temperatures between 3 and 4 °C are suboptimal for krill growth. It is possible that early developmental stages of krill are especially vulnerable to small changes in temperature and so that there is a physiological constraint on successful development to postlarval stages in this area.

The other side of the favourable habitat range seems to be located in the northern Weddell Sea, where the perennial ice exists. Interestingly, field data from the northwestern Weddell Sea showed that hardly any noteworthy concentrations of larvae occurred along transects into water masses flowing out of the Weddell Sea (Hempel and Hempel 1978; Rakusa-Suszczewski 1984; Hempel 1985a; Siegel 2005). Despite this larval densities in the Scotia Sea were high at this time (e.g. in the year 2000) or even exceptionally high (in 1976 and 1981) and the marginal ice zone was a preferred feeding ground for larvae in spring (Siegel et al. 1990). The permanently ice-covered Weddell Sea does not seem to be a major source of krill reproductive output. Melnikov and Spiridonov (1996) reported on the occurrence of low densities of old furcilia and post-larval stages under the permanent sea-ice of the western Weddell Sea. This could have been an effect of a poor year-class, but it may be that the composition of the stock in the Weddell Sea during February–April was very typical for late winter to early spring in a permanently ice covered zone (Daly 1990). Furthermore, these late furcilia did neither grow nor develop further during the study period. Therefore, Melnikov and Spiridonov (1996) concluded that these krill larvae could not belong to the 0+ group originating from the studied season, but must have been born in the summer before and developed extremely slowly under poor feeding conditions under perennial sea-ice. The authors further suggested that the observed larvae in the north-western Weddell Sea could not be krill of local origin, but had arrived by advection from the eastern Weddell Sea where krill were known to be spawning (Hempel and Hempel 1982). These findings contradict the assumptions made by Capella et al. (1992) for modelling the drift of early krill life stages and which resulted in a hypothesis that the western Weddell Sea is an important source of krill larvae in the Bransfield Strait.

To the east of South Georgia similar circumstances to South Georgia have been observed in the past, although data are much sparser. Close to the Polar Front and north and west of Bouvet, several surveys found a "total predominance" of large mature krill (Fevolden 1979; Krafft et al. 2010; Siegel 2012). However, there are no records of early larvae for the Bouvet region, except much later in the season (Marr 1962). The limited data from the 'Discovery' expeditions indicate that the Scotia Sea larval concentrations might disperse further to the east with the progressing season. In autumn to early winter, krill furcilia larvae had been found along 50°–60°S latitude as far as 20°E (Marr 1962; Makarov et al. 1992).

To the south of Bouvet (south of 60°S) first calyptopes were encountered in the central Lazarev Sea in early December with gravid females. Data from Makarov et al. (1992) show a gap in krill larval distribution between 57°S and 63°S, which is mainly an area of small (juvenile) and medium-sized (immature and adult) krill and thus probably a less productive spawning ground. In early autumn calyptopis larvae can occur in fair numbers north and west of Maud Rise and close to the continental shelf (Makarov et al. 1985; Makarov and Menshenina 1992; Siegel 2012). However, mean larval densities for different years (1010 m^{-2} in 1981 and 358 larvae m^{-2} in 2004) recorded by Makarov and Menshenina (1992) and Siegel (2012), respectively, are substantially lower than usually found in the Southwest Atlantic or compared directly with the amount of larvae found during FIBEX in 1981 in the Scotia Sea.

To the west of the Lazarev Sea in the eastern Weddell Sea, *E. superba* larvae were found in the Coastal Current by Fevolden (1979, 1980) and Hempel and Hempel (1982). Although the mean density of 367 calyptopes m^{-2} was similar to the Lazarev Sea densities, it was quite obvious that south of 74°S *E. superba* larvae are very uncommon (Fevolden 1979; Hempel and Hempel 1982). Here the Coastal Current splits into two: one limb turns west into the central permanently ice-covered Weddell Sea along the very deep continental slope with a mean density of 324 krill larvae m^{-2}; the other limb follows along the continental ice shelves into the southernmost Weddell Sea where mean densities did not exceed eight larvae m^{-2} (Fevolden 1980). In the south *E. crystallorophias* larvae dominated the plankton often exceeding several thousand calyptopes and furcilia m^{-2} (Hempel and Hempel 1982).

To the east of the Lazarev Sea between 20° and 80°E krill larvae were mainly caught in a narrow band along the shelf slope where larvae would be transported westwards by the coastal current (Miller 1986; Menshenina et al. 1988; Makarov and Menshenina 1989; Hosie 1991; Kawaguchi et al. 2010a). In January calyptopis stages 1 and 2 were found (Kawaguchi et al. 2010a) which corresponds with larval stages from the adjacent Lazarev Sea (Siegel 2012). While most of the larvae in the eastern section between 70° and 80°E may be transported further west along the Mawson Coast to recruit into the Lazarev Sea population, some of these larvae may potentially be carried back into the Prydz Bay Gyre system and recruit into the local population there.

Pakhomov (2000) presented a generalized diagram for the Cosmonaut and Cooperation Sea (Prydz Bay region) between 30° and 80°E explaining the distribution pattern of larval and spawning stock accumulation. He showed that the spawning stock accumulated along both sides of the Antarctic Divergence. Eggs are released in the oceanic region and hatch in the warm deep waters. When larvae reach the surface in the Divergence region they are transported by surface currents in either a southward or northward direction. Although larvae have occasionally been observed north of the Antarctic Divergence (Hosie et al. 1988; Miquel 1991; Pakhomov and Karpenko 1992) their concentrations were generally low (average 22 larvae m^{-2}).

East of Prydz Bay the narrow band of krill larval distribution extends to the western entrance of the Ross Sea (Witek et al. 1980; Makarov and Menshenina 1989; CPR data see Hosie et al. 2003 and McLeod et al. 2010). East of 120°E, abundances were considerably higher with a mean of 637 larvae m^{-2} (Nicol et al. 2000b). In contrast, the 'Discovery Reports' show a maximum of surface larvae in the East Wind Drift from 120 to 130°E but only moderate abundances in the 90–120°E sector (Marr 1962).

For the north-western Ross Sea, Marr (1962) listed some records of deep and surface larvae, but Sala et al. (2002) did not find any larvae in their plankton samples during January–February. In the northern Ross Sea and Somov Sea around the Balleny Islands, Witek et al. (1980) found just a few calyptopis during January, but most samples yielded no larvae, although later in the season (March/April) Timonin (1987), Makarov and Menshenina (1989) and Makarov et al. (1990b) recorded higher catches regularly exceeding 10^4 larvae m^{-2} off the shelf. Conditions in the southern Ross Sea are probably similar to those in the southern Weddell Sea which also is a preferred habitat for *E. crystallorophias* rather than for *E. superba*.

The Amundsen Sea and Bellingshausen Sea are still seriously under-sampled so that hardly any reliable density figures for krill larvae are available (Marr 1962;

Makarov and Menshenina 1989). However, the high densities of furcilia in autumn in the eastern outflow region of the Bellingshausen Sea off Marguerite Bay suggest that early life stages must be present in greater numbers at least in the Bellingshausen Sea (see also below the modelling exercise by Piñones et al. 2013).

2.3.3 Advection of Larvae

As described in detail above, krill offspring go through an ontogenetic migration, whereby they initially sink as developing eggs and then re-ascend as actively swimming larvae. It has been demonstrated that the rates of sinking and re-ascent are closely related to environmental factors such as water density (Marschall 1983) and temperature (Ross et al. 1988). Hofmann et al. (1992) used published parameters from experimental studies to predict the depth–time trajectory of the early life stages. They found that the presence of the relatively warm Circumpolar Deep Water (CDW) in the Antarctic Peninsula region was instrumental in accelerating development, allowing larvae to re-enter the food-rich surface waters long before energy reserves were exhausted. However, some experimental sinking rates seem to be too low to display properly the deep distribution of eggs and early larvae observed from field samples.

Once calyptopis emerge at the surface they become passive drifters carried away by the major surface currents. Numerical Lagrangian particle tracking experiments were carried out for the region that extends from the western Bellingshausen Sea to the area north of the Antarctic Peninsula (Piñones et al. 2013). The abundance and distribution of Antarctic krill along the western Antarctic Peninsula suggest that the population is maintained by influx via advection from the Bellingshausen Sea. According to the model, particles released in the Bellingshausen Sea were transported to the southern Peninsula shelf in 4 months, which is similar to the time required for krill eggs to develop into furcilia larvae. About one fourth of these particles released along the shelf break will cross the outer shelf by onshore intrusions of the CDW into areas such as the Marguerite Bay area (Dinniman et al. 2011). From drifter trajectories (Beardsley et al. 2004) and geostrophic and ADCP velocities (Klinck et al. 2004) it was argued that loss of larval krill from the shelf could have occurred either to the north or to the south depending on how far inshore the larvae were found. Although results by Piñones et al. (2013) supported the idea that the outer shelf and shelf break larvae will be transported northeastwards with the ACC, the model results also indicated that in addition to the remote input of larvae, local spawning on the mid and inner shelf show low export and higher retention on the shelf but less southward transport of larvae.

An across-shelf gradient is generally observed in larval krill stages and abundance in the Peninsula region, with relatively high abundances of early larvae occurring along the outer shelf (up to a mean of 3885 larvae m^{-2}, see Siegel et al. 2013) and lower densities (average 8–30 larvae m^{-2}) of primarily older furcilia stages occurring on the inner shelf (Daly 2004; Pakhomov et al. 2004; Siegel et al. 2013). A potential mechanism in autumn and winter is the deep water mass which derives from the CDW, and which is present at the outer edge of the south-western Antarctic Peninsula continental shelf at depths of 200–600 m (Hofmann and Klinck 1998; Smith et al. 1999). There is no coastal current front barrier to block flow of CDW from the deep ocean onto the continental shelf (Talbot 1988). Consequently, the outer WAP shelf region is directly affected by the northeastward flowing ACC (Klinck et al. 2004).

Results of a three-dimensional Lagrangian particle tracing model for the greater Bransfield Strait region (Capella et al. 1992) indicated that seasonal changes in the wind stress field influence direction and velocity of surface currents, which is the primary factor influencing the embryo-larva trajectories. Krill eggs released to the north of the South Shetland Islands, west of 62°W, are transported into Drake Passage and carried on by the ACC. Therefore, the transport of krill larvae by the ACC provides an important source to regions downstream farther to the northeast, seeding krill populations along the northern Peninsula and beyond into the Scotia Sea (Siegel 1988, 2005; Hofmann et al. 1998; Fach et al. 2002). Krill larvae transported from the Bellingshausen or Weddell Sea into Bransfield Strait are usually advected westwards.

Further modelling studies by Hofmann et al. (1998) and Fach et al. (2002) used a circulation model to trace the trajectories of krill eggs and larvae across the Scotia Sea region. According to Hofmann et al. (1998) the passive transport of larvae as surface drifters would take on average 140–160 days as they move from the spawning grounds west of the Antarctic Peninsula with the currents associated with the SACCF and Southern Boundary of the ACC (SBF) across the Scotia Sea to South Georgia. Fach and Klinck (2006) reported on WOCE drifters that took 208 days to complete the trajectory from the South Shetland Islands to the east of South Georgia. For all food scenarios tested, krill larvae did not reach growth rates or length compositions observed from field samples. Fach et al. (2002) therefore concluded that interrupted transport in a potentially better food environment (e.g. overwintering in the marginal ice zone) would allow larvae to grow to a size that is observed for the krill population at South Georgia.

The model developed by Hofmann and Hüsrevoglu (2003) indicated that regional variations in bathymetry and the prevalent water masses resulted in different levels of success in the early developmental stages. Although the absolute results (hatching times and depths) may not always match real field observations, it is however interesting to note that the simulated pattern suggests that the circumpolar distribution of hatching depth is not equal for all parts of the Antarctic. The longest times required to ascend to the surface would be in the Amundsen, Ross, and Weddell Seas. According to the model, shortest ascent times occur along the Antarctic Peninsula and in the Bellingshausen Sea. These differences in ascent time result in differences in carbon usage for the non-feeding larval stages, so larvae of the Peninsula/Bellingshausen Sea region ascend quickest and have more energy reserve left when reaching the surface; this then means more time is available for them to encounter suitable feeding conditions. This advantage may be responsible for a better survival rate of larvae and again highlights the importance of the Antarctic Peninsula area for larval development as compared with other regions around the continent. The asymmetry in spawning success around the continent obtained by the model could partly be explained by areas of dependable food supply and according to Hofmann and Murphy (2004) these potential successful spawning habitats are regions where the SBF is near the shelf edge. There will certainly be a need to verify these modelling results with field observations (Siegel 2005).

2.4 The Optimum Zone of Krill Distribution

Figures 2.2 and 2.5 show that the principal postlarval krill concentrations, and in many parts the occurrence of larvae, are mainly located between the SACCF (Orsi et al. 1995), i.e. the southernmost core of the Antarctic Circumpolar Current (ACC) and the Coastal Current. Within the large-scale cold gyres of the Weddell or the Ross Sea, the southernmost shelf areas do not belong to the preferred habitat of Antarctic krill. Only in the western Atlantic sector, both the distribution range of adult krill and the zone of larval abundance, extend beyond the SACCF to the north up to the Polar Front. Meso-scale eddies and meanders associated with the SACCF and the regional, so-called secondary fronts (between the ACC and the large-scale gyres, between the ACC and the Coastal Current (CC), and between the gyres and the Coastal Current) appear to favour the formation of krill concentrations (Makarov et al. 1980; Maslennikov and Popkov 1988). The temperature of the Circumpolar Deep Water (CDW)) in this zone ranges between +0.5 and +2 °C and seems to favour the development of early life stages of Antarctic krill sinking down to greater depth (Hofmann et al. 1992). Seasonal sea-ice cover provides shelter and food resources for overwintering krill larvae and recruits. In spring, sea-ice retreat stimulates phytoplankton development in the water column, which is advantageous for the final maturation process and a successful reproduction of Antarctic krill (Cuzin-Roudy and Labat 1992; Spiridonov 1995; Siegel and Loeb 1995; see also Chap. 3, Reiss 2016). Furthermore, eddies along regional frontal systems (Maslennikov and Popkov 1988) create accumulation and retention areas of phytoplankton concentrations in summer (Heywood and Priddle 1987; Solyankin 1992). Such retention areas are potentially good food environments for a successful growth of calyptopis larvae (Fach et al. 2002). Therefore, the zone between the SACCF and the Coastal Current may be considered as the zone of both climatic and biotic optima for the Antarctic krill distributional range (Spiridonov 1996). In this region the abundance of krill is determined by the optimal conditions for both the biological component of production and survival as well as the physical component represented by retention and export.

2.5 Ocean Basin-Scale Distribution and Biomass

The large scale views of distribution and abundance described in the preceding section were obtained from a variety of surveys and years. They display a very general picture of the krill distribution and the overall mean abundance and show where over time more predictable concentrations of krill may be found. These views however have less ability to show details of year to year changes in krill distribution and abundance. The temporal and spatial scale relationships of biotic and abiotic processes are important to the stock level of Antarctic krill. Seasonal and interannual changes occur at a scale of 100 to >1000 km (Murphy et al. 1988). A closer look into the large and meso-scale survey results will allow for some perspectives on the short and long-term variability of krill stocks within a season and over decades.

2.5.1 Antarctic Peninsula and Scotia Sea

In 1981 the first large-scale population census, known as FIBEX (First International BIOMASS Experiment), was carried out for the Antarctic krill. This multi-national, multi-vessel, quasi-synoptic research activity was part of the research programme BIOMASS (Biological Investigations of Marine Antarctic Systems and Stocks) that set various priorities for krill research (El Sayed 1994) including determining the large-scale biomass of krill, fluctuations in krill abundance, population identity (see Chap. 7, Jarman and Deagle (2016)) and swarming behaviour (see Chap. 8, Tarling and Fielding (2016)). Results of the international effort for the first time led to an acoustic biomass estimate of 26.5 Mt of krill in the surveyed areas of the Southwest Atlantic which covered 0.63 million km^2. A later re-analysis using different target strength information revised the estimate to 30.8 Mt, i.e. 77.6 g m^{-2} (Trathan et al. 1995). These biomass estimates were used by CCAMLR in 1991 when the Commission established the first krill conservation measure for the Southwest Atlantic. The initial maximum annual catch was limited to 1.5 Mt and was adjusted over time with new results coming forward (see Chap. 11, Nicol and Foster (2016).

It took 19 years before another international multi-ship krill "population census" was undertaken in the Scotia Sea in summer 2000 (Watkins et al. 2004). The CCAMLR 2000 Survey covered an area of 2.065 million km^2 of the Scotia Sea and generated an acoustic biomass estimate of 44 Mt (Hewitt et al. 2004). Since 2004 acoustic analysis methods for Antarctic krill have been revised considerably and this led to a series of revised biomass estimates from the CCAMLR 2000 Survey. Finally in 2010 CCAMLR agreed that a biomass of 60.3 Mt (CV 12.8 %) represented the best and most up to date estimate (SC-CAMLR 2010a). This estimate was calculated by the CCAMLR Working Group using a three-frequency identification window and a Stochastic Distorted Wave Born Approximation (SDWBA) model with a cruise specific orientation angle applying an inversion method described by SC-CAMLR (2010b) with additional information in Fielding and Watkins (2011). With this method the target identification was substantially improved (SC-CAMLR 2010b). For details on key stages in the development of this method the reader is referred to papers by McGehee et al. (1998), Demer and Conti (2003a, b, 2004, 2005), and Conti and Demer (2006).

However, this biomass figure of 60 Mt should not be regarded as an increase in krill biomass since the 1981 FIBEX survey. Firstly the area surveyed during CCAMLR 2000 was about five times larger than the FIBEX survey area in 1981. Secondly in terms of analysis procedure, the biomass calculated for FIBEX is comparable with the original CCAMLR 2000 survey but not the updated 2010 value of 60.3 Mt. The original calculated mean density for CCAMLR 2000 was only 21.4 g m^{-2} (scaled up in the re-assessment to 29.2 g m^{-2}) compared to the FIBEX value of 77.6 g m^{-2}. The CCAMLR 2000 Survey also carried out 119 - un-targeted RMT8 hauls, which generated a biomass estimate of 38 Mt, i.e. 18.7 g m^{-2} (Siegel et al. 2004).

In their re-appraisal of the total krill biomass from net sampling data, Atkinson et al. (2009) observed that the CCAMLR 2000 survey area contains 28 % of the total stock. Applying this proportion to the acoustic estimate published by Hewitt et al. (2004), they estimated a total circum-Antarctic krill biomass of 133 Mt in summer 2000. However, now that CCAMLR (2010) has revised this Scotia Sea estimate to 60.3 Mt, then the Atkinson et al. (2009) biomass for the overall stock size in the year 2000 should be revised to 215 Mt. The long-term average net-based biomass of 379 Mt given by Atkinson et al. (2009) is certainly much higher than estimated from the acoustic and net sampling survey in 2000, which may indicate that the CCAMLR 2000 survey hit a period with krill biomass below the long-term average.

A number of other large-scale surveys have been carried out in the Scotia Sea (e.g. Sushin and Shulgovski 1999), and although they have collected valuable quantitative data, these have never been analysed to derive at an overall biomass estimate. The distribution patterns described by these surveys will be discussed below in the section on teleconnections between stocks.

2.5.2 *East Antarctica*

The FIBEX surveys from 1981 also studied the biomass of krill in the area between 30°E and 80°E. Survey results gave an estimated 8.55 Mt of krill biomass for this sector spread over an area of 2.29 million km^2. This estimate was later revised to 11 Mt using new target strength models (Trathan et al. 1992; Trathan and Everson 1994) which relates to 4.8 g m^{-2} (Table 2.2). CCAMLR used the FIBEX survey results to set precautionary limits on the krill fishery and established in 1991 the first krill conservation measure for the Indian Ocean. This limit was later revised due to results from the Australian BROKE-WEST survey in 2006 (Kawaguchi et al. 2010a; see also Chap. 11, Nicol and Foster 2016), that took place in January

Table 2.2 Krill biomass estimates from large-scale acoustic surveys

Year/survey	CCAMLR area	Longitude	Area (10^3 km^2)	Biomass 10^6 t	CV (%)	Density (g m^{-2})	References
1981 FIBEX	SW Atlantic 48.1–48.3	30°W–70°W	471	35.8	13.8	76.0	Trathan et al. (1992, 1995)
1981 FIBEX	SE Atlantic 48.6	15°E–30°E	576	4.6	22.9	8.0	Trathan et al. (1992)
2000 CCAMLR	SW Atlantic 48.1–48.4	20°W–70°W	2065	60.3	12.8	29.2	Hewitt et al. (2002, 2004) [a]
1994 J.C.Ross	South Orkneys 48.2	45°W–48°W	12			1.7–10.70	Brierley and Watkins (1996)
1996 Atlantida	South Orkneys 48.2	30°W–50°W	66	1.7		26.3	Kasatkina et al. (1997)
1981 FIBEX	Indian Ocean 58.4.2	20°E–90°E	1711	3.9	32.0	2.3	Trathan et al. (1992)
1984 SIBEX	Indian Ocean 58.4.2	65°E–75°E	345	5.5		15.9	Shirakihara et al. (1986)
1984 ADBEX	Indian Ocean 58.4.2	58°E–70°E	70	3.5		5.0	Higginbottom et al. (1988)
1985 SIBEX	Indian Ocean 58.4.2	55°E–95°E	1280	3.7		3.2	Higginbottom et al. (1988)
2006 BROKE WEST	Indian Ocean 58.4.2 W	30°E–55°E	681	16.2	18.4	23.8	Jarvis et al. (2010)
2006 BROKE WEST	Indian Ocean 58.4.2 E	55°E–80°E	537	11.6	29.8	21.6	Jarvis et al. (2010)
1991–1993 Aurora Australis	Prydz Bay 58.4.2 E	60°E–80°E	150			7.7–16.6	Pauly et al. (2000)
1996 BROKE	Indian Ocean 58.4.1 W	80°E–115°E	453	3.0	19.0	6.7	Pauly et al. (2000)

(continued)

Table 2.2 (continued)

Year/survey	CCAMLR area	Longitude	Area (10^3 km^2)	Biomass 10^6 t	CV (%)	Density (g m^{-2})	References
1996 BROKE	Indian Ocean 58.4.1E	115°E–150°E	420	1.8	30.0	4.2	Pauly et al. (2000)
1990 Italiantartide	Ross Sea 88.1	175°E–176°W	51/92	0.5/0.9		9.4/10.2	Azzali and Kalinowski (2000)
1994 Italiantartide	Ross Sea 88.1	175°E–176°W	160	3.0/3.1		18.9/19.5	Azzali and Kalinowski (2000)
1992 STERNA	Bellingshausen Sea 88.3	83°W–87°W	18			42.6	Murray et al. (1995)

Numbers for areas are references to CCAMLR Statistical Subareas/Divisions
[a]Biomass given here is for the recalculation undertaken in 2010 (SC-CAMLR 2010a, b; Fielding and Watkins 2011)

to March 2006 and covered an area of approx. 1.3 million km^2 between 30°E and 80°E. The mean density of *E. superba* integrated to 250 m depth across the survey area was 24 g m^{-2}. The total biomass was estimated to be 28.75 Mt (Jarvis et al. 2010). Aggregations of krill were recorded from the coast extending north to 62°S. However, mean densities were highest in the waters to the south of the SBF, while the waters to the north of the SACCF were almost devoid of krill. Half of the entire biomass was found within 80 km of the continental slope and densities peaked within 40 km of the shelf break.

Since BIOMASS, there have been a number of meso-scale (10–100s km) studies in restricted parts of this region, including three zooplankton/krill surveys in Prydz Bay, two krill surveys off Bechervaise Island (Pauly and Higginbottom 1994) and several fishery-related surveys (Dolzhenkov and Timonin 1990; Ichii 1990). Earlier net and acoustic survey activities were carried out by the Soviet Union in the 1970s and 1980s (Pakhomov 1995). However, it should be noted that the coverage of all these surveys was patchy and the survey design was very different to that now accepted by CCAMLR.

In austral summer 1996, the Australian BROKE survey covered the Indian Ocean area between 80°E and 150°E (Nicol 2000). The mean krill density from net samples from this survey in the eastern Indian Ocean was 2.65 krill 1000 m^{-3} which is low compared to other areas that have been surveyed (Table 2.2). The acoustic biomass density was 5.54 g m^{-2} and is at the low end of the range of values reported from the Scotia Sea or even the adjacent Prydz Bay region (Pauly et al. 2000). The biomass of krill originally estimated as 6.67 Mt in the surveyed area (873,000 km^2) was later revised to a biomass of 4.83 Mt (CV 17 %) largely because of a recalculation of the effect of sound absorption during the survey (Pauly et al. 2000). Krill biomass was twice as high in the west of the region (80°E to 120°E) than in the east (120°E to 150°E), which appeared to be related to the large-scale oceanographic conditions of the region, where a southward intrusion of warmer water approached the shelf/slope area. According to the longitudinal differences in krill biomass densities CCAMLR has split the eastern Indian Ocean sector and set separate krill catch limits for the areas east and west of 115°E (see Chap. 11, Nicol and Foster 2016).

2.5.3 *Ross Sea*

First information on krill from the Ross Sea suggested that *Euphausia superba* was almost absent from the area (Marr 1962). This conclusion was later supported by Mackintosh (1973) and Lubimova et al. (1984). It was not until Azzali and Kalinowski (2000) and Sala et al. (2002) summarized their results on krill distribution and biomass from the Italian Antarctic Programme expeditions, that the original view of the Ross Sea as a krill deficient area was changed when dense concentrations of Antarctic krill were recorded as far south as 76°S. The absolute densities of 9.4–19.5 g m^{-2} (Table 2.2) may not be directly comparable with studies in other areas, because of the differing techniques used to classify acoustic backscatter and because their unconventional methodology extrapolated from a single transect to a rectangular area.

However, the relative abundance patterns revealed that the centre of krill biomass was related to the seasonal ice retreat in spring. The progressive northward movement of the ice edge from November to January seems to influence the movement of *E. superba*. The centre of biomass moved from approximate latitude 74°S to around 71.5°S, suggesting that at least a substantial proportion of the stock spreads out into open ocean waters (Azzali et al. 2006). *E. superba* was the main part of the zooplankton biomass in the continental slope area north of 75°S, but its importance diminished to the east while to the south *E. superba* was replaced by *E. crystallorophias* (Faranda et al. 2000) which mirrors the circumstances experienced in the high latitudes of the southern Weddell Sea (Fevolden 1979; Siegel 1982; Boysen-Ennen and Piatkowski 1988).

2.5.4 Other Areas

In general, krill abundance seems to be relatively low in high latitudes of the Southeast Atlantic and East Antarctica when compared with long-term results from the Antarctic Peninsula and Scotia Sea region. All values from the Lazarev Sea of the 2004–2008 seasons are well below the long-term mean density found in the Elephant Island survey, that is, 230 krill 1000 m^{-3} and 39 krill m^{-2}, respectively (data from Table 2.1 in Siegel et al. 2002). During the LAKRIS cruises, numerical or biomass densities in the southern Lazarev Sea never exceeded 7 krill m^{-2} and 2 g m^{-2}, respectively (Siegel 2012). In the area north of 60°S (greater Bouvet Island region), krill abundance seemed to be higher than in the south, but densities still did only reach the average values recorded from the Southwest Atlantic.

The numerical densities and krill biomass recorded from a net sampling survey during the Norwegian AKES study ranged from 0–81.4 krill m^{-2} and 0–63.1 g m^{-2}, respectively (Krafft et al. 2010). However, only the upper limits given above are comparable with the estimates in the Antarctic Peninsula region. Integrated acoustic krill biomass estimates of 4.4 g m^{-2} from the same survey confirmed the overall low average densities in the south-eastern Atlantic (Skaret et al. 2009). Data showed a patchy distribution of krill with only relatively large aggregations of krill around Bouvet Island in the north and Astrid Ridge in the south where krill seem to concentrate along bathymetric and frontal features.

There are little data for large areas of the Bellingshausen, Amundsen and central Weddell Seas. Murray et al. (1995) reported the only acoustic estimate from the central Bellingshausen Sea. The biomass density estimates are comparable with the lower estimates obtained from the FIBEX 1981 survey in the West Atlantic. Mean biomass in the Bellingshausen Sea varied from 20 to 42 g m^{-2} for two consecutive surveys. Biomass densities from the FIBEX survey ranged from 17 g m^{-2} in the oceanic areas to 160 g m^{-2} in the Bransfield Strait.

2.6 Regional Times Series of Biomass and Stock Dynamics

Although the large-scale surveys described above cover huge areas, they represent single point estimates and provide little information on temporal change. However, in the Scotia Sea region several regular standardized meso-scale surveys have been carried out for more than a decade and thus show the dynamics of the population over time. Such surveys of krill distribution and abundance in the Scotia Sea and Antarctic Peninsula region have revealed wide year-to-year variations in krill biomass. A number of other surveys have been repeated at different times of the year, sometimes including both summer and winter sampling : these include BIOMASS SIBEX (Second International BIOMASS Experiment) in 1983–1985, the EPOS expeditions (European Polarstern Studies) in 1986, and the SO-GLOBEC cruises (Lawson et al. 2008) and they give some insight into the seasonal variability of the krill stocks in a given area.

2.6.1 Seasonal Variability

Mackintosh (1973) found evidence for a seasonal southward trend of the outer limit of krill from the southward movement of the whaling fleet with the retreating ice and the high number of net hauls containing no krill. He explained the late summer decline of krill by "not of course active movement southward but through its disappearance in lower latitudes, presumably by consumption". In his opinion whales had much more time in lower latitudes to graze down the krill stock than in seasonally ice covered zones further south. However, Kanda et al. (1982), Siegel (1988), Sprong and Schalk (1992), Lascara et al. (1999), Atkinson et al. (2008), and Krafft et al. (2012) all point to the likelihood of krill undertaking some form of active migration which could also account for such a change in distribution.

The Antarctic Peninsula region is probably the most intensively sampled area and there is good sampling coverage throughout the year. Although the different seasons were usually not sampled within the same annual cycle but in different years, these data offered the opportunity to make some general conclusions about the development of the krill stock. Siegel (1988, 2005) gave a conceptual view of the seasonal variation of krill distribution and abundance in the Antarctic Peninsula and adjacent areas. Krill abundance is usually low in the region during winter and krill occurrence is mainly restricted to the shelf areas, especially the Bransfield Strait which shows relatively high densities in winter and possibly acts as a retention area, while krill were virtually absent in offshore waters of the Drake Passage. From November onwards, krill abundance increases rapidly and krill

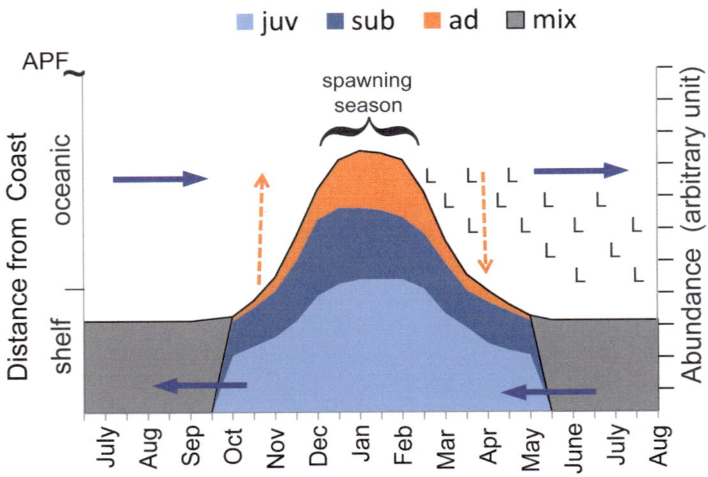

Fig. 2.6 Conceptual view of the seasonal variation in krill stock density and spatial succession of size classes and maturity stages along the Antarctic Peninsula (modified after Siegel 1988, 2000); *solid arrows* indicate passive drift, *broken arrows* indicate seasonal active movement of krill, APF location of Polar Front, L larvae, juv juvenile, sub subadult, ad adult, mix mixture of developmental stages

distribution extends beyond the continental shelf break into oceanic waters. The most northerly distribution limit is reached during the period of maximum abundance in summer which coincides with the main spawning season of the species, mainly within the time window from mid-December to early March (Fig. 2.6).

Although the concept was confirmed in later studies, there are considerable gaps in the understanding of the abundance of krill during winter. The seasonal abandonment of the vast offshore waters, where the majority of the krill biomass resides during austral summer (Atkinson et al. 2009), and shelfward migration suggests that krill density should increase greatly in coastal and near-shore environments during winter (Siegel 1988). Although winter increases in the krill density in coastal waters has not always been detected (Stepnik 1982; Siegel 1988, 2005; Lascara et al. 1999; Nowacek et al. 2011; Zhou et al. 1994) the general concept of seasonal fluctuations in abundance derived from RMT net sampling has been confirmed by acoustic data from the South Shetland Islands region (Kim et al. 1998). During US-AMLR summer surveys, the median krill abundance in the South Shetland Islands area ranged from 0.34 to 0.72 krill m^{-2} (15 years of data, Reiss et al. 2015). In contrast, during the winter, median krill abundance in the same area was an order of magnitude lower, and ranged from 0 to a maximum of just 0.05 krill m^{-2} (3 years of data, Reiss et al. 2015). The LTER and SO-GLOBEC programs further south along the Antarctic Peninsula also confirmed the seasonal trend in krill distribution and biomass (Lascara et al. 1999; Ashijan et al. 2004; Lawson et al. 2008). Mean biomass density in winter was as low as 8 g m^{-2}, increased to 32 g m^{-2} in spring, reaching its maximum of 95 g m^{-2} in summer before declining again to 12 g m^{-2} in autumn. Seasonal changes in krill density have also been

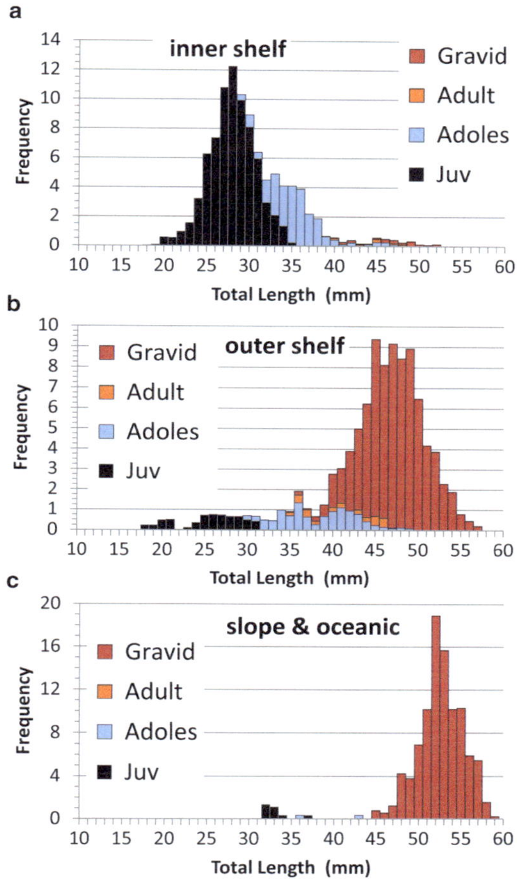

Fig. 2.7 Example of krill length frequency distribution in the WAP region during summer 2011 showing the onshelf – offshelf differences in size composition and maturity stage; *Adoles* adolescent stage (**a**) inner shelf (**b**) outer shelf (**c**) slope and oceanic areas

detected acoustically across the central Scotia Sea. Fielding et al. (2012) carried out small scale acoustic surveys at a series of stations between the South Orkney Islands and South Georgia in spring 2006, summer 2008 and autumn 2009 where autumn krill density was an order of magnitude lower than the spring and summer densities.

In addition to the temporal variability in abundance, spatial patterns of the krill stock also change with season. During austral summer, a spatial succession emerges for krill size classes and developmental stages (Siegel 1988). Juveniles mainly inhabit the shelf area along the Antarctic Peninsula, while on the outer shelf and in the northern Bransfield Strait, subadult krill dominate. Large adults representing the spawning population concentrate in oceanic regions along and beyond the shelf break (Fig. 2.7). A small number of adult krill seem to drift into the Bransfield Strait and spawn along its deep central basin, the total amount, however, remains negligible compared to the oceanic area north of the South Shetland Islands (Jackowska 1980; Kittel 1980; Fevolden and George 1984). After the end of the reproductive season, usually from March onwards, krill abundance declines steadily long before

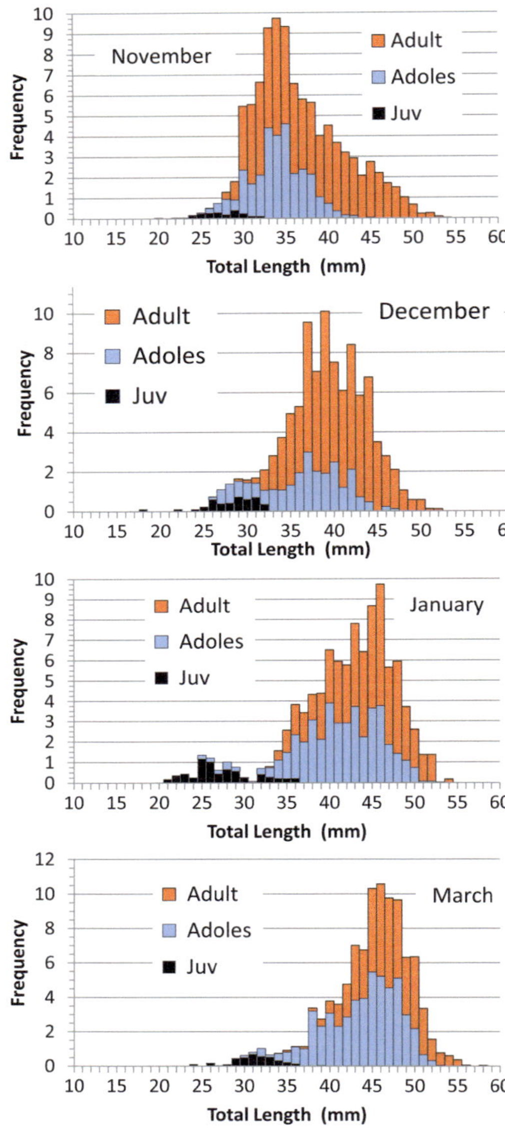

Fig. 2.8 Example for seasonal variation of krill length frequency distributions from Elephant Island from a standard station grid sampled in different months during the 1977/1978 season

the retreat of the winter pack-ice. At the same time absolute post-larval krill abundance becomes extremely low in the oceanic regions. Although the spatial succession of krill size classes fluctuates within the annual cycle, the overall size composition of the stock does not seem to vary substantially. Figure 2.8 indicates that the proportion of juvenile and adult krill is stable during the austral summer period, which is of course of importance for the estimation of the recruitment index (see also Chap. 3, Reiss (2016)). Such information on the length distribution of krill in the catch, and the spatial and temporal patterns of krill length in the population is important for assessing krill population status and dynamics, the potential for achieving the best spatial arrangement for the fishery, and to develop approaches for minimising the impacts of the fishery on krill predators (see also Chap. 11, Nicol and Foster (2016)).

Additional data on seasonal changes in krill abundance are available from the LAKRIS project in the Lazarev Sea. The surveys in autumn 2004, summer 2005/2006 and winter 2006 offered the opportunity to compare results of different types of sampling gear and the development of the krill stock during the same annual cycle. During the surveys Flores et al. (2012a) deployed the Surface and Under Ice Trawl (SUIT), which sampled the 0–2 m surface layer both under sea-ice and in open water. Average surface layer densities ranged between 0.8 krill m^{-2} in summer and autumn, and 2.7 krill m^{-2} in winter. In winter, under-ice densities far surpassed maximum 0–200 m densities obtained from RMT samples (see also Sect. 2.2.5 "Krill and Sea-Ice"). Data analyses of the RMT station grid sampling in the various seasons estimated a biomass of 30,000 t in early summer which increased to 207,000 t in the subsequent winter (Siegel 2012). This relates to mean biomass densities of 0.06 g m^{-2} in summer and 0.56 g m^{-2} in the following winter 7 months later. Obviously, the krill stock in the Lazarev Sea is very dynamic on the seasonal time scale and dependent on relatively high flux rates in and out of the area.

Continuous, year-round measurements of krill density are rare and often confined to observations made from the coastal Antarctic bases (see for example Stepnik 1982). However Saunders et al. (2007) presented acoustic data collected at on-shelf and off-shelf moorings northwest of South Georgia between October 2002 and December 2005. A distinct seasonal pattern of winter low density and summer high density was observed in each year. Moreover the densities estimated from the moored instruments were not significantly different from krill surveys undertaken in the region in spring, summer and autumn each year.

2.6.2 Interannual Variability

All estimates of circumpolar krill biomass are based on composites of data collected over multiple seasons and therefore have temporal and spatial aliasing in the observations. However, both regional monitoring (e.g. Loeb et al. 1997; Siegel 2005) and basin-scale analyses (e.g. Atkinson et al. 2008) show that krill density has varied considerably over the past decades. A series of studies also suggest a decline in krill within the southwest Atlantic sector during the last decades of last century (Loeb et al. 1997; Reid and Croxall 2001; Fraser and Hofmann 2003; Atkinson et al. 2004). Time series of repeated meso-scale surveys provide opportunities to study krill stocks over longer time periods while minimizing the issues of spatial and temporal aliasing. True time series are rare and only the Antarctic Peninsula and the island of South Georgia have been sampled consistently at least every year for more than a decade. The following section will deal with these specific long-term meso-scale survey results to give a better understanding of krill variability in time, both interannual and long-term change, as well as provide an example for the regional variability of the distribution patterns of krill abundance between years.

These regular and standardized surveys of krill abundance in the Atlantic sector have revealed wide year-to-year variations in krill biomass (Siegel and Loeb 1995; Brierley et al. 1999a, b, 2002a; Hewitt et al. 2003; Reiss et al. 2008; Fielding et al. 2014). Only recently Norway has started to conduct standardized krill surveys around the South Orkney Islands using krill fishing vessels as a scientific platform. The vessels were equipped with similar echo sounder systems suitable for quantitative assessments, but the frequencies operated varied between vessels and years. Density estimates from the 5 years of surveys show that, except from the year 2015, krill densities were high, in the range 100–300 g m^{-2} based on the 120 kHz recordings (Skaret et al. 2015).

The only time series from East Antarctica which extended over more than a decade was collected mainly during the 1980s and presented by Pakhomov (2000). Table 2.3 compares krill density estimates for various acoustic and net sampling surveys in different areas. It is obvious from the quantitative data that krill density can vary by more than an order of magnitude between consecutive years. Most of the analyses of these data sets have focused on aspects of variability and change within the areas and this is the focus of our discussion below.

The marine pelagic system around South Georgia is characterised by considerable inter-annual variability in both its physical and biological components. Large-scale climate variability is a major driver of the localised variability at South Georgia (see reviews in Atkinson et al. 2001; Hill et al. 2006; Murphy et al. 2007a, b). Both the El Niño-Southern Oscillation and the Southern Annular Mode are predictors of conditions at South Georgia. Various analyses have examined the links between these indices, sea surface temperature and the availability of Antarctic krill to the major prey species, particularly seal and bird populations during their land-based breeding phase (Trathan et al. 2003, 2006, 2007; Murphy et al. 2007b; Meredith et al. 2008; see also Chap. 9, Trathan and Hill (2016)).

Fielding et al. (2014) summarized results of the long-term data of the British survey at South Georgia from 1997 to 2013. Krill targets were identified in acoustic data using a multi-frequency identification window and converting to krill density using the SDWBA target strength model. In most years, the mean krill density was driven by relatively few very dense swarms. The pattern of interannual variability showed three different periods, with fluctuations every 4–5 years. Years with high krill density >30 g m^{-2} were encountered in 1997–1998, 2001–2003, 2005–2007.

Compiling the data from a range of different studies, we can identify episodic signals with the following years when krill availability was very low off South Georgia: 1969 (Makarov et al. 1980), 1977/1978, 1982/1983, 1990/1991, 1993/1994, (Hempel et al. 1979b; Heywood et al. 1985; Croxall et al. 1999) and 1999/2000, 2004, 2009/2010 (Fielding et al. 2014). The sea surface temperature recorded in early 2009 was above the long term average. The predator data suggested that

Table 2.3 Time series of krill biomass and numerical densities from meso-scale surveys; Acoustic Target Strength model used to estimate krill biomass is identified by SDWBA (see Fielding et al. 2014 for the current CCAMLR recommended implementation of this model) or Greene (Greene et al. 1991)

Split Year	Trawls and acoustic	Acoustic	Acoustic	Acoustic	Acoustic	Net	Net	Net
	$g\ m^{-2}$	$g\ m^{-2}$	$g\ m^{-2}$	$g\ m^{-2}$	$g\ m^{-2}$	$g\ m^{-2}$	$n\ m^{-2}$	$n\ 1000\ m^{-3}$
	Indian Ocean	South Georgia	Elephant Island	Elephant Island	South Orkneys	Elephant Island	Elephant Island	Elephant Island
	Pakhomov (2000)	Brierley et al. (1999a), TS-Greene; Fielding et al. (2014) since 1997	Brierley et al. (1999b), Hewitt and Demer (1994), and Hewitt et al. (2003)	Reiss et al. (2008) and AMLR FSR (2014)	Skaret et al. (2015)	Siegel (2005)	Siegel (2005)	Van Cise (2009) and AMLR FSR (2014)
	TS-unknown, for further details see Table 2.2 of Pakhomov (2000)	TS-SDWBA + 3freq	TS-Greene	TS-SDWBA	120 KHz apart from 2015. TS-SDWBA	TRAWLCI	TRAWLCI	Median
1977	187.7							
1978	65.8					49.70	139.50	
1979	60.7							
1980	20.5							
1981	20	59.7	68.5			30.70	71.72	
1982	22.6	11.7				75.23	336.24	
1983	21.3					9.76	61.26	
1984	17.5		22.5			16.66	37.60	
1985	41.1		8.4			3.39	4.52	
1986	36.6	29.7						
1987	18.3		57.2					
1988	48		41.6			3.53	7.06	
1989	92		82.4			3.18	13.06	

(continued)

Table 2.3 (continued)

Split Year	Trawls and acoustic	Acoustic	Acoustic	Acoustic	Acoustic	Net	Net	Net
	$g\ m^{-2}$	$g\ m^{-2}$	$g\ m^{-2}$	$g\ m^{-2}$	$g\ m^{-2}$	$g\ m^{-2}$	$n\ m^{-2}$	$n\ 1000\ m^{-3}$
	Indian Ocean	South Georgia	Elephant Island	Elephant Island	South Orkneys	Elephant Island	Elephant Island	Elephant Island
	Pakhomov (2000)	Brierley et al. (1999a), TS-Greene; Fielding et al. (2014) since 1997	Brierley et al. (1999b), Hewitt and Demer (1994), and Hewitt et al. (2003)	Reiss et al. (2008) and AMLR FSR (2014)	Skaret et al. (2015)	Siegel (2005)	Siegel (2005)	Van Cise (2009) and AMLR FSR (2014)
	TS-unknown, for further details see Table 2.2 of Pakhomov (2000)	TS-SDWBA + 3freq	TS-Greene	TS-SDWBA	120 KHz apart from 2015. TS-SDWBA	TRAWLCI	TRAWLCI	Median
1990	167	60.1	46.9			2.14	4.06	
1991		6.4	23.8			0.76	1.36	
1992		95	61.2			0.81	3.16	5.7
1993		65.8				3.52	7.46	8.2
1994		4.7	9.6			3.46	5.78	3.1
1995			27.8			2.97	5.00	3.6
1996		33.7	80.8	89.2		4.88	16.54	11.4
1997		31.7	100.47	202.1		46.79	127.52	5.6
1998		38.9	82.3	86.1		3.40	9.61	10.2
1999		9.7	23.7	25.0		0.88	1.55	1.7
2000		2.7		86.4		2.65	3.72	2.1
2001		36.7		41.4		13.04	45.86	3.8
2002		137.0		13.9		40.04	107.08	4.8
2003		84.6		54.2		3.35	12.22	19.9

2004		26.1		10.0		4.26	8.60	2.0
2005		89.4		23.0				9.9
2006		119.1		38.6				7.1
2007		61.1		103.4				21.4
2008				81.1				14.3
2009		28.8		115.1				3.3
2010		15.1		74.9				1.1
2011		59.0		84.1	108.7			
2012		90.1			86.9			
2013		61.8			120.3			
2014					148.3			
2015					7.1			

For the estimation of net densities data were used only from large RMT and IKMT nets; TRAWLCI method to calculate densities from trawl data (see de la Mare 1994); FSR Field Season Report of the US AMLR Program)

2009 was the worst year on record. The acoustic estimates of krill density were also one of the lowest on record (Fielding et al. 2014). There were indications that this occurred during a weak La Niña phase and a positive phase of Southern Annular Mode anomalies. However, cross-correlation analyses of variability in krill density with current and lagged indices of sea-surface temperature (SST), Southern Annular Mode (SAM) and El Niño southern oscillation (ENSO), found the highest correlation between krill density and winter SST from the preceding year (Fielding et al. 2014). The interaction between these two climatic signals is fairly complex at South Georgia, but a warmer-than-average year was expected in 2009 based on the relationships described in Meredith et al. (2008). The evidence suggests that the temperature increase was due to in situ warming rather than advection. Hill et al. (2009) concluded that the resultant warmer waters represent an unfavourable habitat for krill and were associated with reduced krill concentrations over the South Georgia shelf.

Krill at South Georgia are considered to be dependent on periodic recruitment from the Antarctic Peninsula and South Orkney region. Successful recruitment therefore depends on both the strength of recruitment from the upstream sources and also the reliability of the transport mechanism connecting the different areas. Extremely poor krill years at South Georgia are not always reflected in a lack of krill from the upstream areas. Between-year changes in immigration and emigration processes could also cause a high variability in local stock density. Overall krill biomass may not be responsible for the shortage of krill around South Georgia in some years. Variability in ocean current transport rates and mass transport have been described for the Antarctic Circumpolar Current (Naganobu et al. 1999). An example of the variability in distribution is given from net sampling surveys carried out in 1984 and 1988 (Sushin and Shulgovsky 1999). In austral summer 1988, high krill concentrations occurred from the Scotia Sea south of South Georgia to the very north-western end of the island, while in 1984 the situation was completely different, and krill distribution centered mostly across the southern Scotia Sea with very little krill biomass extending into the South Georgia area (Fig. 2.9a, b). A similar situation was observed in 1977/1978 when krill concentrations were high along the WAP and in the Scotia Sea south of 56°S, but krill was almost completely absent from the waters around South Georgia (Hempel et al. 1979a).

Sushin and Shulgovski (1999) argue that the geostrophic current flow was different between 1984 and 1988, with a more isolated and weak current field around South Georgia during years with diminished krill population size around the island. Fluctuations in the input of krill into the South Georgia area can occur as a result of variation in oceanographic variables (e.g. variation in currents, changes in frontal positions). In poor krill years at South Georgia most krill concentrations would pass along further south of the island, while at the same time those reduced stocks around South Georgia may suffer from high temperatures with higher mortality rates, reduced growth rates or poor reproductive success.

In some years, however, shortage of krill around South Georgia can also be caused by changes in biological properties in the upstream areas, e.g. by reproductive success, recruitment failure or low total stock biomass. Extreme events with

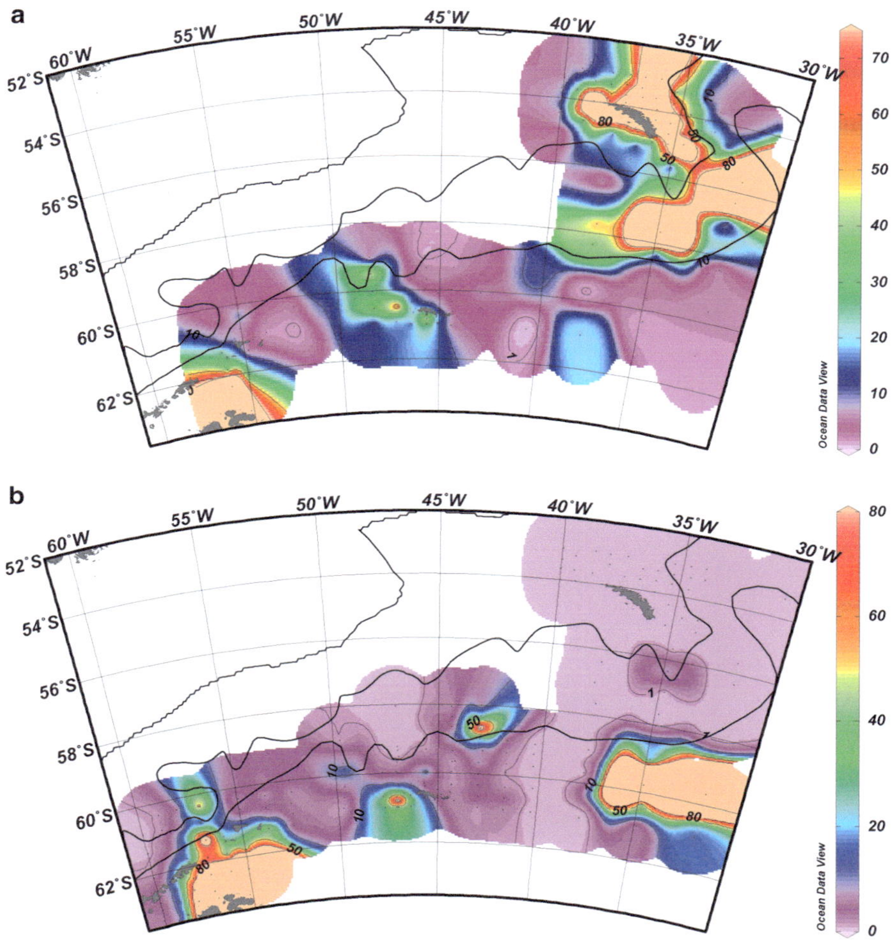

Fig. 2.9 Examples for different large-scale patterns of krill distribution and average density from net sampling surveys (**a**) summer 1984 (**b**) summer 1988; maps are plotted from data provided by the authors of Sushin and Shulgovski (1999)

low or high biomass do not just occur around South Georgia but across the entire stock. Such an event of overall low krill abundance was observed in 1994 (see Table 2.2) which was also one of the most disastrous seasons for krill predators (Croxall et al. 1999). High krill biomass was observed for the Antarctic Peninsula region as well as for South Georgia in the summer seasons of 1981, 1997 and 2002, whereas in 1982 South Georgia did not benefit from the highest krill biomass recorded for the entire time series.

It appears that the survival/mortality rate of the 0 age group over winter could be the dominating proximate cause of recruitment success/failure and consequently greatly impact on stock size in a given year. Siegel and Loeb (1995) found a close

Fig. 2.10 Example of krill length frequency distributions showing interannual variability in krill size composition off the western Antarctic Peninsula; data are from summer months January–February (**a**) 1982 (**b**) 1996 (**c**) 1997 and (**d**) 1985

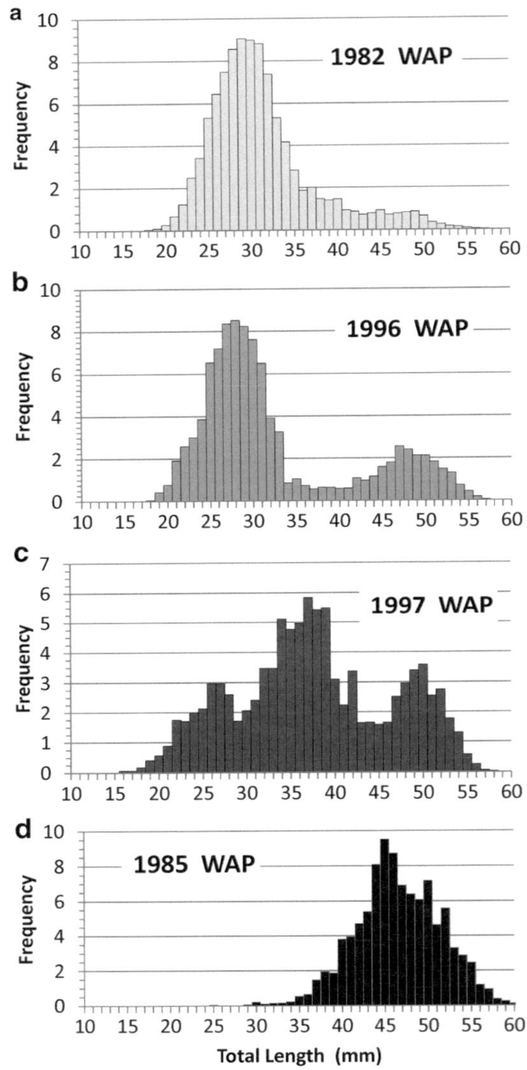

relationship between winter sea-ice cover in the Antarctic Peninsula region and krill recruitment success. Recruitment success becomes obvious in the length frequency data. Examples from different years are given in Fig. 2.10 (LFD in 1982, 1985, 1996, 1997) displaying also the effect of recruitment variability leading to unimodal, bimodal or even polymodal size distributions. According to Siegel and Loeb (1995), long duration and large spatial extent of sea-ice cover result in a high recruitment rate and an early and successful spawning. Model results presented by Fach et al. (2002) show the importance of the timing of spawning, food availability and larval growth. Additionally, studies of the physiological condition of larval krill collected in winters characterized by light and heavy ice showed that the heavy ice favoured higher lipid contents, higher condition factors and growth compared to the light ice conditions (Quetin et al. 1994). On the other hand Daly (2004) indicated that sea-ice biota may not be a primary source of food during winter, but becomes more important in spring, after ice-algae begin to accumulate from mid-September. However, krill larvae were found to contain higher polyunsaturated fatty acids than postlarval krill, which could be traced back to dominant ice algae in the diet of furcilia larvae in winter (Ju and Harvey 2004). Dense and long ice cover may therefore establish the basis for a minimum, but necessary, food resource in late winter for larvae to survive the winter and reduce the risk of starvation and increased mortality (see also Chap. 4, Meyer and Teschke (2016)). Prolonged ice cover may also protect both larvae and older age groups from heavy predation and higher mortality rates. The cumulative effect of these positive influences of more extensive sea-ice results in high krill recruitment leading to high stock abundance. From the above discussion one can draw the conclusion that the krill population along the Antarctic Peninsula is driven by reproductive output and larval survival over winter and the important key variable is sea-ice. The western Antarctic Peninsula region may therefore been seen as a source area, while in contrast, the abundance of krill at South Georgia is primarily driven by influx, retention and export (sink area).

2.6.3 Decadal Changes

A decline in krill abundance between the mid 1980s and mid 1990s periods was first recorded from the Elephant Island/South Shetland Islands area (Loeb et al. 1997; Siegel et al. 1998) and suggest a lower krill abundance in recent years due to reduced ice-cover. Atkinson et al. (2004) confirmed the decline in krill biomass for the entire Scotia Sea region and noted that the krill densities have declined significantly (up to >70 % of that originally observed during the 1920s and 1930s).

The studies by Siegel and Loeb (1995) and Loeb et al. (1997) focused on the 1980s, a time that was strongly influenced by atmospheric warming and dramatic decreases in sea-ice extent in the Antarctic Peninsula region (Smith et al. 1996; Martinson et al. 2008). A study by Loeb et al. (2009) included data from 1990 to 2004, a period characterized by comparatively stable sea-ice conditions and, with the inclusion of additional biological datasets, highlighted more subtle consequences of ENSO variability on biological responses.

White and Peterson (1996) and Jacobs and Mitchell (1996) described regular fluctuations of ocean circulation and sea-ice extent around the Antarctic continent, known as the Circumpolar Wave. Later it has been established that the coupled tropical ocean-atmosphere system known as El Niño–Southern Oscillation (ENSO), drives multiyear cycles of sea-ice extent around the Antarctic Continent, with alternating periods of more and less extensive sea-ice linked to La Niña and El Niño phases of ENSO, respectively (Karoly 1989; White et al. 2002; Carleton 2003; Yuan 2004). ENSO impacts on the Pacific South America (PSA) pattern of Sea Level Pressure (SLP) variability and this signal influences regional climatic conditions in the vicinity of the West Antarctic Peninsula with warm SST anomalies, strong northwesterly wind anomalies and expanded sea-ice extent fluctuating in phase with tropical La Niña. Cool SST anomalies, weak northwesterly wind anomalies and retracted sea-ice extent fluctuate in phase with tropical El Niño (Gloerson and White 2001; Yuan 2004). During La Niña situations the SACCF is displaced south-eastward, the Polar Frontal Jets are intensified and strong northwesterly winds in the South Pacific prevail, reducing the Weddell gyre influence in the South Shetland Island area (Loeb et al. 2010). Loeb et al. (2009) indicate that the temperature time series from Elephant Island indicates shifts in the location of the southern front and boundary of the Antarctic Circumpolar Current, and variable influence of oceanic UCDW and Weddell Sea shelf water, coincidental with changes in the SOI and La Niña and El Niño conditions. Enhanced chlorophyll a concentrations and primary productivity are observed under these La Niña conditions, creating a favourable habitat for the krill population.

Years between 1980 and 1995 were dominated by ENSO-driven sea-ice cycles and apparently low productivity conditions that favoured salps and negatively impacted krill recruitment and abundance. This was also a period characterized by intense and/or prolonged El Niño events. In contrast, the decade since the late 1990s has been dominated by the presence and movements of the SACCF off the South Shetland Islands region and conditions that promote elevated primary and secondary production, including krill recruitment success. This period has been one of primarily La Niña and neutral conditions and significantly increased concentrations of oceanic zooplankton assemblages (Loeb et al. 2009). During the early 1980s krill abundance changed rapidly over relatively short time scales, which was a period marked by strong contrasts between La Niña (1980/1981 and 1984/1985) and El Niño (1983/1984) conditions. However, the last large-scale survey within the Southwest Atlantic took place in 2000, and since then results of new meso-scale surveys showed high interannual variability in krill biomass but no recent evidence for a continuing trend in declining krill biomass (Hill et al. 2015), in krill density (g m^{-2}, e.g. Fielding et al. 2014), or krill abundance (krill caught by research nets, e.g. Atkinson et al. 2014; Steinberg et al. 2015).

2.7 Adult Source Population and Teleconnection Between Areas

We have already seen that krill found at South Georgia are likely to have arisen from the Antarctic Peninsula and South Orkney regions and also that the Antarctic Peninsula region is a key area for krill reproduction. In this section we look in more detail at the seasonal changes in krill abundance and distribution pattern along the Antarctic Peninsula, and then focus on the likely larger scale geographical connections between stocks, their origin and onward transport.

2.7.1 Mechanisms Generating Seasonal Variation in the Antarctic Peninsula Region

As an oceanographic front, the SACCF to some extent restricts krill dispersal to the north while the Weddell and the Ross Gyres and smaller-scale circulations (in the marginal Antarctic seas) bring krill to the optimum habitat zone from the south. Advection and retention by eddies is undoubtedly important for larvae (Marr 1962; Makarov and Maslennikov 1975; Daly 1990), but for adult krill it is likely that passive advection, behaviour and active movement will all have a role to play in determining adult distribution (Atkinson et al. 2008). Adult krill are good swimmers capable of maintaining a swimming speed of 1.5–3.5 body lengths per second indefinitely (Kils 1981, 1983; see also Miller and Hampton 1989 for review of other publications). The observation by Kanda et al. (1982) of a long-distance active migration of krill towards Enderby Land has been considered as a passive drift scenario after further exploration of oceanographic conditions of the area (Bibik et al. 1988). In contrast, a displacement of krill patches in the Coastal Current's direction was well documented during fishing operations (Nasu 1983; Dolzhenkov et al. 1990). Their relatively large size and strong swimming ability suggest that the active movement of krill may play an important role in their spatial re-distribution. Adult krill are certainly micronekton but krill dispersal against the strong jets in the southern ACC or the Coastal Current (with velocities exceeding 10 cm s^{-1}) is improbable. For adult krill we may rather assume retention in certain areas through interactions of krill behaviour with the water column structure; for instance in eddies or areas with weak velocities (Makarov et al. 1980; Witek et al. 1988), or a retarded advection generally following a current direction (Nasu 1983). At the large- and meso-scale there appears to be no alternative but to consider krill as a quasi-passive element subject to advection (Latogursky et al. 1990; Siegel and Kalinowski 1994). Although the power to modify large scale transport through simple behavioural strategies such as vertical migration (Murphy et al. 2004b), or associating with either sea-ice or the underlying water column (Thorpe et al. 2007) has been demonstrated in a number of circulation models. In the following section, we will consider the mechanisms that maintain krill populations within the optimum zone and examine some important key regions as examples.

The concept of seasonal changes in krill abundance along the western Antarctic Peninsula has been described above. This fluctuation of krill abundance over time is accompanied by an onshore-offshore succession of developmental stages, with the spawning stock in the continental slope and oceanic waters. This life history strategy of krill places the developing larvae in locations distinct from the adult winter population which avoids competition for food, but also prevents predation on larval krill by adults. The spatial horizontal succession of size and maturity stages along the Peninsula must at least partly be the result of active migration and not simply be induced by advection (Siegel 1988; Trathan et al. 1993). Movements within the Peninsula region over short north/south distances have major effects on the horizontal drift with the currents (Hofmann et al. 1998; Fach and Klinck 2006) and consequently on the dispersion of the krill population. Such short active migrations would allow krill to move out of the north-east heading ACC into the coastal south-westward moving coastal current and vice versa. A modeling study by Richerson et al. (2015) considered krill life history strategies and distribution together with current velocity, food availability, temperature, and predation risk. When active migration in krill behaviour was included in the model, the resulting distribution patterns were associated with increased survival, growth and reproductive success compared to those results obtained from a purely passive drifting model. Indeed Tarling and Thorpe (2014) found evidence for active movement of krill swarms when analyzing acoustic field data. They demonstrated that instantaneous movement patterns at large scales are responses of krill to local stimuli such as sea-ice and food availability (fluorescence). All these results suggest a strong selective pressure for active krill movement along drift trajectories.

An alternative explanation for the dramatic change in seasonal abundance might be seen in different vertical migration behaviour between summer and winter. However, this does not seem to be the final reason for the low winter biomass estimates of krill in the Peninsula region. Although Lawson et al. (2004) observed a substantial decrease in acoustic scatterings in the upper 300 m along the Peninsula in winter, krill were still mostly present in the upper 200 m. Furthermore, additional deep net tows, although limited in number, indicated that krill along the Antarctic Peninsula may occur deeper in the water column in winter, but below 400 m hardly any krill were caught even in winter. Survey results have shown that the amount of krill captured by nets at depths greater than 200 m was between 0.05 % and 1.4 % of those caught in the upper 200 m (Siegel 1985a). Also Taki et al. (2005) observed a steady increase in fishing depth during autumn and winter indicating a deeper migration of krill; however, mean trawling depth did not exceed 200 m in the South Shetland/ South Orkney region even in winter. At the same time low CPUE values in late autumn and winter support the survey data of an overall lower biomass in winter for this region.

Besides the active inshore/offshore migration of krill, Siegel (1988) hypothesized that seasonal fluctuations of atmospheric pressure and wind stress field and consequently strength of westerly winds over the Drake Passage that cause variability in the velocity of prevailing currents (van Loon 1972; Capella et al. 1992; Naganobu et al. 1999) may have an important impact on the transport rates of krill. The transport of water is significantly higher during summer than in winter (Whitworth 1980). First studies of relative dynamic height indicate changes in geostrophic flow in the northern Bellingshausen Sea (Makarov et al. 1982). Two main current systems occur west of the Peninsula, the ACC and the coastal current. The fronts are "current cores" (Whitworth et al. 1982) with strong horizontal density gradients and considerable variability superimposed on their net flow (Joyce et al. 1978). The location of the frontal zones SACCF and SBF is also influenced by varying high and low pressure zones (Hughes et al. 2003; Meredith et al. 2004; Volkov and Zlotnicki 2012; Thompson and Youngs 2013), with the low pressure zone extending farther north in winter. With krill moving closer to the coast in autumn more krill would be carried to the south-west establishing a winter krill minimum off the Antarctic Peninsula.

The ultimate factors for adults leaving the offshore areas in winter might be seen in the reduction of intraspecific food competition between adults and larvae. Food supply is lower in the oceanic area than in the neritic zone and larger krill have a greater ability to move, are more robust and can switch to a cryopelagic or benthopelagic feeding during winter (Spiridonov et al. 1985; Kawaguchi et al. 1986; Schmidt et al. 2011) or show a shift in their main food items (Schmidt et al. 2014). The interspecific competition is also reduced in oceanic areas during winter, because copepods, for example, are undergoing seasonal vertical migration and are leaving the surface zone, which gives the krill larvae a better chance to survive. This strategy could mean that at least in some areas *E. superba* utilizes its habitat up to the carrying capacity (Siegel 1988).

With the strengthening of the ACC flow in spring and the migration of krill into offshore waters, the seasonal ACC "conveyor belt" would start transporting with increasing amounts of krill to the NE, resulting in high concentrations of the stock in the South Shetland/Bransfield Strait area and beyond into the Scotia Sea (Siegel 1988).

2.7.2 Geographical Connections Across the Scotia Sea

High levels of variation in krill stock size and composition have been repeatedly observed across the Scotia Sea (e.g. Priddle et al. 1988; Fedulov et al. 1996; Murphy et al. 1998, 2004b; Trathan and Murphy 2003). These were explained as partly driven by key variables in the physical environment of the upstream area or around South Georgia itself. In particular the southern portion of the ACC seems to play an important role (Maslennikov and Solyankin 1988; Latogurski et al. 1990; Hofmann et al. 1998; Tynan 1998; Trathan et al. 2003) but does not explain all the observed phenomena in the Scotia Sea. These variations especially in the southern part have led to inconsistent results and provoked questions about spatial origin and means of dispersal.

Three alternative scenarios for the dispersion of krill stocks in the Scotia Sea have been discussed in the past (Siegel 2005)

1. Discrete, self-sustaining krill populations reside in the vicinity of the Antarctic Peninsula, South Orkney Islands and South Georgia with no or little exchange,
2. Krill are transported with the ACC from the Antarctic Peninsula to South Georgia and beyond ("conveyor belt"),

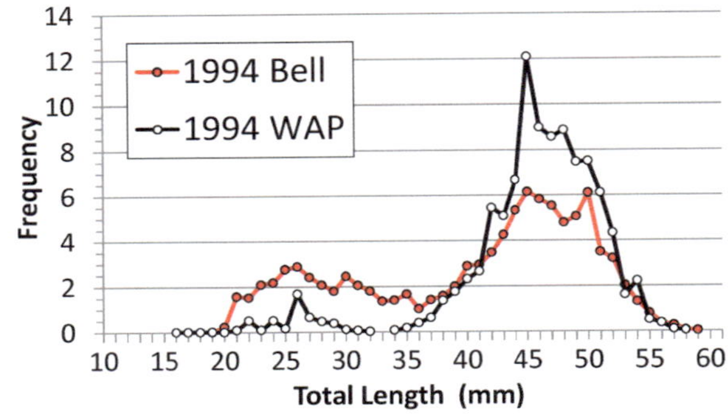

Fig. 2.11 Example of krill length frequency distributions from the central Bellingshausen Sea and northern WAP region in January/February 1995

3. Krill at South Georgia are a mixture of animals transported from the Antarctic Peninsula via the ACC and the eastern Weddell Sea via the Weddell Gyre.

South Georgia is not considered to support a self-sustaining population (Marr 1962; Mackintosh 1973; Murphy 1995). There is a marked seasonal and interannual variability of krill around the island. Hence it has been proposed that krill at South Georgia have their origins at the Antarctic Peninsula and/or in the Weddell Sea (Marr 1962; Maslennikov et al. 1983; Murphy et al. 1998; Watkins 1999; Watkins et al. 1999; Siegel et al. 2003, 2004; Siegel 2012). Similarly there is strong evidence that krill off the South Shetland Islands originates from further south out of the Bellingshausen Sea (Everson 1976; Siegel 1988; Siegel and Loeb 1995; Hofmann et al. 1996; Lascara et al. 1999). Although the size composition and consequently the recruitment index are similar between the Antarctic Peninsula and South Georgia, there is a potential difference between the central Bellingshausen Sea and the western Antarctic Peninsula (see LFD Fig. 2.11). Small 1-year old krill in this example shows a higher proportion in the up-stream region of the Bellingshausen Sea than observed during the entire season at the tip of the Peninsula although this might be explained by the retention and south-west flow of the current on the southern shelf of the Peninsula (see also discussion by Siegel 1988 and below). The Antarctic Peninsula and South Georgia ecosystems are similar in structure and function but are also distinct. The seasonal changes in krill abundance around South Georgia are often accompanied by changes in the population structure which could not simply be explained by variable current flow but may also be influenced by complex interaction of relative year-class strength, timing of immigration into South Georgia, variability in growth rates and dynamic predator-selective effects (Murphy and Reid 2001; Reid et al. 2010). The two systems are part of a continuum from the more intensely seasonal ice covered areas farther south to the more open water regions in the north. The connectivity is generated by the flow of the Antarctic Circumpolar Current. Antarctic krill are central to the food web in both ecosystems while the species composition of zooplankton and predators is different (Murphy et al. 2013).

In a modelling exercise Hofmann at al. (1998) demonstrated that particles released close to Adelaide Island west of the Antarctic Peninsula can be transported to South Georgia within 140–160 days when the wind-induced Ekman drift is combined with the effect of the high speed currents associated with the SACCF. However, according to simulation results by Fach and Klinck (2006) the exact location of the drifter release point is of great importance. Simulated drifters originating in the Bransfield Strait require 263 days to reach South Georgia, while those originating from the continental slope north of the South Shetland Islands only need 211 days. The reason for the difference is due to the time needed to be transported out of the Bransfield Strait and become entrained in the SACCF.

Lagrangian particle tracking for the period prior to the CCAMLR 2000 survey showed that most of the krill were associated with areas to the south of the Antarctic Circumpolar Current in the Weddell–Scotia Confluence (WSC) and farther east in Weddell Sea-influenced waters. The WSC is delimited to the south by the Weddell Front (WF), and to the north by the Southern Boundary of the ACC and creates transient mesoscale eddies within the ACC across the meridional flux in the Scotia Sea (Meredith et al. 2015). Krill present in the high krill biomass regions in January would have come from areas that were covered by sea-ice during late winter/early spring (Murphy et al. 2004b). This study suggests that the summer distribution of krill in the Scotia Sea is connected to the winter sea-ice distribution and probably to the pattern and rate of the spring sea-ice retreat. Many of the krill in the survey region in the summer of 1999/2000 came from under the sea-ice in the eastern Scotia Sea, the southern Scotia Arc, and the northern Weddell Sea. Transport of krill around South Georgia would probably have been in association with the SACCF. Results of a high-resolution dynamic model indicated that variability in the magnitude and spatial distribution of krill was dominated by the proximity and orientation of the SACCF to the South Georgia shelf edge with the main transport pathway off the shelf to the north and northwest (Young et al. 2014). This highlights the potential importance of interaction between large-scale oceanic flows and the shelf edge for on-shelf transport of krill and prolonged retention times in the region. A large proportion of the krill encountered in the eastern Scotia Sea would have exited toward the east, passing north of the South Sandwich Islands, probably in association with the Southern Boundary of the Antarctic Circumpolar Current and Weddell Sea waters that penetrate to the north in this area. The effect of sea-ice was explored by Thorpe et al. (2007) who showed that the association of krill with sea-ice in the southern Scotia Sea likely increases the sources of the krill that arrive at South Georgia, in particular in extreme sea-ice years acting to bring krill out of the Weddell and up into the SACCF.

The two options of Antarctic Peninsula /Bellingshausen Sea and Weddell Sea origin of krill are not mutually exclusive: there are more indications that krill around South Georgia does not originate alone from the western side of the Antarctic Peninsula but may be a mixture of Peninsula and Weddell Sea stocks. This perception would at least partly incorporate the hypothesis by Marr (1962) as well as Maslennikov and Solyankin (1988) and Latogurski et al. (1990) that the large-scale distribution of krill in the Scotia Sea is dominated by the outflows from the Weddell Sea. The general flow pattern in the South Orkney area is strongly influenced by the SACCF and the SBF that flows eastward very close to the edge of the continental shelf west of the Antarctic Peninsula (Orsi et al. 1995). These currents transport krill from the Antarctic Peninsula area to the South Orkneys (Hofmann et al. 1998). Additionally the South Orkney Islands receive waters from the northern periphery of the Weddell Gyre (Kasatkina et al. 1997). The outflow of the Weddell Gyre could be another advective pathway for the krill population through the South Orkney region and transport krill to South Georgia as suggested by Siegel et al. (1990) and Sologub and Remeslo (2011). This hypothesis is supported by the results of Renner et al. (2012) who tracked virtual drifters in an eddy permitting ocean to show that drifters tend to be advected towards South Georgia. Interestingly the modelling exercise of Fach and Klinck (2006) also considers a drift trajectory from the shelf slope of the Weddell Sea to the east of the Antarctic Peninsula. From here krill would be transported to the Weddell Scotia Confluence in an average of 263 days and from there to South Georgia with a total transport time of approximately 20 months.

However, the outflow of the Weddell Gyre in the north-western Weddell Sea seems to be even more complicated. Not all of the particles will drift directly to the South Orkney Islands, some water from the north-western shelf will pass around the tip of the Antarctic Peninsula into Bransfield Strait (Clowes 1934; Deacon 1977) where Antarctic krill occur jointly with ice krill *E. crystallorophias* (Siegel 2013; Guihen et al. 2014). Furthermore, Stein and Heywood (1994) described the advection of water masses in the Bransfield Strait and showed that large-scale changes in the wind field either cause a stronger influx from the Weddell Sea or from the Bellingshausen Sea (flip-flop scenario). In all cases, krill drifting into the southern Bransfield Strait would finally end up in the WSC and the SACCF but due to longer retention times in the Strait would take much longer to reach the eastern Scotia Sea and South Georgia (Fach and Klinck 2006; Murphy et al. 2004b).

Again the exact location of the release point of krill seems to be of great importance for the drift. Krill from the continental slope of the north-western Weddell Sea do not seem to take the route via Bransfield Strait, but enter the Scotia Sea between Elephant Island and the South Orkneys. Thompson and Youngs (2013) explained these different drift trajectories by the presence of the Antarctic Slope Front (ASF) which was observed to bifurcate over the South Scotia Ridge causing drifters to the west to enter Bransfield Strait and drifters to the east to more likely to enter the Scotia Sea with the ASF. Trajectories of surface drifters released in January 2012 in the north-western Weddell Sea suggest that the SACCF acts as a dynamical transport barrier to the drifters (Thompson and Youngs 2013). Specifically, the paper provided the first Lagrangian observations of a direct transport pathway between the Weddell Sea and regions of persistently elevated chlorophyll levels in the Scotia Sea. Interestingly, not all drifters tend to end up at South Georgia, but some pass the island farther south.

It has become obvious that the importance of the two source population for the eastern Scotia Sea /South Georgia region fluctuates over time. Spiridonov (2009) emphasized after his analysis of long-term commercial krill catch data that the dynamics of krill biomass in fishable aggregations in the Antarctic Peninsula region and the South Orkney Islands is not synchronized while there are strong connections to South Georgia. Such a long data base certainly provides a strong argument and it may well be that the influence of the Weddell Sea on the South Georgia krill stock is weaker, however, survey data from the Weddell Sea and even the South Orkney Islands are still sparse and reliable quantitative advection data from the Weddell Sea are difficult to obtain from the perennial sea-ice region. The influence of the Weddell Sea on the eastern Scotia Sea seems to be acceptable in principle, but more work is needed to verify and quantify this source population (see also discussion on the Weddell Gyre below), particularly because of the management implications for the Scotia Sea where the krill fishery is focused and also increasing.

2.7.3 Large Scale Teleconnections in the Weddell Gyre and Beyond

In general around the Antarctic Continent shelf areas are very narrow compared to the Antarctic Peninsula and the krill habitat is characterized by deep oceanic waters. In these areas the concept of seasonal changes in abundance and spatial succession (on/offshelf) does not seem to follow the same pattern as described for the Peninsula region. The fate of krill east of South Georgia and the South Sandwich Islands has hardly been studied but was included in the general discussion by Siegel (2012) on the drift of krill in the larger Weddell Gyre system (Fig. 2.12). Krill distribution

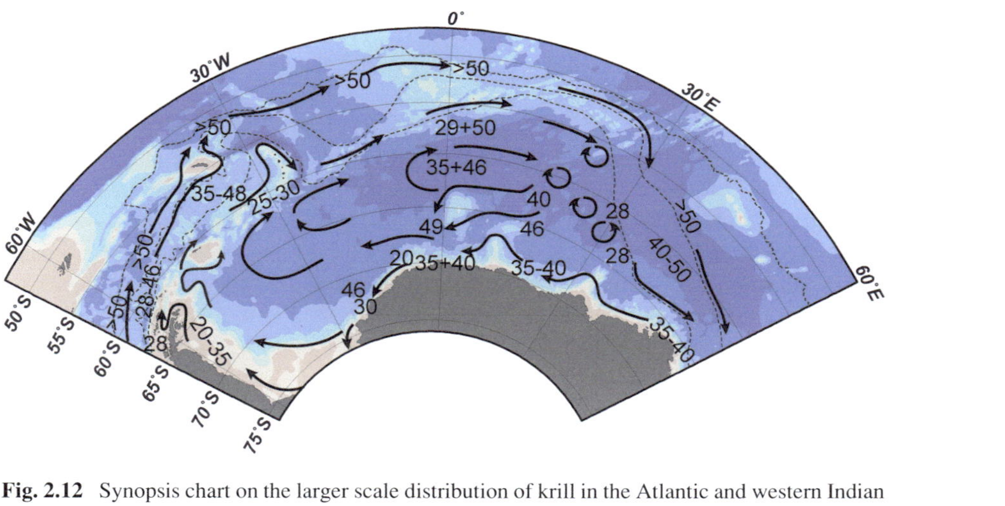

Fig. 2.12 Synopsis chart on the larger scale distribution of krill in the Atlantic and western Indian Ocean sectors; *arrows* display stylised oceanographic flow (for references, see text); *thin dotted lines* mark the three frontal zones from north to south APF, SACCF and SBF (After Siegel 2012 with additional information from Kawaguchi et al. 2010a)

around the 0° meridian in the Southeast Atlantic ranges from the Polar Front at approximately 51°S all the way to the Antarctic continent at 70°S, which is the widest latitudinal coverage in its entire circumpolar distribution. The region north of 60°S, is under the influence of the eastward flowing northern limb of the Weddell Gyre and is downstream of the Scotia Sea krill population (Marr 1962; Tarling et al. 2007; Krafft et al. 2010). Ocean circulation model outputs suggest that possible influx of krill into the area around Bouvet Island could primarily originate from the Antarctic Peninsula via the northern extensions of the South Sandwich Islands along the eastward flow of the ACC (Thorpe et al. 2007), which would classify the Bouvet area as another sink area for the krill population (see also discussion on krill larvae distribution). This view is supported by krill length frequency data that show a high proportion of very large krill in the waters west of the Antarctic Peninsula (Siegel 1987, 1988; Lascara et al. 1999), north of the WSC and South Georgia (Hill et al. 1996 their Fig. 2; Watkins 1999; Reid et al. 2002) and further to the east up to Bouvet Island (Marr 1962; Fevolden 1979; Krafft et al. 2010; Siegel 2012). Another source of drifting particles into the area slightly to the south of Bouvet would be from the Weddell Sea via the northern eastward flow of the Weddell Gyre (Thorpe et al. 2007). These water masses are characterized by a change in krill length composition to small- and medium-sized krill and occurred approximately between 55° and 57°S (Figs. 2.12 and 2.13).

East of Bouvet Island the circulation pattern is undisturbed by any obvious topographic boundary until 30°E (Deacon 1979; Gouretski and Danilov 1993). From here the ACC and frontal zones bend southwards and approach the Antarctic Continent to within as close as a few hundred kilometres (Orsi et al. 1993). East of 30°E and extending as far as 80°E Pakhomov (1995, 2000) and Kawaguchi et al. (2010a) found the large krill again that were recorded in the samples from Bouvet (Siegel 2012 with data from the same summer season). Slightly to the south of the northern current branch, between 15° and 30°E, the Southwest Indian Ridge leads to instabilities in the flow system and to the formation of an intense mesoscale eddy field (Schröder and Fahrbach 1999). The medium to large krill from south of Bouvet will be carried into this eddy field, and finally, these waters are advected to the southwest south of 63°S in the direction of Maud Rise in the Lazarev Sea (Siegel 2012). In the south krill of the Coastal Current originate from the Indian Ocean sector and are of medium to relatively large size. Results presented by Kawaguchi et al. (2010a) indicate that the medium-sized groups occur as far east as 40°E in the Cosmonaut Sea. The central Lazarev Sea south of the Weddell Gyre Centre with its relatively weak-flow velocities (Cisewski et al. 2011) is inhabited by small- and medium-size classes of krill.

These two prevailing currents in the Lazarev Sea take the krill further to the west. In the waters of the Coastal Current krill sizes are similar to those in the adjacent northern area; however, often a larger size mode of 48–50 mm is found for krill in the area slightly to the north of the Coastal Current. These large adult krill never gain the average or maximum size of those large krill regularly observed in the very northern areas of the Scotia Sea and around Bouvet Island (Makarov and Sysoyeva 1985; Siegel 1987, 2012; Hosie et al. 1988; Kawaguchi et al. 2010a). The

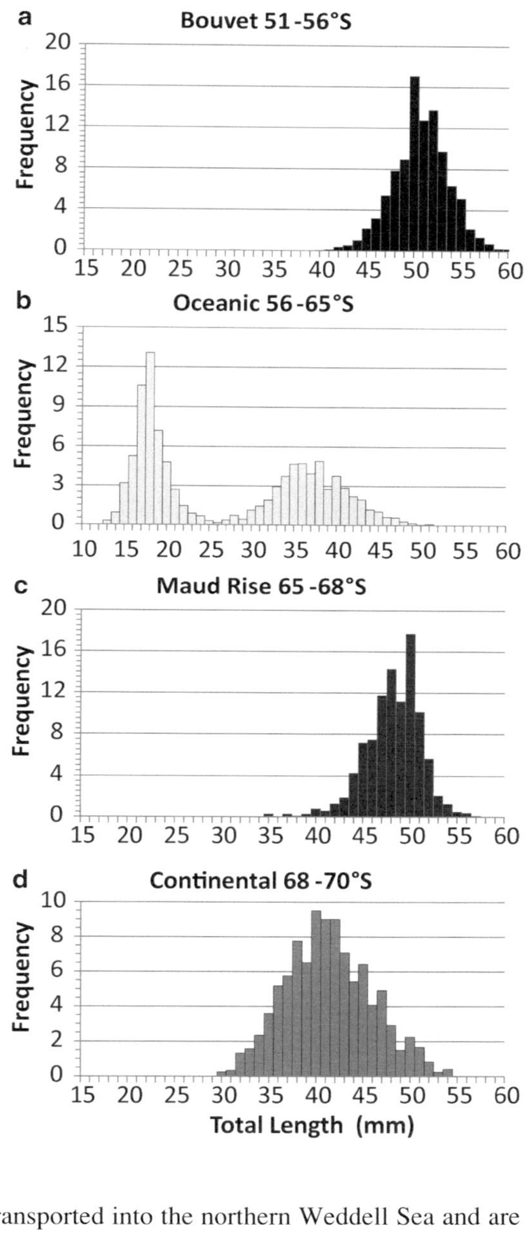

Fig. 2.13 Latitudinal change in composite length frequency distributions in the Eastern Atlantic sector, based on data presented by Siegel (2012)

more offshore water masses are transported into the northern Weddell Sea and are advected to the Scotia Sea mainly between 30° and 45°W that is between the South Orkney and the South Sandwich Islands. The Coastal Current band extends into the eastern Weddell Sea following the continental slope south.

The Coastal Current splits around 74°S into two bands and this bifurcation is believed to be topographically controlled; a slope branch follows the deep slope into the central while the coastal branch flows to the southern Weddell Sea along the Filchner ice-shelf and north again along the eastern side of the Antarctic Peninsula where it finally enters the Scotia Sea south of Elephant Island (Amos 1984). This branch is known for its poor krill abundance (Fevolden 1979; Siegel 1982; Boysen-Ennen and Piatkowski 1988). After entering the Scotia Sea, the Weddell Gyre water meets the water masses from the Antarctic Peninsula forming the WSC (Meredith et al. 2015) and the current system splits into two bands that are deflected eastwards; the northern one can extend as far north as South Georgia, the southern limb flows along the South Sandwich Islands. Both branches seed krill into the larger Bouvet area north of 63°S and complete the long loop around the Weddell Gyre.

The scenario above describes a long-distance connection of krill stocks across ocean boundaries, and the Lazarev Sea acts as a transition zone between the Southwest Atlantic and the Indian Ocean with a high exchange rate in either direction. The southern part of the Lazarev Sea is regarded as a gateway to the Weddell Sea and the entrance of the krill population into the Weddell Gyre system (Mackintosh 1973; Maslennikov 1980) which would contribute to the recorded high krill concentrations in the Scotia Sea (Atkinson et al. 2008). However, when comparing published data Siegel (2012) concluded that the overall abundance of krill and krill larvae in the Lazarev and southern Weddell Seas may on average be lower by almost one order of magnitude than in the Antarctic Peninsula/Scotia Sea region. As a consequence it must be concluded that the influence of the Antarctic Peninsula stock on the Scotia Sea krill aggregations is much greater than the input from the Weddell/Lazarev Sea. From krill biomass estimates recorded above, it is obvious that the biomass density in the Lazarev and eastern Weddell Sea is on average one order of magnitude lower than in the western Peninsula region. It might therefore be considered that the Weddell Sea stock could contribute around 10 % to the krill biomass observed in the Scotia Sea. Since the advection area from the Lazarev Sea into the Weddell Sea extends over roughly 7 degrees of latitude compared to only 3° in the Drake Passage/Peninsula region, the factor for the biomass inflow from the Lazarev Sea into the eastern Weddell Sea would be approx. 2.5 times greater in actual biomass and could therefore contribute up to 25 % of the Scotia Sea krill biomass.

Finally, the question could be raised whether this difference in krill abundance between the two source areas occurs simply because the Antarctic Peninsula is an area with a high reproductive output while the krill stock in the Lazarev Sea is less productive itself and more dependent on seasonal influx from upstream areas. One observation is that krill in the Antarctic Peninsula/Scotia Sea region attain a larger maximum size than krill in the high-latitude areas of the southern Lazarev Sea or Cosmonaut Sea. The occurrence of larger size classes in the northern Scotia Sea also means that the proportion of adult mature specimens increases and because of the exponential relationship between size and fecundity the reproductive output should be substantially greater (Ross and Quetin 1983; Siegel 1985b). Maybe, this difference in additional spawning capacity between the Scotia Sea and Lazarev/

Weddell Sea populations can at least explain part of the discrepancy between the high abundance of krill larvae/recruits in the Antarctic Peninsula and western Scotia Sea source region (Siegel 2012).

2.8 Outlook

The geological history of Antarctica has presented the inhabitants of the Southern Ocean with some demanding challenges to adapt to. Even over the past millennia Antarctic krill, *Euphausia superba*, has experienced dramatic changes in its habitat. During the period of the last glacial maximum ice shelves as well as multi-year sea-ice have probably covered the southern branch of the Weddell Gyre, Prydz Bay, Ross Sea and the south-western Pacific sector which should have led to unfavourable spawning conditions for krill (Spiridonov 1996). On the other hand, during the period of maximum Holocene warmth, the Bellingshausen Sea population was probably larger than today because it approached optimum temperature conditions for Antarctic krill. However, the warming might have also increased the CDW core temperature in the southern ACC up to suboptimal conditions for krill embryonic development (George and Strömberg 1985) which is presently one of the major sources for the entire circum-Antarctic krill population.

Since the 'Discovery' investigations of the 1920s, water temperatures have increased in the Atlantic sector by approx. 1 °C (Mackey et al. 2012). At South Georgia, the mean temperature in the top 100 m of the water column has increased by 0.9 °C in January and 2.3 °C in August over the past 85 years (Whitehouse et al. 2008). In the past three decades the mean extent of sea-ice coverage along the western Peninsula and southern Bellingshausen Sea has declined by 40 % (Jacka 1990; Stammerjohn and Smith 1996; Jacobs and Comiso 1997; Smith and Stammerjohn 2001; Stammerjohn et al. 2008). Although in other regions, particularly the Ross Sea, sea-ice coverage has increased at the same time. In the future, however, the Antarctic Peninsula region especially is predicted to continue to exhibit dramatic changes in sea surface temperature (Anisimov et al. 2007) and in winter sea-ice extent (Stammerjohn et al. 2008). In addition, ocean acidification is likely to magnify the risk of increasing temperatures to the Antarctic ecosystem (Kawaguchi et al. 2010b, 2013).

The observations around South Georgia on the extreme shortage of Antarctic krill in 2009 provided support for the hypothesis that krill is a major mediator of climatic effects on the ecosystem (Forcada et al. 2005; Trathan et al. 2006; Murphy et al. 2007a, b). Murphy et al. (2007b) generated population models derived on the basis of climatically driven oceanic fluctuations to indicate that plausible rates of regional warming of 1 °C over the next 100 years could lead to more than a 95 % reduction in the biomass and abundance of krill across the Scotia Sea by the end of the twenty-first century. Hill et al. (2013) assessed the potential influence of the projected 0.27°–1.08 °C warming of the Scotia Sea on Antarctic krill habitat with a statistical model that links growth to temperature and chlorophyll concentration.

Analysis suggests that the direct effects of warming could reduce the area of krill growth habitat by up to 20 % until the late twenty-first century. It was realized that the effects of a 50 % change in chlorophyll concentration was more significant than the direct effects of a temperature increase. However, even with increasing chlorophyll, the projected warming would cause some degradation of the krill habitat. The reduction in krill habitat could be up to 55 % for northern areas such as South Georgia which is one of the most important breeding grounds for land-based and krill-dependent predators.

Changes in the physical environment will be accompanied by changes in phytoplankton community structure and shifts in biogeochemical processes in the upper water column (Nicol 2006; Hewes et al. 2009). As regional warming reduces both the extent and the duration of sea-ice cover, changes are already evident at most levels of the food web. In some areas, for example, there are indications that diatoms are being replaced by cryptophytes (flagellates) or warm water adapted dinoflagellates. If such a shift spreads across the Antarctic krill habitat, it will surely have observable effects on the energy budget and survival rate of juvenile krill (recruits) congregating under the ice in winter and grazing on the (diatom) ice algae. Habitat loss and physiological stress may be compounded by reduction in food availability through both decreases in total phytoplankton biomass and shifts away from diatom dominance (Kavanaugh et al. 2015). Furthermore, at least in the very important reproductive region of the Antarctic Peninsula, a southward redistribution of spawning grounds is limited by the continental shelf, because the development of krill eggs towards the first feeding stage involves sinking to at least 700–1500 m water depth (Marr 1962; Hempel and Hempel 1986; Marschall 1983). The consequences of such a range displacement and disruption of the oceanic life cycle would also lead to increased competition with *E. crystallorophias* on the Antarctic shelf (Flores et al. 2012b).

These poleward relocation processes could potentially open habitat for species other than those currently dominating the ecosystem (Montes-Hugo et al. 2009). Those warm water species (Mackintosh 1934) usually concentrated in sub-Antarctic and Polar Frontal waters, e.g. *Salpa thompsoni* (Pakhomov et al. 2002; Loeb and Santora 2012), pteropods (Hunt et al. 2008) or omnivorous species that have a relatively broad latitudinal distribution such as the euphausiid *Thysanoessa macrura*, may gain advantage over stenothermic organisms such as *E. superba*. Although salps have fewer predators than krill, they are not a trophic cul-de-sac (Pakhomov et al. 2002). Salps rather represent an example of a different pelagic ecosystem that thrives in "warmer", less productive waters. Therefore, degradation of krill habitat would largely replace one (e.g. krill dominated) community with another (salp dominated) one with subsequent consequences to the top predators and the fishery. Pteropods also have a variety of predators and thus participate in the food chain but are also affected by increasing CO_2 levels and acidification (Hunt et al. 2008). It appears that current climatic trends are likely to be detrimental to crustacean and molluscan members of the zooplankton, while favouring the "gelatinous" organisms such as salps (McClintock et al. 2008).

From these life history traits we can infer that the rapidly reproducing and growing salps or the omnivorous *T. macrura* may have relatively greater plasticity compared to *E. superba* when confronted with a changing pelagic environment. Although the range of habitats that Antarctic krill is now known to exploit, from pelagic to deep-ocean benthos to under ice, suggests that one of the great strengths of *E. superba* has been it high degree of plasticity. Parmesan (2006) and Ji et al. (2010) suggest that greater plasticity may provide organisms an advantage under predicted climate change scenarios (Driscoll et al. 2015), however, it should be kept in mind that even greater plasticity could be outweighted by the short lived and opportunistic pelagic tunicates.

Teleconnections with global climate patterns, such as the ENSO, also act upon the increase in westerly winds (Turner et al. 2005; Harangozo 2006). At the same time, changes in the primary atmospheric climate mode, the Southern Annular Mode (SAM) has strengthened the warm westerly winds (Lovenduski and Gruber 2005). This has driven declines in sea-ice extent and duration in the Antarctic Peninsula increasing the number of low ice and ice free winters (Stammerjohn et al. 2008) potentially impacting overwinter survival, condition and population trends of krill (Atkinson et al. 2004) and krill dependent predators. This increase in the SAM has resulted in poleward heat transport through upwelling of Upper Circumpolar Deep Water (UCDW), as well as in the southward displacement of fronts within the ACC (Gille 2002). While the upwelling of UCDW may have positive effects on the primary productivity and consequently on growth of krill, the displacement of water masses e.g. in the Drake Passage could have immediate negative effects on the dispersion of early life stages and adult krill (Flores et al. 2012b), since it has been shown above that minor changes in the geographical release point will strongly influence the potential drift trajectory and fate of krill downstream in the Scotia Sea and around South Georgia.

Shifts in the pattern of ocean circulation (Bigg 1996; Houghton et al. 2001) may disrupt the established pathways of biological transport, potentially resulting in the collapse of regional food webs (Murphy et al. 2004a). The example around South Georgia and the occasional shortage of krill due the shift in currents and fronts in the central Scotia Sea (see example in Fig. 2.9) demonstrates the dramatic effect that can be expected on the land-breeding predators.

The ecosystems of the Southern Ocean are expected to be some of those most severely affected by ocean acidification (OA) because of the higher solubilities of CO_2 and $CaCO_3$ in cold waters, and because of regional upwelling of hypercapnic deep-sea water (Sabine et al. 2004; Doney et al. 2009). Changes in pCO_2 are already occurring in the Southern Ocean and the physiological cost to krill will rise and so krill vulnerability to stress will increase. Kawaguchi et al. (2013) presented a series of risk maps for Antarctic krill under projected Southern Ocean acidification. This analysis showed that the embryonic development of Antarctic krill in elevated seawater CO_2 levels and successful hatching is impaired at CO_2 levels greater than 1000 μatm. Exposure to elevated CO_2 significantly retards subsequent development. Krill embryos appear more vulnerable to ocean acidification than other pelagic crustaceans such as copepods. The pCO_2 level in the Southern Ocean is predicted to rise to above 1500 μatm in some parts of krill's depth range by the year 2100 unless emissions are mitigated. Risk maps, combining modelled hatch rates and the three-dimensional circumpolar projection of future pCO_2, predict that by 2100 the Weddell Sea and the waters to the east are the highest-risk areas for krill embryos.

It has to be realized that most changes described above are already occurring in the Southern Ocean and the physiological cost of living will rise and so vulnerability to stress will increase. Community structure, life strategies, and adaptive responses of zooplankton species to climate forcing will increase competition among those species. The strong climate links with oceanographic parameters that in turn drive population processes (e.g. reproduction and growth) in krill will – with varying temporal lags – cascade through to higher trophic level predators of krill (Forcada et al. 2005; Trathan et al. 2012) and will be a key factor determining future fisheries (McBride et al. 2014). Such changes highlight the need to think about future decision rules used for the management of the krill fishery as well as the entire ecosystem. Monitoring of stock size and distribution of krill are essential information for assessing the importance of seasonal or long-term climate effects on euphausiid populations because variability is not simply noise and must be well characterized to be able to detect change.

Acknowledgement The authors are very much indebted to Angus Atkinson (Plymouth), Evgeny Pakhomov (University BC, Vancouver), Christian Reiss (NOAA, La Jolla) who gave permission to use KRILLBASE data as well as to all other data supporters from national krill surveys, So Kawaguchi, Graham Hosie, Steve Nicol (AAD, Kingston TAS), Vyacheslav Sushin (AtlantNIRO, Kaliningrad) which enabled us to produce some of the most comprehensive distribution maps for Antarctic krill based on real data.

References

Allen KR (1971) Relation between production and biomass. J Fish Res Bd Can 28:1573–1581

Allen SE, Vindeirinho C, Thomson RE, Foreman MGG, Mackas DL (2001) Physical and biological processes over a submarine canyon during an upwelling event. Can J Fish Aquat Sci 58:671–684

AMLR FSR (2014) 2010/11 field season report U.S. Antarctic Marine Living Resources Program, NOAA-TM-NMFS-SWFSC-524, La Jolla, p 72

Amos AF (1984) Distribution of krill (*Euphausia superba*) and the hydrography of the Southern Ocean: large scale processes. J Crustac Biol 4:306–329

Anisimov OA, Vaughan DG, Callaghan TV, Furgal C, Marchant H, Prowse TD, Vilhjálmsson H, Walsh JE (2007) Polar regions (Arctic and Antarctic). In: Parry ML, Canziani OF, Palutikof JP, van der Linden PJ, Hanson (eds) Climate change. Impacts, adaptation and vulnerability, Contribution of Working Group II to the Fourth Assessment Report of the Intergovernmental Panel on Climate Change. CE Cambridge University Press, Cambridge, pp 653–685

Aseev YP, Vagin AV, Shevzov VV (1988) Distribution of the euphausiids in the south-western cooperation Sea. Complex ecosystem investigations of the pelagial of the Cooperation and Cosmonaut Seas. VNIRO 1988:184–194

Ashijan CJ, Rosenwaks GA, Wiebe PH, Davis CS, Gallager SM, Copley NJ, Lawson GL, Alatalo P (2004) Distribution of zooplankton on the continental shelf off Marguerite Bay, Antarctic Peninsula, during austral fall and winter, 2001. Deep-Sea Res II 51:2073–2998
Ashjian CJ, Davis CS, Gallager SM, Wiebe PH, Lawson GL (2008) Distribution of larval krill and zooplankton in association with hydrography in Marguerite Bay, Antarctic Peninsula, in austral fall and winter 2001 described using the Video Plankton Recorder. Deep-Sea Res II 55:455–471
Atkinson A, Whitehouse MJ, Priddle J, Cripps GC, Ward P, Brandon MA (2001) South Georgia, Antarctica: a productive, cold water, pelagic ecosystem. Mar Ecol Prog Ser 216:279–308
Atkinson A, Siegel V, Pakhomov EA, Rothery P (2004) Long-term decline in krill stock and increase in salps within the Southern Ocean. Nature 432:100–103
Atkinson A, Shreeve RS, Hirst AG, Rothery P, Tarling G, Pond DW, Korb RE, Murphy EJ, Watkins JL (2006) Natural growth rates in Antarctic krill (*Euphausia superba*): II. Predictive models based on food, temperature, body length, sex, and maturity stage. Limnol Oceanogr 51:973–987
Atkinson A, Siegel V, Pakhomov EA, Rothery P, Loeb V, Ross RM, Quetin LB, Schmidt K, Fretwell P, Murphy EJ, Tarling GA, Fleming AH (2008) Oceanic circumpolar habitats of Antarctic krill. Mar Ecol Prog Ser 362:1–23
Atkinson A, Siegel V, Pakhomov EA, Jessopp MJ, Loeb V (2009) A re-appraisal of the total biomass and annual production of Antarctic krill. Deep-Sea Res I 56:727–740
Atkinson A, Hill SL, Barange M, Pakhomov EA, Raubenheimer D, Schmidt K, Simpson SJ, Reiss C (2014) Sardine cycles, krill declines, and locust plagues: revisiting 'wasp-waist' food webs. Trends Ecol Evol 29:309–316
Azzali M, Kalinowski J (2000) Spatial and temporal distribution of krill *Euphausia superba* biomass in the Ross Sea (1989–1990 and 1994). In: Faranda FM, Guglielmo L, Ianora A (eds) Ross sea ecology. Springer, Berlin, pp 433–455
Azzali M, Leonori I, de Felice A, Russo A (2006) Spatial-temporal relationships between two euphausiid species in the Ross Sea. Chem Ecol 22:219–233
Beardsley RC, Limeburner R, Owens WB (2004) Drifter measurements of surface currents near Marguerite Bay on the western Antarctic Peninsula shelf during austral summer and fall, 2001 and 2002. Deep-Sea Res II 51:1947–1964
Bergström BI, Hempel G, Marschall H-P, North A, Siegel V, Strömberg JO (1990) Spring distribution, size composition and behaviour of krill *Euphausia superba* in the western Weddell Sea. Polar Rec 26:85–89
Bibik VA, Maslennikov VV, Pelevin AS, Petrova NG, Samyshev EZ, Solyankin EV, Shevtsov VV (1988) Distribution of *Euphausia superba* DANA in relation to environmental conditions in the Commonwealth and Cosmonauts Seas. In: Makarov RR (ed) Interdisciplinary investigations of the pelagic ecosystem in the Commonwealth and Cosmonauts Seas. VNIRO Publishing, Moscow, pp 109–125 (in Russian)
Bigg G (1996) The oceans and climate. Cambridge University Press, New York, p 266
Bindoff NL, Rosenberg MA, Warner MJ (2000) On the circulation and water-masses over the Antarctic continental slope and rise between 80 and 150° East. Deep-Sea Res II 47:2299–2326
Boysen-Ennen E, Piatkowski U (1988) Meso- and macrozooplankton communities in the Weddell Sea, Antarctica. Polar Biol 9:17–35
Brierley AS, Watkins JL (1996) Acoustic targets at South Georgia and the South Orkney Islands during a season of krill scarcity. Mar Ecol Prog Ser 138:51–61
Brierley AS, Watkins JL (2000) Effects of sea-ice cover on the swarming behaviour of Antarctic krill, *Euphausia superba*. Can J Fish Aquat Sci 57 (Suppl 3):24–30
Brierley AS, Watkins JL, Murray AWA (1997) Interannual variability in krill abundance at South Georgia. Mar Ecol Prog Ser 150:87–98
Brierley AS, Watkins JL, Goss C, Wilkinson MT, Everson I (1999a) Acoustic estimates of krill density at South Georgia, 1981 to 1999. CCAMLR Sci 6:47–57

Brierley AS, Demer DA, Watkins JL, Hewitt RP (1999b) Concordance of interannual fluctuations in acoustically estimated densities of Antarctic krill around South Georgia and Elephant Island: biological evidence of same-year teleconnections across the Scotia Sea. Mar Biol 134:675–681
Brierley AS, Goss C, Grant SA, Watkins JL, Reid K, Belchier M, Everson I, Jessop MJ, Afanasyev V, Robst J (2002a) Significant intra-annual variability in krill distribution and abundance at South Georgia revealed by multiple acoustic surveys during 2000/01. CCAMLR Sci 9:71–82
Brierley AS, Fernandes PG, Brandon MA, Armstrong F, Millard NW, McPhail SD, Stevenson P, Pebody M, Perrett J, Squires M, Bone DG, Griffiths G (2002b) Antarctic krill under sea-ice: elevated abundance in a narrow band just south of ice edge. Science 295(5561):1890–1892
Brinton E (1985) The oceanographic structure of the eastern Scotia Sea – III. Distributions of euphausiid species and their developmental stages in 1981 in relation to hydrography. Deep-Sea Res 32:1153–1180
Brinton E, Huntley M, Townsend AW (1986) Larvae of *Euphausia superba* in the Scotia Sea and Bransfield Strait in March 1984 – development and abundance compared with 1981 larvae. Polar Biol 5:221–234
Capella JE, Quetin LB, Hofmann EE, Ross RM (1992) Models of the early life history of *Euphausia superba* – Part II. Lagrangian calculations. Deep-Sea Res 39:1201–1220
Carleton AM (2003) Atmospheric teleconnections involving the Southern Ocean. J Geophys Res 108(C4):8080. doi:10.1029/2000JC000379
Cisewski B, Strass VH, Leach H (2011) Circulation and transport of water masses in the Lazarev Sea, Antarctica, during summer and winter 2006. Deep-Sea Res I 58:186–199
Clarke A, Tyler PA (2008) Adult Antarctic Krill feeding at abyssal depth. Curr Biol 18:282–285
Clowes AJ (1934) Hydrology of the Bransfield Strait. Discov Rep 9:1–64
Constable AJ, Nicol S (2002) Defining smaller-scale management units to further develop the ecosystem approach in managing large-scale pelagic krill fisheries in Antarctica. CCAMLR Sci 9:117–131
Constable AJ, Nicol S, Strutton PG (2003) Southern Ocean productivity in relation to spatial and temporal variation in the physical environment. J Geophys Res 108:6–21
Conti SG, Demer DA (2006) Improved parameterization of the SDWBA for estimating krill target strength. ICES J Mar Sci 63:928–935
Cook FA (1900) Through the first Antarctic night, 1896–1899: a narrative of the voyage of the Belgica. Among newly discovered lands and over an unknown sea about the South Pole. Doubleday & McClure Co, New York, p 478
Croxall JP, Reid K, Prince PA (1999) Diet, provisioning and productivity responses of marine predators to differences in availability of Antarctic krill. Mar Ecol Prog Ser 177:115–131
Cuzin-Roudy J, Labat P (1992) Early summer distribution of Antarctic krill sexual development in the Scotia-Weddell region: a multivariate approach. Polar Biol 12:65–74
Cuzin-Roudy J, Irisson J-O, Penot F, Kawaguchi S, Vallet C (2014) Southern Ocean Euphausiids. In: De Broyer C, Koubbi P et al (eds) Biogeographic atlas of the Southern Ocean. Scientific Committee on Antarctic Research, Cambridge, pp 309–320
Daly KL (1990) Overwintering development, growth, and feeding of larval *Euphausia superba* in the Antarctic marginal ice zone. Limnol Oceanogr 35:1564–1576
Daly KL (2004) Overwintering growth and development of larval *Euphausia superba*: an interannual comparison under varying environmental conditions west of the Antarctic Peninsula. Deep-Sea Res II 51:2139–2168
Daly KL, Macaulay MC (1988) Abundance and distribution of krill in the ice edge zone of the Weddell Sea, austral spring 1983. Deep-Sea Res 35:21–42
Daly KL, Macaulay MC (1991) Influence of physical and biological meso-scale dynamics on the seasonal distribution and behavior of *Euphausia superba* in the Antarctic marginal ice zone. Mar Ecol Prog Ser 79:37–66
de la Mare WK (1994) Estimating confidence intervals for fish stock abundance estimates from trawl surveys. CCAMLR Sci 1:203–207

Deacon GER (1977) Antarctic water masses and circulation. In: Dunbar MJ (ed) Polar oceans. McGill University Press, Montreal, pp 11–16

Deacon GER (1979) The Weddell Gyre. Deep-Sea Res 26A:981–995

Demer DA (2004) An estimate of error for the CCAMLR 2000 survey estimate of krill biomass. Deep-Sea Res II 51:1237–1251

Demer DA, Conti SG (2003a) Reconciling theoretical versus empirical target strengths of krill; effects of phase variability on the distorted wave Born approximation. ICES J Mar Sci 60:429–434

Demer DA, Conti SG (2003b) Validation of the stochastic distorted-wave born approximation model with broadbandwidth total target strength measurements of Antarctic krill. ICES J Mar Sci 60:625–635

Demer DA, Conti SG (2004) Erratum for: validation of the stochastic distorted wave born approximation model with broadbandwidth total target strength measurements of Antarctic krill. ICES J Mar Sci 61:155–156

Demer DA, Conti SG (2005) New target-strength model indicates more krill in the Southern Ocean. ICES J Mar Sci 62:25–32

Demer DA, Hewitt RP (1995) Bias in acoustic biomass estimates of *Euphausia superba* due to diel vertical migration. Deep-Sea Res 42:455–475

Dinniman MS, Klinck JM, Smith WO Jr (2011) A model study of circumpolar deep water on the West Antarctic Peninsula and Ross Sea continental shelves. Deep-Sea Res II 58:1508–1523

Dolzhenkov VN, Timonin VP (1990) Krill (*Euphausia superba* Dana) resources and distribution in the Wilkes Land area in the seasons 1986 to 1989. Selected scientific papers 1990, CCAMLR, Hobart, pp 149–162

Dolzhenkov VN, Kovalev EN, Spiridonov VA, Timonin VP, Zhigalov IA (1990) Distribution pattern and fishing for Antarctic krill off Wilkes Land and Balleny Islands. SC-CAMLR-WG-Krill 90/16:1–32

Doney SC, Fabry VJ, Feely RA, Kleypas JA (2009) Ocean acidification: the other CO_2 problem. Ann Rev Mar Sci 1:169–192

Driscoll RM, Reiss CS, Hentschel BT (2015) Temperature-dependent growth of *Thysanoessa macrura*: inter-annual and spatial variability around Elephant Island, Antarctica. Mar Ecol Prog Ser 529:49–61

El Sayed SZ (1994) History, organization and accomplishments of the BIOMASS programme. In: El-Sayed SZ (ed) Southern Ocean ecology. The BIOMASS perspective. Cambridge University Press, Cambridge, pp 1–8

Everson I (1976) Antarctic krill: a reappraisal of its distribution. Polar Rec 18:15–23

Everson I (1977) The living resources of the Southern Ocean. FAO GLO/SO/77/1, Rome, p 156

Everson I (2000) Distribution and standing stock. The Southern Ocean. In: Everson I (ed) Krill biology, ecology and fisheries. Blackwell Science, Oxford, pp 63–79

Fach B, Klinck JM (2006) Transport of Antarctic krill (*Euphausia superba*) across the Scotia Sea. Part I: circulation and particle tracking simulations. Deep-Sea Res I 53:987–1010

Fach BA, Hofmann EE, Murphy EJ (2002) Modelling studies of Antarctic krill *Euphausia superba* survival during transport across the Scotia Sea. Mar Ecol Prog Ser 231:187–203

Faranda FM, Guglielmo L, Ianora A (2000) The Italian oceanographic cruises in the Ross Sea (1987–95): strategy, general considerations and description of the sampling sites. In: Faranda FM, Guglielmo L, Ianora A (eds) Ross sea ecology. Springer, Berlin, pp 1–13

Fedulov PP, Murphy EJ, Shulgovsky KE (1996) Environment-krill relations in the South Georgia marine ecosystem. CCAMLR Sci 3:13–30

Fevolden SE (1979) Investigations on krill (Euphausiacea) sampled during the Norwegian Antarctic Research Expedition 1976/77. Sarsia 64:189–198

Fevolden SE (1980) Krill off Bouvetöya and in the southern Weddell Sea with a description of larval stages of *Euphausia crystallorophias*. Sarsia 65:149–162

Fevolden SE, George RY (1984) Size frequency pattern of *Euphausia superba* in the Antarctic Peninsula waters in the austral summer of 1983. J Crust Biol 4(Spec 1):107–122

Fielding S, Watkins J (2011) The ASAM 2010 assessment of krill biomass for Area 48 from the Scotia Sea CCAMLR 2000 synoptic survey. SC-CAMLR-WG-EMM-11/20:1–10

Fielding S, Watkins J, Collins M, Enderlein P, Venables H (2012) Acoustic determination of the distribution of fish and krill across the Scotia Sea in spring 2006, summer 2008 and autumn 2009. Deep-Sea Res II 59–60:173–188

Fielding S, Watkins JL, Trathan PN, Enderlein P, Waluda CM, Stowasser G, Tarling GA, Murphy EJ (2014) Interannual variability in Antarctic krill (*Euphausia superba*) density at South Georgia, Southern Ocean: 1997–2013. ICES J Mar Sci 71:2578–2588

Flores H, van Franeker JA, Siegel V, Haraldsson M, Strass V, Meesters EH, Bathmann U, Wolff WJ (2012a) The association of Antarctic krill *Euphausia superba* with the under-ice habitat. PLoS One 7:1–11

Flores H, Atkinson A, Kawaguchi S, Krafft BA, Milinevsky G, Nicol S, Reiss C, Tarling GA, Werner R, Bravo Rebolledo E, Cirelli V, Cuzin-Roudy J, Fielding S, van Franeker JA, Groeneveld JJ, Haraldsson M, Lombana A, Marschoff E, Meyer B, Pakhomov EA, Van de Putte AP, Rombolá E, Schmidt K, Siegel V, Teschke M, Tonkes H, Toullec JY, Trathan PN, Tremblay N, Werner T (2012b) Impact of climate change on Antarctic krill. Mar Ecol Prog Ser 458:1–19

Forcada J, Trathan PN, Reid K, Murphy EJ (2005) The effects of global climate variability in pup production of Antarctic fur seals. Ecology 86:2408–2417

Fraser FC (1936) On the development and distribution of the young stages of krill (*Euphausia superba*). Discov Rep 14:1–192

Fraser WR, Hofmann EE (2003) A predator's perspective on causal links between climate change, physical forcing and ecosystem response. Mar Ecol Prog Ser 265:1–15

Frazer TK, Quetin LB, Ross RM (1997) Abundance and distribution of larval krill, *Euphausia superba*, associated with annual sea-ice in winter. In: Battaglia B, Valencia J, Walton DWH (eds) Antarctic communities – species, structure and survival. Cambridge University Press, Cambridge, pp 107–111

Garrison DL, Sullivan CW, Ackley SF (1986) Sea-ice microbial communities in Antarctica. BioScience 36:243–250

George RY, Strömberg JO (1985) Development of eggs of Antarctic krill *Euphausia superba* in relation to pressure. Polar Biol 4:125–133

Gille ST (2002) Warming of the Southern Ocean since the 1950s. Science 295:1275–1277

Gloersen P, Campbell WJ, Cavalieri DJ, Comiso JC, Parkinson CL, Zwally HH (1992) Arctic and Antarctic sea-ice, 1978–1987: satellite passive-microwave observations and analysis. National Aeronautics and Space Administration, Washington, DC

Gloerson P, White WB (2001) Re-establishing the circumpolar wave in sea-ice around Antarctica from one winter to the next. J Geophys Res 106:4391–4395

Gouretski VV, Danilov AI (1993) Weddell Gyre structure of the eastern boundary. Deep-Sea Res I 40:561–582

Greene CH, Stanton TK, Wiebe PH, McClatchie S (1991) Acoustic estimates of Antarctic krill. Nature 349:110

Guihen D, Fielding S, Murphy EJ, Heywood KJ, Griffiths G (2014) An assessment of the use of ocean gliders to undertake acoustic measurements of zooplankton: the distribution and density of Antarctic krill (*Euphausia superba*) in the Weddell Sea. Limnol Oceanogr Methods 12:373–389

Gulland JA (1970) The development of the resources of the Antarctic seas. In: Holdgate MW (ed) Antarctic ecology, vol 1. Academic, London, pp 217–223

Gutt J, Siegel V (1994) Benthopelagic aggregations of krill (*Euphausia superba*) on the deeper shelf of the Weddell Sea (Antarctic). Deep-Sea Res 41:169–178

Hamner WM, Hamner PP (2000) Behavior of Antarctic krill (*Euphausia superba*): Schooling, foraging, and antipredatory behavior. Can J Fish Aquat Sci 57:192–202

Hampton I (1985) Abundance, distribution and behaviour of *Euphausia superba* in the Southern Ocean between 15° and 30°E during FIBEX. In: Siegfried WR, Condy PR, Laws RM (eds) Antarctic nutrient cycles and food webs. Springer, Berlin, pp 294–303

Hansen HJ (1908) Schizopoda and Cumacea. Resultats du Voyage du S.Y. Belgica en 1897–1898–1899. Rap Scient Zool Anvers 1–20

Harangozo SA (2006) Atmospheric circulation impacts on winter maximum sea-ice in the west Antarctic Peninsula region (1979–2001). Geophys Res Lett 33:L02502. doi:10.1029/2005GL024978

Hazen EL, Suryan RM, Santora JA, Bograd SJ, Watanuki Y, Wilson RP (2013) Scales and mechanisms of marine hotspot formation. Mar Ecol Prog Ser 487:177–183

Hempel I (1981) Euphausiid larvae in the Scotia Sea and adjacent waters in summer 1977/78. Meeresforsch/Rep Mar Res 29:53–59

Hempel I (1985a) Variation in geographical distribution and abundance of larvae of Antarctic krill, *Euphausia superba*, in the southern Atlantic Ocean. In: Siegfried WR, Condy PR, Laws RM (eds) Antarctic nutrient cycles and food webs. Springer, Berlin, pp 304–307

Hempel I (1985b) Vertical distribution of larvae of Antarctic krill, *Euphausia superba*. In: Siegfried WR, Condy PR, Laws RM (eds) Antarctic nutrient cycles and food webs. Springer, Berlin, pp 308–310

Hempel I, Hempel G (1978) Larval krill (*Euphausia superba*) in the plankton and neuston sampler of the German Antarctic expedition 1975/76. Meeresforsch/Rep Mar Res 26:206–216

Hempel I, Hempel G (1982) Distribution of euphausiid larvae in the southern Weddell Sea. Meeresforschung 29:253–266

Hempel I, Hempel G (1986) Field observations on the developmental ascent of larval *Euphausia superba* (Crustacea). Polar Biol 6:121–126

Hempel G, Sahrhage D, Schreiber W, Steinberg R (1979a) Antarctic expedition 1977/78 of the Federal Republic of Germany. Arch Fisch Wiss 30(Spec Vol 1):1–119

Hempel I, Hempel G, Baker A de C (1979b) Early life history stages of krill (*Euphausia superba*) in Bransfield Strait and Weddell Sea. Meeresforsch/Rep Mar Res 27:267–281

Hewes CD, Reiss CS, Holm-Hansen O (2009) A quantitative analysis of sources for summertime phytoplankton variability over 18 years in the South Shetland Islands (Antarctica) region. Deep-Sea Res I 56:1230–1241

Hewitt RP, Demer DA (1994) Acoustic estimates of krill biomass in the Elephant Island area: 1981–1993. CCAMLR Sci 1:1–5

Hewitt RP, Watkins JL, Naganobu M, Tshernyshkov P, Brierley AS, Demer DA, Kasatkina S, Takao Y, Goss C, Malyshko A, Brandon MA, Kawaguchi S, Siegel V, Trathan PN, Emery JH, Everson I, Miller DGM (2002) Setting a precautionary catch limit for Antarctic krill. Oceanography 15:26–33

Hewitt R, Demer D, Emery JH (2003) An 8-year cycle in krill biomass density inferred from acoustic surveys conducted in the vicinity of the South Shetland Islands during the austral summers of 1991–1992 through 2001–2002. Aquat Living Resour 16:205–213

Hewitt RP, Watkins J, Naganobu M, Sushin V, Brierley AS, Demer D, Kasatkina S, Takao Y, Goss C, Malyshko A, Brandon MA, Kawaguchi K, Siegel V, Trathan P, Emery J, Everson I, Miller D (2004) Biomass of Antarctic krill in the Scotia Sea in January/February 2000 and its use in revising an estimate of precautionary yield. Deep-Sea Res II 51:1215–1236

Heywood RB, Priddle J (1987) Retention of phytoplankton by an eddy. Cont Shelf Res 7:937–955

Heywood RB, Everson I, Priddle J (1985) The absence of krill from the South Georgia zone, winter 1983. Deep-Sea Res 32:369–378

Higginbottom IR, Kerry KR, Wayle SE (1988) Hydroacoustic surveys of the distribution and abundance of krill: Prydz Bay region – FIBEX, ADBEX II and SIBEX II. ANARE Res Notes 62:1–46

Hill HJ, Trathan PN, Croxall JP, Watkins JL (1996) A comparison of Antarctic krill *Euphausia superba* caught by nets and taken by macaroni penguins *Eudyptes chrysolophus*: evidence for selection? Mar Ecol Prog Ser 140:1–11

Hill SL, Murphy EJ, Reid K, Trathan PN, Constable AJ (2006) Modelling Southern Ocean ecosystems: krill, the food-web, and the impacts of harvesting. Biol Rev 81:581–608

Hill S, Belchier M, Collins M, Fielding S, Murphy E, Trathan P, Venables H, Waluda C (2009) Multiple indicators suggest a strong ecosystem anomaly at South Georgia in 2009. SC-CAMLR-WG-EMM-09/23:1–16 (with permission)
Hill SL, Phillips T, Atkinson A (2013) Potential climate change effects on the habitat of Antarctic krill in the Weddell Quadrant of the Southern Ocean. PLoS One 8:e72246. doi:10.1371/journal.pone.0072246
Hill S, Atkinson A, Darby C, Fielding S, Krafft BA, Godø OR, Skaret G, Trathan P, Watkins J (2015) Updating the Antarctic krill biomass estimates for CCAMLR subareas 48.1 to 48.4 using available data. SC-CAMLR-WG-EMM-15/28:1–22 (with permission)
Hofmann EE, Hüsrevoglu YS (2003) A circumpolar modelling study of habitat control of Antarctic krill (*Euphausia superba*) reproductive success. Deep-Sea Res II 50:3121–3142
Hofmann EE, Klinck JM (1998) Thermocline variability of the waters overlying the west Antarctic Peninsula continental shelf. Antarct Res Ser 75:67–81
Hofmann EE, Murphy EJ (2004) Advection, krill, and Antarctic marine ecosystems. Antarct Sci 16:487–499
Hofmann EE, Capella JE, Ross RM, Quetin LB (1992) Models of the early life history of *Euphausia superba*—Part I. Time and temperature dependence during the descent-ascent cycle. Deep-Sea Res 39:1177–1200
Hofmann EE, Klinck JM, Lascara CM, Smith DA (1996) Water mass distribution and circulation west of the Antarctic Peninsula and including Bransfield Strait. Antarct Res Ser 70:61–80
Hofmann EE, Klinck JM, Locarnini RA, Fach B, Murphy E (1998) Krill transport in the Scotia Sea and environs. Antarct Sci 10:406–415
Hosie GW (1991) Distribution and abundance of euphausiid larvae in the Prydz Bay region, Antarctica. Antarct Sci 3:167–180
Hosie GW, Ikeda T, Stolp M (1988) Distribution, abundance and population structure of the Antarctic krill (*Euphausia superba* Dana) in the Prydz Bay region, Antarctica. Polar Biol 8:213–224
Hosie GW, Fukuchi M, Kawaguchi S (2003) Development of the Southern Ocean Continuous Plankton Recorder survey. Prog Oceanogr 58:263–283
Houghton JTY, Diny Y, Giggs DJ, Noguer M, van der Linden PJ, Dai X, Maskell K, Johnson CA (2001) IPCC 2001: climate change 2001—the scientific basis: contribution of working group I to the third assessment report of the Intergovernmental Panel on Climate Change. Cambridge University Press, New York, p 881
Hughes CW, Woodworth PL, Meredith MP, Stepanov V, Whitworth T III, Pyne A (2003) Coherence of Antarctic sea levels. Southern hemisphere annular mode, and flow through Drake Passage. Geophys Res Lett 30:1464. doi:10.1029/2003GL017240
Hunt BPV, Pakhomov EA, Hosie GW, Siegel V, Ward P, Bernard K (2008) Pteropods in Southern Ocean ecosystems. Prog Oceanogr 78:193–221
Huntley ME, Niiler PP (1995) Physical control of population dynamics in the Southern Ocean. ICES J Mar Sci 52:457–468
Ichii T (1990) Distribution of Antarctic krill concentrations exploited by Japanese krill trawlers and Minke whales. Proc NIPR Symp Polar Biol 3:36–56
Jacka TH (1990) Antarctic and Southern Ocean sea-ice and climate trends. Ann Glaciol 14:127–130
Jackowska H (1980) Krill monitoring in Admiralty Bay (King George Island, South Shetland Islands) in summer 1979/1980. Pol Polar Res 1:117–125
Jacobs SS, Comiso JC (1997) Climate variability in the Amundsen and Bellingshausen Seas. Am Meteorol Soc 10:697–709
Jacobs GA, Mitchell JL (1996) Ocean circulation variations associated with the Antarctic Circumpolar Wave. Geophys Res Lett 23:2947–2950
Jarman SN, Deagle BE (2016) Genetics of Antarctic krill. In: Siegel V (ed) The biology and ecology of Antarctic krill, *Euphausia superba* Dana, 1850. Springer, Cham, pp 247–278

Jarvis T, Kelly N, Kawaguchi S, Van Wijk E, Nicol S (2010) Acoustic characterisation of the broad-scale distribution and abundance of Antarctic krill (*Euphausia superba*) off East Antarctica (30–80°E) in January–March 2006. Deep-Sea Res II 57:916–993
Ji R, Edwards M, Mackas DL, Runge JA, Thomas AC (2010) Marine plankton phenology and life history in a changing climate: current research and future directions. J Plankton Res 32:1355–1368
Jiang MS, Charette MA, Measures CI, Zhu Y, Zhou M (2013) Seasonal cycle of circulation in the Antarctic Peninsula and the off shelf transport of shelf waters into Southern Drake Passage and Scotia Sea. Deep-Sea Res I 90:15–30
Joyce TM, Zenk W, Toole JM (1978) The anatomy of the Antarctic polar front in the Drake passage. J Geophys Res 83:6093–6113
Ju SJ, Harvey HR (2004) Lipids as markers of nutritional condition and diet in the Antarctic krill *Euphausia superba* and *Euphausia crystallorophias* during austral winter. Deep-Sea Res II 51:2199–2214
Kanda K, Takagi K, Seki Y (1982) Movement of the larger swarms of Antarctic krill *Euphausia superba* population off Enderby Land during 1976–1977 season. J Tokyo Univ Fish 68:25–42
Karoly DJ (1989) Southern hemisphere circulation features associated with El Niño-Southern Oscillation events. J Climate 2:1239–1252
Kasatkina SM, Shnar VN, Polischuk MI, Abramov AM, Sushin VA (1997) Assessment of krill flux factors in waters of the South Orkney Islands during summer 1996. CCAMLR Sci 4:195–204
Kavanaugh MT, Abdala FN, Ducklow H, Glover D, Fraser W, Martinson D, Stammerjohn S, Schofield O, Doney SC (2015) Effect of continental shelf canyons on phytoplankton biomass and community composition along the western Antarctic Peninsula. Mar Ecol Prog Ser 524:11–26
Kawaguchi S (2016) Reproduction and larval development in Antarctic krill (*Euphausia superba*). In: Siegel V (ed) The biology and ecology of Antarctic krill, *Euphausia superba* Dana, 1850. Springer, Cham, pp 225–246
Kawaguchi K, Ishikawa S, Matsuda O (1986) The overwintering strategy of Antarctic krill (*Euphausia superba* Dana) under the coastal fast ice off the Ongul Islands in Lützow-Holm Bay, Antarctica. Mem Natl Inst Polar Res 44:67–85
Kawaguchi S, Candy SG, King R, Naganobu M, Nicol S (2006) Modelling growth of Antarctic krill. I. Growth trends with sex, length, season and region. Mar Ecol Prog Ser 306:1–15
Kawaguchi S, Nicol S, Virtue P, Davenport SR, Casper R, Swadling KM, Hosie G (2010a) Krill demography and large-scale distribution in the Western Indian Ocean sector of the Southern Ocean (CCAMLR Division 58.4.2.) in Austral summer of 2006. Deep-Sea Res II 57:934–947
Kawaguchi S, Kurihara H, King R, Hale L, Berli T, Robinson JP, Ishida A, Wakita M, Virtue P, Nicol S, Ishimatsu A (2010b) Will krill fare well under Southern Ocean acidification? Biol Lett 25:288–291. doi:10.1098/rsbl.2010.0777
Kawaguchi S, Kilpatrick R, Roberts L, King RA, Nicol S (2011) Ocean-bottom krill sex. J Plankton Res 33:1134–1138
Kawaguchi S, Ishida A, King R, Raymond B, Waller N, Constable A, Nicol S, Wakita M, Ishimatsu A (2013) Risk maps for Antarctic krill under projected Southern Ocean acidification. Nat Clim Chang 3:343–347
Kils U (1981) Swimming behaviour, swimming performance and energy balance of Antarctic krill, *Euphausia superba*. BIOMASS Sci Ser 3:1–122
Kils U (1983) Swimming and feeding of Antarctic krill, *Euphausia superba* – some outstanding energetics and dynamics – some unique morphological details. In: Schnack SB (ed) On the biology of krill *Euphausia superba*. Proceedings of the seminar and report of the Krill Ecology Group, Bremerhaven. Ber Polarforsch 4:130–155
Kim S, Siegel V, Hewitt RP, Naganobu M, Demer DA, Ichii T, Kang SH, Kawaguchi S, Loeb V, Amos SA, Chung KH, Holm-Hanson O, Lee WC, Silva N, Stein M (1998) Temporal changes in marine environments in the Antarctic Peninsula area during the 1994/95 austral summer. Mem Natl Inst Polar Res, Spec Issue 52:186–208

Kittel W (1980) Populational studies on *Euphausia superba* Dana, 1852 (Euphausiacea, Crustacea) in waters of the Admiralty Bay during Antarctic summer of 1978. Pol Arch Hydrobiol 27:267–272

Kittel W, Jazdzewski K (1982) Studies on the larval stages of *Euphausia superba* Dana (Crustacea, Euphausiacea) in the southern Drake Passage and in the Bransfield Strait in February and March 1981 during the BIOMASS-FIBEX expedition. Pol Polar Res 3:273–280

Klinck JM, Hofmann EE, Beardsley RC, Salihoglu B, Howard S (2004) Water-mass properties and circulation on the west Antarctic Peninsula continental shelf in austral fall and winter 2001. Deep-Sea Res II 51:1925–1946

Kottmeier ST, Sullivan CW (1987) Late winter primary production and bacterial production in sea-ice and seawater west of the Antarctic Peninsula. Mar Ecol Prog Ser 36:287–298

Krafft BA, Melle W, Knutsen T, Bagøien E, Broms C, Ellertsen B, Siegel V (2010) Distribution and demography of Antarctic krill in the Southeast Atlantic sector of the Southern Ocean during the austral summer 2008. Polar Biol 33:957–968

Krafft BA, Skaret G, Knutsen T, Melle W, Klevjer T, Soiland H (2012) Antarctic krill swarm characteristics in the Southeast Atlantic sector of the Southern Ocean. Mar Ecol Prog Ser 465:69–83

Krakatitsa VV, Karpenko GP, Pakhomov EA (1993) Distribution peculiarities of the zooplankton in dependence on the temperature stratification of the 1-metre surface water layer in the Cooperation Sea. In: Voronina NM (ed) Pelagic ecosystems of the Southern Ocean. Nauka Press, Moscow, pp 151–157 (in Russian)

La HS, Lee H, Fielding S, Kang D, Ha HK, Atkinson A, Park J, Siegel V, Lee SH, Shin HC (2015) High density of ice krill (*Euphausia crystallorophias*) in the Amundsen sea coastal polynya, Antarctica. Deep-Sea Res I 195:75–84

Lanin VI (1993) Oceanographic bases of the formation functioning and year-to-year variability of the state of ecosystems of the Antarctic Seas. In: Voronina NM (ed) Pelagic ecosystems of the Southern Ocean. Nauka Press, Moscow, pp 63–80 (In Russian)

Lascara CM, Hofmann EE, Ross RM, Quetin LB (1999) Seasonal variability in the distribution of Antarctic krill, *Euphausia superba*, west of the Antarctic Peninsula. Deep-Sea Res 46:951–984

Latogurski VI (1979) Recognizing of independent population of Antarctic krill. Rybnoe Khoziaistvo 1979:12–14 (in Russian)

Latogurski VI, Maklygin LG, Fedotov AS, Makarov RR, Solyankin EV (1990) Distribution and peculiarities of *Euphausia superba* Dana population structure in the area of the Weddell Gyre and adjacent waters. In: Solyankin EV, Danilov AI, Makarov RR (eds) Investigations of the Weddell gyre. Oceanographic conditions and peculiarities of the development of plankton communities. VNIRO Publishing, Moscow, pp 183–206 (in Russian)

Lawson GL, Wiebe PH, Ashijan CJ, Gallager SM, Davis CS, Warren JD (2004) Acoustically-inferred zooplankton distribution in relation to hydrography west of the Antarctic Peninsula. Deep-Sea Res II 51:2041–2072

Lawson GL, Wiebe PH, Ashjian CJ, Stanton TK (2008) Euphausiid distribution along the Western Antarctic Peninsula Part B: distribution of euphausiid aggregations and biomass, and associations with environmental features. Deep-Sea Res II 55:432–454

Loeb VJ, Santora JA (2012) Population dynamics of *Salpa thompsoni* near the Antarctic Peninsula: growth rates and interannual variations in reproductive activity (1993–2009). Prog Oceanogr 96:93–107

Loeb V, Siegel V, Holm-Hansen O, Hewitt R, Fraser W, Trivelpiece W, Trivelpiece S (1997) Effects of sea-ice extent and krill or salp dominance on the Antarctic food web. Nature 387:897–900

Loeb VJ, Hofmann EE, Klinck JM, Holm-Hansen O, White WB (2009) ENSO and variability of the Antarctic Peninsula pelagic marine ecosystem. Antarct Sci 21:135–148

Loeb V, Hofmann EE, Klinck JM, Holm-Hansen O (2010) Hydrographic control of the marine ecosystem in the South Shetland-Elephant Island and Bransfield Strait region. Deep-Sea Res II 57:519–542

Lovenduski NS, Gruber N (2005) Impact of the Southern Annular Mode on Southern Ocean circulation and biology. Geophys Res Lett 32:L11603. doi:10.1029/2005gl022727
Lubimova TG, Makarov RR, Maslennikov VV, Shevtsov W, Shust KV (1984) The ecological peculiarities, stocks, and role of *E. superba* in the trophic structure of the Antarctic ecosystem. Selected papers presented to the Scientific Committee of CCAMLR 1982–1984 Part II, pp 391–505
Mackey AP, Atkinson A, Hill SL, Ward P, Cunningham NJ, Johnston NM, Murphy EJ (2012) Antarctic macrozooplankton of the southwest Atlantic sector and the Bellingshausen Sea: Baseline historical distributions (Discovery investigations, 1928–1935) related to temperature and food, with projections for subsequent ocean warming. Deep-Sea Res II 59–60:130–146
Mackintosh NA (1934) Distribution of the macroplankton in the Atlantic sector of the Antarctic. Discov Rep 9:65–160
Mackintosh NA (1972) Life cycle of Antarctic krill in relation to ice and water conditions. Discov Rep 36:1–94
Mackintosh NA (1973) Distribution of post-larval krill in the Antarctic. Discov Rep 36:95–156
Makarov RR (1982) Ontogenetic migrations in Antarctic euphausiids: "The Antarctic". Soviet Committee of Antarctic Research Report 21, 1982, pp 118–132 (in Russian)
Makarov RR, Maslennikov VN (1975) Investigation of the regularities of distribution and drift of the *Euphausia superba* larvae. Biol Morya 1975:37–43 (in Russian)
Makarov RR, Menshenina LL (1989) On the distribution of euphausiid larvae in the Antarctic waters. Okeanologija Akademija Nauk SSSR, Okeanograficeskaja Komissija, Moskva 29:825–831 (in Russian)
Makarov RR, Menshenina LL (1992) Larvae of euphausiids off Queen Maud Land. Polar Biol 11:515–523
Makarov RR, Shevtsov VV (1972) Some problems in the distribution and biology of Antarctic krill. Israel Prog Sci Transl T.T72(50077), pp 1–10
Makarov RR, Sysoyeva MV (1985) Biology and distribution of *Euphausia superba* in the Lazarev Sea and adjacent waters. Hydrobiol J (Gidrobiologiceskij Zhurnal) Kiev 23:95–99
Makarov RR, Maslennikov VN, Solyankin EV, Shevtsov VV (1980) Peculiarities of quantitative distribution and conditions for formations of *Euphausia superba* aggregations in some areas of the Atlantic and Pacific sector of the Southern Ocean. In: Lubimova TG (ed) Biological resources of the Antarctic krill. VNIRO Publishing, Moscow, pp 114–145 (in Russian)
Makarov RR, Maslennikov VV, Movchan OA, Solyankin EV (1982) Oceanographic conditions and regional peculiarities of seasonal successions in plankton in the coastal waters off the Antarctic Peninsula. Antarktika Dokl Komissii 21:101–117 (in Russian)
Makarov RR, Solyankin EV, Shevtsov VV (1985) Environmental conditions and adaptive features of the biology of *Euphausia superba* Dana in the Lazarev Sea. Antarktika Dokl Komissii 24:158–171 (in Russian)
Makarov RR, Menshenina L, Spiridonov V (1990a) Distributional ecology of euphausiid larvae in the Antarctic Peninsula region and adjacent waters. Proc Natl Inst Polar Res Symp Polar Biol (Tokyo) 3:23–35
Makarov RR, Menshenina LL, Timonin VP, Shurunov NA (1990b) Larval ecology and reproduction of Euphausiidae in the Ross Sea. Biol Morya (Vladivostok) 3:38–45 (in Russian)
Makarov RR, Menshenina LL, Solyankin EV (1992) Euphausiid larvae in eastern part of the Weddell Gyre. Okenaologija 32:276–284 (in Russian)
Marin VH, Brinton E, Huntley M (1991) Depth relationships of *Euphausia superba* eggs, larvae and adults near the Antarctic Peninsula, 1986–87. Deep-Sea Res 38:1241–1149
Marr JWS (1962) The natural history and geography of the Antarctic krill (*Euphausia superba* Dana). Discov Rep 32:33–464
Marschall H-P (1983) Sinking speed, density and size of euphausiid eggs. Rep Mar Res 30:1–9
Marschall H-P (1988) The overwintering strategy of Antarctic krill under the pack-ice of the Weddell Sea. Polar Biol 9:129–135

Martinson DG, Stammerjohn SE, Iannuzzi RA, Smith RC, Vernet M (2008) Western Antarctic Peninsula physical oceanography and spatio-temporal variability. Deep-Sea Res II 55:1964–1987

Maslennikov VV (1980) Modern concepts on the large-scale circulation of the Antarctic waters and the routes of mass drift of the Antarctic krill. In: Lubimova TG (ed) Biological resources of the Antarctic krill. VNIRO Publishing, Moscow, pp 8–27 (in Russian)

Maslennikov VV, Popkov VV (1988) The position of the interaction zone between the high-latitude waters and the ACC as an indication of the northern boundary of the Antarctic krill drift. Antarctica, Moscow 27:134–142 (in Russian)

Maslennikov VV, Solyankin EV (1988) Patterns of fluctuations in the hydrological conditions of the Antarctic and their effect on the distribution of Antarctic krill. In: Sahrhage D (ed) Antarctic ocean and resources variability. Springer, Berlin, pp 209–213

Maslennikov VV, Solyankin EV, Spiridonov VA, Sysoeva MV (1983) A contribution to the study of the origin of *Euphausia superba* Dana occurring off South Georgia. In: Lubimova TG, Makarov RR, Kamenskaya EA (eds) Antarctic krill. The distribution pattern and environment. VNIRO collected papers, Moscow, pp 74–85 (in Russian)

McBride MM, Dalpadado P, Drinkwater KF, Godø OR, Hobday AJ, Hollowed AB, Kristiansen T, Murphy EJ, Ressler PH, Subbey S, Hofmann EE, Loeng H (2014) Krill, climate, and contrasting future scenarios for Arctic and Antarctic fisheries. ICES J Mar Sci 71:1934–1955

McClintock J, Ducklow H, Fraser W (2008) Ecological responses to climate change on the Antarctic Peninsula. Am Sci 96:302–310

McGehee DE, O'Driscoll RL, Martin Traykovski LV (1998) Effects of orientation on acoustic scattering from Antarctic krill at 120 kHz. Deep-Sea Res II 45:1273–1294

McLeod DJ, Hosie GW, Kitchener JA, Takahashi KT, Hunt BPV (2010) Zooplankton atlas of the Southern Ocean: The SCAR SO_CPR survey (1991–2008). Polar Sci 4:353–385

Melnikov IA, Spiridonov VA (1996) Antarctic krill under perennial ice in the western Weddell Sea. Antarct Sci 8:323–329

Menshenina LL (1992) Distribution of euphausiid larvae in the Weddell Gyre in September–October 1989. Symp Polar Biol (Tokyo) 5:44–54

Menshenina LL, Mironenko NI, Samyshev EZ (1988) Euphausiid larvae in plankton of the Cosmonaut and Cooperation seas. In: Makarov RR (ed) Interdisciplinary investigations of pelagic ecosystem in the Cooperation and Cosmonaut seas. VNIRO Publishers, Moscow, pp 143–161 (in Russian)

Menshutkin VV (1993) Imitatsionnaya model populyatsii antarkticheskogo krilya. In: Voronina NM (ed) Pelagicheskie ekosistemy Yuzhnogo Okeana. Nauka, Moskva, pp 182–185 (in Russian)

Meredith MP, Woodworth PL, Hughes CW, Stepanov V (2004) Changes in ocean transport through Drake Passage during 1980s and 1990s, forced by changes in the Southern Annular Mode. Geophys Res Lett 31:L21305

Meredith M, Murphy EJ, Hawker EJ, King JC, Wallace MI (2008) On the interannual variability of ocean temperatures around South Georgia, Southern Ocean: forcing by El Niño/Southern Oscillation and the Southern Annular Mode. Deep-Sea Res II: Top Stud Oceanogr 55:2007–2022

Meredith MP, Meijers AS, Naveira Garabato AC, Brown PJ, Venables HJ, Abrahamsen EP, Messias MJ (2015) Circulation, retention, and mixing of waters within the Weddell-Scotia Confluence, Southern Ocean: the role of stratified Taylor columns. J Geophys Res: Oceans 120:547–562

Meyer B, Teschke M (2016) Physiology of *Euphausia superba*. In: Siegel V (ed) The biology and ecology of Antarctic krill, *Euphausia superba* Dana, 1850. Springer, Cham, pp 145–174

Miller DGM (1986) Results from biological investigations of krill (*Euphausia superba*) in the southern Indian Ocean during SIBEX I. Mem Natl Inst Polar Res (Tokyo) 40:117–139

Miller DGM, Hampton I (1989) Biology and ecology of Antarctic krill (*Euphausia superba* Dana): a review. BIOMASS Sci Ser 9:1–166

Miller DGM, Hampton L, Henry J, Abrams RW, Cooper J (1985) The relationship between krill food requirements and phytoplankton production in a sector of the southern Indian Ocean. In: Siegfried WR, Condy PR, Laws RM (eds) Antarctic nutrient cycles and food webs. Springer, Berlin, pp 362–371

Miller DGM, Barange N, Klindt H, Murray AWA, Hampton I, Siegel V (1993) Antarctic krill aggregation characteristics from acoustic observations in the Southwest Atlantic Ocean. Mar Biol 117:171–183

Miquel JC (1991) Distribution and abundance of post-larval krill (*Euphausia superba* Dana) near Prydz Bay in summer with reference to environmental conditions. Antarct Sci 3:279–292

Montes-Hugo M, Doney SC, Ducklow HW, Fraser W, Martinson D, Stammerjohn SE, Schofield O (2009) Recent changes in phytoplankton communities associated with rapid regional climate change along the Western Antarctic Peninsula. Science 323:1470–1473

Mori M, Butterworth DS (2006) A first step towards modelling the krill-predator dynamics of the Antarctic ecosystem. CCAMLR Sci 13:217–277

Mujica A, Asencio V (1983) Distribution and abundance of krill larvae (*Euphausia superba* Dana). In: Schnack SB (ed) On the biology of krill *Euphausia superba*. Proceedings of the seminar and report of the Krill Ecology Group, Bremerhaven. Ber Polarforsch 4:21–29

Murphy EJ (1995) Spatial structure of the Southern Ocean ecosystem: predator-prey linkages in Southern Ocean food webs. J Anim Ecol 64:333–347

Murphy EJ, Reid K (2001) Modelling Southern Ocean krill population dynamics: biological processes generating fluctuations in the South Georgia ecosystem. Mar Ecol Prog Ser 217:175–189

Murphy EJ, Morris DJ, Watkins JL, Priddle J (1988) Scales of interaction between Antarctic krill and the environment. In: Sahrhage D (ed) Antarctic ocean and resources variability. Springer, Berlin, pp 120–130

Murphy EJ, Watkins JL, Reid K, Trathan PN, Everson I, Croxall JP, Priddle J, Brandon MA, Brierley AS, Hofmann E (1998) Interannual variability of the South Georgia marine ecosystem: biological and physical sources of variation in the abundance of krill. Fish Oceanogr 7:381–390

Murphy EJ, Watkins JL, Meredith MP, Ward P, Trathan PN, Thorpe SE (2004a) Southern Antarctic Circumpolar Current Front to the northeast of South Georgia: Horizontal advection of krill and its role in the ecosystem. J Geophys Res 109:1–10

Murphy EJ, Thorpe SE, Watkins JL, Hewitt R (2004b) Modelling the krill transport pathways in the Scotia Sea: spatial and environmental connections generating the seasonal distribution of krill. Deep-Sea Res II 51:1435–1456

Murphy EJ, Watkins JL, Trathan P, Reid K, Meredith MP, Thorpe S, Johnston N, Clarke A, Tarling G, Collins M, Forcada J, Shreeve R, Atkinson A, Korb, Whitehouse M, Ward P, Rodhouse P, Enderlein P, Hirst A, Martin A, Hill S, Staniland I, Pond D, Briggs D, Cunningham N, Fleming AR (2007a) Spatial and temporal operation of the Scotia Sea ecosystem: a review of large-scale links in a krill centred food web. Phil Trans R Soc B362:113–148

Murphy EJ, Trathan PN, Watkins JL, Reid K, Meredith MP, Forcada J, Thorpe SE, Johnston NM, Rothery P (2007b) Climatically driven fluctuations in Southern Ocean ecosystems. Proc R Soc London Ser B – Biol Sci 274:3057–3067

Murphy EJ, Hofmann EE, Watkins JL, Johnston NM, Piñones A, Ballerini T, Hill SL, Trathan PN, Tarling GA, Cavanagh RA, Young EF, Thorpe SE, Fretwell P (2013) Comparison of the structure and function of Southern Ocean regional ecosystems: the Antarctic Peninsula and South Georgia. J Mar Syst 109–110:22–42

Murray AWA, Watkins JL, Bone DG (1995) A biological acoustic survey in the marginal ice-edge zone of the Bellingshausen Sea. Deep-Sea Res II 42:1159–1175

Naganobu M, Kutsuwada K, Sasai Y, Taguchi S, Siegel V (1999) Relationships between Antarctic krill (*Euphausia superba*) variability and westerly fluctuations and ozone depletion in the Antarctic Peninsula area. J Geophys Res C 104:20651–20665

Nasu K (1983) On the geographic boundary of Antarctic krill distribution. In: Schnack SB (ed) On the biology of krill, *Euphausia superba*. Ber Polarforsch 4:216–222
Nicol S (2000) "BROKE": a multidisciplinary survey of the waters off east Antarctica (80–150°E). Deep-Sea Res II 47:2281–2613
Nicol S (2003) Krill and currents – physical and biological interactions influencing the distribution of *Euphausia superba*. Ocean Polar Res 25:633–644
Nicol S (2006) Krill, currents, and sea ice: *Euphausia superba* and its changing environment. Bioscience 56:111–120
Nicol S, Foster J (2016) The fishery for Antarctic krill – its current status and management regime. In: Siegel V (ed) The biology and ecology of Antarctic krill, *Euphausia superba* Dana, 1850. Springer, Cham, pp 387–422
Nicol S, Constable AJ, Pauly T (2000a) Estimates of circumpolar abundance of Antarctic krill based on recent acoustic density measurements. CCAMLR Sci 7:1–13
Nicol S, Kitchener J, King R, Hosie G, de la Mare WK (2000b) Population structure and condition of Antarctic krill (*Euphausia superba*) off East Antarctica (80–150°E) during the austral summer of 1995/96. Deep-Sea Res II 689:1–29
Nicol S, Pauly T, Bindoff NL, Wright S et al (2000c) Ocean circulation off east Antarctica affects ecosystem structure and sea-ice extent. Nature 406:504–507
Nowacek DP, Friedlaender AS, Halpin PN, Hazen EL, Johnston DW, Read AJ, Espinasse B, Zhou M, Zhu Y (2011) Super-aggregations of krill and Humpback Whales in Wilhelmina Bay, Antarctic Peninsula. PLoS ONE 6:e19173. doi:10.1371/journal.pone.0019173
O'Brien DP (1987) Direct observations of the behaviour of *Euphausia superba* and *Euphausia crystallorophias* (Crustacea: Euphausiacea) under pack ice during the Antarctic spring 1985. J Crust Biol 7:437–448
Orsi AH, Nowlin WD Jr, Whitworth T III (1993) On the circulation and stratification of the Weddell Gyre. Deep-Sea Res I 40:169–203
Orsi AH, Whitworth T III, Nowlin WD Jr (1995) On the meridional extent and fronts of the Antarctic Circumpolar Current. Deep-Sea Res 42:641–673
Pakhomov EA (1995) Demographic studies of Antarctic krill *Euphausia superba* in the Cooperation and Cosmonaut Seas (Indian sector of the Southern Ocean). Mar Ecol Prog Ser 119:45–61
Pakhomov EA (2000) Demography and life cycle of Antarctic krill, *Euphausia superba*, in the Indian sector of the Southern Ocean: long-term comparison between coastal and open ocean regions. Can J Fish Aquat Sci 57(Suppl 3):68–90
Pakhomov EA, Karpenko GP (1992) Oceanographic conditions of Antarctic krill (*Euphausia superba* Dana) reproduction in the Cooperation Sea. In: Rzhonsnitsky VB (ed) Physical oceanography and problems of the biological productivity. Saint Petersburg University Publishers, Saint Petersburg, pp 128–135 (in Russian)
Pakhomov EA, Froneman PW, Perissinotto R (2002) Salp/krill interactions in the Southern Ocean: spatial segregation and implications for the carbon flux. Deep-Sea Res 49:1881–1907
Pakhomov EA, Atkinson A, Meyer B, Oettl B, Bathmann U (2004) Daily rations and growth of larval krill *Euphausia superba* in the eastern Bellingshausen Sea during austral autumn. Deep-Sea Res II 51:2185–2198
Parmesan C (2006) Ecological and evolutionary responses to recent climate change. Ann Rev Ecol Ecol Syst 37:637–639
Pauly T, Higginbottom I (1994) Estimation of the biomass of krill in Prydz Bay during January/ February 1991, February/March 1992 and January/ February 1993 using echo integration. SC-CAMLR-WG-Krill-94/21:1–23
Pauly T, Nicol S, Higginbottom I, Hosie G, Kitchener J (2000) Distribution and abundance of Antarctic krill (*Euphausia superba*) off East Antarctica (80–150°E) during the austral summer of 1995/1996. Deep-Sea Res II 688:1–24
Peck LS, Webb KE, Bailey DM (2004) Extreme sensitivity of biological function to temperature in Antarctic marine species. Funct Ecol 18:625–630

Piñones A, Hofmann EE, Daly KL, Dinniman MS, Klinck JM (2013) Modelling the remote and local connectivity of Antarctic krill populations along the western Antarctic Peninsula. Mar Ecol Prog Ser 481:69–92
Priddle J, Croxall JP, Everson I, Heywood RB, Murphy EJ, Prince PA, Sear CB (1988) Large-scale fluctuations in distribution and abundance of krill – a discussion of possible causes. In: Sahrhage D (ed) Antarctic ocean and resources variability. Springer, Berlin, pp 169–182
Quetin LB, Ross RM (1984) Depth distribution of developing *Euphausia superba* embryos predicted from sinking rates. Mar Biol 79:47–53
Quetin LB, Ross RM, Clarke A (1994) Krill energetics: seasonal and environmental aspects of the physiology of *Euphausia superba*. In: El-Sayed SZ (ed) Southern Ocean ecology. The BIOMASS perspective. Cambridge University Press, Cambridge, pp 165–184
Rakusa-Suszczewski S (1984) Krill larvae in the Atlantic sector of the Southern Ocean during FIBEX 1981. Polar Biol 3:141–147
Reid K, Croxall JP (2001) Environmental response of upper trophic-level predators reveals a system change in an Antarctic marine ecosystem. R Soc 268:377–384
Reid K, Murphy EJ, Loeb V, Hewitt RP (2002) Krill population dynamics in the Scotia Sea: variability in growth and mortality within a single population. J Mar Syst 36:1–10
Reid K, Watkins JL, Murphy EJ, Trathan PN, Fielding S, Enderlein P (2010) Krill population dynamics at South Georgia: implications for ecosystem-based fisheries management. Mar Ecol Prog Ser 399:243–252
Reiss CS (2016) Age, growth, mortality, and recruitment of Antarctic Krill, *Euphausia superba*. In: Siegel V (ed) The biology and ecology of Antarctic krill, *Euphausia superba* Dana, 1850. Springer, Cham, pp 101–144
Reiss CS, Cossio AM, Loeb V, Demer DA (2008) Variations in the biomass of Antarctic krill (*Euphausia superba*) around the South Shetland Islands, 1996–2006. ICES J Mar Sci 65:497–508
Reiss CS, Cossio A, Jones CD, Murray A, Mitchell G, Santora J, Dietrich K, Weiss E, Gimpel C, Walsh J, Watters GM (2015) Winter habitat selection by Antarctic krill will increase krill-predator-fishery interactions during ice free years. SC-CAMLR-WG-EMM-15/05:1–21 (with permission)
Renner AHH, Thorpe SE, Heywood KJ, Murphy EJ, Watkins JL, Meredith MP (2012) Advective pathways near the tip of the Antarctic Peninsula: trends, variability and ecosystem implications. Deep-Sea Res I 63:91–101
Richerson K, Watters GM, Santora JA, Schroeder ID, Mangel M (2015) More than passive drifters: a stochastic dynamic model for the movement of Antarctic krill. Mar Ecol Prog Ser 529:35–48
Ritz DA (2000) Is social aggregation in aquatic crustaceans a strategy to conserve energy? Can J Fish Aquat Sci 57:59–67
Ross RM, Quetin LB (1983) Spawning frequency and fecundity of the Antarctic krill *Euphausia superba*. Mar Biol 77:201–205
Ross RM, Quetin LB, Kirsch E (1988) Effect of temperature on development times and survival of early larval stages of *Euphausia superba* Dana. J Exp Mar Biol Ecol 121:55–71
Ross RM, Quetin LB, Lascara CM (1996) Distribution of Antarctic krill and zooplankton west of the Antarctic Peninsula. In: Ross RM, Hofmann EE, Quetin LB (eds) Foundations for ecological research west of the Antarctic Peninsula. AGU, Washington, pp 199–217
Sabine CL, Feely RA, Gruber N, Key RM et al (2004) The oceanic sink for anthropogenic CO_2. Science 305:367–371
Sala A, Azzali M, Russo A (2002) Krill of the Ross Sea: distribution, abundance and demography of *Euphausia superba* and *Euphausia crystallorophias* during the Italian Antarctic Expedition (January–February 2000). Sci Mar 60:123–133
Santora JA, Sydeman WJ, Schroeder ID, Wells BK, Field JC (2011) Meso-scale structure and oceanographic determinants of krill hotspots in the California Current: implications for trophic transfer and conservation. Prog Ocean 91:397–409

Saunders RA, Brierley AS, Watkins JL, Reid K, Murphy EJ, Enderlein P, Bone DG (2007) Intra-annual variability in the density of Antarctic krill (*Euphausia superba*) at South Georgia, 2002–2005: within-year variation provides a new framework for interpreting previous 'annual' estimates of krill density. CCAMLR Sci 14:27–41
Savidge DK, Amft JA (2009) Circulation on the West Antarctic Peninsula derived from 6 years of shipboard ADCP transects. Deep-Sea Res I 56:1633–1655
SC-CAMLR (2010a) Report of the Working Group on Ecosystem Monitoring and Management. In: Report of the twenty-ninth meeting of the Scientific Committee (SC-CAMLR-XXIX), Annex 6, CCAMLR, Hobart, Australia, pp 173–244
SC-CAMLR (2010b) Report of the fifth meeting of the subgroup on acoustic survey and analysis methods, In: Report of the twenty-ninth meeting of the Scientific Committee (SC-CAMLR-XXIX), Annex 5, CCAMLR, Hobart, Australia, pp 147–172
Schmidt K, Atkinson A (2016) Feeding and food processing in Antarctic krill (*Euphausia superba* Dana). In: Siegel V (ed) The biology and ecology of Antarctic krill, *Euphausia superba* Dana, 1850. Springer, Cham, pp 175–224
Schmidt K, Atkinson A, Steigenberger S, Fielding S, Lindsay MCM, Pond DW, Tarling GA, Klevjer TA, Allen CS, Nicol S, Achterberg EP (2011) Seabed foraging by Antarctic krill: implications for stock assessment, bentho-pelagic coupling, and the vertical transfer of iron. Limnol Oceanogr 56:1411–1428
Schmidt K, Atkinson A, Pond D, Ireland LC (2014) Feeding and overwintering of Antarctic krill across its major habitats: the role of sea-ice cover, water depth, and phytoplankton abundance. Limnol Oceanogr 59:17–36
Schröder M, Fahrbach E (1999) On the structure and the transport of the eastern Weddell Gyre. Deep-Sea Res I 46:501–527
Shirakihara K, Nakayama K, Komaki Y (1986) Acoustic estimation of krill biomass in R.V. Kaiyo Maru SIBEX I survey area (Indian sector of the Southern Ocean). Mem Natl Inst Polar Res (Tokyo) Spec Issue 40:140–152
Siegel V (1982) Investigations on krill (*Euphausia superba*) in the southern Weddell Sea. Meeresforsch/Rep Mar Res 29:244–252
Siegel V (1985a) The distribution pattern of krill, *Euphausia superba*, west of the Antarctic Peninsula in February 1982. Rep Mar Res 30:292–305
Siegel V (1985b) On the fecundity of Antarctic krill, *Euphausia superba* (Euphausiacea). Archiv Fisch Wiss 36:185–193
Siegel V (1986) Investigations on the biology of Antarctic krill, *Euphausia superba*, in the Bransfield Strait and adjacent areas. Mitt Inst Seefisch Hamburg 38:1–244 (in German)
Siegel V (1987) Age and growth of Antarctic Euphausiacea (Crustacea) under natural conditions. Mar Biol 96:483–495
Siegel V (1988) A concept of seasonal variation of krill (*Euphausia superba*) distribution and abundance west of the Antarctic Peninsula. In: Sahrhage D (ed) Antarctic ocean and resources variability. Springer, Berlin, pp 219–230
Siegel V (1989) Winter and spring distribution and status of the krill stock in Antarctic Peninsula waters. Arch Fisch Wiss 39:45–72
Siegel V (2000) Krill demography and variability in abundance and distribution. Can J Fish Aquat Sci 57:151–167
Siegel V (2005) Distribution and population dynamics of *Euphausia superba*: summary of recent findings. Polar Biol 29:1–22
Siegel V (2012) Krill stocks in high latitudes of the Antarctic Lazarev Sea: seasonal and interannual variation in distribution, abundance and demography. Polar Biol 35:1151–1177
Siegel V (2013) Antarctic krill populations in the outflow region of the north-western Weddell Sea. SC-CAMLR-WG-EMM-13/24:1–10
Siegel V, Harm U (1996) The composition, abundance, biomass and diversity of the epipelagic zooplankton communities of the southern Bellingshausen Sea (Antarctic) with special references to krill and salps. Arch Fish Mar Res 44:115–139

Siegel V, Kalinowski J (1994) Krill demography and small-scale processes: a review. In: El-Sayed S (ed) Southern Ocean ecology: the BIOMASS perspective. Cambridge University Press, Cambridge, pp 145–163
Siegel V, Loeb V (1995) Recruitment of Antarctic krill (*Euphausia superba*) and possible causes for its variability. Mar Ecol Prog Ser 123:45–56
Siegel V, Bergström B, Strömberg JO, Schalk PH (1990) Distribution, size frequencies and maturity stages of krill, *Euphausia superba*, in relation to sea-ice in the northern Weddell Sea. Polar Biol 10:549–557
Siegel V, Loeb V, Gröger J (1998) Krill (*Euphausia superba*) density, proportional and absolute recruitment and biomass in the Elephant Island region (Antarctic Peninsula) during the period 1977 to 1997. Polar Biol 19:393–398
Siegel V, Bergström B, Mühlenhardt-Siegel U, Thomasson M (2002) Demography of krill in the Elephant Island area during summer 2001 and its significance of stock recruitment. Antarct Sci 14:162–170
Siegel V, Ross RM, Quetin LB (2003) Krill (*Euphausia superba*) recruitment indices from the western Antarctic Peninsula: are they representative of larger regions? Polar Biol 26:672–679
Siegel V, Kawaguchi S, Ward P, Litvinov F, Sushin V, Loeb V, Watkins J (2004) Krill demography and large-scale distribution in the southwest Atlantic during January/February 2000. Deep-Sea Res II 51:1253–1273
Siegel V, Reiss CS, Dietrich KS, Haraldsson M, Rohardt G (2013) Distribution and abundance of Antarctic krill (*Euphausia superba*) along the Antarctic Peninsula. Deep-Sea Res I 77:63–74
Simard Y, Lavoie D (1999) The rich krill aggregation of the Saguenay–St. Lawrence Marine Park: hydroacoustic and geostatistical biomass estimates, structure, variability and significance for whales. Can J Fish Aquat Sci 56:1182–1197
Skaret G, Krafft BA, Korneliussen R (2009) Krill density estimates in CCAMLR subarea 48.6 based on acoustic data collected during January–March 2008. SC-CAMLR-WG-EMM-09/45:1–11 (with permission)
Skaret G, Krafft BA, Calise L, Watkins J, Pedersen R, Godø OR (2015) Evaluation of Antarctic krill biomass and distribution off the South Orkney Islands 2011–2015. SC-CAMLR-WG-EMM-15/54:1–20 (with permission)
Smith RC, Stammerjohn SE (2001) Variations of surface air temperature and sea-ice extent in the western Antarctic Peninsula region. Ann Glaciol 33:493–500
Smith RC, Stammerjohn SE, Baker KS (1996) Surface air temperature variations in the western Antarctic Peninsula region. In: Ross RM, Hofmann EE, Quetin LB (eds) Foundations for ecological research west of the Antarctic Peninsula. AGU, Washington, DC, pp 105–121
Smith DA, Hofmann EE, Klinck JM, Lascara CM (1999) Hydrography and circulation of the West Antarctic Peninsula continental shelf. Deep-Sea Res 46:925–949
Sologub DO, Remeslo AV (2011) Distribution and size-age composition of Antarctic krill (*Euphausia superba*) in the South Orkney Islands region (CCAMLR Subarea 48.2). CCAMLR Sci 18:123–134
Solyankin EV (1992) Regional peculiarities of biological productivity formation in the Antarctic waters. Antarctica 31:141–157 (in Russian)
Spiridonov V (1995) Spatial and temporal variability in reproductive timing of Antarctic krill (*Euphausia superba* Dana). Polar Biol 15:161–174
Spiridonov VA (1996) A scenario of the Late-Pleistocene-Holocene changes in the distributional range of Antarctic krill (*Euphausia superba*). Mar Ecol 17:519–541
Spiridonov VA (2009) Krill catches indicate the impact of the El Niño – Southern Oscillation related processes on the distribution of krill biomass between subareas of the Atlantic sector of Antarctic. SC-CAMLR-WG-EMM-09/47:1–13
Spiridonov VA, Gruzov EN, Pushkin AE (1985) Observations on schools of the Antarctic *Euphausia superba* (Crustacea, Euphausiacea) under ice. Zool Z 64:1655–1660
Sprong I, Schalk PH (1992) Acoustic observations on krill spring-summer migration and patchiness in the northern Weddell Sea. Weddell Sea ecology. Results of EPOS European "Polarstern" study. Polar Biol 12:261–268

Stammerjohn SE, Smith RC (1996) Spatial and temporal variability of western Antarctic Peninsula sea-ice coverage. In: Ross RM, Hofmann EE, Quetin LB (eds) Ecological research west of the Antarctic Peninsula. AGU, Washington, DC, pp 81–104

Stammerjohn SE, Martinson DG, Smith RC, Yuan X, Rind D (2008) Trends in Antarctic annual sea-ice retreat and advance and their relation to El Niño-Southern Oscillation and Southern Annular Mode variability. J Geophys Res 113:C03S90. doi:10.1029/2007jc004269

Stein M, Heywood RB (1994) Antarctic environment – physical oceanography: the Antarctic Peninsula and southwest Atlantic region of the Southern Ocean. In: El-Sayed SZ (ed) Southern Ocean ecology. The BIOMASS perspective. Cambridge University Press, Cambridge, pp 11–24

Steinberg DK, Ruck KE, Gleiber MR, Garzio LM, Cope JS, Bernard KS, Stammerjohn SE, Schofield OME, Quetin LB, Ross RM (2015) Long-term (1993–2013) changes in macrozooplankton off the Western Antarctic Peninsula. Deep-Sea Res I 101:54–70

Stepnik R (1982) All-year populational studies of Euphausiacea (Crustacea) in the Admiralty Bay (King George Island, South Shetland Islands, Antarctic). Pol Polar Res 3:49–68

Stretch JJ, Hamner PP, Hamner WM, Michel WC, Cook J, Sullivan CW (1988) Foraging behavior of Antarctic krill *Euphausia superba* on sea-ice microalgae. Mar Ecol Prog Ser 44:131–139

Strutton PG, Griffith FB, Waters RL, Wright SW, Bindoff NL (2000) Primary productivity off the coast of East Antarctica (80–150°E): January to March 1996. Deep-Sea Res II 51:2327–2362

Sushin VA, Shulgovsky KE (1999) Krill distribution in the western Atlantic sector of the southern ocean during 1983/84, 1984/85 and 1987/88 based on the results of soviet meso-scale surveys conducted using an Isaacs-Kidd midwater trawl. CCAMLR Sci 6:59–70

Taki K, Hayashi T, Naganobu M (2005) Characteristics of seasonal variation in diurnal vertical migration and aggregation of Antarctic krill (*Euphausia superba*) in the Scotia Sea, using Japanese fishery data. CCAMLR Sci 12:163–172

Talbot MH (1988) Oceanic environment of Georg VI ice shelf, Antarctic Peninsula. Ann Glaciol 11:161–164

Tarling GA, Fielding S (2016) Swarming and behaviour in Antarctic krill. In: Siegel V (ed) The biology and ecology of Antarctic krill, *Euphausia superba* Dana, 1850. Springer, Cham, pp 279–320

Tarling GA, Thorpe SE (2014) Instantaneous movement of krill swarms in the Antarctic Circumpolar Current. Limnol Oceanogr 59:872–886

Tarling G, Cuzin-Roudy J, Thorpe C, Shreeve R, Ward P, Murphy E (2007) Recruitment of Antarctic krill *Euphausia superba* in the South Georgia region: adult fecundity and the fate of larvae. Mar Ecol Prog Ser 331:161–179

Thompson AF, Youngs MK (2013) Surface exchange between Weddell and Scotia Seas. Geophys Res Lett 40:1–6

Thompson AF, Heywood KJ, Thorpe SE, Renner AHH, Trasviña A (2009) Surface circulation at the tip of the Antarctic Peninsula from drifters. J Phys Oceanogr 39:3–26

Thorpe SE, Murphy EJ, Watkins JL (2007) Circumpolar connections between Antarctic krill (*Euphausia superba* Dana) populations: investigating the roles of ocean and sea-ice transport. Deep-Sea Res 1(54):792–810

Timonin VP (1987) The peculiarities of the distribution and the age composition of *Euphausia superba* larvae in the areas of the Antarctic, Pacific and Indian Ocean sectors. In: Makarov RR (ed) Biological oceanographic investigations in the Pacific sector of the Antarctic. VNIRO TINRO, Moscow, p 212

Trathan PN, Everson I (1994) Status of the FIBEX acoustic data from the West Atlantic. CCAMLR Sci 1:35–48

Trathan PN, Hill SL (2016) The importance of krill predation in the Southern Ocean. In: Siegel V (ed) The biology and ecology of Antarctic krill, *Euphausia superba* Dana, 1850. Springer, Cham, pp 321–350

Trathan PN, Murphy EJ (2003) Sea surface temperature anomalies near South Georgia: relationships with the pacific El Niño regions. J Geophys Res 108(C4):SOV2:1–10

Trathan PN, Agnew DJ, Miller DGM, Watkins JL, Everson I, Thorley MR, Murphy E, Murray AWA, Goss C (1992) Krill biomass in area 48 and area 58: recalculations of FIBEX data. SC-CAMLR Select Sci Pap 9:157–181

Trathan PN, Priddle J, Watkins JL, Miller DGM, Murray AWA (1993) Spatial variability of Antarctic krill in relation to mesoscale hydrography. Mar Ecol Prog Ser 98:61–71

Trathan PN, Everson I, Miller DGM, Watkins JL, Murphy EJ (1995) Krill biomass in the Atlantic. Nature 373:201–202

Trathan PN, Brierley AS, Brandon MA, Bone D, Goss C, Grant A, Murphy E, Watkins JL (2003) Oceanographic variability and changes in Antarctic krill (*Euphausia superba*) abundance at South Georgia. Fish Oceanogr 12:569–583

Trathan PN, Murphy EJ, Forcada J, Croxall JP, Reid K, Thorpe SE (2006) Physical forcing in the southwest Atlantic: ecosystem control. In: Boyd IL, Wanless S, Camphuysen CJ (eds) Top predators in marine ecosystems. Cambridge University Press, Cambridge, pp 28–45

Trathan PN, Forcada J, Murphy EJ (2007) Environmental Forcing and Southern Ocean marine predator populations: effects of climate change and variability. Phil Trans R Soc 362:2351–2365

Trathan PN, Ratcliffe N, Masden EA (2012) Ecological drivers of change at South Georgia: the krill surplus, or climate variability. Ecography 35:983–993

Tseitlin VB (1989) Biomassa i produktsiya antarkticheskogo krilya. In: Ponomareva LA (ed) Kompleksnye issledovaniya pelagiali Yuzhnogo Okeana, Institut Okeanology Acad Nauk SSSR, Moskva, pp 118–122 (in Russian)

Turner J, Colwell SR, Marshall GJ, Lachlan-Cope TA et al (2005) Antarctic climate change during the last 50 years. Int J Climatol 25:279–294

Tynan CT (1998) Ecological importance of the southern boundary of the Antarctic circumpolar current. Nature 392:708–710

Van Cise AM (2009) 2008–2009 field season report Antarctic Marine Living Resources program, NOAA-TM-NMFS-SWFSC-445, La Jolla, pp 17–24

van Loon H (1972) Half-yearly oscillations in the Drake Passage. Deep-Sea Res 19:525–527

Volkov DL, Zlotnicki V (2012) Performance of GOCE and GRACE derived mean dynamic topographies in resolving Antarctic Circumpolar Current fronts. Ocean Dyn 62:893–905

von Gyldenfeldt AB, Fahrbach E, Garcia MA, Schröder M (2002) Flow variability at the tip of the Antarctic Peninsula. Deep-Sea Res 49:4743–4766

Voronina NM (1974) An attempt at a functional analysis of the distributional range of *Euphausia superba*. Mar Biol 24:347–352

Voronina NM (1998) Comparative abundance and distribution of major filter-feeders in the Antarctic pelagic zone. J Mar Syst 17:375–390

Ward P, Atkinson A, Peck JM, Wood AG (1990) Euphausiid life cycles and distribution around South Georgia. Antarct Sci 2:43–52

Ward P, Atkinson A, Murray AWA, Wood AG, Williams R, Poulet SA (1995) The summer zooplankton community at South Georgia. Biomass, vertical migration and grazing. Polar Biol 15:195–208

Ward P, Grant S, Brandon M, Siegel V, Sushin V, Loeb V, Griffiths H (2004) Mesozooplankton community structure in the Scotia Sea during the CCAMLR 2000 survey: January–February 2000. Deep-Sea Res II 51:1351–1367

Watkins JL (1999) A composite recruitment index to describe interannual changes in the population structure of Antarctic krill at South Georgia. CCAMLR Sci 6:71–84

Watkins JL (2000) Sampling krill-direct sampling. In: Everson I (ed) Krill biology, ecology and fisheries. Blackwell Science, Oxford, pp 8–19

Watkins JL, Murray AWA, Daly HI (1999) Variation in the distribution of Antarctic krill *Euphausia superba* around South Georgia. Mar Ecol Prog Ser 188:149–160

Watkins JL, Hewitt R, Naganobu M, Sushin V (2004) The CCAMLR 2000 survey: a multinational, multi-ship biological oceanography survey of the Atlantic sector of the Southern Ocean. Deep-Sea Res II 51:1205–1456

Weber LH, El-Sayed SZ (1985) Spatial variability of phytoplankton and the distribution and abundance of krill in the Indian sector of the Southern Ocean. In: Siegfried WR, Condy PR, Laws RM (eds) Antarctic nutrient cycles and food webs. Springer, Berlin, pp 284–293

Weber LH, El-Sayed SZ, Hampton I (1986) The variance spectra of phytoplankton, krill and water temperature in the Antarctic Ocean south of Africa. Deep-Sea Res 33:1327–1343

White WB, Peterson RG (1996) An Antarctic circumpolar wave in surface pressure, wind, temperature and sea-ice extent. Nature 380:699–702

White WB, Chen SC, Allan RJ, Stone RC (2002) Positive feedbacks between the Antarctic circumpolar wave and the global El Niño–Southern Oscillation. J Geophys Res 107:3165. doi:10.1029/2000JC000581

Whitehouse MJ, Meredith MP, Rothery P, Atkinson A, Ward P, Korb RE (2008) Rapid warming of the ocean around South Georgia, Southern Ocean, during the 20th century: forcings, characteristics and implications for lower trophic levels. Deep-Sea Res I 55:1218–1228

Whitworth T (1980) Zonation and geostrophic flow of the Antarctic Circumpolar Current at Drake Passage. Deep-Sea Res 27A:497–507

Whitworth T III, Nowlin WD Jr, Worley SJ (1982) Net transport of the Antarctic Circumpolar Current through Drake Passage. J Phys Oceanogr 12:960–971

Wiebe PH, Ashjian CJ, Gallager SM, Davis CS, Lawson GL, Copley NJ (2004) Using a high-powered strobe light to increase the catch of Antarctic krill. Mar Biol 144:493–502

Wiebe PH, Ashjian CJ, Lawson GL, Piñones A, Copley NJ (2011) Horizontal and vertical distribution of euphausiid species on the Western Antarctic Peninsula U.S. GLOBEC Southern Ocean study site. Deep-Sea Res II 58:1630–1651

Winberg GG (1971) Methods for the estimation of production of aquatic animals. Academic, London, p 175

Witek Z (1979) Phytoplankton distribution and some aspects of the biology of Antarctic krill *Euphausia superba* Dana. ICES Biological Oceanography Cttee CM 1979/L:14, p 19

Witek Z, Kittel W (1985) Larvae of the species of the genus *Euphausia* (Euphausiacea, Crustacea) in the southern part of the Drake Passage and Bransfield Strait during the BIOMASS-SIBEX (December 1983–January 1984). Pol Polar Res 6:117–132

Witek Z, Koronkiewicz A, Soszka GJ (1980) Certain aspects of the early life history of krill *Euphausia superba* Dana (Crustacea). Pol Polar Res 1:97–115

Witek Z, Kalinowski J, Grelowski A, Wolnomiejski N (1981) Studies of aggregations of krill (*Euphausia superba*). Meeresforschung 28:228–243

Witek Z, Kalinowski J, Grelowski A (1988) Formation of Antarctic krill concentrations in relations to hydrodynamic processes and social behavior. In: Sahrhage D (ed) Antarctic ocean and resources variability. Springer, Berlin, pp 237–244

Yamanaka I (1983) Interaction among krill, whales and other animals in the Antarctic ecosystem. Mem Natl Inst Polar Res 27:220–232

Young EF, Thorpe SE, Banglawala N, Murph EJ (2014) Variability in transport pathways on and around South Georgia shelf, Southern Ocean: implications for recruitment and retention. J Geophys Res: Oceans 119:241–252

Yuan X (2004) ENSO-related impacts on Antarctic sea-ice: a synthesis of phenomenon and mechanisms. Antarct Sci 16:415–425

Zhou M, Nordhausen W, Huntley M (1994) ADCP measurements of the distribution and abundance of euphausiids near the Antarctic Peninsula in winter. Deep-Sea Res I 41:1425–1445

Zhou M, Niiler PP, Hu JH (2002) Surface currents in the Bransfield and Gerlache Straits. Deep-Sea Res II 49:267–280

Zhou M, Niiler P, Yiwu Z, Dorland RD (2006) The western boundary current in the Bransfield Strait, Antarctica. Deep Sea Res II 53:1244–1252

Chapter 3
Age, Growth, Mortality, and Recruitment of Antarctic Krill, *Euphausia superba*

Christian S. Reiss

Abstract Laboratory and field studies on the age, growth, mortality, and recruitment dynamics of Antarctic krill (*Euphausia superba*; hereafter "krill") have greatly increased knowledge of its life history over the last 30 years. The development of long, standardized time series of recruitment and abundance have provided data to test and refine conceptual models of krill recruitment, and to examine the role of environmental factors and climatic variability on recruitment and cohort size. Laboratory studies have also greatly increased knowledge about intrinsic and extrinsic factors governing growth and moulting frequency that have been used to develop more robust models for krill growth. These findings have demonstrated the plasticity of krill life history and can provide the foundation for understanding krill population response to future climate change. However, the lack of progress on the development of conservative, direct measures of krill age continue to hamper the ability to properly compare and contrast growth and recruitment across environments.

Keywords Population dynamics • Life history • Vital rates • Southern Ocean • Climate change

3.1 Introduction

Growth, mortality, and recruitment variability are fundamental traits of populations necessary for understanding the life history strategies of animals. Such knowledge is necessary for developing management strategies for harvested populations (Everson 1977a, b, 2000), for understanding the impact of environmental changes on populations (Quetin et al. 2007), and for assessing the impact of harvesting on upper trophic levels. In the Southern Ocean, Antarctic krill (krill) is a key species linking primary production to higher trophic levels (Knox 1994), and also support a major commercial fishery (Nicol et al. 2011). Because of the ecological and

---

C.S. Reiss (✉)
Antarctic Ecosystem Research Division, Southwest Fisheries Science Center, NOAA Fisheries, 8901 La Jolla Shores Dr Rm 333, La Jolla, CA 92037, USA
e-mail: christian.reiss@noaa.gov

V. Siegel (ed.), *Biology and Ecology of Antarctic Krill*, Advances in Polar Ecology,
DOI 10.1007/978-3-319-29279-3_3

economic importance of krill, considerable effort has been made over the last 30 years to understand the causes of annual fluctuations in its abundance and to understand the processes controlling recruitment (Siegel et al. 2003). Long time series of observations (eg, Fielding et al. 2014; Ross et al. 2014), modelling approaches (eg, Fach et al. 2002; Murphy et al. 2004; Thorpe et al. 2007), and the refinement of techniques to measure growth (eg Tarling et al. 2006; Candy and Kawaguchi 2006) have broadened our knowledge of krill ecology and biology.

In this chapter I review studies on the age, growth, mortality, and recruitment of krill from both laboratory and field studies. I focus on aspects of studies on krill biology, behavior, population dynamics and sampling that have both hindered and advanced our ability to reliably estimate these rates, and I consider these rates in the context of assessing climate change impacts on krill populations. I use some unpublished data from the long-term monitoring studies conducted by the U.S. AMLR Program in the northern Antarctic Peninsula and published data from other areas to compare and contrast aspects of krill life history.

3.2 Longevity

Estimating longevity of crustaceans is difficult, as crustaceans lack hard parts that can record chronological age (Hartnoll 2001). Unlike aquatic vertebrates, such as fish, whose otoliths, scales, and other hard parts have been used to accurately estimate daily and annual growth, age, and onset of maturity, to estimate maximum age, longevity of crustaceans has largely been estimated from external attributes of the animal, such as length and maturity.

Inferences about the longevity of krill were first made by examining length-frequency distributions and maturity stages of krill collected during different times of the year and at different latitudes (Ruud 1932; Bargmann 1945). These early studies developed length distributions in order to construct a possible life history model for krill. Ruud (1932) examined length-frequency distributions of krill collected over 2 years, and concluded from the single or double modes in the distributions that krill life span was only 2 years. Bargmann (1945), building on the work of Ruud (1932), examined a large amount of data from the early Discovery cruises and concluded that the maximal age of krill was also 2 years, and that krill died after spawning because the proportion of mature individuals in the population declined after the spawning season. At the time, no one knew that post-larval krill could regress to immature stages during autumn (Makarov 1975), or shrink following ecdysis (Ikeda and Dixon 1982a, b).

As interest in krill increased during the 1960s, considerable work was conducted by Soviet scientists to determine the spatio-temporal patterns in the distribution of fishable quantities of krill. This effort provided an enormous amount of basic biological data that were used to further explore the longevity of krill. Ivanov (1970) examined data from a number of surveys and found that length modes varied among years and areas. In contrast to Ruud (1932), Bargmann (1945), and others (eg, Mackintosh 1972), who viewed krill as short-lived, rapidly-growing organisms, Ivanov (1970) argued that different modes in the length-frequency distributions reflected different krill cohorts and argued that krill could live at least 4 years.

From the late 1960s through the 1980s, a number of important laboratory studies were conducted on krill kept in aquaria during the austral summer research season (eg, Mackintosh 1967; Clarke 1976). These natural maintenance studies documented the length of the intermoult period during summer, and showed that krill could moult approximately every 2 weeks. Makarov (1975), using field data, and McWhinnie and Denys (1978), using animals held in aquaria, described the reversion of mature females to non-mature post-larvae through sequential moults. McWhinnie et al. (1979) further demonstrated that krill did not die after spawning, but instead continued to feed normally. These studies further supported the argument for a longer, more complex life history.

By the mid-1980s, the laboratory work of Ikeda and Dixon (1982a, b) and Thomas and Ikeda (1987) demonstrated clearly that krill could not only revert to immature stages over time, but could shrink when starved for long periods and then recover when food was available again. These authors suggested this was an adaptive strategy for overwintering when food concentrations were low. These studies also showed that krill could live between six and seven years. By the early 2000s, laboratory maintenance studies demonstrated that individual krill could live a decade or more (Pyper 2002). These findings indicate that interpretations on productivity (see also Chap. 2 Siegel and Watkins 2016) and life history as suggested by Ruud (1932) or Bargmann (1945) required substantial revision. The evolution of knowledge of the longevity of krill has greatly impacted the direction of research in this area over the last 30 years.

3.3 Mortality

3.3.1 Mortality Estimation

Age- and stage-specific mortality rates cannot be measured directly; therefore, estimates are necessary to complete the life history models of animal populations, including krill. A number of approaches have been used to constrain estimates of natural mortality in animal populations. Tag-recapture methods are used in some crustaceans to understand both natural and fishing mortality (Claverie and Smith 2007). In larger crustaceans, internal and external tags of various kinds have been successfully used (de Graaf 2007). Unfortunately, the small size, enormous numbers, and lack of both a terminal moult in krill preclude the use of tag-recapture methods for estimating krill mortality. Thus, most estimates of mortality are based on simpler approaches.

The most commonly-used method to infer mortality rates of krill is the linearized catch curve (Siegel 1986, 1989). Other methods have included time (or stage) varying models of mortality, where mortality rates change with age or stages of development (Pakhomov 1995a, b). Still others have used the relationship between the age at maturity and growth rate in the von Bertalanffy function to constrain mortality rate (Basson and Beddington 1989; Basson 1994).

Regardless of the approach to estimate mortality rate, a number of assumptions must be made to constrain estimates. For example, developing mortality estimates based on analyses of length-frequency data requires that the population is relatively closed and individuals of all sizes (ages) are available in the sampling area. If the area of interest cannot be considered relatively closed, then mortality estimates can be biased because of loss of animals to, or gain of animals from, outside the area. This assumption is particularly problematic in the Southern Ocean, where the strong Antarctic Circumpolar Current can advect krill among areas thousands of kilometers apart (Fach et al. 2002, 2006; Hofmann and Murphy 2004). This advection can mix populations of krill which have experienced vastly different environmental conditions that affect growth and mortality rates like water temperature, primary production, and predation rate. Length frequency distributions generated from such mixed populations can result in biased mortality rates when the linearized catch curve is calculated. A further issue in developing consistent mortality estimates is the highly variable recruitment observed in krill populations. Recruitment failures may result in biased mortality estimates due to disproportionately low representation of certain cohorts, violating the assumptions necessary to estimate mortality rate using this technique (Siegel 1991).

Despite these assumptions and caveats, Siegel (1986, 1991) used mixture distribution analysis (MacDonald and Pitcher 1979) to calculate length-frequency distributions from a number of surveys and then estimated natural mortality rates of krill populations in the Weddell Sea and Antarctic Peninsula area. He used a linearized catch curve to estimate the decline in relative abundance with age in the different areas. He assumed that his smallest post larvae in summer represented 1 year old krill, and used this assumption to anchor the length-at-age one. For two seasons of available data, 1987/88 and 1989/90, total mortality, Z, was 0.96 and 0.88, respectively (Fig. 3.1a). These calculated rates suggest an annual survival rate between 38 and 41 %.

A different theoretical approach for estimating krill mortality rates was used by Pakhomov (1995a, b) for surveys conducted in the Indian Ocean, at substantially higher latitude. Instead of assuming a constant mortality rate, Pakhomov used, as a starting point, the work of Zikov and Slepokurov (1982) that suggested using an age-dependent or stage-dependent mortality rate. In this model, mortality rates for larval and juvenile krill are high and decline as krill mature to adulthood, where mortality rates remain relatively low and constant for a number of years (and over some length range). Mortality rates then increase as the population ages or as maximal size is approached (Fig. 3.1b). The results of this technique, when applied to data from the Indian Ocean, show how the pattern of mortality can vary with different assumptions about growth and maturity in this modeling framework. The

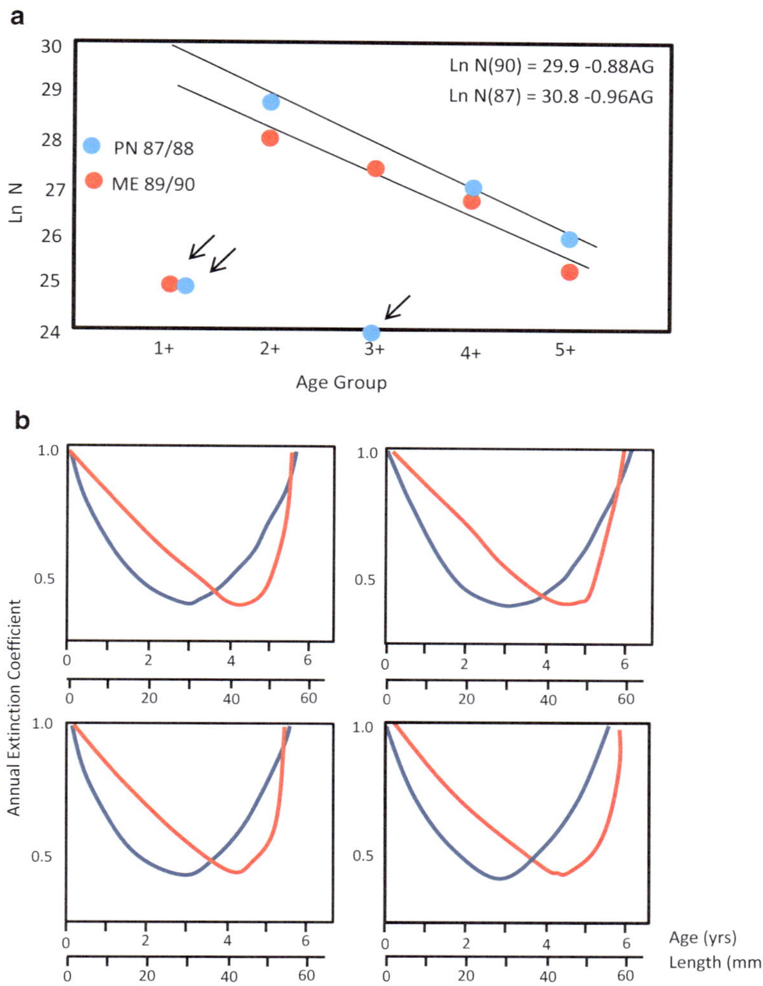

Fig. 3.1 Typical mortality rate functions calculated using two different methods. (**a**) Linearized catch curve from for two survey periods (1987/88 and 1989/90) as calculated by Siegel (1992). Values of age 1 and age 3+ animals are not included. (**b**) Theoretical annual extinction rates for krill in two areas of the Cooperation (*A*, *B*) and Cosmonaut (*C*, D) Seas plotted as a function of age (years; *blue*) and by length (mm; *red*) showing the highest mortality annual extinction rates at small sizes and young ages, minimal mortality rates between 2 and 4 years (30–50 mm) before increasing to maximal values as reported by Pakhomov (1995a, b) based on the Zikov and Slepokurov (1982) technique

age-dependent values of the natural mortality rate (M) calculated using this approach were high (1.1) during the first year of life, minimal (0.52) for animals that had reached maturity, and increased to between 1.29 and 2.41 at maximal age and length, respectively. These mortality rates were significantly higher than estimates by Siegel (1991), but are not directly comparable because of the different assumptions about the life history of krill and its importance to mortality rate.

Natural mortality rate (M) and growth (K) are highly correlated, and it is generally true that animals with high natural mortality rates have high population growth rates (Beverton and Holt 1957). Thus, it is possible to parameterize natural mortality rate with estimates of growth rate and maximal age. Basson (1994) used the age and growth data from Siegel (1987) to estimate a number of natural mortality rates for krill using different model assumptions. Simple estimates of annual natural mortality based on maximal age ranged from 0.38 to 0.77, depending on the proportion of the population alive at maximal age. Basson (1994) also applied the 'Pauly equation' that derives a general relationship between temperature and natural mortality rate using mean maximal size and an estimate of the growth rate (derived from Siegel 1987). Using this model, natural mortality rates ranged from 0.91 to 1.22 for krill in the Antarctic Peninsula, and 1.11 for krill in the Weddell Sea if temperatures were 0 °C. At 3 °C, these natural mortality rates declined to between 0.60 and 0.74 in the Antarctic Peninsula and was estimated to be 0.81 in the Weddell Sea. Other calculations of natural mortality rates show similar results and variability (see Table 4 in Siegel 2000a).

Few data are available on the spatial variability in natural mortality rates. However, mortality rates are likely to vary among areas and times of year because of local and seasonal environmental variability and spatially variable predator distributions. For example, seasonal reproductive patterns in birds and mammals or feeding migrations in others can result in concentration of krill predators (and predation) in nearshore or coastal areas during summer affecting local mortality rates. Estimates of consumption by predators provide estimates of removals. However, estimating mortality rate from consumption is more difficult because of the size selective predation on krill by different predators (see Chap. 9, Trathan and Hill 2016). Mortality rate (rather than estimates of total consumption) is therefore difficult to estimate. In one study, Reid et al. (2002) used a population model to examine krill mortality around South Georgia from the size composition of krill in the diets of Antarctic fur seals. Reid et al. (2002) varied mortality rates and statistically fit the observed length-frequency distributions of krill in the diets of Antarctic fur seals (with their attendant selectivity) at South Georgia to the length-frequency distribution from net tows conducted in the Antarctic Peninsula, a presumed upstream source of krill to South Georgia. They estimated the mortality necessary to account for the changes in length frequency between the Peninsula and South Georgia. To best fit the observed data, natural mortality rates varied from 0.4 to 4.0. These highly variable mortality rates were necessary to account for the changing patterns in the length-frequency distribution in predator diets over the summer, and also to account for unmeasured loss of animals via advection through the system. This one example highlights the difficulty in estimating mortality rate with confidence from predators.

Not only can krill mortality rate vary spatially and temporally, but it may also vary with sex. Virtue et al. (1996) showed that natural mortality rates of male krill were substantially higher than those of female krill collected from the same locations following spawning. Nearly all male krill died within 3 days of capture, and had little to no lipid reserves at the time of death. In contrast, the mortality of female krill was less than 10 % over the same period of time and lipid content implied adequate reserves. The authors inferred from this study that mortality rates of male and female krill differ due to energetic constraints on reproduction. Further evidence in support of sex-specific natural mortality rates comes from a meta-analysis that examined length-specific sex ratios of krill using data from research surveys, the krill fishery, and predator diets. Kawaguchi et al. (2007), using a growth and mortality model, showed that the percent of the population that was male declined with increasing length, especially after 50 mm. These authors suggested that because male krill have higher growth rates and natural mortality rates, large males (>50 mm) are not as prevalent in krill populations. These results support the argument for a tradeoff among growth, maturation, and reproductive success.

These studies are argued to demonstrate that large male krill are under-represented in the population compared to large female krill owing to the costs of reproduction. However, these studies were conducted using data during summer, and say little about size related mortality during the winter. This is important because at intermediate lengths, male krill may be more abundant as a percentage of the population during winter (Daly and Macaulay 1991). In fact, data from three recent winter surveys in the Antarctic Peninsula (Fig. 3.2) conducted by the U.S. AMLR Program between 2012 and 2014 show a similar pattern. And, a recent study combining data from winter and summer fishery and research data from the Southwest Atlantic (Tarling et al. 2016) also support this finding. These winter data suggest that annual natural mortality rates of males and females are more similar than suggested by Kawaguchi et al. (2007), but that the timing of the mortality

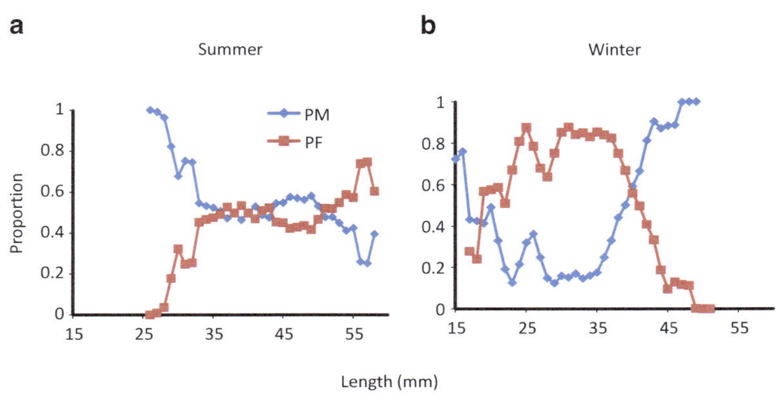

Fig. 3.2 Relative proportion of male (*blue lines*) and female (*red lines*) krill by length class in the South Shetland Islands during summer over 20 years (*left panel*), and winter during three consecutive winters (*right panel*) collected by the US AMLR Program (Data show the switch in dominance of the proportion of male and female krill depending on season)

events differs. One potential interpretation (Volker Siegel, personal communication) is that older male krill die immediately after spawning, while older female krill die later after the spawning season, resulting in the apparent difference in sex ratios observed during summer.

Despite the rather coarse nature of estimates of natural mortality, both Siegel (1986, 1991) and Pakhomov's (1995a, b) values have been used in examinations of krill population dynamics. For example, the development of the General Yield Model for krill (Constable and de la Mare 1996, 2003) has used data from Siegel (1991) and arguments from Basson (1994) to parameterize a fixed mortality rate in developing control rules for krill harvest. Wiedenmann et al. (2008; 2009) used the mortality rates estimated by Pakhomov (1995a, b) in a model to understand recruitment dynamics of krill at two locations along the Antarctic Peninsula in relation to sea ice, transport, and retention. These examples highlight the pragmatic approach to using such coarse estimates to manage fisheries and to understand krill ecology. However, over the last 20 years, progress has stalled on refining estimates of krill natural mortality as the current estimates fit largely within the existing life history models for krill. A substantial obstacle to revising these life history models is the assumption that krill age can be inferred from length, as currently, no definitive method of determining age exists. However, recently-developed techniques to determine age independent of length show promise, and may result in revised models that more accurately estimate natural mortality, and in meaningful comparisons of natural mortality among different areas of the Southern Ocean.

3.4 Ageing

Accurately estimating age in crustaceans is challenging, as crustaceans lack the obvious calcified structures that may accumulate annual growth deposits (eg, otoliths in fish). Without these structures from which to estimate age and growth, age estimates for krill have largely been inferred from length. Studies have demonstrated that growth can fluctuate with environmental conditions (Ikeda and Dixon1982a, b; Quetin and Ross 1991), prompting the investigation into other approaches to generate length-independent methods to estimate crustacean age.

Ettershank (1983) proposed using a fluorescent lipid oxidation byproduct called lipofuscin to estimate age of krill because it is known to accumulate in tissues of all animals linearly with time and may be used as a physiological chronometer (Ettershank et al. 1982; Sheehy 1990). Extracting lipofuscin from krill collected in the Antarctic Peninsula, Ettershank (1983, 1984, 1985) compared his biochemical age distribution of krill with the modal analysis of lengths from the same samples. He was able to resolve six age classes from krill ranging from approximately 16–62 mm. However, despite the initial promise of this method, and because of technical challenges (eg, inability to distinguish lipofuscin fluorescence from other fluorescent tissue compounds, and only a general correlation between lipofuscin and krill age) with interpreting age using this technique, it has not replaced the simplicity of mixture distribution analysis to estimate krill age (Nicol 1987; Sheehy 2008).

Other studies have addressed the possibility of using structures that are largely conserved during growth, despite fluctuations in length, as proxies for age. Sun et al. (1995) examined the relationship between the length and the number of cones in the eye and proposed it could be used as a proxy for age. They modelled the number of crystalline cones (NOCC) vs. body length (BL) using the equation NOCC = 144.4 × BL – 2147. They showed that NOCC of starved krill was greater than predicted for a given length and suggested that the ratio of body length to NOCC could provide an index of shrinkage. Shin and Nicol (2002) further investigated the use of eyeball size as a simple index of age or growth, because a larger number and size of cones would naturally make the eye larger. They examined the changes in eyeball size of krill maintained under two food treatments in the laboratory in order to investigate the relationships among changing feeding conditions, moulting, and eyeball growth relative to changes in body size. For krill fed *ad libitum*, the relationship between change in eyeball size and body length was strong. For krill that were starved, the ratio of eyeball size to body length increased as krill shrunk during ecdysis. When starved krill were then fed *ad libitum*, krill began to grow again, and once positive growth was achieved, between two and three moults (1–1.5 months) were required for body length to catch up to the eyeball size, after which eyeball size and length increased along the same trajectory. These results suggest that eyeball size or cone number can be used to examine the response of krill to recent environmental conditions by looking at deviations in the ratio between eyeball size or cone number and body length. However, as the environment can influence krill growth, eyeball size cannot provide a more robust estimate of krill age compared to length-frequency analysis.

Recently, Kilada et al. (2012) demonstrated that bands form on the eyestalks and gastric mills of some decapod crustaceans. These bands appear to be deposited annually, are conservative, and do not seem to be affected by moulting or growth rate; therefore, they show strong potential with respect to directly measuring age. Preliminary analysis of known-age krill by Reiss et al. (2015) shows the use of this technique to euphausiids. Should this technique prove precise and accurate, comparisons of age and growth between and among areas of the Southern Ocean would be possible, and would allow for the development of age-based krill population assessments.

3.4.1 Length-at-Age

Length-at-age is influenced by variability in the environment (ie, water temperature, primary production) that affect the growth rate over the year, biological interactions that result in changes in length distributions owing to size-dependent predation, and advection to other areas that can mix populations with different growth histories. These physical and biological factors modify the length frequency distributions in the population and complicate the identification of length modes used to estimate age. One of the earlier models that estimated length-at-age to examine population growth was developed by Mackintosh (1972). He examined length modes in samples from a number of other researchers collected in different areas and seasons and argued that the combination of strong seasonality in the environment (that could result in a winter hiatus in growth) and the timing of the spring bloom (fueling strong spring and early summer growth) should result in seasonally varying growth cycle for krill. By pooling and comparing the various datasets that were available at the time (Ruud 1932; Bargmann 1945; Nemoto 1959) along with a re-analysis of the available Discovery data he was able to construct a seasonally varying length-at-age model that reflected the pattern of growth expected of an animal that might live 2–3 years. However, the biologically-based reference point (size at maturity) used to infer age (Bargmann 1945) and the necessary seasonal growth rates (0.66 mm day^{-1}; Nemoto 1959) used to grow to maximum age in 3 years were challenged by other research that argued for a longer lifespan, and thus lower annual growth rates (Ivanov 1970).

Plausible phenomenological models of population growth were also developed to explain the length-at-age with the assumption of a 3 year life span for krill (Astheimer et al. 1985; Astheimer 1986). At the same time, Ettershanks' (1982, 1983, 1984) analysis of the age structure of krill, based on the biochemical lipofuscin approach, suggested a maximal age of at least 6 years, calling into question the earlier work based on less objective and quantitative methods for estimating age. The doubling of maximal age (from 3 to 6 years) requires a much lower annual growth rate than earlier researchers required to estimate growth and length at maximal age. Moreover, the laboratory studies of the effects of starvation on growth and survival that indicated krill could survive overwinter periods of low food availability by shrinking in size and mass (Ikeda and Thomas 1987). This provided a biological mechanism for growth rates of krill to vary seasonally, being zero or negative (shrinkage) during winter but very rapid during summer confounding the expectation of the earlier models of growth used to determine length-at-age.

Rosenberg et al. (1986) constructed a seasonal von-Bertalanffy growth model of krill to examine whether annual growth rates of krill (and length-at-age) might have changed between the early 1920s and the 1980s as a result of the removal of large baleen whales from the Southern Ocean. They used information on length frequencies of krill in whale stomachs over a 10 year period in the 1920s (Mackintosh 1974), the longevity estimates based on the work of Ettershank (1983,1984), and information for the slow or negative growth during winter (Ikeda and Thomas 1987) to compare resulting growth rates during and after whaling. By analysing the length frequency data on a monthly basis for a number of years, across four areas of the Southern Ocean (Scotia sea, the Scotia-Weddell area, an area south of Australia, and the Indian Ocean area), Rosenberg et al. (1986) were able to build a seasonally variable model of growth with parameter estimates that are consistent with contemporary estimates of longevity and a seasonal hiatus in growth. They did not find any significant difference in growth rates and lengths-at-age over time. They also showed that while averaged annual growth rates were 0.116 mm day^{-1}, the fastest growth occurred in just a 3 month window from November through January, consistent with the arguments for strong seasonality in growth in polar environments. Growth parameters (K) of the von Bertalanffy function ranged from 1.28 to 2.18 for the summer growth period, and corresponding summer growth rates averaged 0.148 mm day^{-1} (range 0.105–0.179 mm day^{-1}). Annual growth parameters, estimated from modal analysis of length frequency data averaged 0.41 (range 0.23–0.47). These annual growth coefficients were lowest in the Weddell Sea and similar among the other areas.

Borrowing from length-based fisheries analysis techniques, Siegel (1986, 1987) used mixture distribution analysis to recover age structure from length distributions in different areas of the Antarctic. He developed von-Bertalanffy growth functions to compare length-at-age in some areas of the Southern Ocean and implemented a four stage approach to generate length frequency distributions of krill from a survey area. First, samples from the survey area of similar length frequency were grouped using a clustering procedure. Then, the areal abundance of krill from each cluster was calculated, and a weighted mean abundance was then calculated using standard survey statistics (Saville 1977). In this manner the final length frequency distribution accounted for the relative abundance of krill of different lengths across the survey area. The final stage involved using mixture distribution analysis (MacDonald and Pitcher 1979) to objectively determine nominal age classes from the length frequency distribution. Mixture analysis partitions the length frequency distribution from a survey into component ages assuming that krill lengths at age are normally distributed, separable, and that the means are non-overlapping. By fitting these normal distributions, and their standard deviations using a maximum likelihood approach, length-at-age components of a length frequency distribution can be resolved. Siegel (1987) further argued that the narrow seasonal (December – January) spawning peak (with an assumed December hatch date) observed in the Antarctic Peninsula area indicated that length of the smallest mode in the length frequency distributions represented 1 year old krill, and he used this assumption as an anchor point for determining the length at age-1.

From his analysis, Siegel (1987) found that krill length-at-age varied both spatially and temporally around the Southern Ocean. In the Weddell Sea one-year old krill in late winter averaged 26.4 mm compared to 28.7 mm in early winter in the warmer more productive Antarctic Peninsula. By age three, krill were approximately 44 mm in both locations, while 5+ year old krill averaged 54.2 mm at the onset of winter in the Peninsula, and no krill of this length were present over the study period in the Weddell Sea. The annual von Bertalanffy model constructed for krill in the Weddell Sea and the Antarctic Peninsula areas had growth coefficients (K) of 0.5328, and 0.4781, respectively. As a consequence, mean maximum length (L_{inf}) of krill was lower in the Weddell Sea (56.3 mm) than in the Antarctic Peninsula (61.3 mm), during summer.

Models of length-at-age using this technique and subsequent improvements (de la Mare 1994a, b, c) have subsequently been developed for a number of surveys from around the Southern Ocean (Watkins et al. 1997; Sala et al. 2002; Virtue et al. 2010). These data also show a range in the length-at-age indicating significant variability in growth (Table 3.1, Fig. 3.3). Compilations from earlier work (Pakhomov 2000) and Siegel (2000a, b) provide some context for understanding the variability in length-at-age. At the highest latitudes of the Weddell and Ross Seas, length-at-age was smaller than in more northerly areas like South Georgia. Siegel (1986) reported that 1 year old krill in January in the Weddell Sea were just 23 mm in length. A compilation of data from Pakhomov (2000) based on data in Cooperation and Cosmonauts Seas found that length at age one varied from 19 to 30 mm, depending on the month of the survey. However, the mean length at age one was just 25.45 mm. Sala et al. (2002) did not find 1 year old krill during their survey of the Ross Sea, but age-2 animals were 37.39 mm, similar in length to other age 2 krill from other areas. At South Georgia, Watkins et al. (1997) reported that the mean length for age-1 krill at South Georgia was 30.07 mm, and ranged from 25.2 to 32.9 mm over the 9 years of their study. At the Antarctic Peninsula Siegel (2000a) reported mean lengths at age one of 29.9 mm, while Virtue et al. (2010) reported mean lengths at age one from the area off East Antarctica of 30.9 mm. Recent data for the Antarctic Peninsula for the period from 1995 to 2011 (Table 3.1) show a mean length 29.05 mm at age one. Ross et al. (2014) showed that 1 year old krill in the west Antarctic Peninsula ranged from 22.12 to 28.20 mm and averaged 25.30 mm. Ross et al. (2014) also reported a number of years with a second smaller mode of krill that averaged 19.75 mm, in some years. These smaller krill have

Table 3.1 Results of von Bertalanffy growth models fit through average length-at-age data reported in the literature based on MacDonald and Pitcher (1979) or CMIX (de la Mare 1994c) fits to length frequency modes

Area	Count	R^2	L_{inf} (S.E.)	K (S.E.)	t_0 (S.E.)	Sources
Antarctic Peninsula	27	0.94	55.09	0.44	−0.72	US AMLR PROGRAM (unpublished data)
			(4.41)	(0.13)	(0.35)	
East Antarctica	12	0.97	58.66	0.35	−1.15	Virtue et al. (2010)
			(6.71)	(0.14)	(0.53)	
Indian Ocean	182	0.94	74.19	0.25	−0.69	Pakhomov (2000)
			(5.47)	(0.04)	(0.14)	
Antarctic Peninsula	21	0.98	57.17	0.50	−0.34	Siegel (2000a, b)
			(2.13)	(0.05)	(0.06)	
South Georgia	35	0.96	79.87	0.18	−1.54	Watkins et al. (1997)
			(14.20)	(0.07)	(0.51)	
Weddell Sea	8	0.74	52.86	0.60	−0.51	Siegel (2000a, b)
			(16.17)	(0.57)	(0.52)	
Ross Sea	4	–	–	–	–	Sala et al. (2002)

Count refers to the number of average values for all surveys reported by the source

L_{inf} (is the asymptotic mean maximum length (mm), K is the Brody growth coefficient, and t0 is the theoretical time at zero length. In all cases (S. E.) is the standard error of the parameter estimates. No convergence was possible using the four data points reported in Sala et al. (2002)

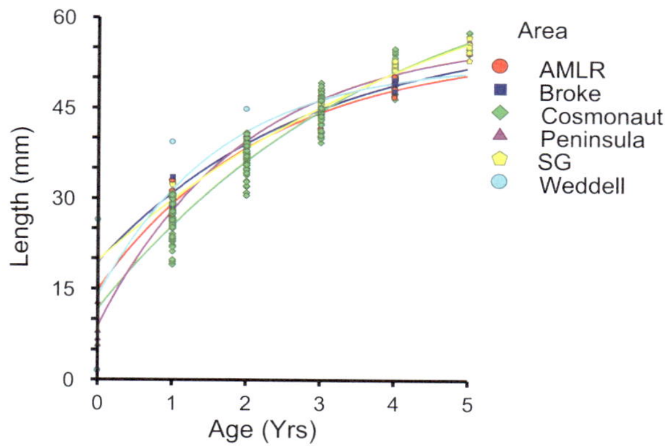

Fig. 3.3 von-Bertalanffy growth curves for mean size at age data from data reported Watkins et al. (1997) for the South Georgia Area, the East Australia area reported by Virtue et al. (2010), the Cosmonaut, Cooperation and Indian Oceans as reported by Pakhomov (2000), the Antarctica Peninsula and the Weddell sea as reported by Siegel (2000a, b) and data form the US AMLR Program collected between 2002 and 2011 (Reiss unpublished)

historically been excluded from estimates of length at age analyses (Siegel 1987). Bigger differences were associated with mean length of 4 year old krill. In the Weddell, Ross and Cosmonaut Seas, krill were 50.10, 47.17, and 51.06 mm. At the Antarctic Peninsula, Siegel (2000a) and the data from the U.S. AMLR Program (Table 3.1) showed that 4 year old krill averaged 50.66 mm and 48.23, respectively. At South Georgia, 4 year old krill were between 2 and 4 mm larger than 4 year old krill elsewhere, and averaged 52.34 mm in length. Thus, there is some evidence for a latitudinal variation in growth with smaller animals at high latitudes where conditions are more strongly seasonal.

3.4.2 *Seasonality*

The strong seasonality in the Antarctic environment led Rosenberg et al. (1986) to model krill growth including a winter growth hiatus. Siegel (1987) furthered this approach by incorporating the potential that krill would not only pause their growth during winter, but instead could shrink. He suggested that modifying the standard von Bertalanffy growth function

$$\mathrm{Lt} = Linf * \left(1 - e^{-(K\,*\,t - t0)}\right) \tag{3.1}$$

to account for seasonal variability in growth using the following equation

$$Lt = Linf * \left(1 - e^{-(K * t - t0) + (C * K/2pi) * sin\,2pi(t - t0)}\right) \quad (3.2)$$

by adding a sinusoidal growth function to the growth coefficient. In this equation, *C* expresses the amplitude of the seasonal growth component and varies between 0 and 1. Values approaching one increase the seasonal amplitude while values close to zero approximate the standard model. He fit this model to data from the Antarctic Peninsula and showed that while the annual growth coefficient was relatively unchanged (K = 0.4728), the seasonal component was fairly large (C = 0.9598) suggesting that at the Antarctic Peninsula seasonal variability in growth rate is very high, and that growth is most rapid during a small part of the year.

3.5 Maturity

Partitioning energy between reproduction and growth is an important factor affecting both the length- and age-at-maturity and the longevity of animals (Stearns 1992). The strong seasonality in primary production in the Southern Ocean leaves krill with a shorter window in which to accumulate energy and allocate it to growth and maturation. So understanding the factors affecting length- and age-at-maturity are critical for understanding the life history of krill and its response future environmental change.

Early work on the maturity of krill was based on microscopic analysis of gonads and provided the first estimates of length-at-maturity (Bargmann 1937). More recent histological and microscopic analysis of gonads (Cuzin-Roudy 1987a, b) has provided considerable insight into the developmental progression of sexual maturation. Using samples collected in Prydz Bay, Cuzin Roudy (1987a, b) demonstrated that sexual differentiation occurred before obvious external characteristics were visible. In fact, female and male krill were sexually differentiable at just 24 mm length when examined histologically. Female krill developed histological evidence of maturation at about 33 mm and males exhibited evidence of maturation at about 36 mm. Importantly she showed that, at least for female krill, there was no evidence of full regression to immature stages after spawning. She concluded that the length at maturity for male and female krill collected in Prydz Bay were similar to length at maturity in other areas in the Antarctic (Bargmann 1945; Poleck and Denys 1982).

Siegel and Loeb (1994) examined the age and size at maturity of krill collected along the Antarctic Peninsula in December 1987 (males and females), December 1989 and January 1990 (males and females), January 1992 (males only), and January 1993 (males and females) (Fig. 3.4). They showed that while both sexes exhibited knife–edge maturity (ie, krill transitioned from juvenile to adult within a single age group), the L_{50} values (the lengths at which 50 % of the krill stock is sexually mature) ranged from 32.9 to 35.9 mm for females, and from 40.20 to 43.71 mm for males. Based on estimated length-at-age from the same data, they

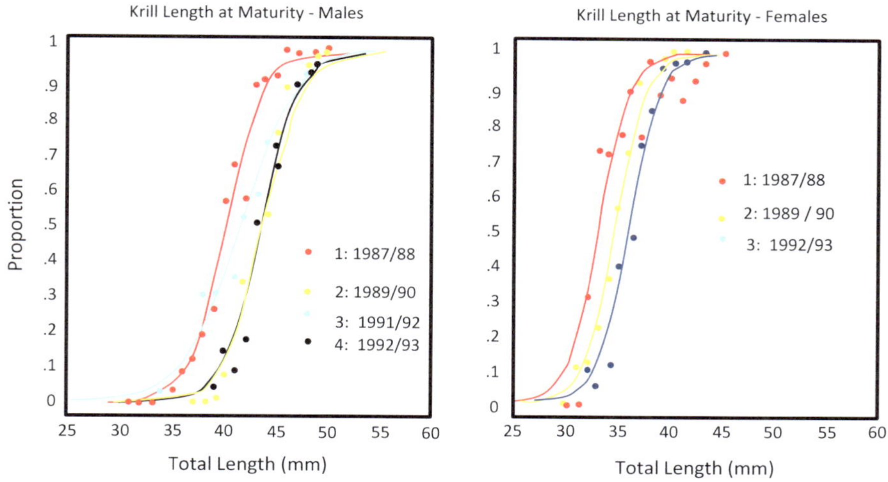

Fig. 3.4 Length at maturity for male and female krill collected in the Antarctic Peninsula and Weddell Sea during summer periods (redrawn from Siegel and Loeb 1994) showing 'knife edge' maturity

concluded that female krill mature during their third year of life, while male krill, because of their later size at maturity, matured during their fourth year of life.

There have been few studies on the variability in length / age at maturity since the work by Siegel and Loeb (1994). However, data collected as part of the long-term monitoring program conducted by the U. S. AMLR Program between 1993 and 2008 (14 years of data) from the Antarctic Peninsula (Table 3.2) can provide some information regarding this variability. Using the approach of Siegel and Loeb (1994) and calculating the length-at-maturity with a logistic regression equation, size-at maturity shows that while the overall pattern of smaller length-at-maturity for female krill is retained, there is a difference in the temporal variability in length-at-maturity for males and females (Table 3.2; Figs. 3.4 and 3.5). Over the time period, length-at maturity for females has declined, averaging 36.5 mm in the 1990s and declining to approximately 33.5 mm in recent years. In contrast, no significant decline in length-at-maturity is found for males. The rate of maturation between males and females varied, and while the slopes of the logistic regressions differed, neither sex showed any temporal trend in the rate of maturation. The overall pattern from these data support previous studies indicating that female krill mature at a smaller size than male krill.

Understanding the factors impacting these maturity schedules is important because changes in length- or age-at-maturity can have significant implications for the dynamics of populations. In fished populations, declines in length-at-maturity are indications of population juvenescence as older slower growing individuals are removed from the population. This can lead to recruitment overfishing. In the Southern Ocean the fishery is just about 200 000–300 000 tonnes, a small fraction of

Table 3.2 Length at which 50 % of the krill stock is mature (L_{50}) for female and male krill at the tip of the Antarctic Peninsula collected during US AMLR surveys in January – February of each austral summer around Elephant Island between 1993 and 2008

	Female			Male		
Year	K	L_{50}	R^2	K	L_{50}	R^2
1993	0.71	36.38	0.99	0.62	43.54	0.99
1994	0.25	34.73	0.49	0.50	45.20	0.97
1995	–	–	–[a]	0.56	38.85	0.96
1996	0.84	36.71	0.98	0.60	42.57	0.98
1997	1.29	36.75	0.99	0.77	41.89	0.79
1998	0.58	37.31	0.99	0.53	42.45	0.99
1999	0.20	28.40	0.21[a]	0.47	40.57	0.97
2001	1.17	34.10	0.98	0.38	40.92	0.91
2003	0.63	36.01	0.94	1.70	39.91	0.86
2004	0.82	35.94	0.98	0.37	43.14	0.96
2005	1.20	34.84	0.99	0.50	41.04	0.95
2006	1.83	33.05	0.92	0.56	40.86	0.98
2007	0.65	33.39	0.87	0.51	42.25	0.98
2008	10.37	31.85	0.89[a]	0.68	44.38	0.94
Average	0.753	31.58	0.81	0.44	42.01	0.90

K represents the slope of the increase in maturation
No convergence in parameters was found for female krill in 2005
[a]data not included in average

the ~60 million tonne estimate of krill biomass for the southwest Atlantic Ocean (Fielding et al. 2011; Nicol et al. 2011; see Chap. 11, Nicol and Foster 2016). This suggests that fishing mortality is not a significant factor explaining the changes in length-at-maturity. In the Antarctic Peninsula area, climate is changing rapidly impacting the physical and biological environments, and this is predicted to have consequences for krill biology and ecology (Quetin et al. 2007). Whether or not these declining female lengths-at-maturity are a response to climatic changes in the region is unknown, regardless, these changes could impact the productivity of krill, and the energetic tradeoffs between growth maturity and fecundity.

3.5.1 Length and Weight

The relationship between length and weight provides fundamental information regarding the condition and growth potential of animals (Ricker 1975). General relationships between length and weight allow comparisons of animal condition and population structure in space and time. And, because of relative ease of this measurement, there are many estimates for krill (see Chap. 1, Siegel 2016; Atkinson et al. 2006; Färber-Lorda 1994; Hewitt et al. 2004; Huntley et al. 1994; Jazdzewski et al. 1978; Hoffman and Lascara 2000; Lockyer 1973; Miller 1983; Morris et al. 1988; Sahrhage 1978; Siegel 1982, 1986, 1989, 1992; Strelnikova 1985).

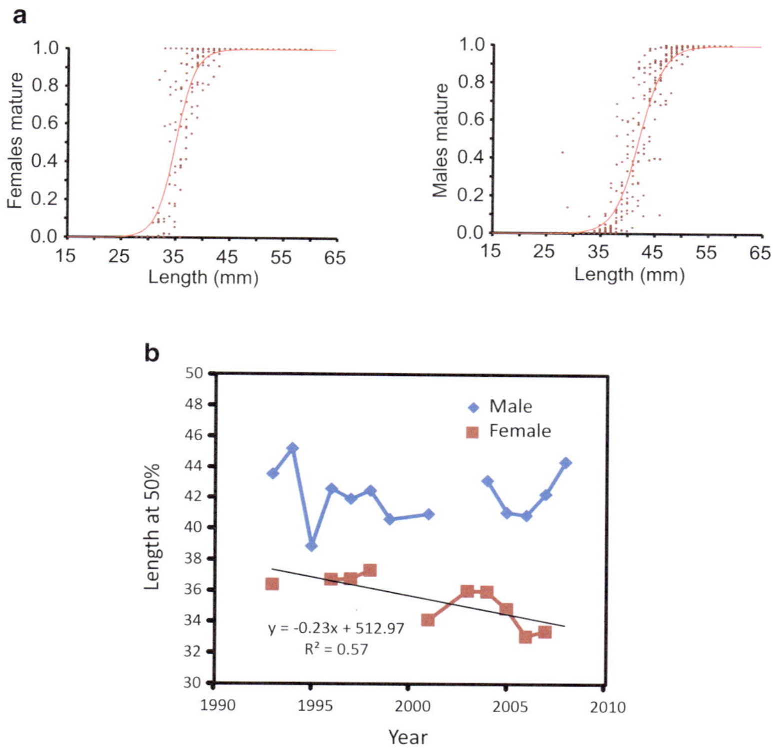

Fig. 3.5 Size at maturity for male (**a**) and female (**b**) krill from 13 cruises conducted between 1996 and 2009 by the US AMLR Program around Elephant Island area of the South Shetland Islands. Temporal trends in predicted size at 50 % maturity (*blue*) and the length based rate of maturation for female (*C*) and male (*D*) krill showing the decline in mean size at 50 % maturity for female krill. No significant trends were found for male krill

With respect to seasonal growth in krill, the length-weight relationships provide an estimate of the potential change in biomass associated with changes in length. For example, Atkinson et al. 2006 found that, on average, krill in the southwest Atlantic increased their mass by 1 % per day. Smaller krill in the colder more southern areas increased their mass more rapidly than the larger krill in the warmer high chl-*a* areas around South Georgia. In some cases swarms of krill increased their mass by more than 5 % d^{-1}, highlighting the variability in growth rates and the resulting changes in seasonal biomass.

There are at times, significant spatial differences in the length-weight, or multivariate body length-weight relationships in krill. These differences may be used to infer the relative condition of krill in different locations or in different circumstances (Färber-Lorda 1994; Färber-Lorda et al. 2009; Nicol 2000). Färber-Lorda (1991, 1994) and Färber-Lorda et al. (2009) developed a differentiation index (DI) based on the ratio of total length to the difference in abdominal and carapace lengths to examine relative condition of krill in different areas of the Indian Ocean. They found that the DI varied with maturity stage of female krill, and between productive and unproductive areas within the Indian Ocean. Lipid content of the different krill was also correlated with location. They suggested that these kinds of relationships could be useful in understanding the life history of krill, but these relationships have yet to be used to understand recruitment probability or feeding success of krill predators.

Importantly, the seasonal change in length weight relationships can provide information regarding the investment in reproduction by female krill. Siegel (2000b) summarized much of the sex and maturity specific krill length-weight data and concluded that both male and female krill have similar length-weight relationships from winter to summer, even as maturation accelerates towards spring (November) and early summer (December) spawning. However, Siegel (2000b) concluded that weight in female krill drops dramatically after post-spawning, altering its length-weight relationship. Because krill of both sexes have similar length-weight relationship at the onset of winter, female krill must gain back weight after spawning, and prior to overwintering. This suggests that after the spawning season female krill forego late-season growth in length and instead accumulate lipid reserves. The differences in length-weight relationships that occur in post-spawning female krill may influence the estimation of biomass of krill significantly (Morris et al. 1988). They argued that accounting for the maturity stage of krill could reduce uncertainty in biomass estimates by 25 %. These large changes in krill mass-at-length impact the calculation of biomass and also impact ideas regarding the productivity of krill. Calculating biomass at the end of the spawning season when the length specific mass of female krill is least will decrease the overall biomass influencing both ecological (energy flow through krill) and fisheries (harvestable biomass) studies.

3.6 Growth

Numerous laboratory and field studies have sought to estimate krill growth over the last 30 years. Modal analysis of length-frequency distributions have been used to examine monthly (Shelton et al. 2013), seasonal (McClatchie et al. 1991), and annual growth (Siegel 1987) by following the progression of length modes over time. The development and subsequent refinement of the Instantaneous Growth Rate (IGR) method (Marr 1962; Mackintosh 1972; Poleck and Denys 1982; Quetin and Ross 1991) provides the most complete understanding of krill growth over shorter time periods (days). The IGR approach has also significantly improved understanding of krill growth to varying environmental (physical and biological) conditions (Atkinson et al. 2006; Tarling et al. 2006; Candy and Kawaguchi 2006).

3.6.1 Estimating Growth from Length Frequency Distributions

The quasi-cyclic nature in fluctuations of krill length and recruitment produces length frequency distributions from sampled populations that are often unimodal or multi-modal length modes (Fig. 3.6). The proportion of individuals in each size class can be tracked over time in both research survey data and in diets of krill predators (Loeb et al. 1997; Saba et al. 2014), providing a window into krill population dynamics. The fairly discrete length modes, especially of young krill, coupled with seasonal or annual sampling strategies have led to the analysis of length frequency distributions to estimate growth rates.

A number of different approaches have been used to calculate age distributions from length frequencies in order to examine growth at the annual scale. Length frequency distributions were initially partitioned by hand (Bargmann 1945; Ivanov 1970; Mackintosh 1972) to generate nominal age distributions without incorporating the spatial variability in abundance generated by biological and physical

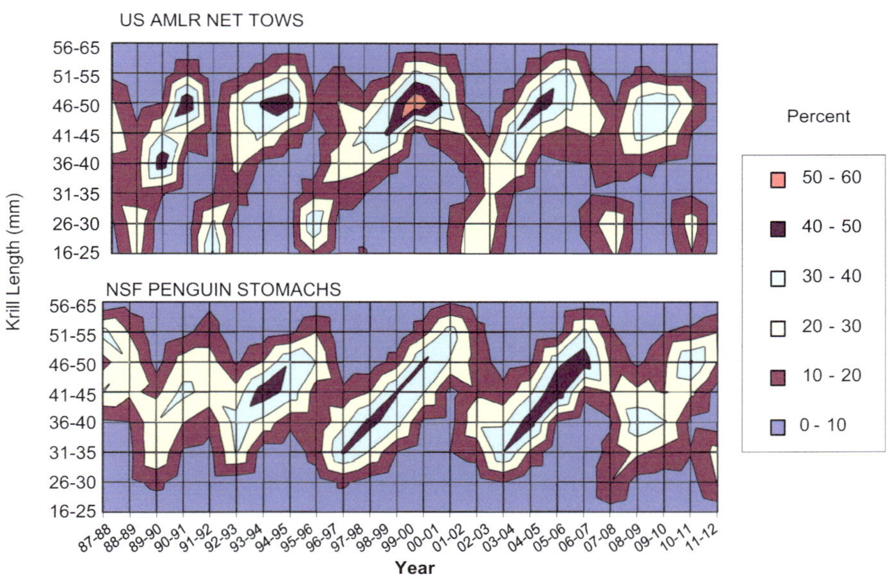

Fig. 3.6 Contour plots of the proportional occurrence of Antarctic krill length classes (5 mm bins) from 1988 to 2012. *Upper panel* is from IKMT net tows conducted around Elephant Island annually during US AMLR summer surveys. *Lower panel* is recalculated from Saba et al. (2014), and represents penguin diet samples collected as part of the National Science Foundation Palmer LTER Program. Diet data were extracted from the NSF Palmer data archive (pal.lternet.edu)

processes. Also, the length-frequency distributions from which these age estimates were generated may have been biased, because they failed to account for the potentially-longer life span of krill (three or more years), the patchy nature of krill associated with schooling and swarming behavior (Hamner 1984), or spatial patterns associated with spawning (Watkins et al. 1992).

Early laboratory and field studies used simple modal analysis of length frequencies to track changes in mean krill length over some period of time to estimate their growth rates. Early laboratory studies (Murano et al.1979; Poleck and Denys 1982; Ikeda et al. 1985; Ikeda and Thomas 1987) found that summer growth rates ranged from approximately 0.024 to 0.156 mm day^{-1}, while early field studies (eg Kanda et al. 1982; Clarke and Morris 1983; McClatchie 1988) found that summer growth rates varied from 0.056 to 0.33 mm day^{-1}. The large variability in summer daily growth rate estimates was less than the variability in winter field estimates for growth conducted around the same time. Daly (1990) found that age-0 krill grew at a rate 0.07 mm day^{-1}, during winter. Others found that growth rates varied from a minimum of 0.01 to 0.045 mm day^{-1}. The overall high variability in growth rates estimated using these techniques suggests that using modal progression of length classes to estimate growth may have limitations.

Ikeda and Dixon (1982a, b) starved krill in a laboratory study to determine how long krill could survive without food. They demonstrated that mortality rates were fairly low, and the starved krill shrank. Over the 211-day experiment, krill lost more than 50 % of their weight and shrunk by an average of 40 % in length. When food was provided again to the starved krill they were able to recover their original body weight. This experiment demonstrated that shrinkage is a fundamental characteristic of the biology of krill, and they concluded that shrinkage during periods of low food (winter) could increase the probability of their survival and was probability an important life history component for krill.

Following this experiment, others sought to determine whether shrinkage was an artifact of confinement in a laboratory, or whether it occurs in the field as well. If shrinkage was also observed to occur in the field, it would further confound estimates of length-at-age, longevity, and would complicate comparisons of life history patterns in different areas of the Southern Ocean or in relation to changing environments. McClatchie (1988) examined krill growth in Admiralty Bay over a year long time series. He suggested that krill in Admiralty Bay were food-limited as the growth rates of age zero, one and two krill were correlated with seasonal chl-*a* concentration. In a follow on study McClatchie et. al. (1991) investigated how krill size varied over the year, and argued that declines in krill size in Admiralty Bay during winter could result from size-dependent predation rather than shrinkage, or from advection of krill into or out of the bay. These field studies and others further highlight the inability of modal analysis or mixture distribution analysis to determine the root cause of seasonally-variable growth rates and length frequencies.

3.6.2 IMP and IGR

Krill, like all crustaceans, must moult in order to grow. This process, known as ecdysis, is under physiological and hormonal control (Buchholz 1989), and the moult cycle and the physiological timing of ecdysis has been investigated by a number of researchers Cuzin Roudy (1987a, b); Buchholz et al. (1989). Buchholz et al. (1989) and Nicol and Stolp (1990) identified stages of this moult cycle using microscopic analysis. They then provided a method to predict timing of ecdysis based on these analyses of the moult stage. However, even prior to understanding of the details of moult cycle, the moulting frequency of krill has been observed and described since krill have been maintained in aquaria during summer for observation (Mackintosh 1967; Clarke 1976; McWhinnie and Denys 1978; McWhinnie et al. 1979; Murano et al. 1979), and experimentation (Poleck and Denys 1982; Ettershank 1983, 1984; Morris and Priddle 1984; Ikeda et al. 1985).

In general, the aforementioned studies generated a range of values for moult frequency and intermoult period (IMP) as environmental conditions varied among the studies. Most of these observational studies found that IMP varied from 12 to 30 days during summer and that krill did not show much growth during moults. These researchers attributed this lack of growth to poor feeding conditions in the aquaria. Early laboratory studies (Ettershank 1983, 1984; Ikeda 1985) supported the findings of these observational experiments, under a number of different laboratory conditions and over a variety of time scales.

Poleck and Denys (1982), following the work of McWhinnie and Denys (1978) and McWhinnie et al. (1979) at Palmer Station on the west Antarctic Peninsula, were among the first to manipulate temperature to examine moulting, growth and maturation of krill. They showed that moult frequency increased with temperature, and that smaller krill moulted more often than larger krill size. These authors suggested that the difference in moult frequency between small and large krill arose because large krill must partition energy between growth and reproduction, whereas small krill may not allocate as much energy to reproduction compared to growth. Poleck and Denys (1982) also showed that moult frequency increased with increasing temperature and ranged from a minimum of 11 d (at 4.48 °C) to 23 d (at 0.12 °C). They found that the growth rate of krill maintained at ~5 °C was almost five times higher (0.031 mm day^{-1}, SD = 0.029) than for krill maintained at 0 °C (0.006 mm day^{-1}, SD = 0.016). These results were supported by the study of Morris and Priddle (1984) around South Georgia that concluded that moult frequencies in winter were much lower than in summer. They also noted that these moult frequencies were variable across the Scotia Sea, but did not assess the intermoult period or any latitudinal pattern in these moult stages.

The data from these studies suggested a general relationship between the environment and the moult frequency of krill. Buchholz et al. (1989) used a combination of laboratory and field data from different areas and times of year to generate a simple equation to predict intermoult period, and then showed that this simple equation could also be used in the field to quantify growth rate over a season. This technique was further refined by Nicol and Stolp (1990) and Buchholz (1991). Buchholz (1991) examined both the moult interval (MI) and percent growth per moult in the laboratory for fed and starved krill and compared the patterns with the moult cycle and histological moult staging from the earlier studies. He found that moult interval was inversely related to food availability (ie, non-starved krill moulted more frequently than starved krill, and therefore had a shorter interval between moults). He also found that that percent growth per moult declined over the course of the experiment. He concluded that growth increment was more sensitive to changes in food availability than moult interval. This explained the earlier findings whereby krill continued to moult in aquaria but exhibited little or no growth. Using data from summarized from the other stduies Buchholz (1991) found a strong negative relationship between moult interval and temperature (Table 6 and Figure 8 in Buchholz 1991), and he noted that moult interval (MI) could be predicted for a given temperature ranging from −0.9 °C to 4.0 °C (Fig. 3.7) by the equation

$$\text{MI (d)} = -3.04\text{T} + 23.21 \tag{3.3}$$

The moulting of crustaceans provides an opportunity to determine their growth rates by comparing the size of the newly moulted individual with the size of the moulted shell. The difference in length is then used as an estimate of growth, and the intermoult period is used to convert the change in length to an instantaneous growth rate (IGR). Different body parts can be used to estimate the IGR. For example, Poleck and Denys (1982) suggested using the difference in the length of the exopodites to estimate the IGR, while Quetin and Ross (1991) used differences in the telson length to estimate IGR. In all cases, these growth rates are dependent on a strong relationship between the length of the krill and the body part of interest.

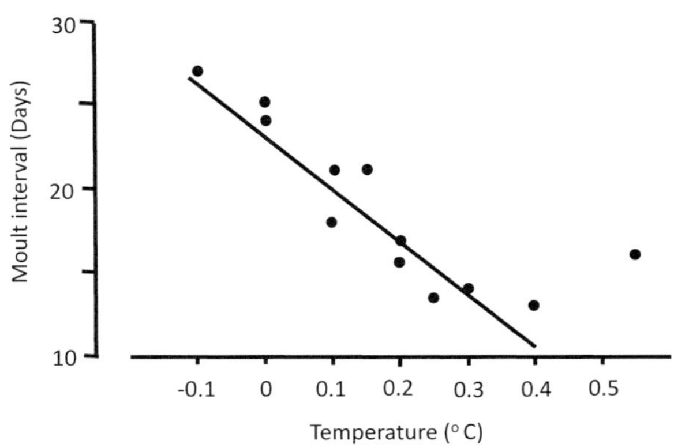

Fig. 3.7 Relationship between moult interval (*MI* days) and water temperature (T °C) derived from 12 field or laboratory maintenance studies. The regression is significant and describes the decrease in intermoult period with increase in water temperature. The equation is MI = −3.04 (T) + 23.21. Redrawn from Buchholz (1991)

Marschoff et al. (2008) modeled the relationship between uropod length and various estimates of krill length to understand whether any biases could impact the estimates of growth rate in krill. He showed that there was considerable variability in some measures of length, with some relationships performing better than others. This suggests that accounting for variability between body length measurements needs to be considered when examining growth rate using the IGR technique. It also suggests that care needs to be exercised when comparing IGR that use different body parts and estimates of length.

A number of studies on the IGR of krill in different areas of the Antarctic showed that IGR can vary seasonally. Quetin and Ross (1991) found that krill grew at a rate of 2–5 % per moult during autumn, but experienced negative growth rates of −0.1–2 % per moult during winter. Around the same time Nicol et al. (1992) examined the IGR of krill over four summers at higher latitudes (Prydz Bay) under food limited conditions. They found variable growth rates that ranged from −2 to 10 % per moult. Ross et al. (2000) also examined growth of late furcilia and young krill using IGR in a number of experiments conducted between 1991 and 1994. They showed that growth rates of these young krill were associated with both the chl-*a* concentration and the phytoplankton species composition. During years with high chl-*a* concentration and a phytoplankton community dominated by diatoms, the IGR ranged from 2 to 10 % of body length per intermoult period (0.02–0.082 mm day^{-1} over the 8 week studies). In contrast, when the phytoplankton community was dominated by prymnesiophytes and cryptophytes, the IGRs were lower, and some krill experienced negative growth. Their study showed that laboratory artifacts, water temperature, and both the type and amount of chl-*a* could significantly influence the intermoult period, and the resulting IGR. Yet, from these IGR studies they concluded that krill growth was food-limited in the west Antarctic Peninsula, as these growth rates were lower than the growth rates of krill maintained in the laboratory by Ikeda and Thomas (1987).

As experiments became more complicated, and more laboratory studies were conducted, a number of biases were revealed that impact the accuracy of the IGR / IMP method (Quetin and Ross 1991). A number of studies noted that growth increments declined and intermoult periods increased for krill captured from the wild and held for extended periods of time without food (eg Nicol et al. 1992). This suggested that accurate estimates of krill could be obtained only for the first couple of moults when krill were responding to their recent feeding environment, rather than conditions in the laboratory. Other studies noted that the calculation of intermoult period could be biased by the manner in which krill are sampled from swarms (Tarling et al. 2006). Because krill from the same swarms were often placed in the experimental chambers at the same time, they had common histories, and so IGRs from krill did not represent independent estimates. Tarling et al. (2006) developed a statistical framework accounting for this sampling factor, and for the number of days since moulting to allow estimates of growth to be generalized to the population.

Most studies of krill IGR were conducted to study growth during individual surveys to understand local growth rates limiting the generality of their results. Candy and Kawaguchi (2006) conducted a meta-analysis of IGR studies from the

Indian Ocean (8 cruises, 51 experiments over 5 months) and the Southwest Atlantic (1 cruise 12 experiments and 4 months), to build a model of krill growth in those areas. They used a linear mixed models (LMM) approach and combined the data with the temperature-dependent moult intervals (based on Buchholz' 1991 summary) into estimates of daily growth. Kawaguchi et al. (2006) showed that growth rates of krill declined with length and varied with season and sex (Fig. 3.8). In the Indian Ocean, juvenile krill (<30 mm) grew fastest in December (0.204–0.279 m d^{-1}) and their growth rates declined through April (to 0.012 and 0.011 mm day^{-1}). Adult female krill (40 mm) followed the same pattern of declining growth rate as juveniles, although rates were lower; growth rate of 40 mm females in December was 0.126 mm day^{-1} and declined to a negative rate of −0.003 mm day^{-1} by April. These growth rates were even lower for 50 mm females (0.083 mm day^{-1} in December to −0.014 mm day^{-1} in April). Adult male krill (40 mm) also followed this pattern, but maintained slightly higher growth rates than females (0.136 mm day^{-1} in December to 0.011 mm day^{-1} in April). Adult male krill 50 mm in length followed a slightly different pattern of growth, although, as with adult females, rates were slower than those for 40 mm krill. Growth in 50 mm males increased from 0.050 mm day^{-1} in December to 0.089 mm day^{-1} in January, and then declined to 0.010 mm day^{-1} in April.

Candy and Kawaguchi (2006) used the same IGR data for juvenile and male krill from the Indian Ocean to model annual growth and develop a von Bertalanffy growth model (female krill were eliminated from model development because of challenges associated with female diversion of energy from growth toward reproduction during summer). Using the same LMM approach, combined with a smoothing spline technique, they used summer growth rate measurements to develop growth trajectories for krill age one through age six. Because of the lack of measured winter growth rates in the Indian Ocean, this study extrapolated winter growth rates and examined whether winter shrinkage should be incorporated into models that estimate growth trajectories. Results indicated that the best-fitting model predicted a seasonal growth period from December to March in the Indian Ocean, and allowed for negative growth between April and November.

Length-at-age estimates derived using this approach were compared to estimates produced by the models of Rosenberg et al. (1986) and Siegel (1986). In general, the growth model of Candy and Kawaguchi (2006) predicted that krill would be slightly smaller at age one and age two compared to the estimates derived using the approaches based on analysis of length modes.

In another series of studies using large amounts of IGR data, Tarling et al. (2006) and Atkinson et al. (2006) examined factors controlling the IMP and growth rates of krill across the southwest Atlantic during two summers. Tarling et al. (2006) examined the effects of temperature and chl-*a* on IMP and developed a method to provide less-biased estimates of IMP. Atkinson et al. (2006) showed that in general, summer growth rates of krill in the Southwest Atlantic could be modeled as a function of temperature, food (ie primary production) and length (Fig. 3.9). They showed that for both water temperature and length, krill growth, as a function of chl-*a* concentration became asymptotic at concentrations greater than

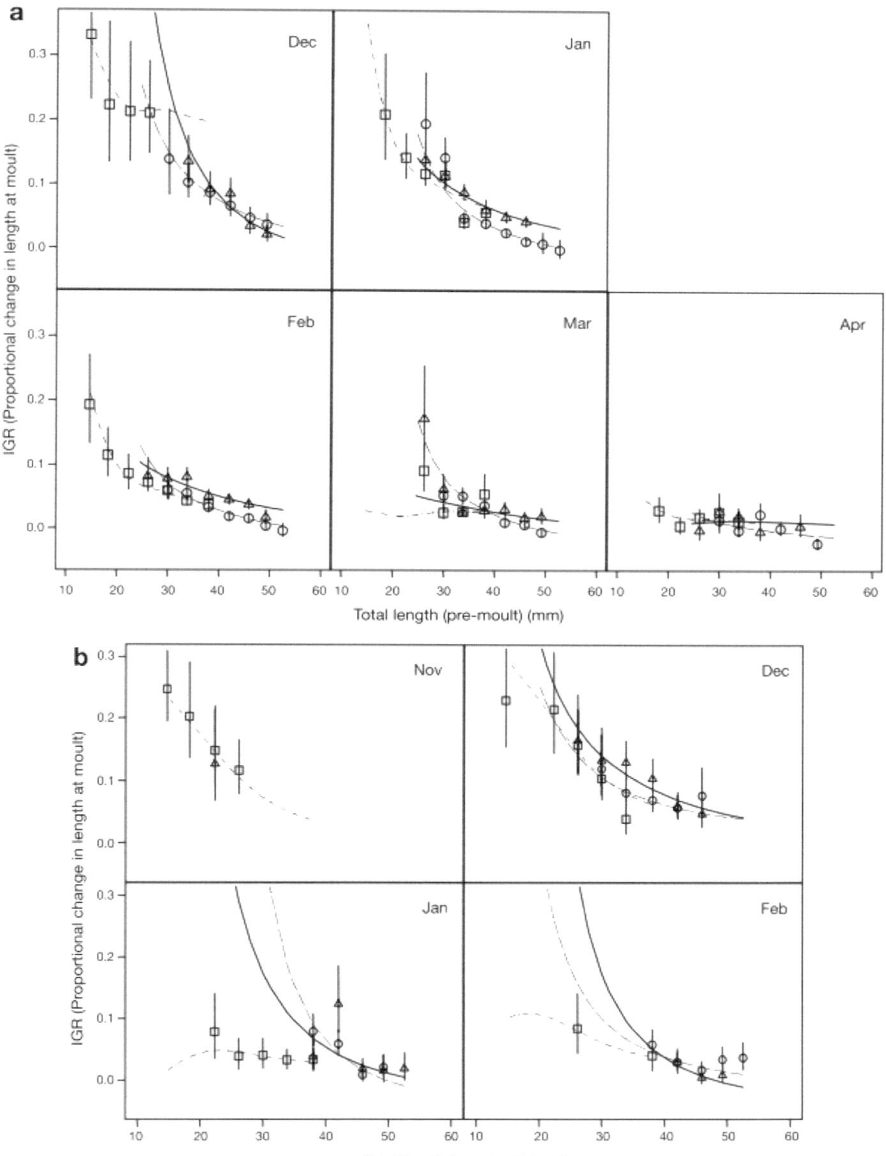

Fig. 3.8 Proportional change in length as a function of length for female (---), male (.-.-.) and juvenile (___) krill during late spring and summer in the Indian Ocean (**a**) and the southwest Atlantic (**b**). Redrawn from Kawaguchi et al. (2006)

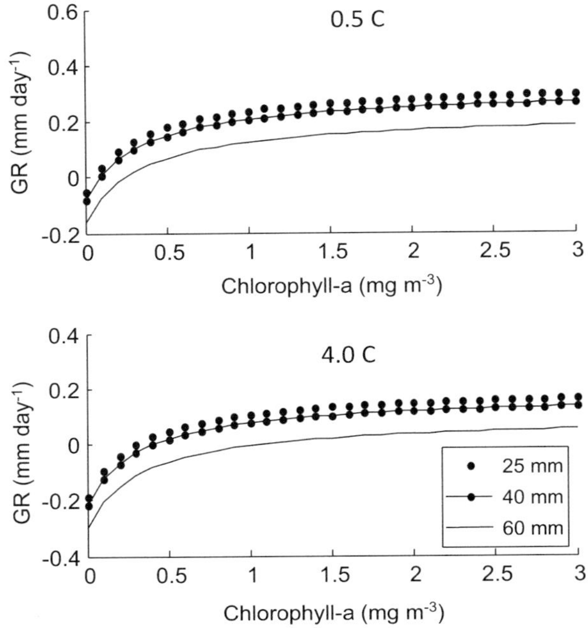

Fig. 3.9 Calculated daily growth rate of krill (mm d^{-1}) of three length classes (25, 40, and 60 mm) in cold (0.5 °C) and warm (4 °C) water, as a function of food concentration (mg chl-*a* m^{-3}) (calculated using equation (3.4); adapted from Atkinson et al. 2006)

approximately 1.0 mg chl-*a* m^{-3}, that food had the strongest effect on growth, and that growth rates were higher at lower temperatures.

They developed a general growth model where

$$Growth = a + b * Length + c * Length^2 + \left[\frac{d * Food}{e + food}\right] + f * Temperature + g * Temperature^2 + \varepsilon \quad (3.4)$$

and length is measured in mm, water temperature (°C) is measured in the upper mixed layer, and food is estimated as µg chl-*a* L^{-1}. The letters *a* through *g* are constants while ε accounts for other random effects. They showed krill sex and maturity stage had smaller influences on growth rate than length, food, or temperature and concluded that this generic model could be used to understand the growth rate of krill under different environmental conditions.

3.6.3 Other Methods to Estimate Growth Rates Using Length Data

Most studies of growth have been conducted over during short (weeks) surveys that are often annually. These studies examine the change in length (age) of cohorts identified by length modes at annual time scales. Other studies have used the IGR method to estimate growth *in situ* (Quetin and Ross 1991) over a cruise (Atkinson et al. 2006; Candy and Kawaguchi 2006). When multiple surveys are conducted within one season, seasonal growth rates may be estimated by examining changes in length between surveys. Such an approach was taken by Shelton et al. (2013) to estimate growth rates during summer (January to March) in the Antarctic Peninsula area. They used kernel density approaches to estimate growth of krill between two summer survey periods approximately a month apart, and over a 19 year time series from data collected as part of the U.S. AMLR Program. This kernel density approach allows for estimates of growth from individual krill in the sampled population over time. It makes several assumptions, including that the same population must be sampled among time periods, and that mortality is size-independent. Meeting these assumptions may be challenging in areas with strong currents, where migration occurs, or where size-based mortality is high. The results of this analysis showed significant interannual variation in summer growth rates, and also significant spatial variability in growth. Median daily growth rates around Elephant Island ranged from 0.00 to 0.17 mm day^{-1}. On the shelf west of Elephant Island, growth rates ranged from −0.06 to 0.22 mm day^{-1}. Mean growth rate over the 19 years of the study was 0.073 mm day^{-1} (CV = 1.1) for a 30 mm krill. Chl-*a* concentration (μg L^{-1}) was the only environmental variable that was correlated with growth rate, supporting the idea that growth may be food limited in the Antarctic Peninsula region. Yet despite the long time series, and the climatic changes occurring in this region, Shelton et al. (2013) were unable to find any trend in the seasonal growth rate underscoring the difficulty in resolving the impact of climate change on the dynamics of Antarctic krill.

3.6.4 Bio-Energetic Models of Growth

Growth can be modeled statistically using the relationships among lengths at different time periods, but growth can also be estimated and modeled numerically using bio-energetic models to drive growth. Astheimer et al. (1985) and Astheimer (1986) were the first to model growth by balancing energy intake with growth. More recently, Hofmann and Lascara (2000) developed a bioenergetic model of krill growth for the west Antarctic Peninsula. They modelled growth across the entire length range and explored the importance of spatially varying food concentrations and winter sea ice algae to the overwinter survival of krill. The results of their model were promising, especially for juvenile krill (<35 mm). They were able to replicate the general pattern of seasonal growth, including winter shrinkage, and were able to establish that models of winter growth rates required both a reduced metabolism component and an additional food source component. This bioenergetics approach to investigating krill growth and bottlenecks was included in a number of subsequent analyses (eg, Fach et al. 2002, 2006, 2008).

3.7 Recruitment

The recruitment (or year class strength) of pelagic spawning species is the result of survival during the first year of life (Hjort 1914), and is usually associated with physical and biological processes during the larval and immediate post-larval stage (Anderson 1988). Thus, depending on environmental conditions and the biological resources available to young life stages, year classes may succeed or fail and affect population structure, regardless of egg production (Cushing 1990; Sinclair 1988). Although it is widely known that many factors can affect recruitment success, for many years the interannual variability in the abundance of krill was largely attributed to oceanographic factors (Priddle et al. 1988). Conclusions from a number of workshops in the 1980s determined that "it appears that the variability [of krill] is generally the result of changes in the krill distribution rather than the result of changes in abundance and biomass... although changes in year classes can also be observed" (Sahrhage 1988). This view was based principally on the observation that krill abundance around South Georgia could be very low in 1 year and very high the following year, and that the knowledge that such rapid changes could not be attributed to biological properties (ie, growth and reproduction) of krill (Siegel 1988). Thus, a central challenge to understanding recruitment variability is to separate the effects of oceanographic and environmental forcing on krill abundance from factors affecting the population dynamics of a species that has a circumpolar distribution, and yet shows little evidence of genetic stock structure (Bortolotto et al. 2011).

In the Southern Ocean, several distinct physical "systems" may confine krill to specific areas or distribute krill to other areas, depending on prevailing circulation or other oceanographic features. Open-ocean island systems, such as South Georgia, that are located in the prevailing circulation are recipients of production from upstream locations. Island-shelf systems that have complicated bathymetric and hydrographic structures, such as the Antarctic Peninsula, are high-production areas and are a source of krill to downstream habitats (Fach et al. 2002). Open-ocean, ice-dominated systems, like the Lazarev Sea system, are dependent on local krill production (Siegel et al. 1990, 2004; Flores et al. 2012). These systems can contribute krill to the general population as the pack ice, which is present throughout most of the year and serves as the primary habitat for krill in these systems, is advected in gyres within the prevailing circulation. Finally, continental shelf – slope systems like the area off east Antarctica (Nicol et al. 2000a, b) or along the coast in the Indian Ocean (Pakhomov 2000) rely on large scale oceanographic features to retain krill, while seasonal sea-ice dynamics affect the primary production and subsequent larval production. In these continental shelf–slope systems, large scale oceanic recirculation (1000s km) maintains krill in regional but "leaky" areas.

In each of these systems recruitment is driven by the same biological processes typical of broadcast spawners, but the magnitude of recruitment is affected by the regional physical processes, including seasonal sea-ice dynamics and variability in ocean circulation patterns. These physical processes operate across large spatial scales because their variability is linked to broad-scale atmospheric forcing. Because recruitment success is strongly linked to the physical environment in polar ecosystems, it is possible that recruitment dynamics of different populations could be correlated over broad spatial and temporal scales potentially linking seemingly different systems.

3.7.1 Defining a Recruitment Index

The use of the mixture distribution approach pioneered by MacDonald and Pitcher (1979) to separate nominal age cohorts from length-frequency distributions provided an important tool for objectively estimating the number of age groups in a sample distribution, and for developing an index of recruitment. The use of the Rosenberg et al. (1986) length-at-age model by the scientific community (Siegel 1987), and fishery resource managers (Everson 1992; Miller and Hampton 1989; de La Mare 1994a, b, c) ushered in a period of increased quantitative study of krill population dynamics because the model accounted for both the longevity of krill and the seasonal growth dynamics. These studies were facilitated by the establishment of several long-term ecosystem-based research programs (eg early German AMLR surveys, U.S. NSF Palmer-LTER, BAS South Georgia monitoring, U.S. AMLR Program) that provided repeated fishery-independent surveys of krill populations in several areas of the Southern Ocean. One key area of progress from these larger studies on krill population dynamics was a more quantitative understanding of recruitment variability.

Although the mixture distribution approach (MacDonald and Pitcher 1979) provided an objective method to extract age cohorts from a length-frequency distribution, it was insufficient to provide an index on the length density from net-based surveys, because several assumptions of the model were difficult to satisfy. Among these assumptions were that length-frequency distributions were obtained from random and independently-distributed animals, and that these animals were completely enumerated and representative of the population (de La Mare 1994a). This approach also did not specifically account for the high probability of a large number of zero-krill hauls, or the highly patchy nature of krill typical of net-based surveys that create complicated statistical properties. de La Mare (1994a, b) developed a method to quantify the gross recruitment rate based on fitting length distributions to survey data. This approach specifically accounted for the proportion of zero hauls in a survey area, using Aitchison's delta distribution as the underlying statistical model for the net tow data (Aitchison 1955; Pennington 1983). Importantly this approach accounted for the probability of having a zero-krill haul, and used a lognormal probability distribution to fit to the highly patchy nature of krill abundance. This approach allowed for the calculation of the length density distribution for any survey. Using this approach he further developed this idea (de La Mare 1994b) and defined a proportional recruitment index, in which the fraction of

1 year old (R1) krill in the survey from the sum of the total abundance of krill age two and greater. A similar index, R2, was defined for surveys where no 1 year old krill were found. He suggested that the ability to separate one or 2 year old krill from the total abundance of krill would be sufficient to determine the relative success of recruitment processes.

The proportional recruitment index is not without shortcomings. Since the index represents a proportion, years of both low and high absolute recruitment may be under-represented, as these years may have the same proportional recruitment. Hewitt (2000) proposed a modification to the proportional recruitment index, to account for cohort mortality, and generated a per capita recruitment index, that has not been used very often to calculate recruitment indices. Thus, despite the limitations and criticisms of the proportional recruitment index, it is used because it is simple and provides useful information about the conditions in which recruitment is successful.

3.7.2 Patterns of Recruitment and the Development of a Conceptual Model

The general notion that ice dynamics were associated with recruitment variability was appreciated as early as the 1930s (Mackintosh 1967), as local krill abundance and size was shown to vary with ice conditions. Siegel (1985, 1986, 1987) was examined the patterns of seasonal krill abundance and commented on the occasional missing year classes which he attributed to recruitment failure. However, connections between krill recruitment and the environment remained largely unexplored until time series could be constructed from combining occasional net surveys (Sahrhage 1988; Siegel and Loeb 1995), fishery independent monitoring and ecological studies in the Antarctic Peninsula (Loeb et al. 1997; Quetin and Ross 2003) and South Georgia (Watkins 1999; Fielding et al. 2014), and also from time series derived from predator analyses (Fraser and Trivelpiece 1995; Fraser et al. 1992; Reid et al. 1999, 2002). These time series have provided data to develop comprehensive conceptual models of krill recruitment to examine how the physical and biological environment drives the population dynamics of krill.

Siegel and Loeb (1995) were the first to develop the first complete conceptual model on krill life history and recruitment based on a 15 year (1977–1994) time-series of krill length frequencies that built on the work by Kawaguchi and Satake (1994) and Fraser and Trivelpiece (1995). The model they developed has four parts that together determine the year class strength by linking the seasonal ice dynamics with spring and summer productivity to drive the recruitment pulses that have observed in a number of different regions (Hosie et al. 1988).

Siegel and Loeb (1995) found a positive correlation between the winter sea-ice extent (km^2) and the proportion of 1 year old krill (R1) the following summer (Fig. 3.10). Additionally, they showed that the abundance of krill larvae was

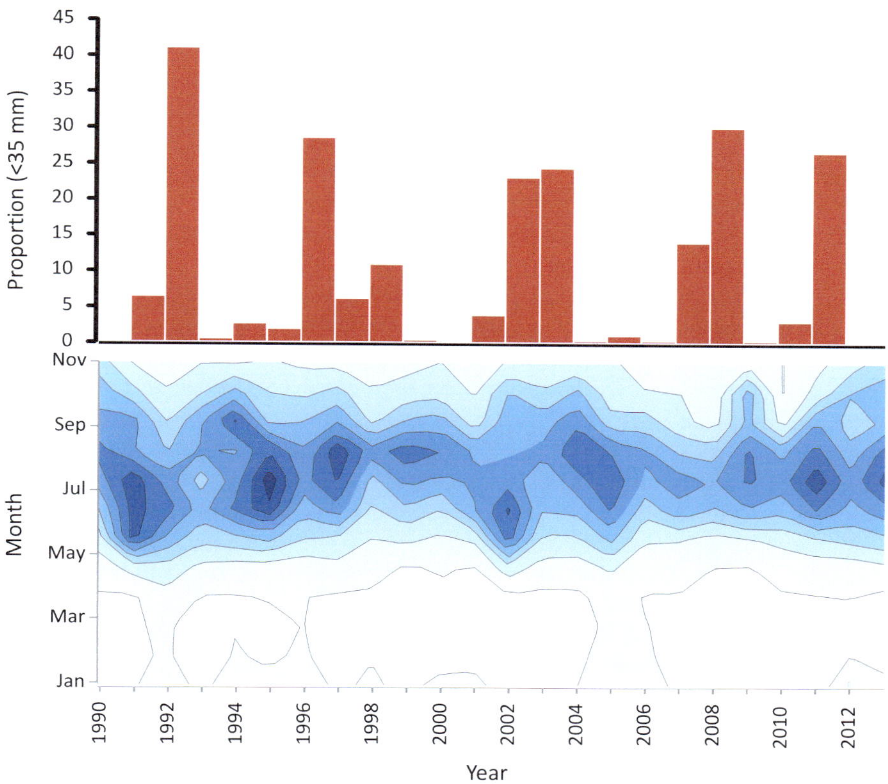

Fig. 3.10 Proportion of krill less than 35 mm in length and seasonal variability in sea ice area (modified and extended from Hewitt and Linen Low 2000) between 1990 and 2011

correlated with the sea-ice extent the previous winter. Their model hypothesized that following sea-ice retreat in spring (that increases stratification and initialises the spring bloom at the ice edge) resulting primary production fuels female reproductive output and, subsequently, survival of the first feeding stages of larvae. After progressing through a number of larval stages during summer, survival of age-0 larvae is controlled by early sea-ice formation during winter. This early winter sea-ice develops a rich sea-ice algal community that provides food, the sea-ice also provides habitat and refuge from predation, and an extensive pack ice zone may provide a more stable habitat, limiting advective loss to the pelagic environment (Meyer 2011). In years with low sea-ice extent or late sea-ice development lower adult reproductive output during spring, and higher larval mortality throughout the year result in recruitment failure. This model identifies the physical environment as a critical mediator in the population dynamics affecting both early life stages and adults.

Loeb et al. (1997) further investigated the recruitment dynamics of krill in relation to environmental forcing, linking warming temperatures to sea-ice variability and to primary production (chl-*a* biomass) around the Antarctic Peninsula. This work firmly established the biological link between atmospheric forcing (air temperature) and sea-ice variability, and this variability in sea ice with recruitment dynamics of krill (Fig. 3.10). Building on this atmospheric connection, Quetin and Ross (2003) investigated the episodic nature of krill recruitment in the western Antarctic Peninsula region in relation to climatic factors. They linked the warming air temperatures and sea-ice variability noted by Loeb et al. (1997) to the El Nino-Southern Oscillation (ENSO) scale of atmospheric climate variability. Examining recruitment variability in their time-series they showed that population abundance in the western Antarctic Peninsula was dominated by occasional episodic recruitment, so that one or two consecutive recruitment events drove the population dynamics until the next set of favorable ice conditions (Fig. 3.6). However, in contrast to the results of Loeb et al. (1997), Quetin and Ross (2003) and Ducklow et al. (2007) argued that ENSO neutral conditions were most favorable to krill recruitment (Fig. 3.11). They found that recruitment was highest during ENSO neutral conditions and recruitment declined on either side of this optimum.

In the Antarctic Peninsula region this sea-ice mediated recruitment relationship has been tested and expanded by a number of follow on studies (Wiedenmann et al. 2008, 2009). Loeb et al. (2009) re-analyzed much of the data collected from the Antarctic Peninsula and tested their original model (Loeb et al. 1997) given the climate driven changes that had occurred over the intervening 20 years. They found that while the recruitment strength could still be predicted from changes in sea-ice extent, the sea-ice dynamics were related less to the ENSO signal and more to the Semi-Annual Mode of climate oscillation. Saba et al. (2014) and Ross et al. (2014) analyzed data for the western Antarctic Peninsula and also showed that the same major climate mode was correlated with the pattern of recruitment. Saba et al. (2014) further showed that winter conditions were directly related to the recruitment potential for krill by affecting the magnitude of spring primary production. Despite the changing ecosystem dynamics at the Antarctic Peninsula, the

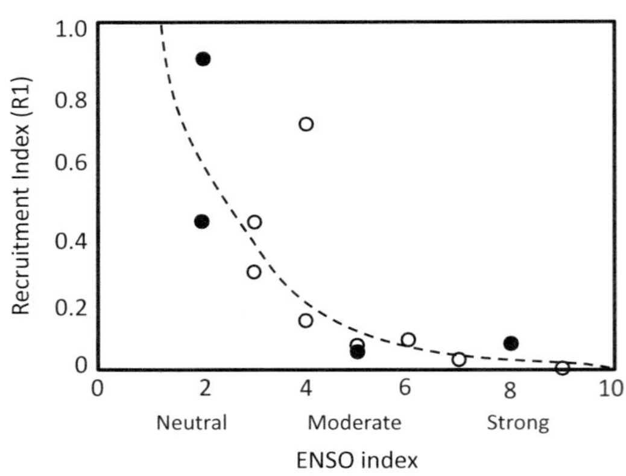

Fig. 3.11 Relationship between ENSO index and the krill recruitment index (*R1*) for the west Antarctic Peninsula area sampled by the USAP, Palmer LTER Program. Redrawn from Ducklow et al. 2007)

continued correlation between the physical environment and recruitment dynamics of krill is important, as most correlational relationships with recruitment fail (Myers 1998).

In other areas of the South Atlantic Ocean, like South Georgia, which is at the most northerern extreme in the distribution of krill (Marr 1962; Mackintosh 1972), there is little evidence for local recruitment. Despite the lack of local recruitment of krill, long term monitoring of upper trophic level diets has documented changes in the size of krill consistent with recruitment variability (eg Reid et al. 1999; Reid 2001). Modelling of krill growth and mortality rates at South Georgia further supports the view that recruitment variability can be resolved in diets of krill predators (Reid et al. 2002). Despite the complex patterns of abundance and recruitment driven by multiple interacting factors, proportional recruitment time series have been developed for some periods of time in this region (Watkins 1999). Although there is sometimes significant production of larvae (Mackintosh 1973; Tarling et al. 2007), these larvae do not recruit to the population. Instead variability in krill abundance of krill at South Georgia area is thought to be driven by recruits produced at the Antarctic Peninsula or from the Weddell Sea and advected to South Georgia (Marr 1962; Mackintosh 1973; Hofmann et al. 1998; Hofmann and Murphy 2004; Murphy et al. 2004; Thorpe et al. 2007). This would suggest a correlation between recruitment indices at the Peninsula and the patterns of recruitment at South Georgia. However, although there has been evidence of concordant fluctuations in biomass between these two regions (Brierley et al. 1999; see Chap. 2, Siegel and Watkins 2016) there are no correlations in recruitment.

There is much less information regarding the causes of fluctuations in krill recruitment in either the Indian Ocean or the Lazarev Sea (Nicol et al. 2000a, b; Pakhomov 2000; Flores et al. 2012; Siegel 2012). It clear that in these regions, sea-ice is critically important to the life history of krill, but there are few time-series of sufficient length to resolve the linkages between krill biology and the physical environment in those environments.

Siegel et al. (2003) investigated the scope of the regional inter-correlation in krill recruitment between areas around the Southern Ocean. They compiled data from a number of sites around the Antarctic (Table 3.3) and showed that the krill recruitment indices (R1) in Bellingshausen Sea and the tip of the Antarctic Peninsula were positively correlated, suggesting that at the scale of 100 s of kilometers krill population dynamics were driven by the same general processes. In contrast, Siegel et al. (2003) found no correlation between recruitment (R1 or R2) patterns in the Indian Ocean and the Bellingshausen Sea suggesting that the fluctuations in recruitment were not in phase at the scale of the Southern Ocean. When the Bellingshausen Sea recruitment time series was correlated with recruitment at South Georgia, a positive correlation was found with the proportion of 1 year old krill (R1) between the Bellingshausen Sea, the Peninsula and South Georgia, supporting the concept that recruitment at South Georgia is driven largely from advection of krill from the southwest Atlantic. Recent publication of recruitment time series for the western Antarctic Peninsula (Ross et al. 2014), and for the tip of the peninsula (Reiss 2008; U.S. AMLR unpublished) shows that recruitment

Table 3.3 Recruitment data updated from Siegel et al. (2003)

Year class	Indian Ocean	Northern Bellingshausen	Elephant Island	South Georgia	US AMLR	LTER
	de la Mare (1994b)	Siegel et al. (2003)	Siegel et al. (2002)	Watkins (1999)	Reiss (unpublished)	Digitized from Ross et al. (2014)
1975/1976						
1976/1977			0.048			
1977/1978						
1978/1979						
1979/1980	0.167		0.559			
1980/1981	0.001		0.757			
1981/1982			0.47			
1982/1983	0.016		0.03			
1983/1984	0.528	0	0.0001			
1984/1985	0.175					
1985/1986	0.025					
1986/1987		0.175 (0.182)	0.156			
1987/1988			0.651			
1988/1989		0	0.057	0.013		
1989/1990	0.314		0.099	0.007		0.82
1990/1991			0.375			0.89
1991/1992	0.064	0.005	0	0.038		0
1992/1993		0.088	0.068	0.075		0.08
1993/1994		0.267 (0.281)	0.046			0.29
1994/1995	0.303a	0.639 (0.852)	0.622	0.682		0.70
1995/1996		0.147 (0.436)	0.198	0.179		0.43

(continued)

Table 3.3 (continued)

	Indian Ocean	Northern Bellingshausen	Elephant Island	South Georgia	US AMLR	LTER
Year class	de la Mare (1994b)	Siegel et al. (2003)	Siegel et al. (2002)	Watkins (1999)	Reiss (unpublished)	Digitized from Ross et al. (2014)
1996/1997		0.095	0.12	0.164	0.25	0.07
1997/1998		0	0		0.47	0.03
1998/1999		0.062 (0.08)	0		0.044	0.03
1999/2000		0.076 (0.056)	0.573		0	0.05
2000/2001		0.748 (0.886)			0.001	0.89
2001/2002					0.64	0.39
2002/2003					0.45	0.12
2003/2004					0.09	0.09
2004/2005					0.008	0.19
2005/2006					0.001	0.86
2006/2007					0.35	0.71
2007/2008					0.61	0.23
2008/2009					0.04	0.38
2009/2010					0.11	0.83

between areas along the Antarctic Peninsula continues to be highly correlated (Table 3.3, Fig. 3.12), despite the changes in climate and sea-ice dynamics in the region (Stammerjohn et al. 2008; Meredith and King 2005).

3.8 Conclusions

The last 30 years of research has provided a wealth of information regarding the growth and recruitment of Antarctic krill. The development of techniques to measure growth *in situ* have led to the accumulation of large data sets collected under varying conditions and across a number of regions in the Southern Ocean.

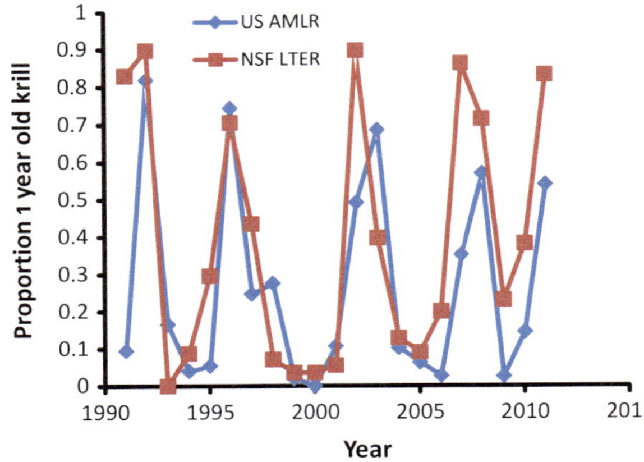

Fig. 3.12 Relative proportional recruitment of krill in the west Antarctic peninsula (*red*) redrawn from Ross et al. (2014) and proportion of krill less than 35 mm in length from the Antarctic Peninsula showing the high correlation between areas over the 21 years of the long-term studies by the USAP and the US AMLR Program

The collection of these data have led to meta-analyses and general relationships that have resulted in a significant increase in understanding the flexibility of growth to different environmental conditions. These general relationships have allowed the development of simulations to numerically test hypotheses that have also broadened our understanding of the links between the physical environment and krill populations, and to examine the projected impacts of climate change on krill habitat and growth.

Long time-series of acoustic biomass and net estimates of abundance have provided data for expanded testing of the link between the physical environment and recruitment and its spatial scale. As a consequence, over the last two decades considerable progress has been made in understanding the climatic links to krill recruitment that may be altered by the climate change occurring in the some parts of the Southern Ocean, like the southwestern Atlantic and South Georgia. In other areas, time-series necessary to understand recruitment, its variability, and its links to climatic forcing are still lacking.

Despite the progress in understanding the variation in growth and recruitment, the continued lack of progress in quantitative comparisons regarding mortality, growth and longevity among areas is directly related to the lack of progress in the direct estimation of age. Modal analysis, whether using body length, eye size, or morphometrics all suffer from the inability to discriminate slow-growing old krill from fast-growing young krill of the same length. Overcoming this limitation is critical to better understanding of the likely impacts of climate change on krill populations. This is because the occasional recruitment and high variability in abundance make detecting changes in the population difficult. Bio-chemical ageing techniques show promise, but have yet to be used to examine time series of krill from surveys, and have been limited to individual studies that document the potential for the technique. Recent development for the direct ageing of crustaceans in general, and krill specifically, are promising but are in their infancy.

Future studies on age, growth, mortality and recruitment of krill will need to focus on the effects of a changing environment and develop methods to directly compare rates amongst areas of the Southern Ocean. Such studies might then be able to validate predictions derived from end-to-end climate models and projections from the IPCC, and lead to better manage Southern Ocean resources in future years.

Acknowledgements I thank Debra Losey and Robin Schnug of the Fisheries Library at the Southwest Fisheries Science Center for meeting each of my many requests for papers with a smile and a wonderful attitude. I thank Jennifer Walsh for editing various parts of this chapter, and I also thank two anonymous reviewers for thoughtful comments for my consideration. The views expressed in this chapter are those of the author and should not be considered as reflecting the views of the Government of the United States.

References

Aitchison J (1955) On the distribution of a positive random variable having a discrete probability mass at the origin. J Am Stat Assoc 50:901–908

Anderson JT (1988) A review of size-dependent survival during pre-recruit stages of fishes in relation to recruitment. J Northwest Atl Fish Sci 8:55–66

Astheimer H (1986) A length class model of the population dynamics of the Antarctic krill *Euphausia superba* Dana. Polar Biol 6:227–232

Astheimer H, Krause H, Rakusa-Suszczewski S (1985) Modelling individual growth of the Antarctic krill *Euphausia superba* Dana. Polar Biol 4:65–73

Atkinson AA, Shreeve RS, Hirst AG, Rothery P, Tarling GA, Pond DW, Korb RE, Murphy EJ, Watkins JL (2006) Natural growth rates in Antarctic krill (*Euphausia superba*): II. Predictive models based on food, temperature, body length, sex, and maturity stage. Limnol Oceanogr 51:973–987

Bargmann HE (1937) The reproductive system of *Euphausia superba*. Discov Rep 4:325–350

Bargmann HE (1945) The development and life-history of adolescent and adult krill *Euphausia superba*. Discov Rep 23:103–176

Basson M (1994) Towards a distribution of M/K for krill (*Euphausia superba*) required for the stochastic krill yield model. CCAMLR WG-Krill-94/11

Basson M, Beddington JR (1989) Preliminary analysis of growth and mortality rates of Antarctic krill (*Euphausia superba*) from length frequency samples. In: Heywood RB (ed) University research in Antarctica: proceedings of British Antarctic Survey Antarctic special topic award scheme symposium. British Antarctic Survey, Cambridge, pp 51–55

Beverton RJH, Holt SJ (1957) On the dynamics of exploited fish populations, vol XIX, Fishery investigations series II. Ministry of Agriculture, Fisheries and Food, London, 533 pp

Bortolotto E, Bucklin A, Mezzavilla M, Zane L, Patarnello T (2011) Gone with the currents: lack of genetic differentiation at the circum-continental scale in the Antarctic krill *Euphausia superba*. BMC Genet 12:32. doi:10.1186/1471-2156-12-32

Brierley AS, Demer DA, Hewitt RP, Watkins JL (1999) Concordance of inter-annual fluctuations in densities of krill around South Georgia and Elephant Islands: biological evidence of same-year teleconnections across the Scotia Sea. Mar Biol 134:675–681

Buchholz F (1989) Moult cycle and seasonal activities of chitinolytic enzymes in the integument and digestive tract of the Antarctic krill, *Euphausia superba*. Polar Biol 9:311–317

Buchholz F (1991) Moult cycle and growth of Antarctic krill *Euphausia superba* in the laboratory. Mar Ecol Prog Ser 69:217–229

Buchholz F, Morris DJ, Watkins JL (1989) Analyses of field moult data: prediction of intermoult period and assessment of seasonal growth in Antarctic krill, *Euphausia superba* Dana. Antarct Sci 1:301–306
Candy SG, Kawaguchi S (2006) Modelling growth of Antarctic krill. II. Novel approach to describing the growth trajectory. Mar Ecol Prog Ser 306:17–30
Clarke A (1976) Some observations on krill (*Euphausia superba* Dana) maintained alive in the laboratory. Bull Br Antarct Surv 43:111–118
Clarke A, Morris DJ (1983) Towards an energy budget for krill; the physiology and biochemistry of *Euphausia superba* Dana. Polar Biol 2:69–86
Claverie T, Smith IP (2007) A comparison of the effect of three common tagging methods on the survival of the galatheid *Munida rugosa* (Fabricius, 1775). Fish Res 86:285–288
Constable AJ, de la Mare WK (1996) A generalised model for evaluating yield and the long term status of fish stocks under conditions of uncertainty. CCAMLR Sci 3:31–54
Constable AJ, de la Mare WK (2003) Generalised yield model, version 5.01b. Australian Antarctic Division, Kingston
Cushing DH (1990) In: Blaxter JHS, Southward AJ (eds) Plankton production and year-class strength in fish populations: an update of the match/mismatch hypothesis in advances in marine biology. Academic, San Diego, pp 250–313
Cuzin-Roudy J (1987a) Sexual differentiation in the Antarctic krill *Euphausia superba* Dana (Crustacea: Euphausiacea). J Crustac Biol 7:518–524
Cuzin-Roudy J (1987b) Gonad history of the Antarctic krill *Euphausia superba* Dana during its breeding season. Polar Biol 7:237–244
Daly KL (1990) Overwintering development, growth, and feeding of larval *Euphausia superba* in the Antarctic marginal ice zone. Limnol Oceanogr 35:1564–1576
Daly KL, Macaulay MC (1991) Influence of physical and biological mesoscale dynamics on the seasonal distribution and behavior of *Euphausia superba* in the Antarctic marginal ice zone. Mar Ecol Prog Ser 79:37–66
de Graaf M (2007) Tag retention, survival and growth of marron *Cherax tenuimanus* (Crustacea: Decapoda) marked with coded micro wire tags. Mar Freshw Res 58:1323–1650
de la Mare WK (1994a) Modelling krill recruitment. CCAMLR Sci 1:49–54
de la Mare WK (1994b) Estimating krill recruitment and its variability. CCAMLR Sci 1:55–69
de la Mare WK (1994c) Estimating confidence intervals for fish stock abundance estimates from trawl surveys. CCAMLR Sci 1:203–207
Ducklow HW, Baker K, Martinson DG, Quetin LB, Ross RM, Smith RC, Stammerjohn SE, Vernet M, Fraser W (2007) Marine pelagic ecosystems: the West Antarctic Peninsula. Philos Trans R Soc B 362:67–94. doi:10.1098/rstb.2006.1955
Ettershank G (1983) Age structure and cyclical annual size change in the Antarctic krill, *Euphausia superba* Dana. Polar Biol 2:189–193
Ettershank G (1984) A new approach to the assessment of longevity in Antarctic krill *Euphausia superba*. J Crustac Biol 4:295–305
Ettershank G (1985) Population age structure in males and juveniles of the Antarctic krill, *Euphausia superba* Dana. Polar Biol 4:199–201
Ettershank G, Macdonnell I, Croft R (1982) The estimation of age in the fleshfly *Sarcophaga bullata* using the age pigment lipofuscin. Aust J Zool 31:131–138
Everson I (1977a) The living resources of the Southern Ocean, GLO/SO/77/1. FAO Southern Ocean Fisheries Survey Programme, Rome, pp 1–156
Everson I (1977b) The living resources of the Southern Ocean. Southern Ocean fisheries survey programme, FAO Document GLO/S0/77/1. FAO, Rome
Everson I (1992) Managing Southern Ocean krill and fish stocks in a changing environment. Philos Trans Roy Soc London B 338:311–317
Everson I (2000) The Southern Ocean food web. In: Everson I (ed) Krill: biology, ecology and fisheries. Blackwell Science, Oxford, pp 182–227

Fach BA, Hofmann EE, Murphy EJ (2002) Modeling studies of Antarctic krill *Euphausia superba* during transport across the Scotia Sea. Mar Ecol Prog Ser 231:187–202

Fach BA, Hofmann EE, Murphy EJ (2006) Transport of Antarctic krill (*Euphausia superba*) across the Scotia Sea. Part II: Krill growth and survival. Deep-Sea Res 53:1011–1043

Fach BA, Meyer B, Wolf-Gladrow D, Bathmann U (2008) Biochemically based modeling study of Antarctic krill *Euphausia superba* growth and development. Mar Ecol Prog Ser 360:147–161

Färber-Lorda J (1991) Multivariate approach to the morphological and biochemical differentiation of *Euphausia superba* and *Thysanoessa macrura*. Deep-Sea Res 38:771–779

Färber-Lorda J (1994) Length–weight relationships and coefficient of condition of *Euphausia superba* and *Thysanoessa macrura* (Crustacea: Euphausiacea) in southwest Indian Ocean during summer. Mar Biol 118:645–650

Färber-Lorda J, Beier E, Mayzaud P (2009) Morphological and biochemical differentiation in Antarctic krill. J Mar Syst 78:525–535

Fielding S, Watkins JL, Cossio A, Reiss C, Watters G, Calise L, Skaret G et al (2011) The ASAM 2010 assessment of krill biomass for area 48 from the Scotia Sea CCAMLR 2000 synoptic survey, WG-EMM 11/20. CCAMLR, Hobart

Fielding S, Watkins JL, Trathan PN, Enderlein P, Waluda CM, Stowasser G, Tarling GA, Murphy EJ (2014) Interannual variability in Antarctic krill (*Euphausia superba*) density at South Georgia, Southern Ocean: 1997–2013. ICES J Mar Sci. doi:10.1093/icesjms/fsu104

Flores H, van Franeker JA, Siegel V, Haraldsson M, Strass V, Meesters EH, Bathmann U, Wolff WJ (2012) The association of Antarctic krill *Euphausia superba* with the under-ice habitat. PLoS ONE 7:2. doi:10.1371/journal.pone.0031775

Fraser WR, Trivelpiece WZ (1995) Palmer LTER: relationships between variability in sea ice coverage, krill recruitment, and the foraging ecology of Adélie penguins. Antarct J US 30:271–272

Fraser WR, Trivelpiece WZ, Ainley DG, Trivelpiece SG (1992) Increases in Antarctic penguin populations: reduced competition with whales or a loss of sea ice due to global warming? Polar Biol 11:525–531

Hamner WM (1984) Aspects of schooling in *Euphausia superba*. J Crustac Biol 4:19–25

Hartnoll RG (2001) Growth in crustacea—twenty years on. Hydrobiologia 449:111–122

Hewitt R (2000) An index of per capita recruitment. CCAMLR Sci 7:179–196

Hewitt RP, Linen Low EH (2000) The fishery on Antarctic krill: defining an ecosystem approach to management. Rev Fish Sci 8:235–298

Hewitt RP, Watkins J, Naganobu M, Sushin V, Brierley AS, Demer D, Kasakina S, Takao Y, Goss C, Malyshko A, Brandon M, Kawaguchi S, Siegel V, Trathan P, Emery J, Everson I, Miller D (2004) Biomass of Antarctic krill in the Scotia Sea in January/February 2000 and its use in revising an estimate of precautionary yield. Deep-Sea Res 51:1215–1236

Hjort J (1914) Fluctuations in the great fisheries of northern Europe, viewed in the light of biological research. Rapp Procés-Verbaux des Réunions du Conseil Permanent International pour l'Exploration de la Mer 20:1–228

Hofmann EE, Lascara CM (2000) Modeling the growth dynamics of Antarctic krill *Euphausia superba*. Mar Ecol Prog Ser 194:219–231

Hofmann EE, Murphy EJ (2004) Advection, krill, and Antarctic marine ecosystems. Antarct Sci 16:487–499

Hofmann EE, Klinck JM, Locarini RA, Fach B, Murphy EJ (1998) Krill transport in the Scotia Sea and environs. Antarct Sci 10:406–415

Hosie GW, Ikeda T, Stolp M (1988) Distribution, abundance and population structure of the Antarctic krill (*Euphausia superba* Dana) in the Prydz Bay region, Antarctica. Polar Biol 8:213–224

Huntley ME, Nordhausen W, Lopez MDG (1994) Elemental composition, metabolic activity and growth of Antarctic krill *Euphausia superba* during winter. Mar Ecol Prog Ser 107:23–40

Ikeda T (1985) Life history of Antarctic krill *Euphausia superba*: a new look from an experimental approach. Bull Mar Sci 37:599–608

Ikeda T, Dixon P (1982a) Body shrinkage as a possible overwintering mechanism of the Antarctic krill, *Euphausia superba* Dana. J Exp Mar Biol Ecol 62:143–151
Ikeda T, Dixon P (1982b) Observations on moulting in Antarctic krill (*Euphausia superba* Dana). Aust J Mar Freshwat Res 33:71–76
Ikeda T, Thomas PG (1987) Moulting interval and growth of juvenile Antarctic krill (*Euphausia superba*) fed different concentrations of the diatom *Phaeodactylum tricornutum* in the laboratory. Polar Biol 7:339–343
Ikeda T, Dixon P, Kirkwood J (1985) laboratory observations of moulting, growth and maturation in Antarctic krill (*Euphausia superba* Dana). Polar Biol 4:1–8
Ivanov BG (1970) On the biology of the Antarctic krill *Euphausia superba*. Mar Biol 7:340–351
Jazdzewski K, Dzik J, Porebski J, Rakusa-Suszczewski S, Witek A, Wolnomiejski N (1978) Biological and populational studies on krill near South Shetland Islands, Scotia Sea and South Georgia in the summer 1976. Pol Arch Hydrobiol 25:607–631
Kanda K, Takagi K, Seki Y (1982) Movement of the larger swarms of Antarctic krill *Euphausia superba* population off Enderby Land during 1976–1977 season. J Tokyo Univ Fish 68:25–42
Kawaguchi S, Satake M (1994) Relationship between recruitment of the Antarctic krill and the degree of ice cover near the South Shetland Islands. Fish Sci 60:123–124
Kawaguchi S, Candy SG, King R, Naganobu M, Nicol S (2006) Modelling growth of Antarctic krill. I. Growth trends with sex, length, season, and region. Mar Ecol Prog Ser 306:1–15
Kawaguchi S, Finley LA, Jarman S, Candy SG, Ross RM, Quetin LB, Siegel V, Trivelpiece W, Naganobu M, Nicol S (2007) Male krill grow fast and die young. Mar Ecol Prog Ser 345:199–210
Kilada R, Sainte-Marie B, Rochette R, Davis N, Vanier C, Campana S (2012) Direct determination of age in shrimps, crabs, and lobsters. Can J Fish Aquat Sci 69:1728–1733
Knox GA (1994) The biology of the Southern Ocean. Cambridge University Press, New York, 444 p
Lockyer C (1973) Wet weight, volume and length correlation in the Antarctic krill *Euphausia superba*. Appendix to N.A. Mackintosh: distribution of post-larval krill in the Antarctic. Discov Rep 36:152–155
Loeb V, Siegel V, Holm-Hansen O, Hewitt R, Fraser W, Trivelpiece W, Trivelpiece S (1997) Effects of sea ice extent and krill or salp dominance on the Antarctic food web. Nature 387:897–900
Loeb V, Hofmann EE, Klinck JM, Holm-Hansen O, White WB (2009) ENSO and variability of the Antarctic Peninsula pelagic marine ecosystem. Antarct Sci 21:135–148
MacDonald PDM, Pitcher TJ (1979) Age-groups from size-frequency data: a versatile and efficient method of analysing distribution mixtures. J Fish Res Board Can 36:987–1001
Mackintosh NA (1967) Maintenance of living *Euphausia superba* and frequency of moults. Norsk Hvalfangsttid 56:97–102
Mackintosh NA (1972) Life cycle of Antarctic krill in relation to ice and water conditions. Discov Rep 36:1–94
Mackintosh NA (1973) Distribution of post-larval krill in the Antarctic. Discov Rep 36:95–156
Mackintosh NA (1974) Sizes of Krill eaten by whales in the Antarctic. Discov Rep 36:157–178
Makarov RR (1975) Study of the repeated maturation of female euphausiids (Eucarida, Euphausiacca). Zoologichesky Zhurnal 54(5):1975
Marr JWS (1962) The natural history and geography of the Antarctic krill (*Euphausia superba* Dana). Discov Rep 32:33–464
Marschoff E, Rombolá E, Coria N (2008) The uropod as a proxy for total length distribution in Antarctic krill: an assessment of different models. Polar Biol 31:717–724
McClatchie S (1988) Food-limited growth of *Euphausia superba* in Admiralty Bay, South Shetland Islands, Antarctica. Cont Shelf Res 8:329–345
McClatchie S, Rakusa-Suszczewski S, Filcek K (1991) Seasonal growth and mortality of *Euphausia superba* in Admiralty Bay, South Shetland Islands, Antarctica. ICES J Mar Sci 48:335–342

McWhinnie MA, Denys CJ (1978) Biological studies of Antarctic krill, austral summer, 1977–78. Antarct J US 13:133–135

McWhinnie MA, Denys CJ, Parkin R, Parkin MA, Marciniak P (1979) Biological investigation of *Euphausia superba* (krill). Antarct J US 14:163–164

Meredith MP, King JC (2005) Rapid climate change in the ocean west of the Antarctic Peninsula during the second half of the 20th century. Geophys Res Lett 32:L19604. doi:10.1029/2005GL024042

Meyer B (2011) The overwintering of Antarctic krill, *Euphausia superba* from an ecophysiological perspective. Polar Biol 35:15–37. doi:10.1007/s00300-011-1120-0

Miller DGM (1983) Variation in body length measurement of *Euphausia superba* Dana. Polar Biol 2:17–20

Miller DGM, Hampton I (1989) Biology and ecology of the Antarctic krill (*Euphausia superba* Dana): a review. BIOMASS Sci Ser 9:1–166

Morris DJ, Priddle J (1984) Observations on the feeding and moulting of the Antarctic Krill, *Euphausia superba* Dana, in winter. Br Antarct Surv Bull 65:57–63

Morris DJ, Watkins JL, Ricketts C, Buchholz F, Priddle J (1988) An assessment of the merits of length and weight measurements of Antarctic krill *Euphausia superba*. Br Antarct Surv Bull 79:27–50

Murano M, Segawa S, Kato M (1979) Moulting and growth of the Antarctic krill in the laboratory. Trans Tokyo Univ Fish 3:99–106

Murphy EJ, Thorpe SE, Watkins JL, Hewitt R (2004) Modeling the krill transport pathways in the Scotia Sea: spatial and environmental connections generating the seasonal distribution of krill. Deep-Sea Res 51:1435–1456

Myers RA (1998) When do environment-recruit correlations work? Rev Fish Biol Fish 8:285–305

Nemoto T (1959) Food of baleen whales in the northern Pacific. Sci Rep Whales Res Inst Tokyo 12:33–89

Nicol S (1987) Some limitations on the use of the lipofuscin ageing technique. Mar Biol 93:609–614

Nicol S (2000) Understanding krill growth and aging: the contribution of experimental studies. Can J Fish Aquat Sci 57:168–177

Nicol S, Foster J (2016) The fishery for Antarctic krill – its current status and management regime. In: Siegel V (ed) The biology and ecology of Antarctic krill, *Euphausia superba* Dana, 1850. Advances in polar science. Springer, Cham, pp 387–422

Nicol S, Stolp M (1990) A refinement of the moult-staging technique for Antarctic krill (*Euphausia superba*). Mar Biol 104:169–173

Nicol S, Stolp M, Cochran T, Geijsel P, Marshall J (1992) Growth and shrinkage of Antarctic krill *Euphausia superba* from the Indian Ocean sector of the Southern Ocean during summer. Mar Ecol Prog Ser 89:175–181

Nicol S, Pauly T, Bindoff NL, Wright S, Thiele D, Hosie GW, Strutton PG, Woehler E (2000a) Ocean circulation off east Antarctica affects ecosystem structure and sea-ice extent. Nature 406:504–507

Nicol S, Kitchener J, King R, Hosie G, de la Mare WK (2000b) Population structure and condition of Antarctic krill (*Euphausia superba*) off East Antarctica (80-1501E) during the Austral summer of 1995/1996. Deep-Sea Res II 47:2489–2517

Nicol S, Foster J, Kawaguchi S (2011) The fishery for Antarctic krill- recent developments. Fish Fish 13:30–40. doi:10.1111/j.1467-2979.2011.00406.x

Pakhomov EA (1995a) Demographic studies of Antarctic krill *Euphausia superba* in the cooperation and Cosmonaut Seas (Indian sector of the Southern Ocean). Mar Ecol Prog Ser 119:45–61

Pakhomov EA (1995b) Natural age-dependent mortality rates of Antarctic krill *Euphausia superba* Dana in the Indian sector of the Southern Ocean. Polar Biol 15:69–71

Pakhomov EA (2000) Demography and life cycle of Antarctic krill, *Euphausia superba*, in the Indian sector of the Southern Ocean: long-term comparison between coastal and open-ocean regions. Can J Fish Aquat Sci 57:68–90
Pennington M (1983) Efficient estimators of abundance, for fish and plankton surveys. Biometrics 39:281–286
Poleck TP, Denys CJ (1982) Effect of temperature on the moulting, growth and maturation of the Antarctic krill *Euphausia superba* (Crustacea: Euphausiacea) under laboratory conditions. Mar Biol 70:255–265
Priddle J, Croxall JP, Everson I, Heywood RB, Murphy EJ, Prince PA, Sear CB (1988) Large-scale fluctuations in distribution and abundance of krill – a discussion of possible causes. In: Sahrhage D (ed) Antarctic Ocean, resources variability. Springer, Berlin, pp 169–182
Pyper W (2002) Age riddle confounds fair fishing for krill. ECOS 111:3
Quetin LB, Ross RM (1991) Behavioural and physiological characteristics of the Antarctic krill, *Euphausia superba*. Am Zool 31:49–63
Quetin LB, Ross RM (2003) Episodic recruitment in Antarctic krill *Euphausia superba* in the Palmer LTER study region. Mar Ecol Prog Ser 259:185–200
Quetin LB, Ross RM, Fritsen CH, Vernet M (2007) Ecological responses of Antarctic krill to environmental variability: can we predict the future? Antarct Science 19:253–266
Reid K (2001) Growth of Antarctic krill *Euphausia superba* at South Georgia. Mar Biol 138:57–62
Reid K, Watkins JL, Croxall JP, Murphy EJ (1999) Krill population dynamics at South Georgia 1991–1997, based on data from predators and nets. Mar Ecol Prog Ser 177:103–114
Reid KR, Murphy EJ, Loeb V, Hewitt RP (2002) Krill population dynamics in the Scotia Sea: variability in growth and mortality within a single population. J Mar Sci 36:1–10
Reiss CS (2008) Updated krill recruitment data for the Elephant Island region of the South Shetland Islands, Antarctica; 2002–2008. WG-EMM-08/41
Reiss CS, Kilada R, Kawaguchi S (2015) Direct ageing of Antarctic Krill (*Euphausia superba*) – potential utility of eyestalk sections for age determination. CCAMLR WG-EMM 15/46
Ricker WE (1975) Computation and interpretation of the biological statistics of fish populations. Bull Fish Res Board Can 191:1–382
Rosenberg AA, Beddington JR, Basson M (1986) Growth and longevity of krill during the first decade of pelagic whaling. Nature 324:152–154
Ross RM, Quetin LB, Baker KS, Vernet M, Smith RC (2000) Growth limitation in young *Euphausia superba* under field conditions. Limnol Oceanogr 45:31–43
Ross RM, Quetin LB, Newberger T, Shaw CT, Jones JL, Oakes SA, Moore KJ (2014) Trends, cycles, interannual variability for three pelagic species west of the Antarctic Peninsula 1993–2008. Mar Ecol Prog Ser 515:11–32
Ruud JT (1932) On the biology of southern Euphausiidae. Hvalradets Skr 2:1–105
Saba GK, Fraser WR, Saba VS, Iannuzzi RA, Coleman KE, Doney SC, Ducklow HW, Martinson DG, Miles TN, Patterson-Fraser DL, Stammerjohn SE, Steinberg DK, Schofield OM (2014) Winter and spring controls on the summer food web of the coastal West Antarctic Peninsula. Nat Commun 5:4318. doi:10.1038/ncomms5318
Sahrhage D (1978) Zur Laengen-Gewichts-Beziehung beim antarktischen Krill (*Euphausia superba*). Meeresforschung 26:47–49
Sahrhage D (1988) Some indications for environmental and krill resources variability in the Southern Ocean. In: Sahrhage D (ed) Antarctic Ocean and resources variability. Springer, Berlin/Heidelberg/New York, pp 33–40
Sala A, Azzali M, Russo A (2002) Krill of the Ross Sea: distribution, abundance and demography of *Euphausia superba* and *Euphausia crystallorophias* during the Italian Antarctic Expedition (January–February 2000). Sci Mar 66:123–133
Saville A (1977) Survey methods of appraising fisheries resources, FAO fisheries technical paper 171. FAO, Rome, 76 p
Sheehy MRJ (1990) Widespread occurrence of fluorescent morphological lipofuscin in the crustacean brain. J Crustac Biol 10:613–622

Sheehy MRJ (2008) Questioning the use of biochemical extraction to measure lipofuscin for age determination of crabs: Comment on Ju et al. (1999, 2001). Mar Ecol Prog Ser 353:303–306

Shelton OA, Kinzey D, Reiss C, Munch S, Watters G, Mangel M (2013) Among-year variation in growth of Antarctic krill *Euphausia superba* based on length-frequency data. Mar Ecol Prog Ser 481:53–67

Shin H-C, Nicol S (2002) Using the relationship between eye diameter and body length to detect the effects of long-term starvation on Antarctic krill *Euphausia superba*. Mar Ecol Prog Ser 239:157–167

Siegel V (1982) Relationship of various length measurements of *Euphausia superba* Dana. Meeresforschung 29:114–117

Siegel V (1985) The distribution pattern of krill, *Euphausia superba*, west of the Antarctic Peninsula in February 1982. Meeresforschung 30:292–305

Siegel V (1986) Investigations on the biology of Antarctic krill *Euphausia superba*, in the Bransfield Strait and adjacent areas (In German). Mitt Inst Seefisch Hamb 38:1–244

Siegel V (1987) Age and growth of Antarctic Euphausiacea (Crustacea) under natural conditions. Mar Biol 96:483–495

Siegel V (1988) A concept of seasonal variation of krill (*Euphausia superba*) distribution and abundance west of the Antarctic Peninsula. In: Sahrhage D (ed) Antarctic ocean and resources variability. Springer, Berlin, pp 219–230

Siegel V (1989) Winter and spring distribution and status of the krill stock in Antarctic Peninsula waters. Arch Fisch-wiss 39:45–72

Siegel V (1991) Estimation of krill (*Euphausia superba*) mortality and production rate in the Antarctic Peninsula region, SC-CAMLR-SPP/8. CCAMLR, Hobart, pp 159–176

Siegel V (1992) Assessment of the krill (*Euphausia superba*) spawning stock of the Antarctic Peninsula. Arch Fisch 41:101–130

Siegel V (2000a) Krill (Euphausiacea) demography and variability in abundance and distribution. Can J Fish Aquat Sci 57:151–167

Siegel V (2000b) Krill (Euphausiacea) life history and aspects of population dynamics. Can J Fish Aquat Sci 57:130–150

Siegel V (2012) Krill stocks in high latitudes of the Antarctic Lazarev Sea: seasonal and interannual variation in distribution, abundance and demography. Polar Biol 35:1151–1177

Siegel V (2016) Introducing Antarctic krill *Euphausia superba* Dana, 1850. In: Siegel V (ed) The biology and ecology of Antarctic krill, *Euphausia superba* Dana, 1850. Advances in Polar Science. Springer, Cham, pp 1–20

Siegel V, Loeb V (1994) Length and age at maturity of Antarctic krill. Antarct Sci 6:479–482

Siegel V, Loeb V (1995) Recruitment of Antarctic krill *Euphausia superba* and possible causes for its variability. Mar Ecol Prog Ser 123:45–56

Siegel V, Watkins J (2016) Distribution, biomass and demography of Antarctic krill, *Euphausia superba*. In: Siegel V (ed) The biology and ecology of Antarctic krill, *Euphausia superba* Dana, 1850. Advances in Polar Science. Springer, Cham, pp 21–100

Siegel V, Bergström B, Strömberg JO, Schalk P (1990) Distribution, size frequencies and maturity stages of krill, *Euphausia superba*, in relation to sea ice in the northern Weddell Sea. Polar Biol 10:549–557

Siegel V, Bergström B, Mühlenhardt-Siegel U, Thomasson M (2002) Demography of krill in the Elephant Island area during summer 2001 and its significance for stock recruitment. Antarct Sci 14:162–170

Siegel V, Ross RM, Quetin LB (2003) Krill (*Euphausia superba*) recruitment indices from the western Antarctic Peninsula: are they representative of larger regions? Polar Biol 26:672–679

Siegel V, Loeb V, Bergstrom B, Scholing S, Haraldsson, M, Kitchener J, Vortkamp M (2004) Demography of Antarctic krill in the Lazarez Sea (Subarea 48.6) and the Elephant Island area (Subarea 48.1) in 2004. SC-CAMLR WG-EMM-04/23

Sinclair M (1988) Marine populations: an essay on population regulation and speciation. Washington Sea Grant University, Washington Press, Seattle, 252 p

Stammerjohn SE, Martinson DG, Smith RC, Iannuzzi RA (2008) Sea ice in the western Antarctic Peninsula region: spatio-temporal variability from ecological and climate change perspectives. Deep-Sea Res II 55:2041–2058
Stearns SC (1992) The evolution of life histories. Oxford University Press, London, 249 pp
Strelnikova VM (1985) Weight-length relationships and calorific value of Antarctic krill. Oceanolgy 5:45–49
Sun S, de la Mare W, Nicol S (1995) The compound eye as an indicator of age and shrinkage in Antarctic krill. Antarct Sci 7:387–392
Tarling GA, Shreeve RS, Hirst AG, Atkinson A, Pond D, Murphy EJ, Watkins JL (2006) Natural growth rates in Antarctic krill (*Euphausia superba*): I. Improving methodology and predicting intermoult period. Limnol Oceanogr 51:959–972
Tarling GA, Cuzin-Roudy J, Thorpe SE, Shreeve RS, Ward P, Murphy EJ (2007) Recruitment of Antarctic krill *Euphausia superba* in the South Georgia region: adult fecundity and the fate of larvae. Mar Ecol Prog Ser 331:161–179
Tarling GA, Hill S, Peat H, Fielding S, Reiss C, Atkinson A (2016) Sex-dependent growth and shrinkage in Antarctic krill (*Euphausia superba* Dana). Mar Ecol Prog Ser 547:61–78
Thomas PG, Ikeda T (1987) Sexual regression, shrinkage, rematuration and growth of spent female *Euphausia superba* in the laboratory. Mar Biol 95:357–363
Thorpe SE, Murphy EJ, Watkins JL (2007) Circumpolar connections between Antarctic krill (*Euphausia superba* Dana) populations: Investigating the roles of ocean and sea ice transport. Deep-Sea Res I 54:792–810
Trathan PN, Hill S (2016) The importance of krill predation in the Southern Ocean. Siegel V (ed) The biology and ecology of Antarctic krill, *Euphausia superba* Dana, 1850. Advances in Polar Science. Springer, Cham, pp 321–350
Virtue P, Nichols PD, Nicol S, Hosie G (1996) Reproductive trade-off in male Antarctic krill *Euphausia superba*. Mar Biol 126:521–527
Virtue P, Kawaguchi S, McIvor J, Nicol S, Wotherspoon S, Brown M, Casper R, Davenport S, Finlay L, Foster J, Yoshida T, Yoshiki T (2010) Krill growth and condition in Western Indian Ocean sector of the Southern Ocean 30–801E in austral summer 2006. Deep-Sea Res 57:948–955
Watkins J (1999) A composite recruitment index to describe interannual changes in the population structure of Antarctic krill at South Georgia. CCAMLR Sci 6:71–84
Watkins JL, Buchholz F, Priddle J, Morris DJ, Ricketts C (1992) Variation in reproductive status of Antarctic krill swarms; evidence for a size-related sorting mechanism? Mar Ecol Prog Ser 82:163–174
Watkins JL, Murray AWA, Daly HI (1997) Variation in the distribution of Antarctic krill *Euphausia superba* around South Georgia. Mar Ecol Prog Ser 188:149–160
Wiedenmann J, Cresswell K, Mangel M (2008) Temperature-dependent growth of Antarctic krill: predictions for a changing climate from a cohort model. Mar Ecol Prog Ser 358:191–202
Wiedenmann J, Cresswell K, Mangel M (2009) Connecting recruitment of Antarctic krill and sea ice. Limnol Oceanogr 54:799–811
Zikov LA, Slepokurov VA (1982) The equation for the estimation of the natural mortality of fishes (for example fish from Lake Endir). Ribnoe chozjaistvo 3:36–37 (in Russian)

Chapter 4
Physiology of *Euphausia superba*

Bettina Meyer and Mathias Teschke

Abstract Since the 1920s, *E. superba* is one of the best studied species in the Southern Ocean in terms of their general biology. The main driver for this research focus has been the fisheries' requirements for stock forecasting and conservation measures. Nowadays this is joined by concerns over climate change effects and the requirement to take a more holistic view to understand food web structures. So far, however, we do not have a clear understanding of the physiological response of krill and hence their adaptability to cope with ongoing environmental changes, caused by the anthropogenic carbon emissions. This is due to the extreme lack of intense studies on krill physiology, especially of their larval stages in relation to their seasonal environment. A major aim of this book chapter is on the one hand to summarize how physiological functions such as lipid accumulation and utilisation, metabolic activity and growth change with ontogeny and season and to demonstrate which environmental factors are the main drivers for seasonal variability of these functions in adult and larval krill. On the other hand, we draw the attention to the importance of photoperiod (day length) as an entrainment cue for endogenous rhythms and clocks in the life cycle of krill. Furthermore, we give an overview of the current knowledge on the impact of elevated seawater temperature and ocean acidification on krill.

Keywords Krill physiology • Larval krill physiology • Krill growth • Photoperiod

---

B. Meyer (✉)
Alfred Wegener Institute, Helmholtz Centre for Polar and Marine Research, Section Polar Biological Oceanography, Am Handelshafen 12, D-27570 Bremerhaven, Germany

Institute for Chemistry and Biology of the Marine Environment, Carl von Ossietzky University of Oldenburg, Carl-von-Ossietzky-Straße 9-11, D-26111 Oldenburg, Germany
e-mail: bettina.meyer@awi.de

M. Teschke
Institute for Chemistry and Biology of the Marine Environment, Carl von Ossietzky University of Oldenburg, Carl-von-Ossietzky-Straße 9-11, D-26111 Oldenburg, Germany
e-mail: mathias.teschke@awi.de

V. Siegel (ed.), *Biology and Ecology of Antarctic Krill*, Advances in Polar Ecology,
DOI 10.1007/978-3-319-29279-3_4

4.1 General

E. superba is one of the best studied Antarctic species in terms of general biology and ecology, with emphasis on growth, biochemical composition, reproduction and on their distribution and abundance in the Southern Ocean. The main drivers for these research foci have been the fisheries' requirements for stock forecasting and conservation measures. Nowadays, this is joined by concerns over climate change effects and the requirement to take a more holistic view to understand food web structures (Quetin and Ross 1991; Teschke et al. 2008; Meyer 2012). While there is considerable scientific knowledge about krill's biology and ecology, relative little is known how physiological functions are mediated in the different ontogenetic stages in krill.

Physiology aims to understand the mechanisms how organisms are functioning at different organismic levels (genes, cells, the whole organism) and how they adapt to vastly different environmental condition.

Table 4.1 summarises physiological studies that reviewed previous and recent findings and mirrored the emphasis that physiological studies had at this time. In the 1980s, the focus was to calculate an energy budget of adult krill in summer (Clarke and Morris 1983; Miller and Hampton 1989), whereas in the 1990s the winter period was identified as important part in krill's annual cycle to understand its recruitment success as well as the relation of adult and larval krill with the winter sea ice (Quetin and Ross 1991; Ross and Quetin 1991; Quetin et al. 1994). Two studies, focusing on long term abundance data of adult krill in relation to environmental parameters, strongly influenced the scientific view of the sensitivity of krill to climate induced environmental changes and influenced the further directions in physiological krill research. Siegel and Loeb (1995) and Loeb et al. (1997) demonstrated first that the recruitment success and, hence population size of krill, are

Table 4.1 Overview of physiological papers that summarised recent findings at the time they were published

Researcher	Topic addressed
Clarke and Morris (1983)	Adult krill in summer: oxygen consumption, excretion and filtration rates, growth, moulting and reproduction
Miller and Hampton (1989)	Adult krill in summer: metabolic activity at different temperatures, excretion, filtration and feeding, growth, moulting, reproduction, biochemistry of egg production
Quetin and Ross (1991)	Adult krill: foraging and gathering ability, overwintering mechanisms
Ross and Quetin (1991)	Larval krill: starvation tolerance, growth, lipid content and shrinkage in winter
Quetin et al. (1994)	Adult krill: filtration and digestion, reproduction, seasonal growth, respiration, and lipid content, role of sea ice in krill energetics
Meyer (2012)	Adult and larval krill: seasonality of metabolic and feeding activity, biochemical composition, and growth, overwintering strategies, the importance of sea ice

linked to extent and duration of winter sea ice cover, with low ice years related to poor recruitment. The investigations of Atkinson et al. (2004) showed a krill decline of almost 80 % since the mid 1970s in the SW Atlantic sector in concert with a decline in winter sea ice extent and duration (see also Chap. 2, Siegel and Watkins 2016). Both studies made us aware that sea ice seems to be important for the recruitment success of krill but the mechanistic relationship between a successful larval krill development to 1 year juveniles during winter and the winter sea ice coverage is still lacking. Therefore, it is essential to go beyond correlative studies towards a mechanistic understanding of the linkage of all life stages of krill with their environment.

Today, krill physiology aims to understand the performance of all ontogenetic stages of krill to environmental stressors caused by anthropogenic warming. In this respect it is important to investigate the seasonality of important physiological life cycle functions such as body lipid dynamic, metabolic activity, and growth in adults and larvae from different regions in the Southern Ocean in relation to environmental variables such as food availability, photoperiod (day length) and winter sea ice coverage (Meyer 2011). Krill live in a highly seasonal environment in terms of sea ice cover, food availability, and photoperiod (day length) and their annual life cycle has been synchronized to this seasonality over millions of years. However, to understand the adaptability of krill to the ongoing environmental changes caused by the global warming and to make reliable predictions as to how the population might change in the future, we have to understand the cues that are related (e.g. food quantity and quality, timing of phytoplankton spring bloom) and unrelated (photoperiod) to climate induced environmental changes and their impact on krill's annual cycle. Due to logistical constrains to obtain sufficient ship time and to keep krill in captivity for long periods over month or even over 1 year, seasonal investigations on physiological function of all ontogenetic stages of krill are rare. Table 4.2 gives an overview of seasonal investigations on larval and adult krill and illustrates why our knowledge is still limited to understand the principle mechanisms behind important life cycle functions of krill.

In the present chapter, we summarise the current knowledge on seasonal metabolic activity, feeding and growth of larval and adult krill with emphasis on the winter period and larval krill. The impact of global warming on the density of Antarctic krill suggests that the overwintering success especially of the larval stages (the new recruits) is regarded as the major single factor that dictates recruitment success, and hence population size of krill in the Southern Ocean (Siegel and Loeb 1995; Loeb et al. 1997; Atkinson et al. 2004). However, the potential overwintering mechanisms of krill, especially of their larval stages, are still poorly researched (Meyer 2011). In addition, we draw the attention on the impact of rising seawater temperature and ocean acidification on physiological functions and Krill's ontogeny, respectively and to the important role of endogenous rhythms and clocks in the function of important life cycle functions of krill.

As outlined above, the understanding of how krill survive the winter season in general and of their interaction with the sea ice in particular, is crucial to predict the impact of climatic warming on krill. It has been demonstrated that adult krill

Table 4.2 Summary of seasonal investigations of physiological functions and/or biochemical composition of larval and adult krill

Parameter analysed	Region	Season	References
Adult krill			
	Bransfield Strait, north of South Shetland Islands	Feb–Apr 1984, 1985, Aug–Sept 1985	Quetin and Ross (1991)
Growth, feeding activity (gut fluorescence, faecal pellet production), lipid content, metabolic activity	Northern Antarctic Peninsula, Eastern and Western Weddell Sea, Lazarev Sea	Oct, Nov 1983, Jan, Feb, 1985, July, Aug 1986, Oct, Nov 1986, Apr, May 1992	Hagen et al. (2001)
Metabolic and feeding activity, elemental and biochemical composition, growth	Lazarev Sea	Mar–May 2004, Nov–Jan 2005/2006, June–Aug 2006	Meyer et al. (2010)
Feeding activity	Scotia Sea, Bransfield Strait	Jan, Feb 2002, 2003, 2005, 2006, Mar 2004, Apr 2007, June–Aug 2004, July–Aug 2005, 2006, Nov 2006	Schmidt et al. (2012)
Feeding activity	Lazarev Sea, Bransfield Strait, Scotia Sea, South Georgia	Jan–Feb 2002, 2003, 2005, 2006, Mar 2004, Apr 2004, July 2005, June–Aug 2006, Nov 2006, Dec 2005	Schmidt et al. (2014)
Larval krill			
Morphometrics, growth	Bransfield Strait, Marguerite Bay	June, July 1987, July 1989, May, Sept 1991, Apr–Sep 1993, June, July 1994, June 1999	Quetin et al. (2003)
Morphometrics, elemental composition, feeding activity, growth	Marguerite Bay	Apr–June 2001, July, Aug 2002	Daly (2004)
Metabolic and feeding activity, elemental and biochemical composition, growth	Lazarev Sea	Mar–May 204, June–Aug 2006	Meyer et al. (2009)

Modified according to Meyer (2011)

employ a suite of overwintering mechanisms, which provide considerable flexibility in their response to winter conditions. Proposed mechanisms of adult krill fall into two major categories: (a) Non-feeding strategies and (b) Utilisation of food sources other than phytoplankton in the water column. The first category consists of the reduction in metabolic rates (Kawaguchi et al. 1986; Quetin and Ross 1991; Torres et al. 1994a), the utilisation of body lipids (Quetin and Ross 1991; Hagen et al. 2001), and shrinkage in size (Ikeda and Dixon 1982; Quetin and Ross 1991). Alternative food sources might be zooplankton (Huntley et al. 1994) and seabed detritus in shallow regions (Kawaguchi et al. 1986). The utilisation of ice algae by krill during winter was discussed (e.g. Hamner et al. 1983; Marshall 1988; Spirodonov 1992).

Larval krill appear during summer, develop during Antarctic winter and recruit to the postlarval population during the following spring. Recruitment success depends both that potential recruits enter winter in an advanced stage (reflecting reproductive output and growth and survivorship of larvae during summer and autumn) and on larval growth and survival during their first winter (Quetin and Ross 1991; Siegel 2005). However, up to recently, most studies on larval krill have focused on their distribution and abundance (Hempel 1981; Hempel and Hempel 1978, 1982; Makarov and Sysoyeva 1985; Hubold and Hempel 1987; Daly and Macaulay 1988, 1991; Makarov and Maslennikov 1992; Menshenina 1992; Siegel et al. 1992; Melnikov and Spirodonov 1996; Frazer et al. 1997; Siegel 2005, see also Chap. 2, Siegel and Watkins 2016). Until the end of the 1990s, we had only anecdotal information on larval krill physiology and development through winter. At this time, it was not clear if larval krill have adapted similar overwintering mechanism as the adults, although it was presumed that dependence on ice becomes progressively less with ontogeny (Quetin et al. 1996). The review of Meyer (2011) on the overwintering of krill demonstrates to what extent seasonal variation in larval and adult krill physiology is mediated by those environmental factors that show a strong seasonality in the Southern Ocean, such as food supply or day length, and which physiological functions are adopted by adults and larvae to survive the winter season.

4.2 Seasonal Physiological Functions in Adults and Larvae and the Critical Winter Period

4.2.1 Metabolic Activity

Adult krill show a clear seasonal pattern in their metabolic activity (Fig. 4.1), with highest values in late spring and summer. In autumn and winter oxygen consumption rates were 50 % and 30 % of summer rates, respectively (Meyer et al. 2010). The activities of the enzymes citrate synthase (CS) and malate dehydrogenase (MDH), proxies for metabolic activity, mirror the pattern of the oxygen uptake rates shown in Fig. 4.1 (Meyer et al. 2010). In autumn and winter, respectively, the CS and MDH activities in krill were 47 % and 39 % (CS), and 77 and 66 % (MDH) of those in summer (Meyer et al. 2010).

In contrast to the adult stage, larvae from different regions showed no significant differences in metabolic activity between summer and autumn, ranging from 0.7 to 1.4 µl O_2 mg DW h^{-1} (Ikeda 1981; Meyer et al. 2002a, b, 2003, 2009), whereas in winter the respiration rates were only half of those in summer (Daly and Macauley 1991; Daly 2004; Frazer et al. 2002a; Meyer et al. 2009), with 0.95 ± 0.28 mL O_2 mg^{-1} DW h^{-1} and 0.54 ± 0.19 mL O_2 mg^{-1} DW h^{-1} in autumn and winter,

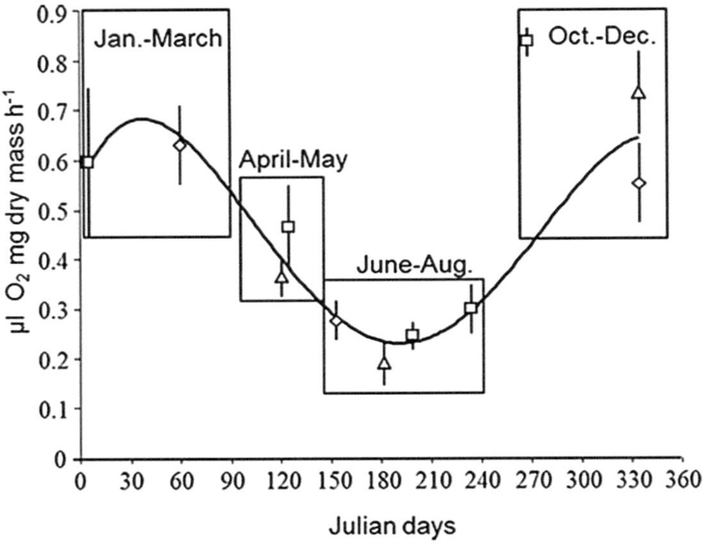

Fig. 4.1 From Meyer 2011: seasonal respiration rates of adult krill from different study sites. △ Lazarev Sea (Meyer et al. 2010), ◇ Southern Scotia-Northern Weddell Sea Torres et al. (1994b), □ Lütz-Holm Bay (Kawaguchi et al. 1986)

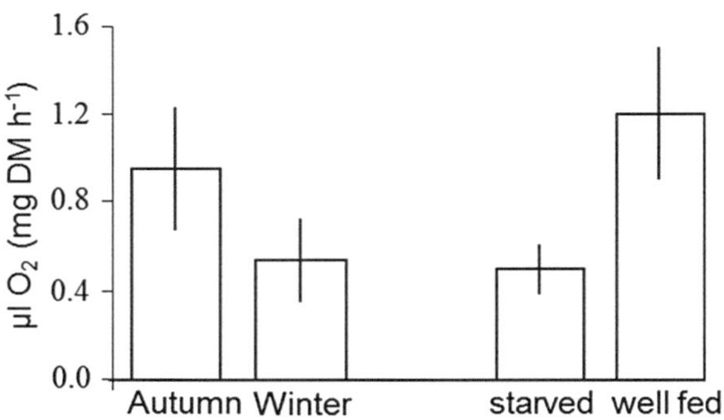

Fig. 4.2 Mean oxygen uptake rates of freshly caught furcilia larvae in autumn (1999, 2000) and winter (2006) in the Lazarev Sea and after 1 week starvation or exposure to high food concentrations

respectively. However, shipboard experiments demonstrate that the oxygen uptake rates of winter larvae are comparable with rates of starved furcilia (Meyer et al. 2002a; Frazer et al. 2002a) and winter larvae from open water areas in pack ice regions (Fig. 4.2), which is supposed to be an unfavourable feeding ground for krill larvae in winter (Quetin et al. 2003).

Based on the correlation between the size and consequently age of krill and their corresponding individual oxygen uptake rates (Fig. 4.3), it appears that the clear

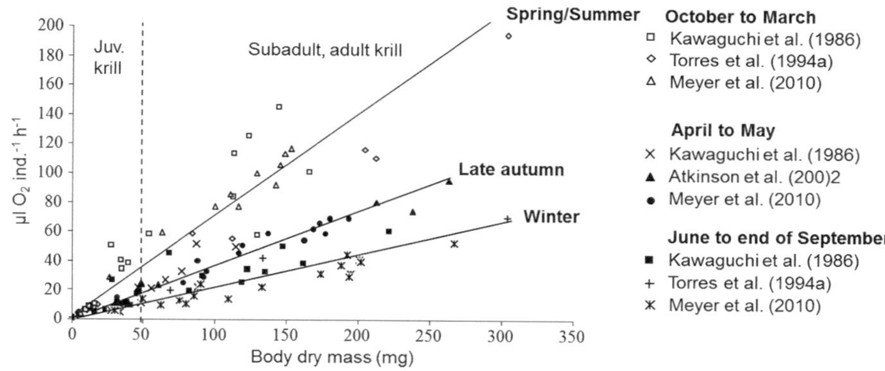

Fig. 4.3 From Meyer (2011): relationship between body dry mass and individual oxygen uptake rates in different seasons and latitudinal regions in the Southern Ocean. Juvenile, subadult and adult krill were defined according to Siegel (1987). The equations are as follows: Spring/Summer: $y = 0.69x$, $r^2 = 0.86$, $n = 36$; late autumn: $y = 0.37x$, $r^2 = 0.93$, $n = 29$; winter: $y = 0.22x$, $r^2 = 0.78$, $n = 49$. The regions were the oxygen uptake rates were measured are as follows: Lützow-Holm Bay (Kawaguchi et al. 1986), Southern Scotia-Northern Weddell Sea (Torres et al. 1994a), Lazarev Sea (Atkinson et al. 2002; Meyer et al. 2010)

seasonal pattern in metabolic activity varies during krill ontogenesis. The regression lines demonstrate that the clear seasonal pattern in metabolic activity is most pronounced in the adult stages. It is unknown to date, whether the shift of metabolic activity between seasons takes place abruptly with the final larval moult or if there is a subtle transition period during juvenile stage. Until the end of the 1990s, the prevailing view was that the differences in metabolic activity between seasons are the result of the low food supply during winter. Other studies, however, have mentioned the possibility that the seasonal changes in day light duration in the Southern Ocean might also be able to cause the seasonal variability in metabolic activity (Kawaguchi et al. 1986; Torres 1994a). We will take up this view in more detail in Sect. 4.2.3.

4.2.2 Feeding

The feeding activity of krill follows the highly seasonal pattern in food availability with high food intake during spring and summer, whereas in the winter month of low food supply, the feeding activity is reduced. The reduced feeding activity of adults in the field during winter has been indicated by a low stomach- and/or gut fullness (Morris and Priddle 1984; Kawaguchi et al. 1986; Buchholz 1989a; Lancraft et al. 1991; Daly and Macaulay 1991; Nishino and Kawamura 1994) and reduced digestive gland size compared with summer animals (Fig. 4.4). In the Western Antarctic Peninsula (WAP) region, the rates of fecal pellet production

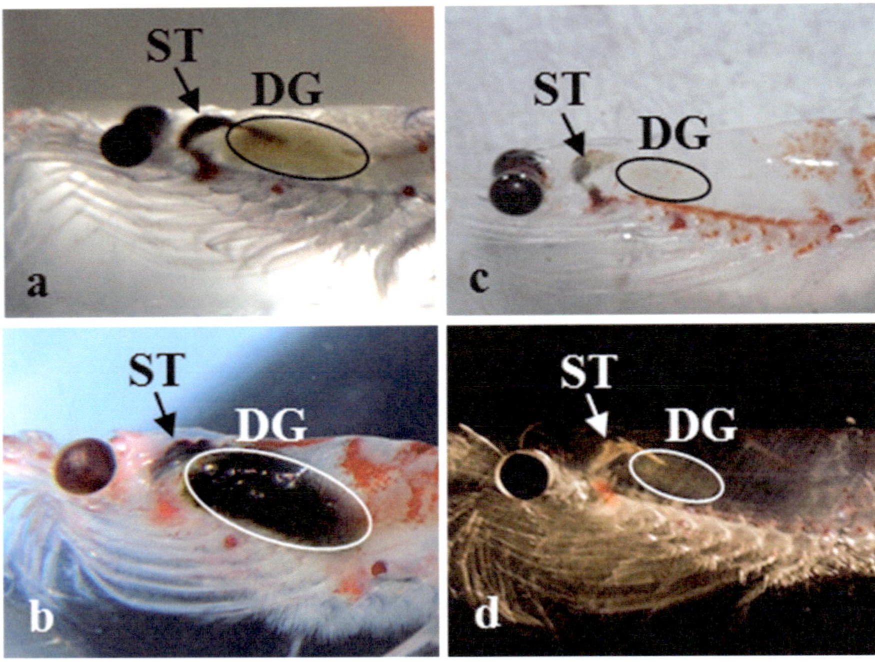

Fig. 4.4 From Meyer et al. (2010): carapace with digestive gland (DG) and stomach (ST) of freshly caught adult krill. (**a**) *Yellow* and (**b**) *green-black* DG in spring, and (**c**) *milky white* and (**d**) *pale yellow* DG in winter

and ingestion of phytoplankton of krill in winter were less than 2–3 % of summer rates (Quetin and Ross 1991). Recent investigations on the seasonal feeding activity of adult krill show that, in winter, the feeding activities seem to differ with the latitudinal location of krill. The stomach fullness of krill from the Bransfield Strait and the Lazarev Sea was only 77 % and 10 % of the ones from South Georgia respectively (Schmidt et al. 2014). The quality of food ingested by krill in the different seasons will be discussed in detail in Chap. 5, (Schmidt and Atkinson 2016).

Shipboard experiments in different seasons demonstrated that the functional response of krill to increasing food supply was significantly different between seasons (Fig. 4.5). In early December, late spring krill displayed a clear functional response with increasing food availability. In autumn and winter, when adults were exposed to similar food concentration than in late spring, they ingested only 20 % and 14 % of the maximum daily ration determined in late spring. These findings, in combination with the winter feeding activity of krill from different latitudes (Schmidt et al. 2014), indicate that another factor than food supply must be responsible for the low oxygen uptake rates during winter in adults. The impact of photoperiod on physiological functions of adult krill will be discussed in detail in the following paragraph (Sect. 4.2.3).

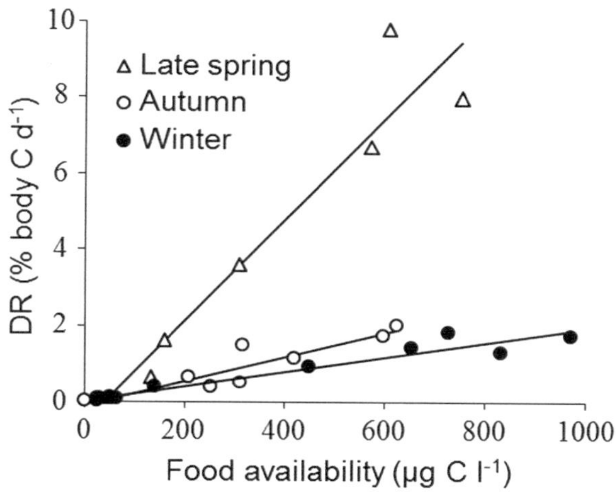

Fig. 4.5 From Meyer et al. (2010). Seasonal daily C ration (DR) as a function of food availability. Data points represent individual experiments

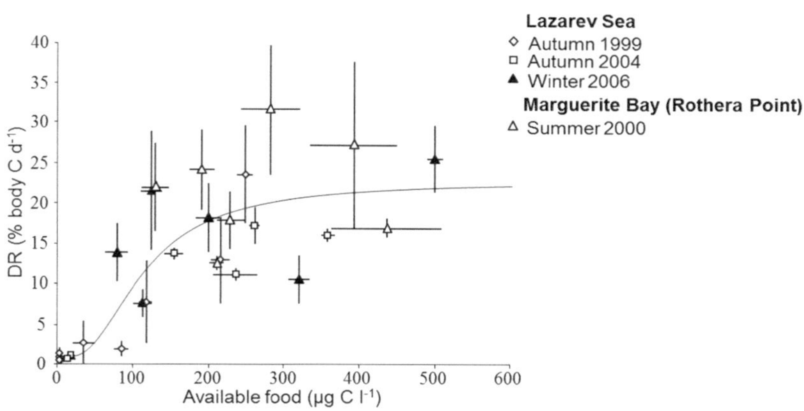

Fig. 4.6 Mean daily ration (DR) of furcilia larvae in different seasons and regions when exposed to different concentration of natural food assemblages

In contrast to adult krill, autumn and winter larvae showed a positive functional response in metabolic (Fig. 4.2) and feeding activities (Fig. 4.6) when exposed to increasing food concentrations (Meyer et al. 2002b, 2003, 2009). A maximum daily food intake was reached in larval krill at a food concentration of ca. 200 µg C l^{-1}, which corresponded to a Chl *a* concentration of 3–4 µg l^{-1} (Ross et al. 2000; Meyer et al. 2009). These findings, in combination with the respiration rates of starved and well fed larvae shown in Fig. 4.2, suggest that the low oxygen uptake rates measured in winter larvae may be a flexible adaptive behaviour to cope with the low food supply.

4.2.3 The Role of Photoperiod in Controlling Physiological Functions in Krill

Biological timing that guarantees regulation of krill's physiology and behaviour in reaction to annual fluctuations of biologically significant factors seems of particular advantage and the ability of krill to anticipate and prepare for forthcoming environmental changes and synchronize seasonal events (e.g. reproduction) to environmental fluctuations is likely to be a major factor behind its success in the Southern Ocean. Results from previous and recent studies on seasonal feeding and metabolic activity on krill point to the impact of photoperiod (day length) in mediating physiological functions in krill.

In the field Kawaguchi et al. (1986) demonstrated a decline of feeding and metabolic activity from April to May to the end of September and a slow but steady increase thereafter, at a time when food was still scarce in the water column. In addition, freshly caught krill feed on natural food assemblages in autumn and winter are unable to respond to high food concentrations despite exposure to abundant food for almost 2 weeks (Atkinson et al. 2002; Meyer et al. 2010). The maximum feeding activity of krill was only 20 % (autumn) and 14 % (winter) of summer rates (Meyer et al. 2010). Both studies were the first that were indicating a food independent component in the regulation of metabolic activity in krill and therefore suggesting that other environmental factors such as photoperiod might play a functional role. Further experiments to test this hypothesis in terms of krill's behaviour, specific enzyme activities, and gene expression levels were therefore needed.

Long-term experiments with krill in the laboratory demonstrated that feeding and metabolic activity was affected by different simulated light-dark cycles (LD) irrespective of food supply suggesting that light simulations of prolonged photoperiod can force animals into a state of increasing metabolic activity and vice versa (Teschke et al. 2007). In addition, it was demonstrated that seasonal patterns of maturity are maintained, independent of direct control by factors such as food, light or temperature (Thomas and Ikeda 1987; Kawaguchi et al. 2006; Brown et al. 2013), indicating that krill passes through a fundamental inherent seasonal transition in maturity development. Other experiments showed that light conditions of prolonged photoperiod can force krill into maturity while shortened photoperiods force animals into regression of maturity faster than under a natural light-dark regime (Teschke et al. 2008; Brown et al. 2011). These experiments strongly indicate that the overt cycles of maturity and metabolic activity represent endogenous seasonal rhythms, controlled by an endogenous timing system in krill (endogenous clock) in which photoperiod act as a main *Zeitgeber* (entrainment cue), synchronizing the clock with the natural year.

Indeed, molecular analyses (Seear et al. 2009) revealed differential gene expression of target genes (i.e. involved in metabolism) in response to photoperiodic changes indicating a signaling cascade that link the photoperiod cue to the target response. The experiments demonstrated that autumn krill exposed to a winter photoperiod showed a pattern of gene expression consistent with known features of regression and quiescence, such as the slowing of moult rate, the lowering of activity levels, and the reduction in metabolic rate while under the natural light cycle in autumn gene expression levels were significantly higher (Seear et al. 2009). The molecular response became evident after just 7 days of exposure to winter light conditions. The results indicate that krill can switch rapidly between active and quiescent states and that the photoperiodic cycle plays a key role in this process. Seear et al. (2009) also found a down-regulation of myosin when krill were exposed to the winter photoperiod, suggesting that krill are less active in winter. Together with a present study where the author makes similar findings within a freshly captured wild krill population caught during summer and winter near the Antarctic Peninsula (60° S) (Seear et al. 2012), these results indicate that wintertime metabolic depression is accompanied by lower activity levels consistent with krill entering a winter quiescent state, where foraging decreases and the level of muscular activity becomes reduced. If seasonal metabolic depression were simply a matter of specific dynamic activity in terms of feeding and digestion, it is likely that muscle use would be similar year-round, since foraging for patches of food would need to continue.

Antarctic krill with its circumpolar distribution shows a latitudinal range from approx. 51 to 74° S meaning that the seasonal photoperiodic regime as well as the accompanied seasonal food pattern experienced by krill may be very different at a given point during the season. An individual krill may be capable of travelling across many degrees of latitude within one season (Thorpe et al. 2007). Taking this into consideration a central question is whether photoperiod as a dominant *Zeitgeber* would leave krill too inflexible for the array of environmental conditions it encounters? In this sense, a comparison between two different overwintering locations for krill, namely the Antarctic Peninsula (62° S) and South Georgia (54° S) showed no differential expression levels of genes involved in respiration, motor activity or vitellogenesis, suggesting that both populations of krill are undergoing a phase of physiological quiescence at the same time of the year and thereby coordinating their general life-cycle despite the very different seasonal cycles of photoperiod prevailing at the two different latitudes (Seear et al. 2012). However, winter krill from the South Georgia region, with favorable feeding conditions even in winter, showed an up-regulation of genes involved in feeding, digestion, and immunity, indicating that krill are capable of adjusting specific functional physiological pathways to prevailing environmental conditions.

The results may also indicate that the power of photoperiod as an environmental *Zeitgeber* may lose strength towards lower latitudes (Helm et al. 2013). At lower latitudes a hierarchy of external signals or interplay of different environmental cues with day length may become more important to modulate the life-cycle of krill, and food in particular may play a prominent role as modulating factor. The importance of temporal synchronization through internal clocks in adult krill, triggered by photoperiod and food, could be recently demonstrated by use of an individual-based model (Groeneveld et al. 2015). To conclusively understand a potential interplay of photoperiod and food supply as environmental *Zeitgebers*, more complex, two-factor experiments have to be conducted in the future. These studies will have to combine different photoperiods and different levels of food supply at the same time and in which krill can feed freely. Moreover, such experiments should be performed at different times of the year to take into account the potential seasonal effects.

The clock mechanism that seems to be involved in the modulation of krill's seasonal cycles is yet not understood. Critical molecular components of an endogenous circadian clock, such as a cry-gene, have been identified in Antarctic krill and evidence of its functional significance in the clock machinery has been found (Mazzotta et al. 2010). The authors observed daily changes in krill *cry2* mRNA throughout a 24 h cycle and proposed an endogenous circadian time-keeping system in krill. A similar conclusion resulted from a laboratory study on circadian behavioural patterns of krill (Gaten et al. 2008). The first report of an endogenous circadian timing system in Antarctic krill and its link to metabolic key processes was demonstrated by Teschke et al. (2011). The study found that expression levels of the canonical clock gene *cry2* were highly rhythmic both in a light-dark cycle and in constant darkness. An ongoing rhythm even in constant darkness is evidence for the presence of an endogenous clock in krill. Surprisingly, the authors found a remarkable short circadian period, which they interpreted as a special feature of the krill's circadian clock that helps to entrain the circadian system to the extreme range of photoperiods krill are exposed to throughout the year. Furthermore, important key metabolic enzymes of krill showed bimodal circadian oscillations (9–12 h periods) in transcript abundance and enzymatic activity. Oxygen consumption of krill showed (9–12 h) oscillations that correlated with the temporal activity profile of key enzymes of aerobic energy metabolism. The 12 h bimodal oscillatory pattern in aerobic energy metabolism may reflect distinctive 12 h behavioural patterns in locomotor activity of krill. Indeed, Gaten et al. (2008) identified a 12 h rhythmic component in the diel vertical migration (DVM) of Antarctic krill during laboratory experiments, which correlate with findings from the field (Godlewska 1996), where a 12 h component in DVM of krill becomes predominant at certain times during the year when food is scarce. Often, these rhythms have been interpreted as bimodal patterns of one circadian-behavioural output.

De Pittà et al. (2013) performed the first diurnal transcriptome characterization of krill in natural conditions on a summer day in Antarctica. About 609 genes were found to be periodically expressed across the time points analysed and many of these showed a cyclic sinusoidal expression. Bimodal oscillatory patterns showed 60 % of the transcripts (371 out of 609) with a periodicity of approximately 12 h, as recently reported for key enzymes involved in aerobic energy metabolism (see above, Teschke et al. 2011). The majority (56.5 %) of the transcripts characterized by a 12 h periodicity showed two high expression peaks at 10:00 h and consequently 22:00, after the morning and evening light transitions represented, respectively, by the increase (10:00, 2.98 W/m^2) and the decrease (06:00, 0.36 W/m^2) of light irradiance at the fishing depth. About 40 % (238 out of 609) of periodically regulated genes showed only one peak of expression during the 24-h cycle with a significant enrichment (47.8 %) in the early day (01:00–06:00). Interestingly, genes belonging to a number of specific functional categories showed peak expressions limited to distinct temporal intervals, confining specific biological processes to particular times of the day. Thus, the breakdown of energy-yielding nutrients (glycolysis, the Krebs cycle and the electron transport chain) and energy storage pathways (glycogen synthesis and fatty acid synthesis) are specifically activated in the early morning, while glycogen mobilization, gluconeogenesis and fatty acids catabolism are used as a stored energy source in the evening and throughout the night. Together these results indicate the presence of a functional circadian clock in krill that controls a chronological progression of biochemical and physiological events throughout the 24-h cycle. This system is most likely of essential importance for krill as it facilitates synchronization of its physiology and behaviour to daily environmental cycles. Future studies are necessary, to define the circadian transcriptome of *E. superba* analyzing the expression levels of light: dark and dark: dark (LD/DD) entrained krill samples to get a better understanding of the physiological processes that are regulated by krill's circadian clock.

Generally, our understanding of how circadian clocks of high latitude organisms such as krill might have adapted to the strong variability in annual day length that at the extreme may ranges from constant darkness in winter to constant light in the summer is small. The results of De Pittà et al. (2013) indicate that krill do not become arrhythmic under constant photoperiodic conditions and we have some evidence that at least *cry*2 levels are still showing daily oscillations under summer conditions of constant light (Mazzotta et al. 2010). A central question in this context is the putative role of the krill's circadian clock in photoperiodic time measurement. There is some evidence that organisms might be using components of the circadian system to sense changes in day length and to mediate the photoperiodic response (Oster et al. 2002; Schultz and Kay 2003). What will be the plasticity of krill's circadian clock in the context of different seasons? Determining transcript levels of canonical clock genes in correlation with key physiologicall target genes around the circadian cycle under different LD conditions throughout simulated seasonal courses of photoperiod will potentially answer this question and should be the focus of future laboratory and field studies.

The phenology of environmental conditions to which the seasonality of krill is adapted is changing, due to the increasing anthropogenic CO_2 emission, whereas the Zeitgeber cue (photoperiod) will remain unchanged. For making reliable prediction how the krill population might change in the future we have to identify how important life cycle parameters in all developmental stages of krill are mediated by environmental cues such as photoperiod via endogenous timing systems (circadian/circannual). A prerequisite for future research in this respect is the functional characterization of the endogenous clock machinery in krill itself. The identification of the molecular components of the *E. superba* clock machinery is currently under way by combining the classical gene candidate approach (primer design on orthologous genes, 5′ and 3′ RACE) with next generation sequencing data (ESTs: (De Pittà et al. 2008; Seear et al. 2010); 454 sequencing data: (Clark et al. 2011; De Pittà et al. 2013; Meyer et al. 2015).

4.2.4 Growth and Shrinkage

In krill physiology, growth, estimated by the Instantaneous Growth Rate (IGR) method' is used as a parameter to quantify the condition of adult and larval krill in different seasons (Ross et al. 2000; Daly 2004; Meyer et al. 2009, 2010). One of the main assumptions of the IGR method is that the growth increment at moult (GI), measured in the first few days, reflects the in situ conditions experienced by krill in the previous intermoult period (Tarling et al. 2006). The IGR method was first described by Quetin and Ross (1991) and modified by Nicol et al. (1992). Large scale investigations, that related growth to environmental variability, are rare (see also Chap. 3, Reiss 2016). They investigated natural growth rates of juvenile krill in spring and summer at the WAP region (Ross et al. 2000), growth rates on adults in summer across the Southwest Atlantic sector of the Southern Ocean (Arnold et al. 2004; Atkinson et al. 2006) and in East Antarctica (Nicol et al. 2000). Krill maintain high growth in summer to take advantage of short-term fluctuations in food (Arnold et al. 2004). The variability in growth rates at this time of the year fit best with food concentration in terms of Chl*a* (Atkinson et al. 2006; Ross et al. 2000). During the highly productive summer month in the Scotia Sea growth rates in adult krill were between 0.013 and 0.32 mm d^{-1}, with mean values of 0.1 mm d^{-1} (Atkinson et al. 2006). An average growth rate of 0.1 mm d^{-1} in adult krill during summer can be found in different regions such as around South Georgia (Arnold et al. 2004), in the Lazarev Sea (Meyer et al. 2010) and off East Antarctica (Nicol et al. 2000). In juvenile krill, during a 4 years study, growth rates ranged on average between 0.072 mm d^{-1} and 0.082 mm d^{-1} and reaching maximum growth rates above a critical concentration of about 3.5 μg Chl*a* l^{-1} (Ross et al. 2000).

In austral winter, the low feeding activity in adults is accompanied by zero to low individual growth rates (Meyer et al. 2010) up to shrinkage (Quetin and Ross 1991). Previous studies, which estimated growth at the population level, reported similar results compared with the IGR method, namely zero to low growth during winter (Mackintosh 1972; Stepnik 1982; Morris and Priddle 1984; Kawaguchi et al. 1986; Buchholz et al. 1989b).

Growth rates of larval krill measured by the IGR method are calculated as the % change in uropod or telson length, based on the GI at moult, these values are highly variable from the onset of winter in April until its end in September. They range on average from 15 % growth in autumn (Pakhomov et al. 2004) to extreme low (Daly 2004; Meyer et al. 2009) or negative values (body shrinkage) in mid-winter (Quetin et al. 2003; Ross et al. 2004). A comparison of growth rate data from autumn and winter larvae from different regions have demonstrated that larval krill follow a specific growth pattern from late autumn to winter, reflecting the high variability in food supply from late autumn, during winter, until the end of winter (Fig. 4.7). The larvae had a clear positive growth in April, a steady decrease in growth until a minimum in mid-winter (from June to August) and a recurring increase of their growth rates in later winter (end September). A summary of daily larval growth rates from laboratory experiments and field data are shown in Table 4.3.

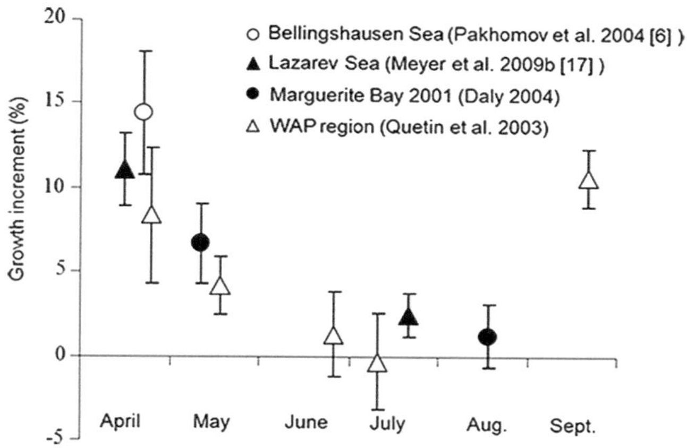

Fig. 4.7 From Meyer (2011): average growth rates of larval krill, measured with the instantaneous growth rate (IGR) method, from different regions and years given as growth increment (GI) in % change in uropod or telson length at moulting (Meyer et al. 2009). WAP means different study regions west of the Antarctic Peninsula. The data from Fig. 4.5 in Quetin et al. (2003) were extracted with the image processing program Image. The number of growth data were as follows: Pakhomov et al. (2004), April: n = 9; Meyer et al. (2009), April: n = 6, July: n = 5; Daly (2004), May: n = 7, August: n = 6; Quetin et al. (2003), April n = 3, May: n = 8, June: n = 13, July: n = 3, September: n = 16

Table 4.3 Average daily growth data of larval krill from laboratory experiments and different field seasons. WAP means Western Antarctic Peninsula

Laboratory/field/ season	Region	Daily growth rates (mm)	References
Laboratory/summer food condition		0.06–0.11	Ikeda (1984, 1985)
Laboratory/winter food condition		0.02	Ross and Quetin (1991)
Field/summer		0.06	Witek et al. (1980) and Huntley and Brinton (1991)
Field/autumn	WAP	0.013–0.036	Daly (2004)
	Lazarev Sea	0.006–0.015	Meyer et al. (2009)
Field/mid-winter (June–August)	WAP	0.017	Ross and Quetin (1991)
	WAP	0.00–0.013	Daly (2004)
	Lazarev Sea	0.001–0.002	Meyer et al. (2009)
	WAP	0.0005–0.002	Quetin et al. (2003)
Field/late winter (Sept./Oct.)	WAP	0.05 Pack ice region	Quetin et al. (2003)
	Northern Weddell Sea	0.07 Pack ice region	Daly (1990)

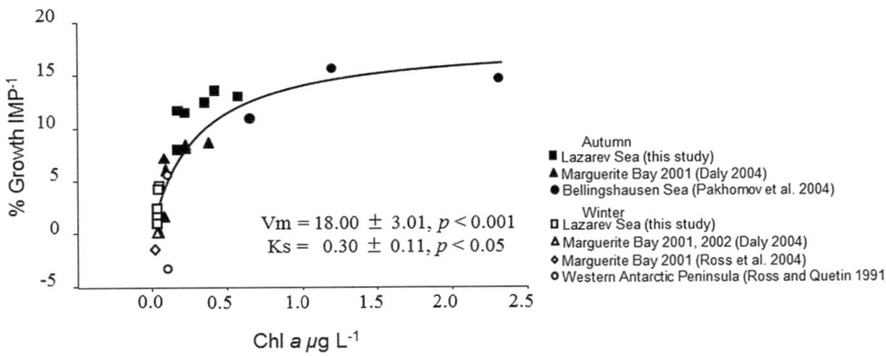

Fig. 4.8 From Meyer et al. (2009): relationship between mean chlorophyll *a* (Chl *a*) concentration and % growth per intermoult period (IMP^{-1}) of larval krill from autumn and winter in different regions in the Southern Ocean. Data are expressed as a Michaelis-Menten uptake function as follows: % growth $IMP^{-1} = 18.00 \times [\text{Chl } a/(0.30 + \text{Chl } a)]$, $r^2 = 0.68$, $n = 25$. V_m and K_s are constants representing, respectively, maximum growth and the Chl *a* concentration at which growth is half the maximum. K_s reflect the ability to grow at low food concentrations

When plotting larval growth rate data from different regions and seasons versus food supply in terms of ambient Chl *a* concentration in the surface layer, the Chl *a* is a surprisingly good overall predictor of larval growth (Fig. 4.8). This basic relationship, however, only holds because winter Chl *a* concentrations and larval growth rates are much lower than those during autumn. Therefore, the use of water column Chl *a* concentration as a food proxy to predict growth is most precise in autumn, but less reliable in winter when growth varies greatly, e.g. from 1 % to 4 % in uropod length at moulting (Daly 2004; Meyer et al. 2009), despite very low Chl *a* concentrations (<0.03 μg Chl*a* l^{-1}). Thus, some other energy sources than algal diet must be used by the larvae to explain the high variability in their growth rates during winter. Stomach analyses demonstrated that larval krill made use of heterotrophic diet during winter (Meyer et al. 2009; Töbe et al. 2010) In autumn, maximum growth was reached by the larvae at a Chl *a* concentration >1 μg l^{-1}.

During winter, larvae not only showed a high variability in their growth rates, but also a large flexibility in their growth pattern following direct and indirect developmental pathways and prolonged intermoult period (IMP). During direct development, larvae moult to the following ontogenetic stage, whereas during indirect development, they moult to the same stage (delayed development) or to an intermediate form. During both developmental pathways, winter IMP was mainly that of summer and autumn (Quetin et al. 2003; Meyer et al. 2009). Consequently, the amplitude in seasonal variation in daily growth rates (in mm per day, defined as the GI divided by the IMP) would be roughly twice that depicted in Fig. 4.7.

Indirect development was mainly observed in older furcilia stages. In mid-winter, the majority of FVI larvae moulted to the same stage, whereas FIV and FV larvae moulted to an intermediate from where an indirect developmental pathway was observed (Daly 2004; Meyer et al. 2009). This might be a strategy to survive the severe winter condition in pack ice regions with low food supply because moulting to the next advance stage means higher energy demands, which have to be covered to survive.

The debate over whether adult krill regularly shrink in the field is still ongoing, ever since the long term krill starvation study of Ikeda and Dixon was published in 1982. This study demonstrated that krill are able to shrink when food is scarce or absent. Based on this study, it was hypothesised that krill do not grow in winter (Ikeda 1985; Ikeda and Thomas 1987). However, the results available so far are controversial. In the Lazarev Sea, the low feeding activity in winter was accompanied by low individual growth rates but not shrinkage (Meyer et al. 2010), whereas shrinkage was determined in the Bellingshausen Sea (Quetin and Ross 1991). Other studies, which measured growth by traditional length-frequency analysis, reported zero or low growth (Mackintosh 1972; Stepnik 1982; Morris and Priddle 1984; Kawaguchi et al. 1986; Siegel 1987; Buchholz et al. 1989b) as well as shrinkage (Ettershank 1983). Determining shrinkage of krill by length-frequency analysis at the population level is problematic though. Different size structures of krill within a population might occur due to selective size mortality (predation) and/or due to selective size immigration/emigration and exchange of water masses (see also Chap. 3, Reiss 2016).

Shrinkage has been recorded in the laboratory and in nature in various decapod species such as crabs (Guyselman 1953; Skinner 1985; Hopkins 1985; Asakura 1992), shrimps (Lloyd and Yonge 1947), lobsters (Marshall 1945; Lindberg 1955; Sweat 1968; Little 1972) and spiny lobsters (Cockcroft and Goosen 1995), euphausiid crustaceans in general (Lasker 1966; Hosie and Ritz 1989; Nicol et al. 1992) and *E. superba* in particular (Ikeda and Dixon 1982). Such negative growth has been associated with adverse experimental conditions such as starvation (Ikeda and Dixon 1982; Nicol et al. 1992) or unfavourable environmental conditions (Cockcroft and Goosen 1995; Atkinson et al. 2006) and/or severe physiological conditions in terms of body lipid levels (Nicol et al. 2004; Meyer et al. 2010). Atkinson et al. (2006) have demonstrated that unfavourable environmental conditions in terms of Chl *a* concentration do not result in negative growth per se as models predict (e.g. Hofmann and Lascara 2000; Fach et al. 2008). The available data suggest that adult krill might be able to shrink in the field during winter but the data also suggest that such behaviour may not be the rule but rather the exception, due to unfavourable conditions. During future seasonal growth studies, condition parameters of krill such as body carbon and lipid content, or length-mass relationships should be analysed in addition to environmental conditions such as food supply in terms of Chl *a*. This would certainly help to explain when and why shrinkage in krill occurs in the field. In growth experiments with larval krill in different regions during winter 13 to up to 40 % of larvae were shrinking (Ross and Quetin 1991; Quetin et al. 2003; Ross et al. 2004; Daly 2004; Meyer et. al. 2009). In some studies shrinkage was related to very low body lipid levels ($\leq$5 % DM^{-1}) of larvae (Meyer et al. 2009), suggesting that shrinkage, similar to adult krill, is an indicator of unfavourable food conditions and hence severe physiological condition in terms of body lipid reserves.

4.2.5 Energy Provision

Energy must be provided in order for the organism to function. In this respect, the winter season is most critical due to the absence of autotrophic food in the water column for several months. There seem to be two adaptations to accomplish survival: (1) accumulation of large lipid reserves during summer for winter consumption and (2) an omnivorous feeding at low rates during winter. In the following we will outline in more detail the utilisation of energy reserves by krill, whereas their feeing behaviour during winter will be discussed in Chap. 5, (Schmidt and Atkinson 2016).

The body lipid contents of adult krill from different regions show a strong seasonality, with highest levels in late autumn (35–48 % DM^{-1}) and minimum values in mid spring (5–20 % DM^{-1}, Hagen et al. 2001), demonstrating the accumulation of large lipid reserves until the onset of winter and a depletion of these reserves during winter each year. The accumulation of energy reserves during the feeding season depends on the quantity and quality of phytoplankton during the preceding summer and autumn (Hagen et al. 2001). The lipid depletion in krill from the WAP region (Quetin and Ross 1991), the southern Scotia-northern Weddell Sea region (Torres et al. 1994b), the eastern and western Weddell Sea (Hagen et al. 2001) and from the Lazarev Sea (Meyer et al. 2010) correspond to a utilisation of body lipids in krill at a rate of 10 % DM^{-1} $month^{-1}$ from their initial values in April/May to October/November. The molar ratio of oxygen uptake to ammonia excreted (O:N ratio) provide information at which relative proportion lipids, proteins, and carbohydrates are being utilized as metabolic substrates (Ikeda 1974). When zooplankton utilizes proteins as the only metabolic substrate, the O:N ratio is generally about 8, but the ratio increased greatly the more lipids are metabolised. An O:N ratio of 24 is observed when equal amounts of protein and lipid are metabolised (Ikeda 1974; Omori and Ikeda 1984). The O:N ratio of adult summer krill is between 12 and 17 (Ikeda and Mitchell 1982; Ishii et al. 1987). The low O:N ratio of late autumn krill (average 15) indicates that they had not resorted to burning their reserves at the onset of winter (Atkinson et al. 2002), while the high O:N ratios of adult winter krill (average 66) suggests that they were relying on body lipids only until the forthcoming spring (Meyer et al. 2010). In addition, the metabolic enzyme 3-hydroxyacyl-CoA dehydrogenase (HOAD), an indicator for lipid catabolism (turnover, Auerswald et al. 2009), showed significantly higher activities in winter krill than in those from summer and autumn. This highlights the importance of lipid utilization of krill during winter (Meyer et al. 2010).

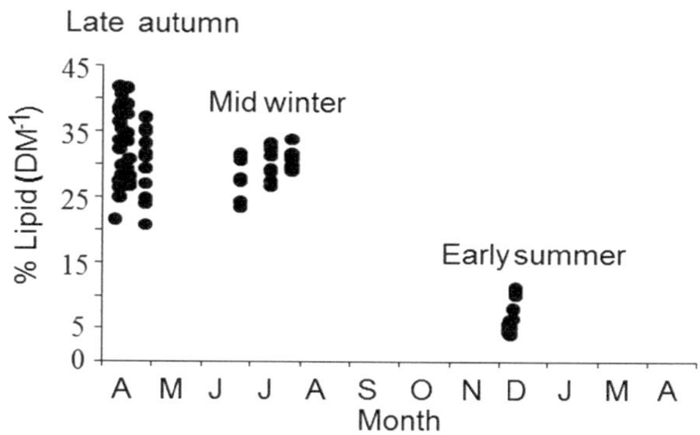

Fig. 4.9 Seasonal lipid levels of adult krill in the Lazarev Sea (Modified after Meyer et al. 2010)

The remaining lipid content of 5 % DM^{-1} in December 2005 of adults from the Lazarev Sea (Fig. 4.9) is very low and on the border of what is deemed essential for the functioning of bio-membranes and hence survival (Saether et al. 1985). Such low reserves in krill were documented previously in late October/mid November (Hagen et al. 2001). However, survival of winter and reproductive success in the subsequent summer depend on the accumulated energy reserves at the onset of winter. Completely depleted body lipids at the start of the reproductive season would affect the onset and maintenance of reproduction in krill negatively and hence their reproductive success (Clarke and Morris 1983; Cuzin-Roudy and Labat 1992; Quetin et al. 1994). This is particularly true when the spring phytoplankton bloom is delayed (Auerswald et al. 2009). The importance of a sufficient accumulation of lipids until the commencement of winter is also supported by results of Teschke et al. (2008), indicating that development of external maturation may be fuelled preferentially by lipid reserves, whereas the following internal maturation is fuelled by grazing on large diatom blooms (Schmidt et al. 2012). Compared with lipids, proteins play a minor role as an energy provision for overwintering. Krill depleted protein levels from March to August by 16 % (Torres et al. 1994a, b). Such utilization is minor (ca. 3 % $month^{-1}$) compared to the lipid depletion during winter as outlined above.

In contrast to adults, larvae have no seasonal pattern in their lipid dynamics and the same larval stages show a high variability in their lipid level within one season and year (Fig. 4.10) and between years (Table 4.4, Deibel and Daly 2007). The average body lipid content in larvae increased with ontogenetic stage (CIII: 7 %, FI: 10 %, FII; III: 15 % DM^{-1}), and hence their tolerance to short starvation periods from days to a few weeks (Meyer and Oettl 2005; Daly 2004; Meyer et al. 2009). In late autumn, the body lipid content of larval krill do not exceed 20 % DM^{-1} compared with 40 % DM^{-1} in adult krill. This implies that, unlike adults, no larval stage could survive several months or even the entire winter without food.

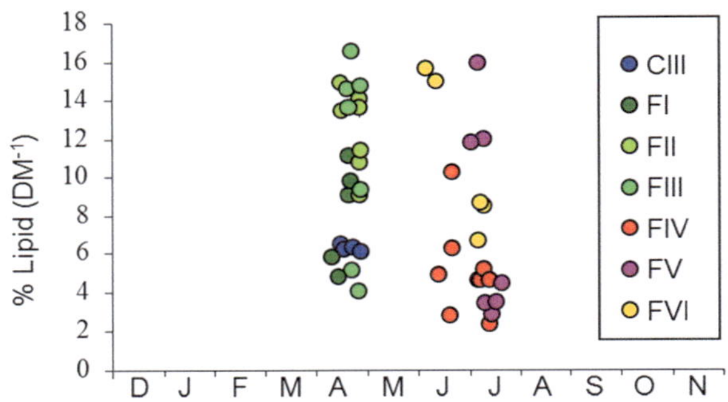

Fig. 4.10 Variability of body lipid content of larval stages in late autumn (April) and mid-winter (June, July) in the Lazarev Sea (Modified after Meyer 2011)

Table 4.4 Body lipid content per dry mass (DM) in different ontogenetic stages of krill and different regions

Season	Larval stages	Lipid (% DM^{-1})	Region	Month and year	References
Mid-summer	CI, II	12.3 ± 1.4 (8)	Weddell Sea	Jan–Feb 1986	Hagen et al. (2001)
	CIII	15 (1)	Weddell Sea	Apr–May 1992	
Late autumn	CIII	6.6 ± 0.1 (5)	Lazarev Sea	Apr–May 2004	Meyer et al. (2009b)
	FI–FIII	12.4 ± 2.2 (14)		Apr–May 1999	Meyer et al. (2002a)
	FIII	15.5 ± 4.3 (18)			
	FI–FIV	18.4 ± 5.4 (12)	Weddell Sea	Apr–May 1992	Hagen et al. (2001)
Winter	FIV–FVI	7.4 ± 3.8 (19)	Lazarev Sea	July 2006	Meyer et al. (2009b)
Mid spring	FV	7.3 ± 1.6 (2)	Lazarev Sea	Oct–Nov 1986	Hagen et al. (2001)
	FV, FVI	9.7 ± 1.2 (5)	Antarctic Peninsula	Oct–Nov 1983	

Number of replicates in *brackets*
C calyptopis, *F* furcilia

4.2.6 Energy Demand of Krill in Winter

While there is considerable knowledge of krill's high energy demands in summer (see Table 4.1) in the following section we will focus on krill's energy demands in winter.

According to the seasonal physiological functions of adults outlined above the energy demand was calculated for adult krill (Fig. 4.11) for the winter period, i.e. from the start of April to the end of September (183 days). According to recent results on the seasonal metabolic activity of krill from the Lazarev Sea

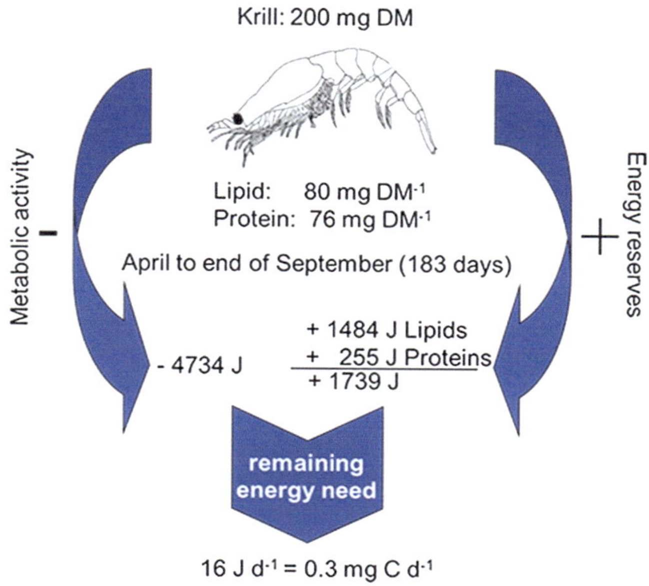

Fig. 4.11 From Meyer (2011): energy demand of adult krill of 200 mg dry mass (DM) from the onset of winter to beginning of spring (begin of April until the end of September) south of 66°S from the Lazarev Sea

(Meyer et al. 2010), metabolic depression was already initiated in April, whereas at the end of September, i.e. in early spring, adult krill reverted to an active mode again in late autumn and winter. For calculation of the daily energy requirements from the start of April to the end of September, the equations of the regressions in Fig. 4.3 were used (April and May, 61 days: µl O_2 ind. $h^{-1} = 0.37 \times$ DM in mg; June until end of September, 122 days: µl O_2 ind. $h^{-1} = 0.22 \times$ DM in mg). The resulting calculated respiration rates were converted to energy consumption rates, assuming an equivalent of 19.4 J ml O_2^{-1} (Brett and Groves 1979). Hence, a 200 mg model krill consumes about 1.8 ml O_2 daily in April and May and 1.1 ml from June to the end of September. This translates into a total demand of 4,734 J to cover metabolic activity for the 6 month of winter "hibernation".

Previous and recent investigations as well as our own data have shown that krill use high amounts of body lipid and moderate amounts of body protein reserves at rates of 10 % and 3 %, month $^{-1}$, respectively (Lit. summarised in Meyer 2011). During the 6-month of metabolic winter depression, krill consume 37 mg of body lipids and 13 mg of body proteins, which correspond to an energy yield of 1,484 J from lipids and 255 J from proteins, assuming conversion factors of 39.6 J mg^{-1} for lipid and 20.1 J mg^{-1} for proteins (Brett and Groves 1979). Consequently, the total energy krill obtain from both stores amounts to 1,739 J. At the end of "hibernation" (end of September), krill still contained 42 mg lipids and 63 mg proteins, which correspond to ca. 20 % and 30 % DM^{-1}, respectively. This is a reasonable amount to start the spring season in favourable physiological conditions.

Our calculation above revealed a difference of 2,995 J between the energy consumed and that provided from body reserves of lipids and proteins, corresponding with a lack of energy of 16 J d^{-1} for the period from April to the end of September. This daily energy requirement equates to 0.3 mg C, assuming an energy yield of 45.7 J mg^{-1} C (Salonen et al. 1976). According to microscopic analyses of the stomach and gut contents of winter krill (see Chap. 6), 0.3 mg C d^{-1} is an amount of energy that krill can gain by occasional feeding, even at low rates. This budget of the energy demand in winter demonstrates that the low feeding activity during winter is an important part for a successful overwintering of krill. It enables krill to start the spring season with a reasonable lipid level to survive when, after the metabolic winter depression, a high energy demand is needed by krill and the spring diatom blooms might by delayed or less extent.

In larval krill, lipid levels are highly variable, reflecting the high variability in food availability, outlined above. Therefore food quality and quantity is the major factor needed to cover the energy demands for their metabolic activity and growth. According to seasonal investigations on larval krill by Meyer et al. (2009) it can be summarised:

1. In autumn, the larvae from within the ice were in better physiological condition than those from open water.
2. Within the ice, autumn larvae were in better physiological state than winter larvae.
3. Different food sources were used in autumn and winter. While in autumn larvae were mainly feeding on autotrophic food sources, during winter heterotrophic organisms (small copepods, protozoans, or both) were important food items for larvae.
4. High amounts of larvae are observed in rafted sea ice regions with high structured under ice topography. This might dictate whether larval krill can exploit the food associated with sea ice or are drifted away from suitable feeding habits.

In addition, this habitat might not only favourable food regions but also protected areas due to predation. The measured growth and respiration rates of furcilia in autumn (0.95 ± 0.28 µl O_2 mg^{-1} DW h^{-1}, 0.011 mm d^{-1}) and winter (0.54 ± 0.19 µl O_2 mg-1 DW h^{-1}, 0.001 mm d^{-1}) in the Lazarev Sea, correspond to a C allocation of 2.5 % of body C d^{-1} into growth in autumn and 1.2 % in winter and 3.3 % body C d^{-1} to fuel respiration in autumn and 1.6 % in winter (Meyer et al. 2009), with a respiratory quotient of 0.97 (Ikeda et al. 2000). In winter, this corresponds for a slow growing furcilia VI (0.001 mm d^{-1}) of 1 mg DW and 37 % body C to an energy demand of 0.01 mg C. The average POC concentration in the water directly under sea ice was 0.07 ± 0.04 mg C L^{-1} and corresponds to the low growth rates and even shrinkage of larval krill in winter.

4.3 The Impact of Elevated Seawater Temperature and Ocean Acidification on Krill

The life cycle of krill is characterized by a strong interaction between endogenous physiological functions and seasonal environmental factors (Teschke et al. 2011). Therefore, it is crucial to understand how important physiological life cycle functions are affected by stressors such as seawater temperature rise, increasing ocean acidification and decreasing salinity due to glacier melt caused by anthropogenic CO_2 emission. Several studies have linked climate-induced environmental changes (temperature; pH; ice cover) to krill habitat (Hill et al. 2013) and krill distribution, recruitment and survival (Atkinson et al. 2004; Kawaguchi et al. 2013; Quetin et al. 2007). However, studies on the performance of krill to these stressors are rare and focused mainly on temperature. Previous field and laboratory studies have demonstrated an increasing moulting frequency with rising temperatures (Brown et al. 2010; Buchholz 1991) and accelerated maturation (McWhinnie and Marciniak 1964), increasing respiration rates with temperature up to 5 °C (McWhinnie and Marciniak 1964; Rakus-Suszczewski and Opaliński 1978; Opaliński et al. 1993; Opaliński 1991), altered lipid metabolism at different temperatures between seasons (Mayzaud et al. 2000) and declining growth rates at temperatures between 3 and 4 °C (Atkinson et al. 2006). Some of these findings suggest that krill seem to be relatively sensitive to rising temperatures. However, previous studies also show that krill are able to tolerate temperatures up to 10 °C for short periods of time (Opalinski 1991; Hirche 1984).

Only few studies exist so far on the impact of ocean acidification on krill. Laboratory experiments show that early larval stages seem to be most affected by increasing pCO_2 and especially in combination with rising temperatures (Kawaguchi et al. 2011, 2013).

A powerful approach to examine organismal responses to environmental change is by combining physiology performance indicators with transcriptomic changes, as demonstrated by recent characterization of the optimal thermal window for Antarctic fish (Windisch et al. 2011, 2014), showing a hepatic metabolic reorganization, indicating an alteration from a lipid-based metabolic network to pathways associated with carbohydrate metabolism. This picture of cellular adjustments to the warmth illustrate that we have to take a holistic view by identifying molecular networks rather than single genes to understand marine ectotherms capacities to cope with environmental change caused by the anthropogenic CO_2 emission (e.g. elevated seawater temperature, ocean acidification or reduced salinity due to glacier melt etc.). In adult Antarctic krill, a shift in metabolic pathways as shown for the Antarctic eelpout would have profound implications for krill's overwintering and spawning activity in the forthcoming spring. Krill build up considerable amount of body lipid reserves during the austral summer for their utilization during winter (Hagen et al. 2001). Increasing energy demands due to a warming environment (Pörtner and Farrell 2008) may impede the buildup of sufficient reserves during summer to allow survival of the winter season (Hagen et al. 2001) and to fulfill the external maturation process (Teschke et al. 2008).

Long-term experiments focusing on the impact of krill to rising seawater temperature, ocean acidification and reduced salinity, using physiological performance indicators with transcriptomic changes are needed to understand the flexibility of krill to the ongoing environmental changes and to make reliable predictions how the krill population might alter in the future. The only means by which an organism may respond to temperature change are physiological flexibility in the short-term, adaptation in the long-term or migration to a more favourable environment as an evasion strategy (Peck et al. 2010). In terms of krill we have only a rudimentary knowledge in this respect.

References

Arnold KH, Shreeve RS, Angus A, Clarke A (2004) Growth rates of Antarctic krill, *Euphausia superba*: comparison of the instantaneous growth rate method with nitrogen and phosphorus stoichiometry. Limnol Oceanogr 49:2152–2161

Asakura A (1992) Population ecology of the sand-dwelling hermit crab *Diogenes nitidimanus* Terao. 5. Ecological implications in the pattern of moulting. J Crust Biol 12:537–545

Atkinson A, Meyer B, Stübing D, Hagen W, Bathmann UV (2002) Feeding and energy budgets of Antarctic krill *Euphausia superba* at the onset of winter – II. Juveniles and adults. Limnol Oceanogr 47:953–966

Atkinson A, Siegel V, Pakhomov EA, Rothery P (2004) Long-term decline in krill stock and increase in salps within the Southern Ocean. Nature 432:100–103

Atkinson A, Shreeve RS, Hirst AG, Rothery P, Tarling GA, Pond DW, Korb RE, Murphy EJ, Watkins JL (2006) Natural growth rates in Antarctic krill (*Euphausia superba*): II: predictive models based on food, temperature, body length, sex, and maturity stage. Limnol Oceanogr 51:973–987

Auerswald L, Pape C, Stübing D, Lopata AL, Meyer B (2009) Effect of short-term starvation of adult Antarctic krill, *Euphausia superba*, at the onset of summer. J Exp Mar Biol Ecol 381:47–56

Brett JR, Groves TDD (1979) Physiological energetics. In: Hoar WS, Randall DJ, Brett JR (eds) Fish physiology 8. Academic, New York, pp 280–352

Brown M, Kawaguchi S, Candy GS, Virtue P (2010) Temperature effects on the growth and maturation of Antarctic krill (*Euphausia superba*). Deep Sea Res Part 2 Top Stud Oceanogr 57:672–682

Brown M, Kawaguchi S, King R, Virtue P, Nicol S (2011) Flexible adaptation of the seasonal krill maturity cycle in the laboratory. J Plankton Res 33:821–826. doi:10.1093/plankt/fbq123

Brown M, Kawaguchi S, Candy S, Yoshida T, Virtue P, Nicol S (2013) Long-term effect of photoperiod, temperature and feeding regimes on the respiration rates of Antarctic krill (*Euphausia superba*) in the laboratory. Open J Mar Sci 3:40–51. doi:10.4236/ojms.2013.32A005

Buchholz F (1989) Moult cycle and seasonal activities of chitinolytic enzymes in the integument and digestive tract of the Antarctic krill, *Euphausia superba*. Polar Biol 9:311–317

Buchholz F (1991) Moult cycle and growth of Antarctic krill *Euphausia superba* in the laboratory. Mar Ecol Prog Ser 3:217–229

Buchholz F, Morris DJ, Watkins JL (1989) Analyses of field moult data: prediction of intermoult period and assessment of seasonal growth in Antarctic krill, *Euphausia superba* Dana. Antarct Sci 1:301–306

Clark MS, Thorne MA, Toullec JY, Meng Y, Guan LL, Peck LS, Moore S (2011) Antarctic krill 454 pyrosequencing reveals chaperone and stress transcriptome. PLoS One 6:e15919. doi:10.1371/journal.pone.0015919

Clarke A, Morris DJ (1983) Towards an energy budget for krill: the physiology and biochemistry of *Euphausia superba* Dana. Polar Biol 2:69–86

Cockcroft AC, Goosen PC (1995) Shrinkage at moulting in the rock lobster *Jasus Lalandii* and the associated changes in the reproductive parameters. S Afr J Mar Sci 16:195–203

Cuzin-Roudy J, Labat JP (1992) Early summer distribution of Antarctic krill sexual development in the Scotia-Weddell region: a multivariate approach. Polar Biol 12:65–74

Daly KL (1990) Overwintering development, growth, and feeding of larval *Euphausia superba* in the Antarctic marginal ice zone. Limnol Oceanogr 35:1564–1576

Daly KL (2004) Overwintering growth and development of larval *Euphausia superba* an interannual comparison under varying environmental conditions west of the Antarctic Peninsula. Deep Sea Res II 51:2139–2168

Daly KL, Macaulay MC (1988) Abundance and distribution of krill in the ice edge zone of the Weddell Sea, austral spring 1983. Deep Sea Res 35:21–41

Daly KL, Macaulay MC (1991) Influence of physical and biological mesoscale dynamics on the seasonal distribution and behaviour of *Euphausia superba* in the Antarctic marginal ice zone. Mar Ecol Prog Ser 79:37–66

De Pittà C, Bertolucci C, Mazzotta GM, Bernante F, Rizzo G, De Nardi B, Pallavicini A, Lanfranchi G, Costa R (2008) Systematic sequencing of mRNA from the Antarctic krill (*Euphausia superba*) and first tissue specific transcriptional signature. BMC Genomics 28:9–45

De Pitta' C, Biscontin A, Albiero A et al (2013) The Antarctic krill *euphausia superba* shows diurnal cycles of transcription under natural conditions. PLoS ONE 8(7):e68652. doi:10.1371/journal.pone.0068652

Deibel D, Daly KL (2007) Zooplankton of the southern ocean. In: Smith WO, Barber DG (eds) Zooplankton processes in Arctic and Antarctic Polynyas, chapter 9. Elsevier oceanography series 74. Elsevier BV, pp 287–322. doi:10.1016/S0422-9894(06)74009-0

Ettershank G (1983) Age structure and cyclical annual size change in the Antarctic krill, *Euphausia superba* Dana. Polar Biol 2:189–193

Fach BA, Meyer B, Wolf-Gladrow D, Bathmann U (2008) Biochemically base modelling study of Antarctic krill *Euphausia superba* growth and development. Mar Ecol Prog Ser 360:147–161

Frazer TK, Quetin LB, Ross RM (1997) Abundance and distribution of larval krill, *Euphausia superba*, associated with annual sea ice in winter. In: Battaglia B, Valencia J, Walton DWH (eds) Antarctic communities: species, structure and survival. Cambridge University Press, pp 107–111

Frazer TK, Quetin LB, Ross RM (2002) Energetic demands of larval krill, *Euphausia superba*, in winter. J Exp Mar Biol Ecol 277:157–171

Gaten E, Tarling G, Dowse H et al (2008) Is vertical migration in Antarctic krill (*Euphausia superba*) influenced by an underlying circadian rhythm? J Genet 5:473–483

Godlewska M (1996) Vertical migrations of krill (*Euphausia superba* Dana). Pol Arch Hydrobiol 43:9–63

Groeneveld J, Johst K, Kawaguchi S, Meyer B, Teschke M, Grimm V (2015) How biological clocks and changing environmental conditions determine local population growth and species distribution in Antarctic krill (*Euphausia superba*): a conceptual model. Ecol Model 303:78–86

Guyselmann JB (1953) An analysis of the moulting process in the fiddler crab, *Uca pugilator*. Biol Bull Mar Boil Lab Woods Hole 104:115–137

Hagen W, Kattner G, Terbrüggen A, Van Vleet ES (2001) Lipid metabolism of the Antarctic krill *Euphausia superba* and its ecological implications. Mar Biol 139:95–104

Hamner WM, Hamner PP, Strand SW, Gilmer RW (1983) Behaviour of Antarctic krill *Euphausia superba*: chemoreception, feeding schooling, moulting. Science 220:433–435

Helm B, Ben-Shlomo R, Sheriff MJ, Hut RA, Foster R, Barnes BM, Dominoni D (2013) Annual rhythms that underlie phenology: biological time-keeping meets environmental change. Proc R Soc B 280:20130016, http://dx.doi.org/10.1098/rspb.2013.0016

Hempel I (1981) Euphausiid larvae in the Scotia Sea and adjacent waters in summer 1977/1978. Meeresforsch 29:53–59

Hempel I, Hempel G (1978) Larval krill (*Euphausia superba*) in the plankton and neuston samples of the German Antarctic Expedition 1975/1976. Meeresforsch 26:206–216

Hempel I, Hempel G (1982) Distribution of euphausiid larvae in the southern Weddell Sea. Meeresforsch 29:253–266

Hill SL, Phillips T, Atkinson A (2013) Potential climate change effects on the habitat of Antarctic krill in the Weddell quadrant of the southern ocean. PLoS One 8:e72246

Hirche H (1984) Temperature and metabolism of plankton—I. Respiration of Antarctic zooplankton at different temperatures with a comparison of Antarctic and nordic krill. Comp Biochem Physiol A Physiol 77:361–368. doi:10.1016/0300-9629(84)90074-4

Hoffmann EE, Lascara CM (2000) Modeling the growth dynamics of Antarctic krill *Euphausia superba*. Mar Ecol Prog Ser 194:219–231

Hopkins PM (1985) Regeneration and relative growth in the fiddler crab. In: Crustacean. Issues 3

Hosie GW, Ritz DA (1989) Body shrinkage in the sub-tropical euphausiid *Nyctiphanes australis* GO Sars. J Plankt Res 11:595–598

Hubold G, Hempel I (1987) Seasonal variability of zooplankton in the Southern Weddell Sea. Meeresforsch 31:185–192

Huntley M, Brinton E (1991) Mesoscale variation in growth and early development of *Euphausia superba* Dana in the western Bransfield Strait region. Deep Sea Res 38:1213–1240

Huntley M, Nordhausen W, Lopez MDG (1994) Elemental composition, metabolic activity and growth of Antarctic krill *Euphausia superba* during winter. Mar Ecol Prog Ser 107:23–40

Ikeda T (1974) Nutritional ecology of marine zooplankton. Mem Fac Fish Hokkaido Univ 22:1–97

Ikeda T (1981) Metabolic activity of larval stages of Antarctic krill. Antarct J US 16:161–162

Ikeda T (1984) Development of the larvae of the Antarctic krill (*Euphausia superba* Dana) observed in the laboratory. J Exp Mar Biol Ecol 75:107–117

Ikeda T (1985) Metabolic rate and elemental composition (C and N) of embryos and non-feeding early larval stages of (*Euphausia superba* Dana). J Exp Mar Biol Ecol 90:119–127

Ikeda T, Dixon T (1982) Body shrinkage as a possible overwintering mechanism of the Antarctic krill, *Euphausia superba* Dana. J Exp Mar Biol Ecol 62:143–151

Ikeda T, Mitchell AW (1982) Oxygen uptake, ammonia excretion and phosphate excretion by krill and other Antarctic zooplankton in relation to their body size and chemical composition. Mar Biol 71:283–298

Ikeda T, Thomas PG (1987) Longevity of the Antarctic krill (*Euphausia superba* Dana) based on a laboratory experiment. Proc NIPR Symp Polar Biol 1:56–62

Ikeda T, Torres JJ, Hernandez-Leon S, Geiger SP (2000) Metabolism. In: Harris RP, Wiebe PH, Lenz J, Skjoldal HR, Huntley M (eds) ICES zooplankton methodology manual. Academic, London, pp 455–532

Ishii H, Omori M, Maeda M, Watanabe Y (1987) Metabolic rates and elemental composition of Antarctic krill *Euphausia superba* Dana. Polar Biol 7:379–382

Kawaguchi K, Ishikawa S, Matsuda O (1986) The overwintering strategy of Antarctic krill (*Euphausia superba* Dana) under the coastal fast ice off the Ongul Islands in Lützow-Holm Bay, Antarctica. Mem Nat Inst Polar Res Spec Iss 44:67–85

Kawaguchi S, Yoshida T, Finley L, Cramp P, Nicol S (2006) The krill maturity cycle: a conceptual model of the seasonal cycle in Antarctic krill. Polar Biol 30:689–698. doi:10.1007/s00300-006-0226-2

Kawaguchi S, Kilpatrick R, Roberts L, King R, Nicol S (2011) Ocean-bottom krill sex. J Plank Res 33:1134–1138. doi:10.1093/plankt/fbr006

Kawaguchi S, Ishida A, King R et al (2013) Risk maps for Antarctic krill under projected Southern Ocean acidification. Nat Clim Chang 3:843–847. doi:10.1038/NCLIMATE1937

Lancraft TM, Hopkins TL, Torres JJ, Donelly J (1991) Oceanic micronektonic/macrozooplanktonic community structure and feeding in ice covered Antarctic waters during winter (AMERIEZ 1988). Polar Biol 11:157–167

Lasker R (1966) Feeding growth, respiration and carbon utilization of a euphausiid crustacean. J Fish Res Bd Can 23:1291–1317

Lindberg RG (1955) Growth, population dynamic, and field behaviour in the spiny lobster, *Panulirus interruptus* (Randall). Univ Calif Publs Zool 59:157–231

Little EJ (1972) Tagging of spiny lobster (*Panulirus argus*) in the Florida Keys, 1967–1969. Spec Sci Rep Fla Dept Nat Resour Mar Res Lab 31:1–23

Lloyd AJ, Yonge CM (1947) The biology of *Crangon L.* in the Bristol Channel and Severn Estuary. Mar Biol Ass UK 26:626–661

Loeb V, Siegel V, Holm-Hansen O, Hewitt R, Fraser W, Trivelpiece W, Trivelpiece S (1997) Effects of sea-ice extent and krill or salp dominance on the Antarctic food web. Nature 387:897–900

Mackintosh NA (1972) Life cycle of Antarctic krill in relation to ice and water conditions. Discov Rep 36:1–94

Makarov RR, Maslennikov VV (1992) Larvae of euphausiids off Queen Maud Land. Polar Biol 11:515–523

Makarov RR, Sysoyeva MV (1985) Biology and distribution of *Euphausia superba* in the Lazarev sea and adjacent waters. In: Legkaya i pishevaya promyshlennost (ed) Antarktichestkiy Kril Osobennosti raspredeleniya I Sreda, Moscow, pp 110–116

Marschall HP (1988) The overwintering strategy of Antarctic krill under pack-ice of the Weddell Sea. Polar Biol 9:245–250

Marshall N (1945) The moulting without growth of spiny lobsters *Panurilus argus* kept in a live car. Trans Am Sish Soc 75:267–269

Mayzaud P, Albessard E, Virtue P, Boutoute M (2000) Environmental constraints on the lipid composition and metabolism of euphausiids: the case of *Euphausia superba* and *Meganyctiphanes norvegica*. Can J Fish Aquat Sci 57:91–103

Mazzotta GM, De Pittà C, Benna C, Tosatto SCE, Lanfranchi G, Bertolucci C, Costa R (2010) A cry from the Krill. Chronobiology 27:425–445

McWhinnie MA, Marciniak P (1964) Temperature responses and tissue respiration in Antarctic crustacea with particular reference to the krill *Euphausia superba*. In: Biology of the Antarctic seas. American Geophysical Union, Wiley, pp 63–72. doi: 10.1029/AR001p0063

Melnikov IA, Spiridonov VA (1996) Antarctic krill under perennial sea ice in the western Weddell Sea. Antarct Sci 8:323–329

Menshenina L (1992) Distribution of Euphausiid larvae in the Weddell Gyre in September–October 1989. Polar Biol 5:44–54

Meyer B (2012) The overwintering of Antarctic krill, *Euphausia superba*, from an ecophysiological perspective, – a review. Polar Biol 35:15–37

Meyer B, Oettl B (2005) Effects of short-term starvation on composition and metabolism of larval Antarctic krill, *Euphausia superba*. Mar Ecol Prog Ser 292:263–270

Meyer B, Atkinson A, Stübing D, Oettl B, Hagen W, Bathmann UV (2002a) Feeding and energy budgets of Antarctic krill *Euphausia superba* at the onset of winter — I. Furcilia III larvae. Limnol Oceanogr 47:943–952

Meyer B, Saborowski R, Atkinson A, Buchholz F, Bathmann U (2002b) Seasonal differences in citrate synthase and digestive activity in larval and postlarval Antarctic krill, *Euphausia superba*. Mar Biol 141:855–862

Meyer B, Atkinson A, Blume B, Bathmann UV (2003) Feeding and energy budgets of larval Antarctic krill, *Euphausia superba*, in summer. Mar Ecol Prog Ser 257:167–178

Meyer B, Fuentes V, Guerra C, Schmidt K, Atkinson A, Spahic S, Cisewski B, Freier U, Olariaga A, Bathmann U (2009) Physiology, growth and development of larval krill *Euphausia superba* in autumn and winter in the Lazarev Sea, Antarctica. Limnol Oceanogr 54:1595–1614

Meyer B, Auerswald L, Siegel V, Spahic S, Pape C, Fach B, Teschke M, Lopata A, Fuentes V (2010) Seasonal variation in body composition, metabolic activity, feeding, and growth of adult krill *Euphausia superba* in the Lazarev Sea. Mar Ecol Prog Ser 398:1–18

Meyer B, Martini P, Biscontin A et al (2015) Pyrosequencing and de novo assembly of Antarctic krill (*Euphausia superba*) transcriptome to study of the adaptability of krill to climate induced environmental changes. Mol Ecol Resour 15:1460–1471. doi:10.1111/1755-0998.12408

Miller DGM, Hampton I (1989) Biology and ecology of the Antarctic krill (*Euphausia superba* Dana): a review. Biol Investig Mar Antarct Syst Sticks (BIOMASS) Sci Ser 9:63–85

Morris DJ, Priddle (1984) Observations on the feeding and moulting of the Antarctic Krill, *Euphausia superba* Dana, in winter. Br Antarct Surv Bull 65:57–63

Nicol S, Stolp M, Cochran T, Geijsel T, Marshall J (1992) Growth and shrinkage of Antarctic krill, *Euphausia superba*, from the Indian Ocean sector of the Southern Ocean during summer. Mar Ecol Prog Ser 89:175–181

Nicol S, Kitchener J, King R, Hosie G, de la Mare WK (2000) Population structure and condition of Antarctic krill (*Euphausia superba*) off east Antarctica (80–150°E) during the Austral summer of 1995/1996. Deep-Sea Res Part II 47:2489–2517

Nicol S, Virtue P, King R, Davenport SR, McGaffin AF, Nichols P (2004) Condition of *Euphausia crystallorophias* off East Antarctica in winter in comparison to other seasons. Deep-Sea Res II 51:2215–2224

Nishino Y, Kawamura A (1994) Winter gut contents of Antarctic krill (*Euphausia superba* Dana) collected in the South Georgia area. Proc NIPR Symp Polar Biol 7:82–90

Omori M, Ikeda T (1984) Methods in marine zooplankton ecology. Wiley, New York, 332 pp

Opaliński KW (1991) Respiratory metabolism and metabolic 632 adaptations of Antarctic krill *Euphausia superba*. Pol Arch Hydrobiol 38:183–263

Opaliński KW, Klekowski RZ, Kidawa A (1993) Respiratory metabolism of *Euphausia superba*: effect of temperature. the second Polish-Soviet Antarctic symposium, pp 183–189

Oster H, Maronde E, Albrecht U (2002) The circadian clock as a molecular calendar. Chronobiol Int 19:507–516

Pakhomov EA, Atkinson A, Meyer B, Oettl B, Bathmann U (2004) Daily rations and growth of larval krill *Euphausia superba* in the Eastern Bellingshausen Sea during austral autumn. Deep-Sea Res II 51:2185–2198

Peck LS, Morley SA, Clark MS (2010) Poor acclimation capacities in Antarctic marine ectotherms. Mar Biol 157:2051–2059

Pörtner H-O, Farrell A (2008) Physiology and climate change. Science 322:690–692

Quetin LB, Ross RM (1991) Behavioural and physiological characteristics of the Antarctic krill, *Euphausia superba*. Am Zool 31:49–63

Quetin LB, Ross RM, Clarke A (1994) Krill energetics: seasonal and environmental aspects of the physiology of *Euphausia superba*. In: El-Sayed SZ (ed) Southern ocean ecology: the BIOMASS perspective. Cambridge University Press, Cambridge, pp 165–184

Quetin LB, Ross RM, Frazer TK, Habermann KL (1996) Factors affecting distribution and abundance of zooplankton, with an emphasis on Antarctic krill, *Euphausia superba*. Ant Res Ser 70:357–371

Quetin LB, Ross RM, Frazer TK, Amsler MO, Wyatt-Evens C, Oakes SA (2003) Growth of larval krill, *Euphausia superba*, in fall and winter west of the Antarctic Peninsula. Mar Biol 143:833–843

Quetin LB, Ross RM, Fritsen CH, Vernet M (2007) Ecological responses of Antarctic krill to environmental variability: can we predict the future? Antarct Sci 19:253–266

Rakusa-Suszczewski S, Opaliński KW (1978) Oxygen consumption in *Euphausia superba*. Polskie Archiwum Hydrobiologii 25:633–641

Reiss CS (2016) Age, growth, mortality, and recruitment of Antarctic Krill, *Euphausia superba*. In: Siegel V (ed) The biology and ecology of Antarctic krill, *Euphausia superba* Dana, 1850. Springer, Cham, pp 101–144

Ross RM, Quetin LB (1991) Ecological physiology of larval euphausiids, *Euphausia superba* (Euphausiaceae). Mem Queensl Mus 31:321–333

Ross RM, Quetin LB, Baker KS, Vernet M, Smith RS (2000) Growth limitation in young *Euphausia superba* under field conditions. Limnol Oceanogr 45:31–43

Ross RM, Quetin LB, Newberger T, Oakes SA (2004) Growth and behaviour of larval krill (*Euphausia superba*) under the ice in late winter 2001 west of the Antarctic Peninsula. Deep Sea Res II 51:2169–2184

Saether O, Ellingsen TE, Mohr V (1985) Lipids of North Atlantic krill. J Lipid Res 27:274–285

Salonen K, Sarvala J, Hakala I, Viljanen M-L (1976) The relation of energy and organic carbon in aquatic invertebrates. Limnol Oceanogr 21:724–730

Schmidt K, Atkinson A (2016) Feeding and food processing in Antarctic krill (Euphausia superba Dana). In: Siegel V (ed) The biology and ecology of Antarctic krill, *Euphausia superba* Dana, 1850. Springer, Cham, pp 175–224

Schmidt K, Atkinson A, Venables H, Pond DW (2012) Early spawning of Antarctic krill in the Scotia Sea is fuelled by 'superfluous' feeding on non-ice associated phytoplankton blooms. Deep Sea Res II 59:159–172

Schmidt K, Atkinson, Pond DW, Ireland LC (2014) Feeding and overwintering of Antarctic krill across its major habitats: the role of sea ice cover, water depth, and phytoplankton abundance. Limnol Oceanogr 59:17–36

Schultz TF, Kay SA (2003) Circadian clocks in daily and seasonal control of development. Science 301:326–328

Seear PJ, Geraint AT, Teschke M, Meyer B, Thorne MAS, Clark MS, Gaten E, Rosato E (2009) Effects of simulated light regimes on gene expression in Antarctic krill (*Euphausia superba* Dana). J Exp Mar Biol Ecol 381:57–64

Seear PJ, Tarling GA, Burns G, Goodall-Copestake WP, Gaten E, Ozkaya O, Rosato E (2010) Differential gene expression during the moult cycle of Antarctic krill (*Euphausia superba*). BMC Genomics 11:582

Seear PJ, Goodall-Copestake WP, Fleming AH, Rosato E, Tarling GA (2012) Seasonal and spatial influences on gene expression in Antarctic krill *Euphausia superba*. Mar Ecol Prog Ser 467:61–75

Siegel V (1987) Age and growth of Antarctic Euphausiaceae (Crustacea) under natural conditions. Mar Biol 96:483–495

Siegel V (2005) Distribution and population dynamics of *Euphausia superba*: summary of recent findings. Polar Biol 29:1–22

Siegel V, Loeb V (1995) Recruitment of Antarctic krill *Euphausia superba* and possible causes for its variability. Mar Ecol Prog Ser 123:45–56

Siegel V, Watkins J (2016) Distribution, biomass and demography of Antarctic Krill, *Euphausia superba*. In: Siegel V (ed) The biology and ecology of Antarctic krill, *Euphausia superba* Dana, 1850. Springer, Cham, pp 21–100

Siegel V, Skibowski PA, Harm U (1992) Community structure of the epipelagic zooplankton community under the sea-ice of the northern Weddell Sea. Polar Biol 12:15–24

Skinner DM (1985) Moulting and regeneration. In: Bliss DE, Mantel LH (eds) The biology of crustacea. 9. Integument, pigments, and hormonal processes. Academic, New York, pp 43–146

Spiridonov VA (1992) Distribution and biological properties of the Antarctic krill *Euphausia superba* Dana during winter Weddell Gyre study (WWGS) 1989. Proc NIPR Symp Polar Biol 5:55–64

Stepnik R (1982) All-year population studies of Euphausiaceae (Crustacea) in the Admiralty Bay (King George Island, South Shetland Island Antarctica). Pol Polar Res 3:49–68

Sweat DE (1968) Growth and tagging studies on *Panulirus argus* (Latreille) in the Florida Keys. Tech Ser Fla St Bd Conserv 57:30

Tarling G, Shreeve RS, Hirst AG, Atkinson A, Pond DW, Murphy EJ, Watkins JL (2006) Natural growth rates in Antarctic krill (*Euphausia superba*): I. Improving methodology and predicting intermoult period. Limnol Oceanogr 51:959–972

Teschke M, Kawaguchi S, Meyer B (2007) Simulated light regimes affect feeding and metabolism of Antarctic krill, *Euphausia superba*. Limnol Oceanogr 52:1046–1054

Teschke M, Kawaguchi S, Meyer B (2008) Effects of simulated light regimes on maturity and body composition of Antarctic krill, *Euphausia superba*. Mar Biol 154:315–332

Teschke M, Wendt S, Kawaguchi S, Kramer A, Meyer B (2011) A circadian clock in Antarctic krill: an endogenous timing system governs metabolic output rhythms in the Euphausid species *Euphausia superba*. PLoS ONE 6:e26090. doi:10.1371/journal.pone.0026090

Thomas PG, Ikeda I (1987) Sexual regression, shrinkage, re-maturation and growth of spent female Euphausia superba in the laboratory. Mar Biol 95:357–363

Thorpe SE, Murphy EJ, Watkins JL (2007) Circumpolar connections between Antarctic krill (*Euphausia superba* Dana) populations: investigating the roles of ocean and sea ice transport. Deep-Sea Res I 54:792–810

Töbe K, Meyer B, Fuentes V (2010) Detection of zooplankton items in the stomach and gut content of larval krill, *Euphausia superba*, using a molecular biology approach. Polar Biol 33:407–414

Torres JJ, Aarset AV, Donnelly J, Hopkins TL, Lancraft TM, Ainley DG (1994a) Metabolism of Antarctic micronektonic Crustacea as a function of depth of occurrence and season. Mar Ecol Prog Ser 113:207–219

Torres JJ, Aarset AV, Donnelly J, Hopkins TL, Lancraft TM, Ainley DG (1994b) Proximate composition and overwintering strategies of Antarctic micronektonic Crustacea. Mar Ecol Prog Ser 113:221–232

Windisch HS, Kathöver R, Pörtner H-O et al (2011) Thermal acclimation in Antarctic fish: transcriptomic profiling of metabolic pathways. Am J Physiol Regul Integr Comp Physiol 301:R1453–R1466

Windisch HS, Frickenhaus S, John U et al (2014) Stress response or beneficial temperature acclimation: transcriptome signatures in Antarctic fish (*Pachycara brachycephalum*). Mol Ecol 23:3469–3482

Witek Z, Koronkiewicz A, Soszka GJ (1980) Certain aspects of the early life history of krill, *Euphausia superba* Dana (Crustacea). Pol Polar Res 1:97–115

Chapter 5
Feeding and Food Processing in Antarctic Krill (*Euphausia superba* Dana)

Katrin Schmidt and Angus Atkinson

Abstract *Euphausia superba* is exceptional among euphausiids for the large filtering surface of the feeding basket and its fine mesh size (2–3 μm), which remain into adulthood. This enables them to feed efficiently on nano- and microplankton, and to reach substantial growth rates with food concentrations as low as 0.5 μg Chlorophyll *a* L^{-1}. Even though phytoplankton – in particular diatoms – are their staple food, protozoans and small copepods are ingested simultaneously and represent an important supplementary food source year-round. However, krill feeding behaviour is more complex than just filter-feeding in the water column, it includes raptorial capture of larger zooplankton, handling of 'giant' diatoms, scraping algae from beneath sea ice and lifting detritus from the seabed. High mobility and physiological robustness enable krill to explore three feeding grounds – the water column, the sea ice and the benthos. Variability in access and productivity of these feeding grounds leads to fundamental differences in krill overwintering across their habitats. Gut passage time, absorption efficiency and fecal pellet density vary with food concentration and nutritional needs. Therefore krill fecal pellets have a dual role; some promote the export of carbon and nutrients while others facilitate the recycling of material in the upper water column. Krill grazing can suppress phytoplankton blooms, but this tends to be a localised phenomenon where krill abundances are exceptionally high. Conversely, krill appear to have major conditioning effects due to nutrient supply (e.g. ammonium, iron), although their role in Southern Ocean biogeochemical cycles is only starting to be discovered.

Keywords Diet • Ingestion rates • Stable isotopes • Polyunsaturated fatty acids • Fecal pellets

K. Schmidt (✉)
Sir Alister Hardy Foundation for Ocean Science, The Laboratory, Citadel Hill, The Hoe, Plymouth PL1 2PB, UK
e-mail: katsch@sahfos.ac.uk

A. Atkinson
Plymouth Marine Laboratory, Prospect Place, The Hoe, Plymouth PL1 3DH, UK
e-mail: aat@pml.ac.uk

V. Siegel (ed.), *Biology and Ecology of Antarctic Krill*, Advances in Polar Ecology,
DOI 10.1007/978-3-319-29279-3_5

5.1 Introduction

Nutrition provides the building blocks of all organisms and fuels the dynamic interactions between them (Simpson and Raubenheimer 2012). Most aspects of the biology of *Euphausia superba* indeed hinge in some way around feeding or avoiding being eaten. For example the amount and quality of food sets the ingestion rate, gut retention time and absorption efficiency. Those in turn affect the rest of the krill physiology, including excretion, respiration, mobility, growth and gonad production (Ikeda and Dixon 1984; Ross et al. 2000; Pond et al. 2005). Finding food requires aggregation in the water column, or visits to sea ice or the seabed according to region, season and life stage, with specialised feeding mechanisms on each substrate. Conceptual models of distribution, from the circumpolar scales to that of an individual within a school, involve food acquisition and predator avoidance as well as ocean physics. Feeding is therefore a central thread of krill biology.

Studies of krill feeding have a long history and the slowness in progress reflects the great difficulty in obtaining real-world data on how they feed, what they eat and how much. On one hand, krill are large and well suited for laboratory maintenance and experimentation (Ross and Quetin 2003). On the other hand, their complex foraging behaviour in association with schooling and extensive vertical and horizontal migration make it difficult to recreate natural conditions in the laboratory.

Early studies of krill nutrition examined their gut contents (Barkley 1940) and mouthpart morphology (Kils 1983; Suh and Nemoto 1988; Hamner and Hamner 1988) and established that the species is well adapted to feeding on phytoplankton. This earlier work contains the most thorough observations of the mouthpart morphology and feeding mechanisms made to date. It is surprising that the great improvements in filming technology in the last 30 years have not been applied to krill feeding.

In the 1980s, research on krill rapidly expanded with new studies covering all aspects of their ecology. Many of the early krill incubation experiments used small (<10 L) containers and derived daily rations that were only a few percent of the total body carbon per day (see Perissinotto et al. 1997; Pakhomov et al. 2002). With the increase in incubation volume, the application of through-flow systems and the use of new, more in-situ related approaches, krill clearance- and ingestion rates started to reflect those of micronekton (Clarke et al. 1988). Nevertheless, it was still not properly known what they eat and where they feed. Three breakthrough findings followed, however. (1) Both larval and postlarval krill were filmed feeding on the underside of sea ice (Stretch et al. 1988; Marschall 1988; Hamner et al. 1989), (2) they were found to readily eat copepods (Price et al. 1988), and (3) they were seen in aggregations at the seabed (Gutt and Siegel 1994), providing yet another potential food source. It is only now, three decades later, that we are starting to quantify the relative roles of these food substrates.

The 1990s and 2000s were perhaps a less glamorous time to study krill feeding. These marked a period of infilling the above breakthroughs to provide solid quantitative information. To do this the methods needed to be used carefully and in combination (see Sect. 5.4). For instance the application of isotopic- and fatty acid trophic markers increased, and gut fluorescence and fecal pellet egestion estimates were used alongside traditional feeding incubations and gut content analysis.

The last 5 years have made us realise that some of the previous controversies about krill feeding have in fact reflected real differences, either in terms of region, season or krill ontogeny. For instance the feeding on copepods is very region-specific and so are overwintering strategies (Schmidt et al. 2014), and the reliance on ice algae decreases with ontogeny (Quetin et al. 1994; Meyer 2012). There has also been a resurgence of studies that examine the biogeochemical ramifications of krill feeding, for instance in carbon export via their fecal pellets (Atkinson et al. 2012; Manno et al. 2015), or in controlling phytoplankton blooms and regenerating nutrients (Tovar-Sanchez et al. 2007; Schmidt et al. 2011; Whitehouse et al. 2011).

The sections below review these developments in krill feeding. The topic has been incorporated previously by Clarke and Morris (1983), Miller and Hampton (1989), Knox (1994), and Quetin et al. (1994) in their wider-ranging reviews of krill ecology. This chapter highlights the key developments in the study of krill feeding, namely how they feed, what they feed on and at what rates, before putting krill feeding into a wider context. This context includes the energy budget of krill, the comparison with other Southern Ocean euphausiids, their role in the food web and in biogeochemical cycling. While we describe larval feeding, most emphasis is on post-larvae, given the separate chapter dedicated to larval krill biology (see Chap. 6, Kawaguchi 2016).

5.2 Feeding Apparatus

In euphausiids, feeding involves the integrated action of externally located thoracopods and mouthparts and the internally placed armature of the stomach wall (Hamner and Hamner 1988; Suh and Nemoto 1988). The thoracopods gather the food, the mouthparts handle it and the mandibles pierce, cut and grind the items before they are swallowed. The internal armature of the stomach breaks the food into even smaller pieces until they are fine enough to enter the digestive gland for final digestion and absorption (Suh 1996). The principal filtering apparatus of *Euphausia* species is a feeding basket collectively formed by six pairs of thoracopods (Barkley 1940; Kils 1983; McClatchie and Boyd 1983). The fine structure of this basket is a three-dimensional filter of primary, secondary and tertiary setae (Suh and Nemoto 1987). Primary setae are present on the ischium and merus of the thoracopods. Along the primary setae, there are two rows of secondary setae inserted at an angle of 90° and again a single row of tertiary setae along the secondary setae (Suh and Nemoto 1987). The terminal segments of the thoracopods bear comb setae with a comb-like device at their ends. The filter area and the intersetal distance determine the filtering efficiency of the feeding basket, which can differ between species and size classes (Boyd et al. 1984; Suh and Nemoto 1987).

In *E. superba*, the filtering area of the basket increases from ~70 mm^2 in juveniles (20 mm body length) to ~277 mm^2 in adults (50 mm body length), while the minimum spacing between tertiary setae remains small (2–3 μm, Suh and Nemoto 1987). In other Southern Ocean euphausiids, the filter area of the basket is only 20–30 % of that of *E. superba* with the same body length, the minimum spacing between tertiary setae is larger (>8 μm) and the maximum spacing between primary setae is similar or slightly larger (see Sect. 5.12). McClatchie (1985) compared the filtering area of *E. superba* with that of Northern krill, *Meganyctiphanes norvegica*, and suggested that if the two species apply their feeding baskets in the same way and at the same rate, 22–39 mm long *E. superba* would have a three-times higher clearance rate than *M. norvegica* of similar size (McClatchie 1985). This indicates that the feeding basket of *E. superba* is exceptional among euphausiids for its large filter area, the fine mesh sizes and the nearly consistent minimum spacing between tertiary setae from juvenile to adult.

Once the food is enclosed in the feeding basket, various mouthparts (mandibular palps, maxillules and maxillae) are involved in passing it on to the oral cavity built by the upper lip (labrum), mandibles and lower lip (labia). The mandibles are hard and have strong cusps, the *pars incisive*, in the ventral region and grinding surfaces, the *pars molaris*, in the dorsal region. Large grinding areas of the mandible are associated with a phytophagous tendency, whereas a pronounced cutting region indicates carnivorous feeding (Nemoto 1967; Mauchline 1989). Although the large grinding region of the *E. superba* mandible is characteristic of an herbivorous euphausiid (Nemoto 1967), the marginal teeth are spine-like as in omnivorous-carnivorous species (McClatchie and Boyd 1983). These spines may facilitate piercing of animal prey, functioning as an adjunct to the *pars incisive* (McClatchie and Boyd 1983). The grinding region is differentiated into very rugged cusped transitional areas and a broad plate-like region, which indicates specialization for fracturing hard tests and for finer grinding of particles (McClatchie and Boyd 1983). Ridges on the plate-like region are spaced at ~5-μm intervals, suggesting that particles smaller than 10 μm can be ground up efficiently (McClatchie and Boyd 1983). After crushing and grinding by the mandibles, the stomach is an additional organ for the maceration of food particles (Suh 1996). The gastric mill is the main grinding region within the stomach. *E. superba* have a well-developed gastric mill with strong cuticular structures, lateral teeth and cluster spines, which act in crushing hard food items such as diatom frustules (Suh and Nemoto 1988; Ullrich et al. 1991). A complex system of muscles enables movements of the stomach wall, which compress the food between the armoured areas.

In summary, *E. superba* is equipped to feed on a wide range of food items. Two different types of filter nets are formed by the setal arrangements of their feeding basket. First, a very fine net of secondary and tertiary setae, which allows *E. superba* to filter nanoflagellates and small resuspended particles, and second, a coarse net formed by primary and comb setae suitable to retain larger items such as diatoms. Other Southern Ocean euphausiids are not equipped to feed efficiently on items as small as 2–3 μm, which gives *E. superba* a considerable competitive advantage (Suh and Nemoto 1987). The elaborated *pars molaris* and the well-developed gastric mill enable *E. superba* to break and macerate strongly-silicified diatoms. Handling of larger animal prey is supported by spine-like marginal teeth on their mandibles.

5.3 Feeding Mechanisms

5.3.1 Filter Feeding

Euphausiids are able to sense odours of phytoplankton along a diffusion gradient (Price 1989; Hamner and Hamner 2000). In a positive response, the scent trail is tracked; filtration rates increase and the krill try to remain within the area of highest phytoplankton concentration (Price 1989; Hamner and Hamner 2000). Negative responses to chemoreception include the rejection and avoidance of particles in the water. Surprisingly few studies examined the actual mechanism of food- and water flow through the feeding appendages in close-up detail. Of these, the study by Kils (1983) stands out since it details quite different ways of moving the feeding appendages that could explain some of the contrasting results on food size selection (see Sect. 5.9).

Euphausia superba has a chamber-like food basket which acts as a pressure-pumping mechanism (Hamner et al. 1983). Kils (1983) suggested from observations of wild and captive krill that the most common movement of these feeding appendages was in compression filtration (he also termed this mode "pump filtration" although subsequent authors have described these as if they were separate modes). This energy-intensive opening and closing of the filtering basket is synchronised with the pleopod swimming beat to maintain steady forwards motion. To open the basket, the paired thoracic legs move downward and outward in a metachronal rhythm which creates a pressure gradient that sucks water and particles into the basket from the front (McClatchie and Boyd 1983; Hamner and Hamner 1988). Food collects inside the basket while the euphausiid expels water laterally through the setae by rapidly compressing the thoracopods. The filter setae are then scraped and cleaned by a second set of comb setae and the particles passed forward to the mouth (Hamner and Hamner 1988). The metachronal rhythm of the thoracic legs during expansion and contraction of the basket enables the passage of food from posterior to anterior legs (Mauchline 1989). After the mouthparts gather a food bolus the feeding bouts cease, the mandibular palps press the bolus against the mandibles and ingestion ensues (Hamner et al. 1983).

The distribution of intersetal distances in the feeding basket determines the minimum size of particles retained. When the thoracopods are closed and pressed to the ventral side of the body, water is rapidly compressed within the feeding basket. This increases water velocity and Reynolds number, and decreases the width of the boundary layer around the setules, so that water is forced through the very finely spaced (2–3 μm) tertiary setae, termed microsetules (McClatchie and Boyd 1983). A fundamental difference between the relatively small maxillary filter of copepods and the large thoracic filter of krill is that only in the latter the filtering area is great enough for the water volume to be passed through the meshes (McClatchie and Boyd 1983).

5.3.2 Feeding on Ice Algae

Krill's ability to feed on ice algae was initially encountered in the laboratory (Hamner et al. 1983), but has subsequently also been observed in situ (Spiridonov et al. 1985; O'Brien 1987; Stretch et al. 1988; Marschall 1988). Stretch et al. (1988) found that krill foraging near ice floes exhibits two distinct behaviour patterns. When stimulated by algae released from melting ice they show area-intensive foraging. This behaviour is characterised by high speed swimming and rapid turning, accompanied by fast opening and closing of the feeding basket. Thereafter krill often orientate themselves with the ventral side towards the under surface of the ice and scrape algae with the tip of their fully-extend thoracopods. The terminal segments of their thoracopods (dactylopodites) are well-suited for this behaviour as they have rake-like structures which are much stronger and thicker than normal setae (Kils 1983). Observations by Marschall (1988) using a remotely operated vehicle showed that krill were rare under smooth-bottomed ice, but reached high densities under rugged ice. Individuals close to the ice had often dark green digestive glands due to the intensive uptake of chlorophyll pigments.

5.3.3 Feeding at the Seabed

While krill were previously known to associate with the seabed (Gutt and Siegel 1994; Ligowski 2000), it was only in summer 2006/2007 that adult krill were for the first time observed feeding at the benthos. This was off the western Antarctic Peninsula in water depths ranging from 500 to 3500 m (Clarke and Tyler 2008). Typically, the krill would dive head first into the sediment from a height of <1 m above the seabed. This would raise a small volume of sediment into the water column, and the krill would then swim rapidly upward and filter the resuspended material with characteristic movements of their feeding baskets. It seemed that krill were resuspending sediment to extract phytodetritus or other food material (Clarke and Tyler 2008). To date, there are more than 30 studies that bring evidence of krill-seabed-interactions. These comprise direct observations of krill at the seabed, their entrapment in epibenthic sampling equipment, their presence in the stomachs of benthic predators or the encounter of seabed material in their own stomachs. This shows that feeding at the seabed may be a common behaviour of this species across Southern Ocean habitats, with major implications for the food web and the vertical transport of nutrients (Schmidt et al. 2011).

5.3.4 Feeding on Copepods

In situ, krill feed on a range of copepod species and size classes (prosome length: 0.1–2.2 mm; modal prosome length: 0.27 mm; Schmidt et al. 2014). The following taxa were identified in the stomach content of freshly caught krill: *Oithona* spp., *Ctenocalanus citer*, *Drepanopus forcipatus*, *Stephos longipes*, *Microcalanus pygmaeus*, *Metridia* spp., *Calanoides acutus* and *Calanus propinquus* (Hopkins and Torres 1989; Lancraft et al. 1991; Schmidt et al. 2014). It has been suggested that small copepods such as *Oithona* spp. are passively caught when euphausiids filter-feed on phytoplankton, since they are unable to withstand the negative pressure caused by the feeding beats (Barange et al. 1991; Gibbons et al. 1991; Schmidt 2010). In contrast, raptorial feeding on larger copepods entails the complex succession of detection, attack, capture and finally ingestion. Euphausiids can use vision, mechanoreception and chemoreception to detect prey (Hamner et al. 1983; Torgersen 2001; Abrahamsen et al. 2010). So far, raptorial feeding mechanisms have not been described for *Euphausia superba*. However, it has been suggested that northern krill, *Meganyctiphanes norvegica* use mechanoreception, not vision, as the main sensory mode in near-field prey detection (Browman 2005; Abrahamsen et al. 2010). Browman (2005) clarified that the morphology of the *M. norvegica* eye does not enable sufficient spatial resolution to detect small objects at close range. However, additional clues such as movements, changes in light intensity or bioluminescence may enable krill vision to pick up clusters of prey at a distance. The process of copepod capture by *M. norvegica* has been described as follows: Attack responses are initiated well before the copepod reaches the feeding appendances. During an attack, antennae move towards the target, followed by propulsion and opening of the feeding basket (Abrahamsen et al. 2010). If successful, the copepod is sucked into the basket with the inward flow of water. Once the copepod is captured, the cusps of the mandibles and the spines of the maxillules can pierce the integument (Mauchline and Fisher 1969). It has been reported that some euphausiid species only extract the soft internal tissue of the copepods and discard the remains (Beyer 1992, and references therein). *E. superba* seem to ingest copepods completely (Atkinson, personal observations), although it is not yet known whether the mechanisms of copepod capture are the same as those described above for *M. norvegica*.

5.4 Methods to Study Krill Feeding

Like no other aspect of animal biology, diet and feeding have provoked the development of a large array of study methods. For krill at least 13 different approaches have been applied, covering various aspects of 'feeding' from morphological adaptations and feeding mechanisms to diet, trophic level, food selectivity and feeding rates. Each of the approaches has its strengths and limitations, but in their sum, they have supplied a large amount of information on krill feeding (Table 5.1). Many of the early studies were carried out in the laboratory, keeping krill in confinement. Even though these studies show what krill are doing under specific conditions, the results may not reflect their in situ behaviour. For instance, krill's preference and high intake of animal food in the laboratory (McWhinnie and Denys 1978; Boyd et al. 1984; Price et al. 1988; Nordhausen et al. 1992; Granéli et al. 1993; Kawaguchi and Takahashi 1996; Atkinson and Snÿder 1997) is not confirmed by trophic level estimates or stomach content analysis on freshly caught krill (Table 5.1). The latter has shown that krill can feed on a range of copepod species and size classes (see Sect. 5.4), but the number of copepod mandibles found in their stomachs is low (mean: 1 ± 2, max: 18 mandibles stomach^{-1}, Schmidt et al. 2014) compared to the more carnivorous euphausiid *Meganyctiphanes norvegic a* (mean: 48 ± 41, max: 151 mandibles stomach^{-1}, Schmidt 2010; see also Båmstedt and Karlson 1998). Therefore krill ingestion rates of over 700 copepods d^{-1} (Nordhausen et al. 1992) seem to be a laboratory artefact. Likewise, even though krill feed readily on moults, euphausiid ommatidia have not been found in their stomachs (Hopkins et al. 1993b; Schmidt et al. 2006), suggesting that cannibalism (McWhinnie and Denys 1978) is not a common feeding strategy. Finally, in the field, the likelihood of predator-prey-interactions between krill and larger copepods or salps is reduced as the latter divert into deeper water to avoid vertical overlap with krill swarms (Atkinson et al. 1999; Pakhomov et al. 2002).

Krill feeding is a complex interaction between nutritional requirements, swimming, swarming, and vertical and horizontal migration, which cannot easily be recreated in the laboratory. Therefore, laboratory experiments may be suitable to study specific aspects of feeding (e.g. the mechanics of food capture), while for diet studies the use of in situ-based approaches is recommended. Here, three in situ approaches are highlighted:

1. *Direct observations in the field.* Krill feeding underneath ice and at the seabed are aspects that require better regional coverage and more detailed understanding. While the initial studies have suggested that this behaviour is stimulated by the available food such as seabed phytodetritus (Clarke and Tyler 2008) or abundant ice algae (Marschall 1988), we still do not know what percentage of the local population is involved in this behaviour and what krill are gaining from this diet. Therefore it would be useful to extend the in situ observations over longer time-scales and different regions, and to collect krill directly from these surfaces (e.g. by pump suction or with epibenthic sledges) to examine their diet and body stores in comparison to krill concurrently sampled in the open water away from sea ice or the seabed.
2. *Evaluating krillfeeding activityin relation to surface chlorophyll a concentrations and krill swarm characteristics.* Most juvenile and adult krill live in the open ocean (Atkinson et al. 2008) with reduced access to a food-rich seabed or ice habitat. Therefore the diet of these krill may differ from those living over shelf-areas and needs separate consideration. As krill spend a large part of their

Table 5.1A Overview of methods to study krill feeding behaviour and diet

Method	Principle	Studied aspect	Uses	Limitations
Functional morphology studies	Cross-species comparison of the morphology of feeding-related structures (e.g. feeding basket, mandibles, gastric mill).	Lower size limit of food items. Existence of structures to fracture and grind diatoms, or pierce copepods.	Indicates ontogenetic- and species-specific differences in diet or food-size spectrum	Mandibles and gastric mill are similar for herbivorous and omnivorous species as the feeding on hard-shelled food items imposes a high demand on their mechanical capabilities
Direct in situ observations	Observing krill in their natural environment by remotely operated vehicles, moored cameras or scuba divers.	Feeding behaviour	Helps to realise the full range of krill feeding behaviour under natural conditions.	Does not give information about 'how typical' this behaviour is for the population. Observations may be specific to certain locations.
Assessing the fullness of the krill stomach and gut	Freshly caught krill are scored for the fullness of their stomach/gut as an index of their feeding activity	Feeding activity	If temporal or spatial coverage is high, feeding activities can be related to e.g. time of the day, season or sampling depth.	A sufficient number of krill needs to be scored from each station. Gives no information on ingestion rates.
Assessing the colouration of the digestive gland	Freshly caught krill are scored for the 'greenness' of their digestive gland, as a measure of ingested phytoplankton pigment.	Feeding on phytoplankton	If temporal or spatial coverage is high, feeding activities can be related to in situ phytoplankton abundance, size-class structure or taxonomic composition.	This is only a crude indication of carbon intake, as pigment-to- carbon ratios are variable. A sufficient number of krill needs to be scored from each station. Gives no information on ingestion rates.
Gut fluorescence analysis	In situ ingestion rates can be estimated from phytoplankton pigment content in freshly-caught krill, gut passage time and pigment-to-carbon ratio.	Feeding activity and phytoplankton ingestion rates	Insights into short-term variability in feeding activity (i.e. gut fullness). Phytoplankton ingestion rates and overall krill grazing impact can be estimated.	Calculations of ingestion rates have to account for pigment destruction during gut passage, changes in gut passage time with no/or alternative food, and variable pigment-to-carbon ratios.

(continued)

Table 5.1A (continued)

Method	Principle	Studied aspect	Uses	Limitations
Fecal pellet egestion rates	Freshly-caught krill are incubated onboard ship to measure their egestion rates.	Carbon egestion rates (with possible extrapolation to ingestion rates)	Reflects in situ feeding activity. Measurement of carbon egestion, fecal pellet density and sinking rates give insights into food processing and potential fate of pellets.	Conversion to ingestion rates requires knowledge of the carbon absorption efficiency (literature values vary from 42 to 94 % for krill).
Visual identification of the stomach content	Items in the stomach of freshly-caught krill are identified and measured under the microscope	Krill diet (restricted to food with solid remains; e.g. diatoms, thecate dinoflagellates, tintinnids, foraminifera, copepods, cnidaria, lithogenic particles)	Size-measurements of food items and subsequent volume calculations allow a crude estimation of the relative importance of main food categories (diatoms, thecate protozoans, copepods) in the krill diet.	No information on soft-shelled or very small food items (e.g. athecate protozoans, bacteria). Partial digestion and maceration of solid items can hinder identification to species level.
Molecular identification of the stomach content	Items in the stomach of freshly-caught krill are identified via their DNA using polymerase chain reaction (PCR)	Krill diet	Soft- and very small food items can be targeted. Theoretically, all items can be identified to species-level.	Relies on complete nucleotide databases or the ability to develop species-specific primers. No information on size of items. Fast DNA-digestion limits quantitative estimates
Immuno-chemical analysis of stomach content	Stomach extracts from freshly-caught krill are tested for anti-serum activity to a specific target species with an enzyme-linked immuno-sorbent assay (ELISA)	Krill diet	Gives unambiguous information about the ingestion of items of particular interest under natural conditions, e.g. krill feeding on *Phaeocystis* spp.	Requires additional laboratory experiments to calibrate anti-serum responses against known proportions of the target species in the stomach extract
Fatty acid (FA) biomarkers	Some organisms produce unique FA that can be traced in their consumers, e.g. in their stomach content (recent diet) or tissue (integrated diet).	Krill diet (restricted to food sources with indicative FA, e.g. diatoms, flagellates, bacteria, wax-ester-rich copepods)	Certain food sources can be compared for their relative importance in the krill diet, e.g. diatoms-*vs*-flagellates, or diatoms-*vs*-copepods.	Dietary effects can be obscured by different total lipid content of krill, as storage lipids (triacylglycerol) contain generally more saturated FA and

				structure lipids (phospholipids) more unsaturated FA.
Bulk stable isotopes	Naturally occurring isotopes of nitrogen ($^{15}N/^{14}N$) show an enrichment of ~3.4 ‰ from prey to consumer. The trophic level of krill can be calculated as $TL_{krill} = [(\delta^{15}N_{krill}-\delta^{15}N_{primary\ producer})/3.4]+1$.	Trophic level (TL)	Stable isotope measurements can be carried out in a similar manner across the whole food web; so the TL of krill can directly be compared to that of other species.	The interpretation of small-scale differences between stations may be hindered if krill feed in multiple habitats and $\delta^{15}N$ values in krill and their food reflect different integration periods.
Compound-specific stable isotopes	Some amino acids are strongly fractionated when passing through the food web (e.g. Glu), while others remain unchanged (e.g. Phe). This gives an internal index of trophic position: $\Delta\delta^{15}N_{Glu\text{-}Phe}$: + 7‰ per trophic level	Trophic level	Small-scale differences in trophic level can be assessed.	Heterotrophic food sources (protozoans, copepods) remain unidentified. Some studies suggest that microbes fractionate nitrogen in a similar manner to metazoan, while others found that feeding on protozoans is isotopically invisible.
Incubation experiments	Krill are incubated in laboratory containers, and food is offered in a range of concentrations or mixtures. The removal of food is assessed.	Krill diet, food selectivity, feeding rates	Suitable to study selective feeding on phytoplankton assemblages and seasonal differences in feeding activity.	In confinement, krill and their prey may divert from their natural behaviour due to wall effects, lack of turbulence, hindered vertical migration and swarming, lack of escape options for prey.

Table 5.1B Key findings revealed by each of the methods in Table 5.1A

Method	Important findings
Functional morphology studies	Suh and Nemoto (1987): The lower size limit of filterable particles is 2–3 μm for juvenile and adult *E.s.* Compared to other SO euphausiids, the feeding basket of *E.s.* has the largest filtering area and smallest mesh sizes. *E.s.* are the most efficient filter feeder on small items.
	McClatchie and Boyd (1983): Among euphausiids, the mandibles of *E.s.* have the largest *pars molaris* (= grinding region), indicating herbivory. But in common with omnivorous-carnivorous species, *E.s.* mandibles have spine-like marginal 'teeth', facilitating a piercing of animal prey.
	Ullrich et al. (1991): The gastric mill of *E.s.* has strong cuticular lateral teeth and numerous cluster spines, which are suitable to crush hard-shelled food items, e.g. diatom frustules.
Direct in situ observations	Hamner et al. (1983): When krill orientate in schools and swim rapidly, their feeding basket remains closed (i. e. schooling and intensive feeding are incompatible).
	Marschall (1988): Krill aggregate under rugged sea ice with high algal growth and feed by continuously scraping the ice surface with the tips of their thoracopods while opening and closing their feeding basket.
	Clarke and Tyler (2008): Krill swim or sink to the seabed as deep as 3500 m and feed on phytodetritus by nose-diving into the sediment and filtering the resuspended material.
Assessing the fullness of the krill stomach and gut	Morris and Ricketts (1984): There are more krill with full stomachs during or immediately after the hours of darkness, probably reflecting an increased level of feeding activity at night.
	Schmidt et al. (2014): Krill show regional differences in their feeding activity during winter. Individuals with empty stomachs are common in the Lazarev Sea (~80 %), but rare at South Georgia and in the Bransfield Strait (20 %).
Assessing the colouration of the digestive gland	Kawaguchi et al. (1999): There is in situ evidence that krill are feeding on micro- (>10 μm) and nanophytoplankton (2–10 μm), but not on picophytoplankton (<2 μm).
Gut fluorescence analysis	Priddle et al. (1990): Differences in the gut fullness of 3264 krill from 38 swarms reflect the patchiness of phytoplankton, while intrinsic properties (i.e. krill body length, sex-and-maturity stage), swarm characteristics (i.e. packing density, cross-sectional area), sampling time and depth are of minor importance.
	Perissinotto et al. (1997): Krill ingest ~13 % body carbon day^{-1} in a dense phytoplankton bloom. Gut passage times are 3–4.5 h for juveniles and 6–10 h for adults. Krill populations can consume 0.1–51 % of the local primary production day^{-1}.

Fecal pellet egestion rates	Clarke et al. (1988): High fecal production rates and high C content in pellets (5–14 % of dry mass) suggests that krill energy intake may be 17–28 % body carbon day^{-1}, far exceeding estimates with other methods.
	Atkinson et al. (2012): At high food concentrations, krill feed 'superfluously', which facilitates maximum rates of nutrient uptake – i.e. high feeding rates and fast gut passage ensure a constantly high substrate concentration in the gut and therefore a high rate of nutrient gain. This can lead to low absorption efficiencies and therefore high carbon content in krill fecal pellets (up to 30 % of dry mass).
Visual identification of the stomach content	Barkley (1940): During the summer months, stomachs of *E.s.* contain a large amount of diatoms. Species belong mainly to the following groups: Fragilariopsis, Coscinodiscus, Actinocyclus, Asteromphalus and Biddulphia.
	Hopkins and Torres (1989), Hopkins et al. (1993a, b): Protozoa (dinoflagellates, tintinnids, foraminifera, radiolaria, heliozoa) and metazoa (copepods, cnidaria) are regularly found in krill stomachs in autumn, winter and spring.
	Schmidt et al. (2014): Across 92 stations, algae comprised 71 ± 29 %, protozoans 17 ± 21 % and metazoans 12 ± 25 % of the identifiable food volume in krill stomachs. Copepods and protozoans supplement the krill diet year-round and can occasionally be the dominant food source. The relative importance of copepods is highest in ice covered regions during autumn and winter.
Molecular identification of the stomach content	Martin et al. (2006): Krill ingest a broad diversity of taxa, including diatoms, dinoflagellates, cryptomonads, prasinophytes, ciliates, cercozoans, choanoflagellates, turbellarians and sponge larvae. Diatoms are the most prevalent group.
Immuno-chemical analysis of stomach content	Haberman et al. (2002): Krill from the western Antarctic Peninsula feed on the prymnesiophyte *Phaeocystis antarctica*, especially in nearshore areas. However, overall *P. antarctica* comprises a small proportion of the phytoplankton intake by krill.
Fatty acid (FA) biomarkers	Pond et al. (2005): In the vicinity of South Georgia, high growth rates of krill co-occur with intensive feeding on diatoms.
	Schmidt et al. (2006): In a non-bloom, flagellate-dominated system krill are able to sustain moderate growth rates when feeding on athecate dinoflagellates.
	Schmidt et al. (2014): During winter, krill larvae contain variable but sometimes very high proportions of diatom-markers and moderate proportions of copepod-markers in their tissues. In postlarval krill, the role of diatoms and flagellates decreases, and the role of copepods increases with body length.

(continued)

Table 5.1B (continued)

Method	Important findings
Bulk stable isotopes	Stowasser et al. (2012): In the pelagic food web of the Scotia Sea, the TL of *E.s.* is higher than that of salps, similar or lower than that of common pelagic copepods, and lower than that of other euphausiids. In total the system has 4 TL, with krill occupying level 2.5 (i.e. omnivorous).
	Polito et al. (2013): Adult krill (both males and females) have a broader dietary niche and occupy a higher TL than juveniles. While phytoplankton remains life-long an important food source, krill feed on average more carnivorously as their body size increases.
Compound-specific stable isotopes	Schmidt et al. (2006): During spring-summer, krill occupied a higher TL at South Georgia (TL 2.8) and in an ice edge bloom (TL 2.7), than in non-bloom regions of the Scotia Sea (TL 2.5).
Incubation experiments	Haberman et al. (2003): Krill grazing is negligible in cryptophyte-dominated assemblages. In phytoplankton mixture, krill select diatoms over prymnesiophytes (*Phaeocystes*) even when both are of similar size.
	Wickham and Berninger (2007): Krill furcilia larvae can feed on a range of ciliates and on the small copepod *Oithona similis*. In natural autumn assemblages, ciliates are preferential ingested over phytoplankton.
	Meyer et al. (2009), Meyer (2012): Krill ingestion rates show a positive linear response to the available food concentration. However, while for furcilia larvae this functional response has a similar slope year-round (i.e. no seasonal change in feeding activity), adult krill from the Lazarev Sea show a steep slope during summer but gentle slopes in autumn and winter (i.e. seasonal reduction in feeding activity).

lives within swarms, feeding and swarming are necessarily interlinked (see Chap. 8, Tarling and Fielding 2016). High travelling speed and large swarm size may enhance the likelihood of finding patchy food; on the other hand, krill in large dense swarms may be more prone to starvation if food is scarce. Therefore it would be instructive to relate the average krill feeding activity in different swarms to in situ food availability and swarm characteristics such as speed, size, shape and density (Priddle et al. 1990). With indices of feeding activity such as gut fluorescence, stomach/gut fullness or colour of the digestive gland a large number of animals can be assessed in relatively short time and therefore a sufficient temporal-spatial coverage can be achieved. Combined data from different regions and seasons can give an overview about the in situ frequency of 'high', 'moderate' and 'low' feeding activity. This may show that even during summer 'high' feeding activities are relatively rare in the open Southern Ocean.

3. *Relating krill diet to performance indices*. To date there is no study which has quantified all major food items in the krill diet. Visual stomach content analysis gives only information about food sources with digestion-resistant remains and therefore misses out on the numerous soft items such as naked flagellates, athecate ciliates and some metazoa. Ratios of fatty acid trophic markers can indicate changes in the relative importance of food sources (e.g. diatoms vs. flagellates, or diatoms vs. copepods), but conversion to carbon intake is difficult. Newly emerging molecular approaches may offer a way forward (see Sect. 5.14), however, rapid or differential digestion of prey DNA remains an obstacle (Troedsson et al. 2009; see Chap. 7, Jarman and Deagle 2016). The estimation of feeding rates is likewise problematic. Gut fluorescence and fecal pellet egestion of freshly-caught krill are considered to reflect in situ feeding rates but both approaches have severe limitations (Table 5.1). Therefore valuable insights into the effects of food quantity and quality may arise when krill diet is related to performance indices such as egg production rate, lipid content or instantaneous growth rate (Ross et al. 2000; Schmidt et al. 2006; see also Chap. 3, Reiss 2016 and Chap. 4, Meyer and Teschke 2016). In this way the net benefit of a certain feeding environment can be established.

5.5 Ontogenetic Changes in Feeding Activity and Diet

Female krill spawn during summer (November–March) and their offspring develop through a succession of larval stages until they become juveniles the following spring (see Chap. 6, Kawaguchi 2016). After one whole season of growth as juveniles, krill reach adulthood at the beginning of their second year. During this period the body length increases from ~0.6 to ~30 mm and the facility to resist starvation increases. Thus, juvenile and adult krill deposit large amounts of lipid before the winter (~40 % of dry mass, Hagen et al. 2001) and can survive over 200 days without food (Ikeda and Dixon 1982). In contrast, larvae contain small lipid stores (6–25 % of dry mass, Hagen et al. 2001; Meyer et al. 2002a) and require regular food intake (Meyer and Oettl 2005).

The ontogenetic differences in body size and food requirements are reflected in the feeding activity and diet of juvenile and adult krill: First, younger krill have overall a higher feeding activity than adults due to higher metabolic rates and lower lipid stores (Fig. 5.1a). Second, adult krill are stronger swimmers (Huntley and Zhou 2004), which allows them to explore a wider range of habitats (water column, sea ice, seabed) and increases the ability to capture motile prey. Consequently, adult krill have a wider dietary niche than juveniles and are on average more carnivorous (Fig. 5.1b, Polito et al. 2013; Schmidt et al. 2014). Finally, during winter, larval krill are more closely associated with sea ice than adults. This spatial segregation most likely reflects the different balance between the need to feed (high in larvae, lower in adults) and the risk of predation under the ice (low in larvae, high in adults; Quetin et al. 1994). Therefore, larvae krill feed on ice algae and associated heterotrophs during winter, while adults are more likely to starve or occasionally catch larger copepods (Fig. 5.1c, Töbe et al. 2010; Schmidt et al. 2014).

Gravid male and female krill differ in their energy expenses and lipid metabolism (Clarke and Morris 1983; Pond et al. 1995; Virtue et al. 1996); however, it seems that their overall food intake and diet remain similar (Priddle et al. 1990; Schmidt et al. 2004, 2006; Polito et al. 2013). Nevertheless, concurrent stable isotope- and fatty acid measurements on individuals from the same swarm indicate the existence of dietary 'preferences' (Schmidt et al. 2006). Neighbouring krill can differ by 1–2 ‰ in their $\delta^{15}N$ values (~0.5 trophic level) unrelated to sex, maturity stage or body length (Schmidt et al. 2006, Polito, personal communication). This may reduce within-swarm competition for food, but also indicates the complexity of krill feeding behaviour.

5.6 Seasonality of Feeding

It has often been suggested that krill feed on phytoplankton during summer, and when it becomes scarce in autumn and winter they switch to heterotrophic food, benthic material or ice algae, or they cease feeding completely. Now we know that this view is too simplistic. Benthic feeding can occur year-round (Ligowski 2000; Clarke and Tyler 2008; Schmidt et al. 2011). Heterotrophic food such as copepods and protozoans are often abundant within phytoplankton blooms (Leakey et al. 1994) and therefore supplement the diet even in spring and summer (Hopkins et al. 1993a; Schmidt et al. 2006). Depending on the latitude, phytoplankton may be available in winter (Morris and Priddle 1984) and ice algae may be available in summer (Brierley et al. 2002). The krill habitat is highly variable in terms of day-length, ice cover, ocean productivity, water depth, convolution of the coastline and seabed, and therefore seasonal aspects in krill feeding are specific to the region they are living in (see Sect. 5.7). In essence, krill encounter two broadly-defined periods in a seasonal cycle – one where the phytoplankton abundance is sufficient to

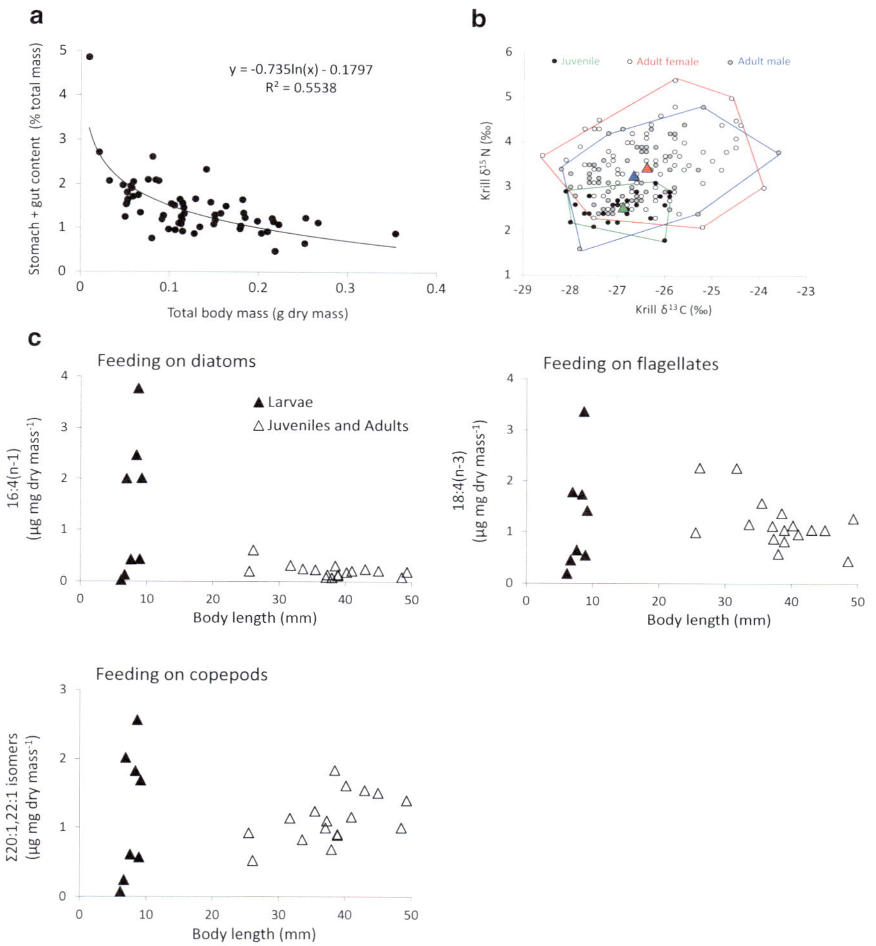

Fig. 5.1 Ontogenetic changes in feeding activity and diet. (**a**) The relative mass of the stomach- and gut content as a function of the total body mass. The panel indicates that a swarm of small krill (0.1 g dry mass) contains on average about twice as much total ingested food than a swarm of large krill (0.3 g dry mass) with the same biomass. Each of the symbols represents a pooled sample of 10 krill with the same body length. The data derived from two summer cruises in the Scotia Sea and at South Georgia (Schmidt, unpublished). (**b**) Trophic level differences between juvenile and adult krill. The plot indicates individual krill (circles); mean values for juveniles (green triangle), adult females (red triangle) and adult males (blue triangle) and the total isotopic niche area for each of them (solid line convex hulls). The data derived from two summer cruises near the South Shetland Islands and the northern Antarctic Peninsula (Polito et al. 2013). (**c**) The amount of fatty acid trophic marker in larval and postlarval krill during a winter cruise in the Lazarev Sea. The data show that feeding conditions for larvae were highly variable, but at some stations superior to postlarval krill. In postlarval krill, ingestion of diatoms and flagellates decreased with body length, while the ingestion of copepods increased. Each of the symbols represents a pooled sample of 10 krill with the same body length (Modified after Schmidt et al. 2014)

grow and spawn (chlorophyll *a*: ≥1 μg L^{-1}, Ross et al. 2000; Atkinson et al. 2006), and one where it is not sufficient. The length of the favourable period varies from several months at South Georgia to a few weeks at Bouvet Island or East Antarctica (see Sect. 5.7).

To fulfil their life-cycle, krill have firstly to make maximal use of the phytoplankton blooms to fuel growth, reproduction and the build-up of body reserves. Secondly they need to avoid excessive loss of body condition during non-bloom periods. The first is most likely facilitated by their ability to ingest and process diatoms and co-occurring heterotrophs efficiently (see Sect. 5.11, Pond et al. 2005; Schmidt et al. 2012). They achieve the second by a combination of using lipid stores and feeding on alternative food sources (flagellates, copepods, seabed material, see Sect. 5.7). The seabed for instance can act as a 'food bank', where seasonally high fluxes of phytoplankton and fecal pellets arrive, become buried and degrade only slowly (Smith et al. 2006). Studies on benthic deposit feeders have shown that high-quality organic matter can be available at the seabed even in winter (Smith and DeMaster 2008).

It has also been suggested that krill enter a stage of inactivity (quiescence) during winter, where they reduce their metabolic rates to save energy (Quetin and Ross 1991; Torres et al. 1994; Meyer et al. 2002b; Meyer 2012). This change in behaviour may be mediated by photoperiod (Teschke et al. 2007). However, an alternative explanation is possible. Rather than being an 'overwintering strategy', lower respiration rates during winter may simply reflect lower food intake. Respiration rates of adult krill are on average ~3 times lower in winter compared to summer (Meyer 2012), well within the >fourfold differences seen between fed and non-fed copepods in the laboratory (Kiørboe et al. 1985). Feeding is associated with energy-demanding processes such as absorption of food and biosynthesis of new tissue, which explains the causal link between starvation and reduced respiration rates (Kiørboe et al. 1985). Longer-term starvation may have additional side-effects such as a reduction of the gut surface area and a drop in the activity of digestive enzymes, which will affect subsequent ingestion rates (Kreibich et al. 2008; Wirtz 2013). This may explain why winter krill do not respond to excess food in the laboratory in the same manner as well-fed summer krill (Meyer 2012). There is evidence that krill feeding activity during winter differs between regions of similar latitude (e.g. Lazarev Sea vs. Bransfield Strait, Schmidt et al. 2014) and therefore seasonal feeding behaviour may not be triggered solely by photoperiod.

5.7 Regional Differences in Krill Feeding

The circumpolar habitat of Antarctic krill spans about 19 million km^2, with the islands of South Georgia and Bouvet as the northern limit (~53°S) and the pack ice zone of the southeastern Weddell Sea as the southern limit (~75°S) (see Chap. 2,

Siegel and Watkins 2016). Environmental conditions clearly differ across these regions (Table 5.2). South Georgia, for instance, has an extended shelf habitat with summer surface temperatures of up to 5 °C. Here, the phytoplankton bloom lasts for several months and the area is ice-free even in winter. The Lazarev Sea is another extreme: deep-oceanic, ice covered for 4–9 months year^{-1}, temperatures rarely exceed ~0 °C and the phytoplankton bloom is as short as ~1 month year^{-1}.

These regional differences are reflected in krill feeding activity and diet. At South Georgia, most krill engage in feeding even during winter, they ingest high amounts of lithogenic particles year-round due to both feeding at the seabed and by uptake of glacial flour in the water column. Copepods and protozoans are an important supplementary food source in summer, autumn and winter (Fig. 5.2a, Morris and Priddle 1984; Schmidt et al. 2014). In the Lazarev Sea, feeding during winter seems less common for postlarval krill (Schmidt et al. 2014). However, the occasional consumption of copepods represents a considerable food intake (Fig. 5.2a). During both autumn and winter, copepods contribute substantially to the diet of adult krill in the Lazarev Sea (Fig. 5.2a, b). This is in contrast to the Scotia Sea and Bransfield Strait, where krill feeding on copepods was rarely found, neither when using visual stomach content analysis nor fatty acid trophic markers (Fig. 5.2a, b). At South Georgia, krill feed mainly on small copepods such as *Oithona* spp., whereas in the Lazarev Sea the large winter-active *Calanus propinquus* is a common prey (Schmidt et al. 2014). While *Oithona* spp. is most likely caught when krill are filter-feeding on suspended material, the capture of *C. propinquus* may require directed raptorial behaviour. The relative importance of diatoms vs. flagellates in the krill diet also differs between regions. At South Georgia krill feed mainly on diatoms, but proportions are shifted towards flagellates in the Scotia Sea, Bransfield Strait and Lazarev Sea, especially during summer and autumn (Fig. 5.2b). Long-lasting diatom blooms at South Georgia are favoured by high nutrient supply from the island and shelf area.

The comparison between South Georgia and the Lazarev Sea suggests that in productive shelf areas postlarvae krill are likely to supplement their diet with seabed material, while in the deep ice-covered ocean they feed more carnivorously. Additional, more subtle factors may also play a role, such as the local copepod abundance and species composition (Rudjakov 1996; Atkinson and Sinclair 2000), the type and 'age' of the sea ice (Marschall 1988), the overall productivity of the area and the seabed morphology (Wakefield et al. 2012). To resolve such issues, more cross-regional studies are required. Therefore it is important that krill feeding is examined in a consistent manner, i.e. using the same study method or set of methods. Table 5.2 gives an overview of environmental conditions across the main krill habitats, and available information on krill observations under the ice or at the seabed. This table also indicates that there are regional differences in krill body condition and size structure of the population, which most likely reflect their nutrition. Understanding these regional differences is essential to predict fluctuations in the circumpolar krill stock and to guide a sustainable krill fishery.

Table 5.2 Characteristics of the circumpolar habitats of Antarctic krill

Region	Water depth (m)	Average seasonal length (days) of ice cover (white), phytoplankton bloom (black) and non-bloom (grey)	Krill onshore-offshore distribution	Max. krill body length (mm, 95 % percentile)	Autumn/Winter lipid content (% dry mass)	Krill observed under sea ice	Krill-seabed interactions
Lazarev Sea 57–70°S; 5°W-5°E	3000–5000	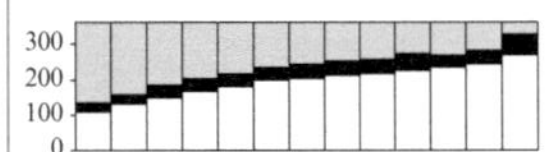	Mainly oceanic	52	47	Sp, S, A, W	
Cosmonaut Sea 57–70°S; 30–50°E	1000–5000	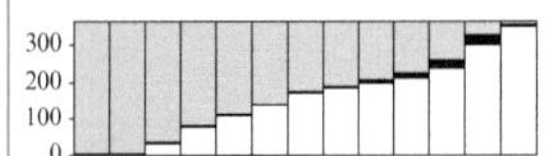	Shelf break and off-shelf	51		Sp, S, W	S, A, W
Cooperation Sea, Prydz Bay 57–70°S; 70–80°E	1000–3000	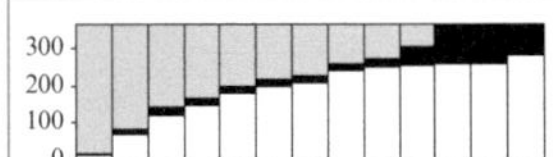	Shelf break and off-shelf	55			S, A
Mawson Sea 57–66°S; 95–120°E	1000–4000	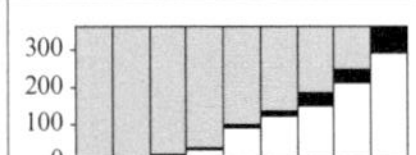	Shelf break and off-shelf	50		Sp	
Somov Sea 58–71°S; 160–170°E	1000–3000	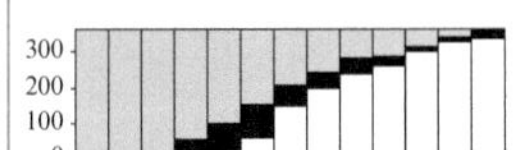	Shelf break	51		S	S
Western Antarctic Peninsula 57–70°S; 65–75°W	500–3000	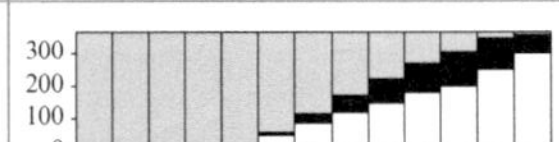	Inner shelf	56	20	Sp, S, A, W	S, A, W

Bransfield Strait 62–64°S; 58–61°W	100–700	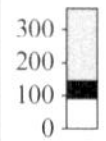	Shelf	54	48	W	S, W
Scotia Sea 53–61°S; 40–50°W	500–3000	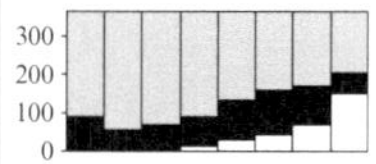	Oceanic	53		Sp, A, W	Sp, S
Weddell Sea 78°S; 30–60°W	1000–5000	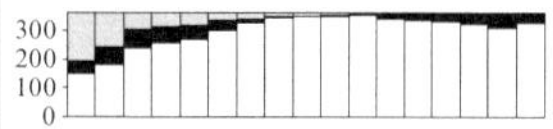	Unknown	unknown	40	Sp, S, A, W	S
South Georgia 53–54.4°S; 36–41°W	200–1000	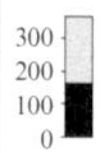	Mainly shelf	58	36	–	Sp, S, A, W

(1) Annual length of ice cover, phytoplankton bloom and non-bloom as in Schmidt et al. (2014). Histogram bars represent 1° latitudinal steps from the north (left) to the Antarctic continent (right), except for Bransfield Strait and South Georgia where data are integrated across the whole latitudinal range given
(2) Maximum krill body length (i.e. body length that includes 95 % of the ≥40 mm population); Schmidt et al. (2014)
(3) Krill winter lipid content; Hagen et al. (2001), Ju and Harvey (2004), and Schmidt et al. (2014)
(4) Krill observed under sea ice; Guzmán (1983), Naito et al. (1986), Kawaguchi et al. (1986), O'Brien (1987), Kottmeier and Sullivan (1987), Marschall (1988), Stretch et al. (1988), Daly (1998), Hamner et al. (1989), Siegel et al. (1990), Daly and Macaulay (1991), Melnikov and Spiridonov (1996), Brierley et al. (2002), O'Brien et al. (2011), and Flores et al. (2012b)
(5) Krill-seabed interactions, Schmidt et al. (2011)
Sp spring, *S* summer, *A* autumn, *W* winter

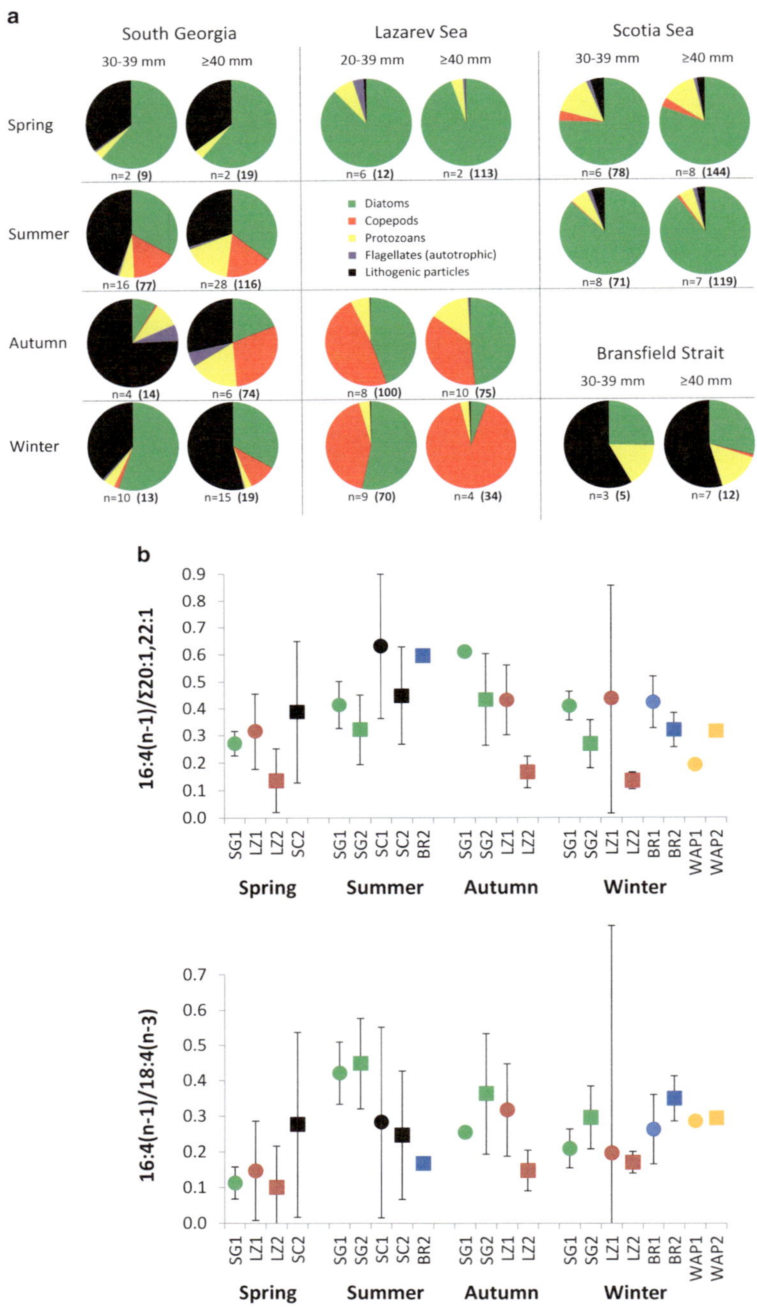

Fig. 5.2 Regional differences in krill diet. (**a**) Krill stomach content at South Georgia (spring, summer, autumn, winter), in the Lazarev Sea (spring, autumn, winter), Scotia Sea (spring, summer) and Bransfield Strait (winter) (Original data in Schmidt et al. 2014). Two size classes of krill were considered, ≥ 40 mm (mainly adults) and < 40 mm (mainly juveniles). n number of

5.8 Ingestion Rates and Functional Response

5.8.1 Ingestion Rates

A previous review of krill energetics (Quetin et al. 1994) concluded that: "We now view *Euphausia superba* as an active organism, perhaps more like a small schooling fish than a scaled-up copepod. *E. superba* has a high energy throughput, perhaps 20 % of body carbon per day or higher, sustained by a high and effective rate of filtration". This radically different view questioned previous approaches whereby feeding rates were derived from incubations in small bottles. However it begs the question of just how to derive real-world feeding rate measurements for krill, and 20 years later this has still not been resolved.

Several authors have compiled daily rations of krill estimated with a wide range of direct and indirect methods (Knox 1994; Perissinotto et al. 1997; Pakhomov et al. 2002). These values range from a few percent to ~28 % with little consensus on what the maximum daily ration might realistically be. In Table 5.3, we compile a series of measurements that may represent maximum ingestion rates of larvae and postlarvae. These values were all derived in bloom conditions in summer, based either on bottle incubations (larvae), gut fluorescence (juveniles) or faecal egestion (adults). With the latter method exceptionally high values were estimated, 17–28 % body C d^{-1}. Here ingestion rates were calculated from fecal pellet egestion rates assuming a carbon absorption efficiency of 70–85 % (Clarke et al. 1988).

However daily rations much greater than 20 % may be unrealistically high for postlarval krill for three reasons: Firstly, krill are known to feed "superfluously" in high food concentrations with fast gut throughput and high egestion rates, which leads to low absorption efficiencies and carbon-rich fecal pellets (Atkinson et al. 2012; Schmidt et al. 2012). Thus, some of the high ingestion rates calculated from carbon egestion and literature values on absorption efficiency may have been overestimates, with the real carbon absorption efficiency being lower. Secondly, maximum in situ growth rates of postlarval krill are equivalent to ~5 % of body C d^{-1} (Clarke and Morris 1983; Atkinson et al. 2006). Such growth rates require a daily ration of no more than 16–19 % when assuming a gross growth efficiency of 0.26–0.32 (Lasker 1960, 1966). The third reason is that, in line with allometric scaling expectations, the mass of the krill stomach content as a percentage of the total

Fig. 5.2 (continued) stations. The number in brackets is the total volume of items identified in the stomach (× 10^6 μm^3) for individuals that had been feeding. (**b**) The ratio of fatty acid trophic markers in krill tissue from South Georgia (SG, green), the Lazarev Sea (LZ, brown), Scotia Sea (SC, black), Bransfield Strait (BR, blue) and the western Antarctic Peninsula (WAP, yellow) during different seasons. Juveniles (1, circles) and adults (2, squares) are presented separately. The following marker fatty acids were used 16:4(n−1) for diatoms, Σ20:1, 22:1 isomers for copepods and 18:4(n−3) for flagellates. The plots are based on data presented in Cripps and Atkinson (2000), Atkinson et al. (2002), Ju and Harvey (2004), Pond et al. (2005), Schmidt et al. (2014), and Reiss et al. (2015)

Table 5.3 Compilation of studies conducted during summer bloom periods in which high (possibly near maximum) daily rations of krill life stages have been determined

Stage	Maximum daily ration (% body C d^{-1})	References
Calyptopis III	25.9	Meyer et al. (2003)
Calyptopis III – Furcilia I	17.8	Huntley and Brinton (1991)
Furcilia I	26.2	Meyer et al. (2003)
Furcilia I–II	8.5	Huntley and Brinton (1991)
Furcilia II	14.6	Meyer et al. (2003)
Juvenile	13	Perissinotto et al. (1997)
Adult	17–28	Clarke et al. (1988)

body mass declines with increasing krill size (Fig. 5.1a). This suggests that on average daily rations of adults are lower than those of larvae (see Table 5.3).

In Fig. 5.3, these maximum daily ration estimates are plotted against a compilation of maximum ingestion rates of pelagic invertebrates ranging from nanoflagellates to fish (Kiørboe and Hirst 2014). For all life-cycle stages of krill the maximum rates exceed those predicted from the regression line, being particularly evident for the postlarvae. Clearly there are caveats with any such attempt to put krill feeding rates into wider context, for example the values have all been adjusted to a temperature of 15 °C using the same Q_{10} value of 2.8 (Kiørboe and Hirst 2014). Nevertheless, even when based on the lower rations for postlarvae of 13–17 % (Table 5.3) and notwithstanding uncertainty over temperature conversions, *E. superba* postlarvae seem to achieve exceptionally high food intake for their size (see Fig. 5.3). In summary, while it seems unlikely that maximum daily rations of postlarvae greatly exceed 20 %, krill are clearly a species with high energy throughput.

5.8.2 Functional Response

Functional responses of *Euphausia superba* feeding rates have been measured in the laboratory both for adults and larvae. A common finding is that feeding rates do not saturate even at high food concentrations (Price et al. 1988; Atkinson and Snÿder 1997; Ross et al. 1998; Meyer 2012; see Chap. 4, Meyer and Teschke 2016). This is in contrast to results of growth studies using the instantaneous growth rate (IGR) method, where modest half saturation concentrations of 0.33 μg Chl *a* L^{-3} (Atkinson et al. 2006) and 0.50 μg Chl *a* L^{-3} (Ross et al. 2000) have been determined.

Several factors may contribute to this finding. First, it may represent "superfluous" feeding (Schmidt et al. 2012) whereby at high food concentrations, feeding rates increase and gut transit time and absorption efficiency decrease, allowing increased total absorption rates of essential molecules. This has been established for

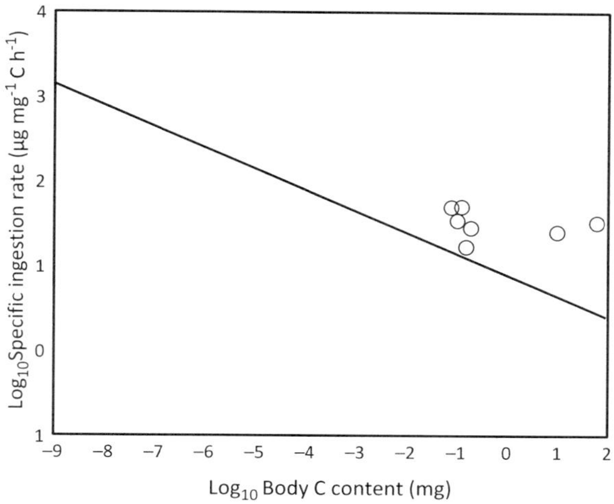

Fig. 5.3 Ingestion rates and functional response. Comparison of high (likely maximum) specific ingestion rates of ontogentic stages of krill derived in summer bloom conditions (Table 5.3) with a literature best-fit regression (Kiørboe and Hirst 2014) across a wide range of pelagic organisms, denoted by the solid line. Krill source data were from the publications in Table 5.3, adjusted to a common reference temperature of 15 °C using a Q_{10} of 2.8 in common with Kiørboe and Hirst (2014). A conservative value for the maximum adult ration (17 %) is plotted for reasons presented in the text. The regression line (slope −0.25) derived from a meta-analyses of 327 maximum ingestion rates for protozoans, flagellates, ciliates, copepods, other crustaceans, euphausiids, amphipods, chaetognaths, cnidarians, ctenophores, tunicates and fish (Kiørboe and Hirst 2014)

copepods (Thor and Wendt 2010) where maximum ingestion rates are controlled by food processing in the gut rather than by food concentration or the ability to feed rapidly (Wirtz 2013). For krill, fast gut throughput rates (Clarke et al. 1988; Pond et al. 1995) and high carbon- and fatty acid concentrations in rapidly egested pellets support the notion of 'superfluous' feeding (Atkinson et al. 2012; Schmidt et al. 2012). Secondly, functional response experiments are seldom run for a full 24 h-period, so longer experiments would be needed to test whether the high ingestion rates, measured for instance over 12 h, sustain commensurately high daily ration values. Finally, the low saturation concentration for growth may reflect only the growth in length (mm d^{-1}) measured by the IGR technique; while additional capacity for growth may be channelled into the build-up of gametes or lipid reserves (Hagen et al. 2001). In any case, the discrepancy between functional responses for somatic growth and feeding requires consideration when quantifying the energy budget of krill.

5.9 Effects of Temperature, pCO_2 and Food Size

In future climate scenarios, increased water temperature, reduced pH, increased freshwater run-off and increased abundance of small cryptophytes relative to diatoms are commonly cited combinations of conditions (Flores et al. 2012a;

Kawaguchi et al. 2013; Ducklow et al. 2013; Constable et al. 2014). Even though several of these apply only to part of the krill habitat (e.g. Western Antarctic Peninsula) this section examines the effects of these potential stressors on feeding.

To our knowledge, the direct effect of increased temperature on krill feeding rates has not been measured, probably because it is hard to do so in a laboratory setting due to artefacts associated with containment and temperature acclimation. The only experiment to date examining direct effects of pCO_2 on krill metabolism is by Saba et al. (2012). At pCO_2 concentrations of ~700 ppm, a stress-type response of elevated feeding-, respiration- and excretion rates was found when compared to rates at ambient pCO_2 concentrations.

The notion that krill feed inefficiently on nano-sized particles while salps benefit from decreased food sizes is mentioned increasingly in the context of climate change stressors (see Sailley et al. 2013; Constable et al. 2014). This concept may be based on the observation of low krill feeding rates when a cryptophyte-dominated diet was offered in the laboratory (Haberman et al. 2003). However, it counters other studies which suggest that krill can feed efficiently on small items and have a large predator-prey size ratio (Kils 1983; Suh and Nemoto 1987; Kawaguchi et al. 1999). In Fig. 5.4, available experimental and in-situ studies are combined to derive their food-size spectrum. A major drawback of such studies is that they invariably test only a minor component of the full food spectrum. When offered just one or a few food sources in the laboratory, feeding behaviour can differ substantially from that in natural mixtures in the sea (Wirtz 2014). Nevertheless, these studies document krill's ability to ingest cells as small as 3–4 μm as well as copepods as large as ~3 mm.

While the potential food size ranges across three orders of magnitude, the preferred size is still not clear. Some studies suggest it to be ~20–30 μm (Boyd et al. 1984), or >40 μm (Quetin and Ross 1985), while a laboratory study with a mixture of copepods, algae and protozoans suggested maximal clearance rates were on copepods of ~1 mm (Atkinson and Snÿder 1997). Subsequent field studies, however, suggest that this latter result (and indeed several others in which laboratory-held krill ingested copepods in preference to algae) are laboratory artefacts. Analysing the gut content of krill from a variety of regions and seasons, the modal copepod prosome length based on mandible widths was only 267 μm (Schmidt et al. 2014). This supports the concept that *E. superba* are mainly a filter feeding species that only occasionally catches some larger metazoans.

The optimum prey size for krill is likely somewhere within the range 40–300 μm, and their predator-prey size ratios are not unusually high, compared to other suspension feeders of equivalent size. Krill have perhaps been described as having an exceptionally high predator-prey size ratio because they are often compared with copepods, whose ratios are mainly in the range 10–50 (Hansen et al. 1994). Figure 5.5 illustrates the fact that, among microplankton feeders, predator-prey size ratios tend to increase strongly with grazer size. This brings the optimum predator-prey size ratio for krill to ~1000, in line with other large microphages such as salps and planktivorous fish.

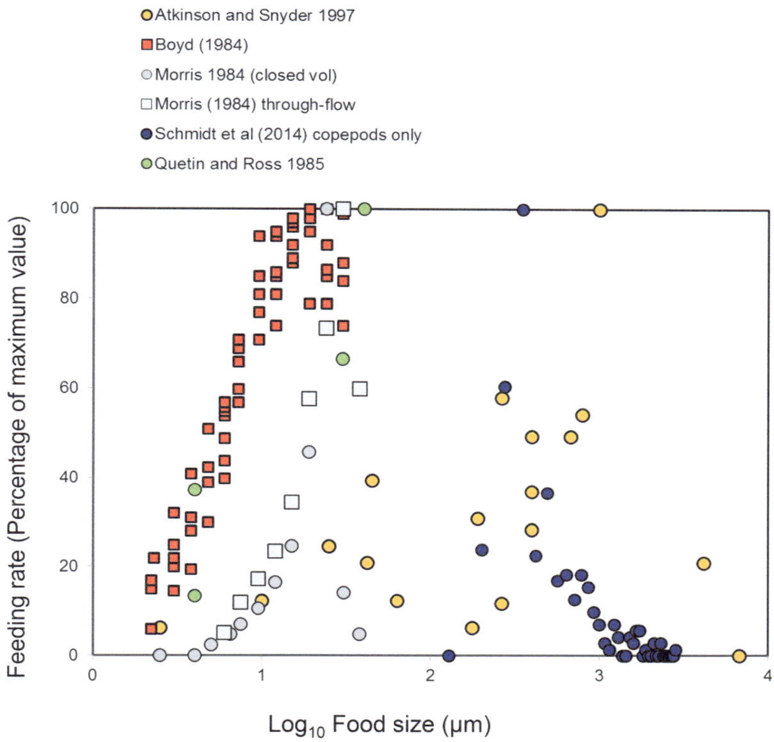

Fig. 5.4 Effects of food size. Compilation of studies that examine *E. superba* feeding rates across a spectrum of food sizes. Results are normalised by expressing them as percentages of the maximum value obtained in each study. For Schmidt et al. (2014) the results are from multiple regions and seasons (see Fig. 5.2a), with copepod prosome lengths calculated from mandible width using Karlson and Båmstedt (1994) and converted to total body length assuming this is 1.3× prosome length (a total of 253 mandibles were recorded and measured). Values in each size category are expressed as percentages of the maximum value, which is 71 mandibles recorded within the category of 350 μm copepod total length. Other values pertain to clearance or ingestion rates (for mixed assemblages of phyto- and zooplankton in Atkinson and Snÿder 1997, and solely phytoplankton in the remaining studies). Food size (x-axis) refers to the maximum linear dimension

In addition to the large predator-prey size ratio, another important trait of krill is their wide range in potential predator-prey size ratios (roughly 20–20,000, by body length). Figure 5.5 shows that *E. superba* can access food items spanning about three orders of magnitude in equivalent spherical diameter. While salps and sardines also display this ability to a lesser extent, the range of available food sizes is far smaller in many copepods (Fig. 5.5). Therefore, in a given volume of water, krill encounter a larger amount of suitable food items than copepods. Based on Southern Ocean biomass spectra there are similar amounts of plankton biomass within equal logarithmic intervals of mass (Tarling et al. 2012). This would imply that for krill

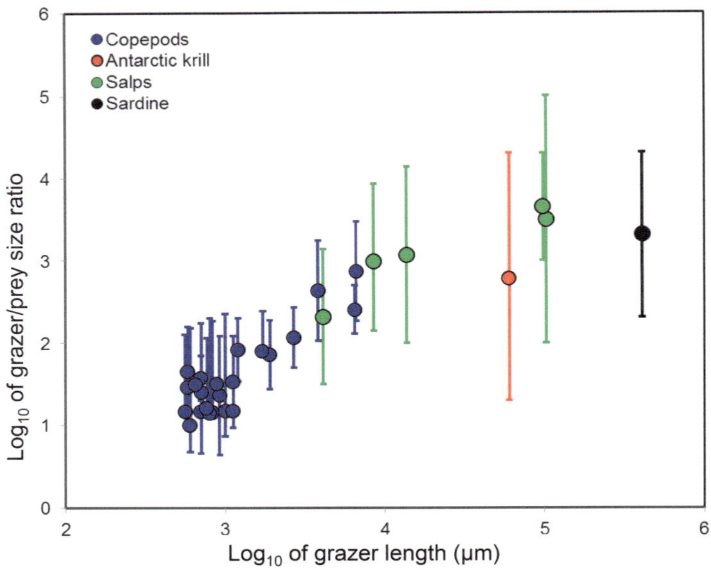

Fig. 5.5 Effects of food size. Compilation of data on optimum and range in predator-prey size ratio (in terms of length or equivalent spherical diameter), plotted against grazer length. Non-krill data are compiled from Hansen et al. (1994), Fuchs and Franks (2010), Wirtz (2012), Saiz et al. (2014), and Nikiloudakis et al. (2012) and references therein. Krill data are derived from Fig. 5.4

with a thousand-fold range in food size the available food concentration could be ~50 % higher than for copepods with a hundred-fold range in food size.

With the rapid warming at the Western Antarctic Peninsula there have been predictions of increased meltwater run-off that may favour the occurrence of cryptophytes over diatoms (Ducklow et al. 2013; Mendes et al. 2013). Diatoms are generally considered of higher food quality for zooplankton than cryptophytes or prymnesiophytes (Ross et al. 2000). Indeed, fast gonad development in krill co-occurs with the spring diatom bloom (Cuzin-Roudy and Labat 1992; Schmidt et al. 2012) and krill are more enriched in polyunsaturated fatty acids (PUFA) when feeding on diatoms rather than copepods (Schmidt et al. 2014). This may imply that the lack of diatoms can have adverse effects on krill development and their nutritional quality for higher predators. However, krill occurrence at the inner shelf is a phenomenon specific to the Western Antarctic Peninsula, while in other regions they have a more oceanic distribution and are therefore less affected by melting glaciers (Atkinson et al. 2008). Moreover, food quality depends on a range of factors such as species composition and status of the bloom. In a study in the Scotia Sea for instance, diatom-dominated diets led to moderate- or high growth rates in krill depending on whether it was a spring bloom near the ice edge or a summer bloom at South Georgia, while a nanoflagellate-dominated diet in the open ocean did likewise support moderate growth rates (Schmidt et al. 2006).

5.10 Feeding Selectivity

Feeding selectivity in krill can occur in several forms. In terms of food size, the mechanical characteristics of the feeding basket will help to set an upper and lower possible size of particles that can be eaten (see Sect. 5.9). Superimposed on that, the feeding basket can be used in different ways (see Sect. 5.3) to achieve different effective mesh sizes depending on food concentration and size (Kils 1983). However, this physics-based selectivity does not explain the finding of Haberman et al. (2003), where laboratory-acclimated krill selected diatoms over *Phaeocystis antarctica* of similar size. Remote chemical detection is a possible selection mechanism employed by copepods (Kiørboe 2011). Alternatively, tasting and post capture rejection may lead to selectivity. However, given the great size of the krill feeding basket compared to the volume of water entrained in a copepod's feeding current, rejection or ingestion of small cells on an individual basis seems unlikely.

In the field, the large behavioural repertoire of krill allows them to switch between food substrates – the water column, the sea ice, the sea bed. Given the mobility of krill, they may be caught in different vertical horizons from where they have been recently feeding. Thus individuals caught from the upper water column can retain tracers of feeding activity either from the overlying sea ice (Meyer et al. 2002a; Stübing et al. 2003) or from the seabed (Ligowski 2000; Schmidt et al. 2014). Such mobility makes it very hard to disentangle the various forms of selectivity that krill may use. With the improvements in filming techniques since the 1980s, further in situ studies may reveal more detail of the actual mechanisms of food selection.

5.11 Food Processing

The stomach and digestive gland are sites of food digestion. In the anterior region of the stomach (cardia) the food is ground and mixed with digestive enzymes. In the posterior part of the stomach (pylorus) the crushed food is filtered and the filtrate is pumped into the digestive gland (Ullrich et al. 1991). The digestive gland is a system of blind-ending tubules consisting of a uni-layer epithelium with special cells for enzyme synthesis and nutrient resorption (Sabarowski and Buchholz 1999). Digestive enzymes are released directly into the lumina of the tubules to act upon the filtered chymus. Coarse food residues are transported into the hindgut, where material is packed into fecal pellets for egestion.

A number of digestive enzymes have been identified from the digestive tract of *E. superba*, including glucanases (e.g. laminarinase, amylase, cellulose, galactosidase), proteases (e.g. trypsin, chymotrypsin) and chitinases (e.g. endo-chitinase, *N*-acetyl-β-D-glucosaminidase) (Mayzaud et al. 1985; Sabarowski and Buchholz 1999; Saborowski 2012). Chitin consists of amino sugar and is therefore of considerable nutritive value. There are various potential sources of chitin in the krill diet, e.g. copepods, diatoms of the genus *Thalassiosira* and moulted cuticles of their fellows. The latter are regularly found in stomachs of freshly-caught krill (Hopkins et al. 1993a, b; Schmidt et al. 2006) and krill have been observed to skilfully handle and ingest moults during laboratory incubations (Hamner et al. 1983). Saborowski (2012) compared the proteolytic activities of *E. superba* with those of *Meganyctiphanes norvegica*, and found more complex digestive properties in *E. superba*, e.g. the expression of four rather than one trypsin isoforms and elevated proteinase activities in the stomach. The author explained these differences with the more herbivorous diet of *E. superba* (Saborowski 2012). Phytoplankton contains less protein than zooplankton, therefore the protein uptake has to be optimised, for example by increasing the digestive enzyme activities and by using multiple endopeptidases, each with slightly different substrate specificities.

Algae are generally considered 'less digestible' or 'low energy' food, while copepods and athecate protozoans are more 'easily digestible' prey. For decapods it has been found that herbivorous larvae adjust to their low energy food with high enzyme activities, rapid gut passage and low absorption efficiency (Le Vay et al. 2001). In contrast, carnivorous larvae show lower levels of enzyme activities but compensate by longer gut retention and higher absorption efficiency (Le Vay et al. 2001). This fits with observations on *E. superba*, where carbon absorption efficiency usually ranges from 72 to 94 % (Kato et al. 1982), although values as low as 42 % have been reported (Schnack 1985a). When feeding on phytoplankton blooms, krill often produce large amounts of relatively carbon-rich, loosely-packed fecal pellets indicating fast gut passage and low absorption efficiency (Atkinson et al. 2012). The benefit of this 'superfluous' feeding is that substrate concentrations are constantly high in the stomach, which combined with high digestive enzyme activities leads to high absolute rates of nutrient gain (e.g. mol nutrient absorbed hour^{-1}) (Jumars 2000). Thus, even extensive uptake of 'indigestible' lithogenic particles (Schmidt et al. 2011) does not necessarily indicate inefficient feeding, because high enzyme activities and fast gut passage can ensure that significant amounts of the associated organic matter are extracted.

For adult krill, estimates of gut passage time vary from 47 min to ~10 h (Pond et al. 1995; Clarke et al. 1988; Pakhomov et al. 1997; Perissinotto et al. 1997), reflecting differences in food quantity and quality (Pond et al. 1995; Perissinotto and Pakhomov 1996). To evaluate the relative absorption efficiency for specific food components, their ratios can be compared in the diet and the fecal pellets. During a multi-seasonal study in the Scotia Sea, the C:N mass ratio of krill fecal pellets ranged from 4.9 to 13.2 (median 7.8), which was higher than values in krill tissue (3.9) or their food (5.4), pointing to preferential uptake of nitrogen over carbon (Atkinson et al. 2012). There is also evidence that krill can vary the absorption efficiency for individual fatty acids according to their nutritional needs (Schmidt et al. 2012). At an initial stage after the winter, fatty acid absorption was most efficient for the essential PUFAs 20:5(n−3) and 22:6(n−3), while during vitellogenesis when oocysts are supplied with lipidic yolk and grow in size, the uptake of 14:0 and 16:1(n−7) was favoured (Schmidt et al. 2012). The preferential absorption of PUFAs and 16:1(n−7) was also found in laboratory feeding experiments with postlarval krill (Stübing et al. 2003).

5.12 Trophic Overlap with Other Southern Ocean Euphausiids

Seven species of euphausiids occur in the Southern Ocean and the region just north of the Polar Front, with their habitats spanning different latitudes (John 1936; Everson 2000; Mackey et al. 2012). *Euphausia crystallorophias* prefers neritic waters and is the most common euphausiid on the Antarctic continental shelf. It is the only species that prevails in the permanent pack ice zone. *E. superba* and *Thysanoessa macrura* are found from the seasonal pack ice zone to the Antarctic Polar Front. *Euphausia triacantha* and *E. frigida* are distributed from north of the continental shelf break to the Antarctic Polar Front, *Thysanoessa vicina* is dominant in a narrow band on both sides of the Polar Front and *E. vallentini* occurs north of it. While *E. superba* often occupies the upper ~100 m water column, the populations of other species spread from the subsurface down to variable, but usually much greater water depths (Lancraft et al. 1991; Haraldsson and Siegel 2014).

There are only a few studies which have compared the diet of these species using the same methodological approach (Table 4.4). Examinations of the feeding apparatus have shown that the basket of *E. superba* has a larger filtering area and finer mesh sizes than that of the other euphausiids. The feeding baskets of *E. vallentini*, *E. crystallorophias* and *E. frigida* have medium-size meshes, and that of *E. triacantha* has coarse meshes. Therefore, *E. superba* seems better adapted to filter-feed on small particles than the other euphausiids. In agreement with this morphology, *E. superba* contained more diatom indicating fatty acids in their tissues and occupied a lower trophic level when analysed together with other species (Table 4.4). According to their fatty acid composition, *E. triacantha* and *T. macrura* are the most carnivorous species. However, *E. crystallorophias* has also been described to feed on metazoans, including polychaetes, pteropods and coelenterates (Hopkins 1987). At the same time, relatively high proportions of the fatty acids 16:1(n − 7) and 18:4(n − 3) indicate that *E. crystallorophias* also graze on phytoplankton (Kattner and Hagen 1998). Despite living in the pack-ice region, *E. crystallorophias* seems not to inhabit ice crevices or to scrape algae from beneath the ice (O'Brien 1987; Nordhausen 1994), unlike *E. superba*. However, *E. crystallorophias* have been caught and filmed at the seabed (Atkinson, unpublished observations) and benthic diatoms were found in their stomachs (Kittel and Ligowski 1980). Given the overlapping habitats of the Southern Ocean euphausiids, further comparative studies of their diet and food processing would be valuable.

5.13 Ecosystem Implications of Krill Feeding

5.13.1 Grazing Impact on Phytoplankton

Primary production in the Southern Ocean may be limited by three factors: nutrients, light and grazing (Venables and Moore 2010). During a scientific cruise in the Scotia-Weddell Sea it was observed that a krill swarm grazed down a diatom dominated bloom within a few hours (Smetacek and Veth 1989). This is not surprising because large krill swarms may contain up to 10,000–30,000 individuals m^{-3} (Hamner et al. 1983), each capable of clearing several litres per hour (Quetin et al. 1994). However, since swarms are local phenomena it is not to be expected that krill can graze down phytoplankton across large areas (Atkinson et al. 2014). Thus, krill grazing impact is highly variable, even within the same region and season (*see* Pakhomov et al. 2002, their Table 5.4). High rates (40–420 % of daily primary production) have been observed near the South Shetland Islands (Holm-Hansen and Huntley 1984), in the Lazarev Sea (Perissinotto et al. 1997), in the Bransfield Strait (von Bodungen 1986), at the western Antarctic Peninsula (Ross et al. 1998), in the Prydz Bay Region (Samyshev 1991) and at South Georgia (Pakhomov et al. 1997).

At the eastern side of South Georgia, a negative relationship was found between krill density and phytoplankton abundance (Whitehouse et al. 2009). Calculations confirmed that krill grazing rates exceeded the phytoplankton growth rates in this area (Whitehouse et al. 2009). Further downstream, along the island, ample supply of micro- and macronutrients promoted higher primary production rates, while lower krill densities had less grazing impact. Here the relationship between krill- and phytoplankton abundance was positive (Whitehouse et al. 2009). Both high krill densities and low chlorophyll *a* concentrations are a recurring phenomenon at the eastern side of South Georgia (Fig. 5.6), which indicates that in this region phytoplankton is often under 'top-down' control by krill.

In addition to reducing phytoplankton stocks, krill can also modify its species composition. Near the Antarctic Peninsula, Kopczynska (1992) found strong evidence that both deep mixing and krill grazing act to suppress diatom blooms and cause flagellates to dominate. Likewise, Jacques and Panouse (1991) found in the Weddell/Scotia Confluence area a rapid change from a high biomass netplankton community to a nanoplankton system and interpreted this as an effect of krill grazing. Larval krill may also be important phytoplankton grazers as a study in the Bellingshausen Sea has shown (Pakhomov et al. 2004).

5.13.2 Nutrient Recycling and Mobilisation

Traditionally, microheterotrophs are considered the main agents in nutrient recycling. However, Antarctic krill play a crucial role in some parts of the Southern Ocean due to their large biomass, intensive feeding and access to food sources

Table 5.4 Dietary differences between *E. superba* and other Southern Ocean euphausiids

Feeding indicator	Characteristics	E. s.	T. m.	E. c.	E. f.	E. t.	E. v.	References
Body length (mm)		60	37	41				
Morphology of the feeding basket (adults)	Lower limit of filterable particle sizes (μm)	2–3		16–23	15–19	27–39	8–11	1
	Filter area (mm^2)	277		45	15	45	28	1
Fatty acid trophic marker (autumn)	Diatoms – 16:4(n−1) (% of total FA)	0.7	0.2		0.3	0.3		2, 3
	Flagellates – 18:4 (n-3) (% of total FA)	2.0	1.0/ 1.6	3.4	0.9	1.1		2–4
	Copepods – 20:1 (n-9) (% of total FA)	1.1	1.1/2.3	1.3	1.4	2.1		2–4
Trophic level based on $\delta^{15}N$ values		2.5	3.1	3.2	3.0	3.3		5
Stomach content	Spring	Phytoplankton, protozoans	Phytoplankton, invertebrate eggs		Phytoplankton, invertebrate eggs			6
	Summer		Metazoans, phytoplankton, protozoans	Metazoans, phytoplankton, protozoans				7

(continued)

Table 5.4 (continued)

Feeding indicator	Characteristics	E. s.	T. m.	E. c.	E. f.	E. t.	E. v.	References
	Autumn	Phytoplankton, protozoans, metazoans	Metazoans, phytoplankton, protozoans					8, 9
	Winter	Phytoplankton, protozoans, metazoans						10

E. s. *Euphausia superba*, T. m. *Thysanoessa macrura*, E. c. *Euphausia crystallorophias*, E. f. *Euphausia frigida*, E. t. *Euphausia triacantha*, E. v. *Euphausia vallentini*

(1) Suh and Nemoto (1987), (2) Stübing and Hagen (2003), (3) Kattner and Hagen (1998), (4) Hagen and Kattner (1998), (5) Stowasser et al. (2012) and Stowasser unpublished, (6) Hopkins et al. (1993a), (7) Hopkins (1987), (8) Hopkins (1985), (9) Hopkins and Torres (1989), (10) Hopkins et al. (1993b)

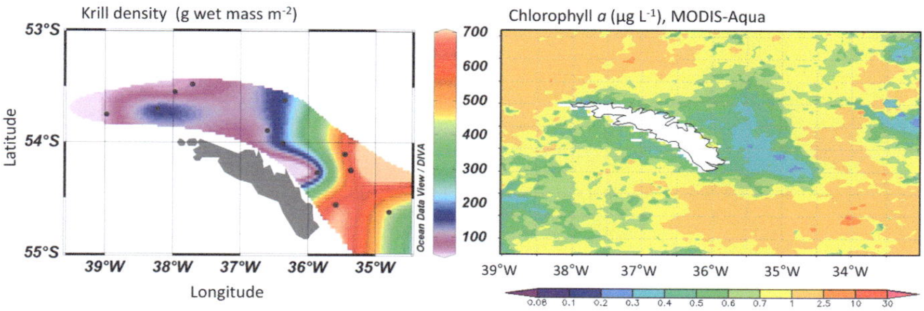

Fig. 5.6 Krill grazing impact on phytoplankton. Spatial overlap between the region of high krill density (left) and low chlorophyll a concentrations (right) at the eastern side of South Georgia. Here the phenomenon is illustrated with data from summer 2010/2011 (Fielding and Schmidt, unpublished), but has also been encountered during other seasons (e.g. Whitehouse et al. 2009)

which are less efficiently used by microbes, e.g. lithogenic particles or large, silicified diatoms (Hamm et al. 2003). While krill grazing reduces phytoplankton stocks, their simultaneous regeneration of nutrients promotes new growth. One example is their excretion of ammonium. Many phytoplankton species prefer the uptake of ammonium over nitrate for energetic reasons (Dortch 1990), and the amount of ammonium released by krill covers a significant part of the primary producers requirements at South Georgia (Atkinson and Whitehouse 2001; Whitehouse et al. 2011) and at the Western Antarctic Peninsula (Lehette et al. 2012).

Another example is the mobilisation of iron by krill, which is often a limiting nutrient in the Southern Ocean. Krill not only recycle iron when grazing on phytoplankton (Tovar-Sanchez et al. 2007), they also introduce new iron into the foodweb when feeding on lithogenic particles from the seabed or glacial outlets (Schmidt et al. 2011). Acidic digestion and mechanical impact during gut passage mobilise some of the iron attached to the lithogenic particles (Lewis and Syvitski 1980). Dissolved iron released by krill can cover >30 % of the iron demand during a phytoplankton bloom on the north-western shelf of South Georgia (Schmidt et al. unpublished data).

Krill feeding is also a major source of dissolved organic carbon in the Southern Ocean, which stimulates microbial activity and bacteria-mediated nutrient recycling (Ortega-Retuerta et al. 2009; Ruiz-Halpern et al. 2011; Arístegui et al. 2014). In common with releases of ammonium and iron, DOC excretions can lead to higher primary production rates. These examples illustrate that ocean productivity is a complex phenomenon, initially set by physical and geochemical nutrient supply, but enhanced in intensity and duration by biological processes which facilitate the mobilisation and recycling of these nutrients.

5.13.3 Repackaging of Material into Fecal Pellets

During krill gut passage, undigested components are packaged into fecal pellets. Depending on diet and food processing, these fecal pellets are highly variable in size, carbon content, density and therefore sinking rate (Atkinson et al. 2012). The carbon content of krill fecal pellets varies from 0.8 to 29 % of dry mass (median ~10 %) and the sinking rates from 16 to 1218 m d^{-1} (McDonnell and Buesseler 2010; Atkinson et al. 2012).

The fate of these pellets depends on a number of factors, including the depth of release, water column mixing, structure and abundance of the mesopelagic community. Broadly, there are three scenarios: Some pellets are eaten and remineralised within the surface layer (Cadée et al. 1992; Gonzalez 1992). Others sink out of the mixed layer and provide nutrient-rich food to bacteria, protozoans, copepods and other scavengers in the ocean's twilight zone (von Bodungen 1986; Steinberg et al. 2008; Manno et al. 2015). The remainder reach the deep ocean or seabed and contribute to the long-term sequestration of atmospheric carbon (Wefer et al. 1988; Manno et al. 2015).

Paradoxically, it is krill pellets that are mostly found in sediment traps even though salps can be just as numerous in the Southern Ocean and their pellets tend to sink even faster (Pakhomov et al. 2006). One suggested mechanism is that krill occur in swarms and produce a 'rain' of pellets which can exceed the repackaging abilities of scavengers. Significant particle export via krill fecal pellets has been reported from sediment trap studies at the Western Antarctic Peninsula (McDonnell and Buesseler 2010; Gleiber et al. 2012), Bransfield Strait (von Bodungen 1986; Wefer et al. 1988), near the South Shetland Islands (Schnack 1985b), Weddell-Scotia area (Cadée et al. 1992; Gonzalez 1992), Weddell Sea (Bathmann et al. 1991; Gonzalez et al. 1994), Prydz Bay (Whiteley 2003), Davis Sea (Suzuki et al. 2003) and near South Georgia (Priddle et al. 1995; Manno et al. 2015). Further studies are required to clarify the dual role of krill fecal pellets promoting either the recycling of carbon and nutrients in surface waters or their export to depth.

5.13.4 Krill as a Food Source

The high abundance, relatively large body size (4–6 cm), and appearance as dispersed individuals as well as dense swarms makes krill a favourable food source for a range of predators, including squid, fish, benthic fauna, sea birds, penguins, seals and whales (see Chap. 9, Trathan and Hill 2016). Krill are considered a high quality lipid and protein source, with unique abundance of the oxmega-3 polyunsaturated fatty acids (PUFA) eicosapentaenoic acid (EPA) and docosahexaenoic acid (DHA) (Gigliotti et al. 2011). The latter are attributed to krill consuming marine micro-algae (Kolakowska et al. 1994), with most of the EPA deriving from diatoms and DHA from dinoflagellates (Dalsgaard et al. 2003).

Both lipid- and PUFA content in krill vary with season, region, body size and large-scale climatic conditions (Hagen et al. 2001; Ruck et al. 2014; Schmidt et al. 2014; Reiss et al. 2015). High krill lipid content coincides with ice cover and low temperatures (Ruck et al. 2014), while high PUFA levels are associated with a diatom diet (Schmidt et al. 2014). Differences can be considerable, for instance, a 20 % reduction in lipid content co-occurred with a 1–2° increase in temperature (Ruck et al. 2014), and the PUFA content was ~50 % lower in krill feeding on copepods rather than diatoms (Fig. 5.7). Therefore, rising water temperatures and the replacement of diatoms by other species, as predicted for Southern Ocean climate change scenarios (Vaughan et al. 2003; Mendes et al. 2013) may lead to significant reductions in krill lipid- and PUFA content. This has implications both for krill and their dependant predators, as lipids form energy stores for overwintering, and PUFA have key structural and regulatory roles in organisms (Dalsgaard et al. 2003; Trumble and Kanatous 2012).

High fluoride levels in Antarctic krill have attracted attention because they are toxic to many terrestrial vertebrates. Thus, to allow consumption by domestic animals or humans, fluoride has to be removed and those expenses adversely affected the economics of krill exploitation in the past. Natural predators of krill seem to be immune to these toxic effects, and build up exceptionally high fluorine concentrations in their bones (Schneppenheim 1980). In turn, this has led to the application of fluorine as a biotracer of krill in penguin diets (Thomas et al. 2013). High fluoride levels are not restricted to *E. superba* but are found in several Southern Ocean euphausiids and across a range of other Antarctic invertebrate taxa (Sands et al. 1998). There appears to be a process of active fluorine uptake but the mechanisms and reasons for this remain elusive. One possible explanation is that fluoride is taken up as an exoskeleton hardener. This is supported by Sands

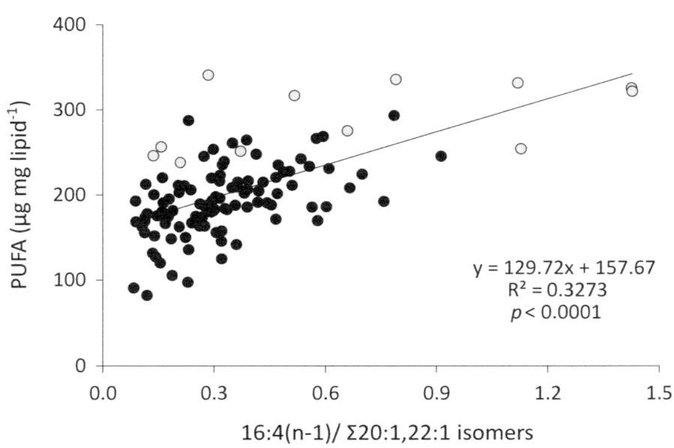

Fig. 5.7 Krill as a food source. Krill PUFA content as a function of their feeding on diatoms vs. copepods. The fatty acid 16:4(n−1) is used as a marker for diatoms and the Σ20:1, 22:1 isomers as marker for copepods. *Grey dots* indicate larval krill and *black dots* postlarval krill. Each of the *symbols* represents a pooled sample of 10 krill with the same body length (Modified after Schmidt et al. 2014)

et al. (1998) who found the highest fluoride concentrations in krill mouthparts, which need strengthening to handle large, heavily silicified diatoms (Hamm et al. 2003).

5.14 Future Prospects

The process of feeding and nutrition shapes almost all aspects of krill biology in some way or other. However, this chapter has shown that the topic of feeding is so hard to study without artefacts that progress has been slow over the last century. So what major knowledge gaps remain, which methods have told us most in the past, and are there any promising new techniques on the horizon?

Krill nutrition is included either directly or indirectly in many conceptual and numerical models of Southern Ocean food webs and biogeochemical cycles. Probably the major single challenge is to encapsulate, in a numerical way, the enormous flexibility of krill feeding behaviour. For instance the interaction of ice type, water depth and water column food levels might dictate the time budgets for krill life stages between ice, seabed and water column. A further facet to this is the great regional variability in the biology of *Euphausia superba* around Antarctica. For example their basic onshelf-offshelf distribution differs radically between the Western Antarctic Peninsula, the Scotia Sea and the Indian sector (Atkinson et al. 2008) and this is corresponds with fundamental regional differences in feeding and overwintering strategy (Schmidt et al. 2014). Clearly we should not generalise about krill from studies in just one place, but rather understand what is causing these differences. Do they represent highly flexible behaviour of a single population or specific subpopulation-level responses?

Looking back at the observations relating to krill feeding over the last century, most progress has been made, in our opinion, by in-situ – based approaches. While laboratory incubations of krill are suitable to study certain aspects of their behaviour and the processing of food, other topics such as food selectivity on mobile prey or feeding rates are certainly affected by the confinement and pre-conditioning of krill. The "natural" approach includes the breakthrough photographic observations of Marschall (1988), Hamner and Hamner (2000), and Clarke and Tyler (2008) which have changed the way we think about krill, even though they are not always accompanied by a weight of numerical data. Authors have repeatedly stressed that a social species like krill needs to be studied in situ, within its natural schooling element (Ritz et al. 2011). However, this in no way restricts us to acoustic-, photographic- or diver observations. Many of the references cited in this chapter in fact entail "in-situ" based approaches. Their authors have caught krill from a multitude of distinct schools and then either instantly frozen them for biochemical or microscopical analyses, or immediately incubated them to determine excretion, egestion or moulting rates that still reflect the in situ feeding conditions.

Several new opportunities and technologies may help us to progress our understanding of krill feeding and nutrition. High-throughput molecular techniques are advancing, and Next Generation Sequencing (NGS) is starting to allow us to read thousands of recovered gene sequences from plankton samples (Lindeque et al. 2013; see also Chap. 7, Jarman and Deagle 2016). DNA digests rapidly in the stomach (Troedsson et al. 2009) and differential digestion of prey is a potential bias. However this approach is not limited to prey with identifiable hard parts and unlike some of the targeted Polymerase Chain Reaction (PCR) approaches (Töbe et al. 2010), the detectable suite of prey items are not restricted to those previously selected for amplification (Pinol et al. 2014). This ability of NGS to cover the diversity of potential prey items is a great advantage, notwithstanding the fact that many of these sequences may not yet be in any molecular reference database (Lindeque et al. 2013). NGS methods can be standardised to make them transferable between laboratories, which would allow a large scale study of krill feeding behaviour that is so valuable.

A message emerging from this chapter is that no single diet method is suitable on its own, and combining multiple methods in a standardised manner across multiple years, seasons or regions is needed. NGS is only just starting to be used to examine marine invertebrate diets (O'Rorke et al. 2012), and while it will not replace existing methods, it promises to be a highly complementary approach. As an example, fatty acid markers and lipid content of krill are commonly measured, and these would provide time-integrated indices of diet and feeding performance as an ideal complement to NGS-type snapshots of diet. Alongside in-situ indices of performance such as the Instantaneous Growth Rate (IGR) method or morphometric condition indices, it will be possible to make the link between available food, the feeding process and the value of this food for krill.

Another development that may help us to understand krill feeding and nutrition may come from the krill fishery (Kawaguchi and Nicol 2007). The current krill fishery is dominated by Norway, who target high quality products for omega-3 food supplements (Nicol et al. 2012). These vessels are intensive samplers of krill; for example they have supplied frozen krill for diet studies to cover the winter period that is poorly accessible to science cruises (Schmidt et al. 2014). Like krill predators the fishery benefits from specimens with high PUFA-content. As this quality is continuously monitored aboard ship during year-round fishing, this could provide insights into nutrient transfer through the food web.

In summary, krill nutrition and feeding behaviour form the link between the biogeochemical part of the food web and fisheries- or predator-based models (Murphy et al. 2007, 2012; Hill et al. 2012). Even though krill feeding is not parameterised explicitly in some of these models it is still important. For example changing sea ice concentrations can affect krill recruitment, but while this is likely via larval feeding success, the mechanisms are debated (Lowe et al. 2012). Other potential stressors of climate change, such as warming or decreased pH, may mean increased food requirements to compensate for higher metabolic costs (Saba et al. 2012). Likewise, increased glacial meltwater and cryptophytes may have affects through feeding mechanisms or food quality. In turn, krill have been suggested to exert "wasp-waist" control on the rest of the food web, namely top-down on levels below them and bottom up on their predators, although the mechanisms and strengths of these controls are still unclear (Atkinson et al. 2014). Understanding such dynamics, and incorporating them into emerging modelling approaches (Litchman et al. 2013; Murphy et al. 2012) requires new and existing methods to be combined, and applied across larger spatial and temporal domains.

Acknowledgments We thank V. Siegel for the invitation to write this book chapter and a reviewer for helpful comments. Numerous colleagues and friends have provided stimulating discussions on the subject of krill feeding and food processing over the years: especially E. Pakhomov, S. Kawaguchi, B. Meyer, D. Pond, R. Ross, L. Quetin, J. Cuzin, V. Smetacek, D. Stübing, H. Flores, J. van Franeker, J. McClelland, D. Steinberg, A. Clarke, S. Fielding, G. Tarling, W. Hagen, M. Voss and C. Reiss. KS was funded by the Natural Environment Research Council (NERC) grant NE/F01547X/1. AA was funded by the "Marine Ecosystems" Programme of NERC and the Department for Environment, Food and Rural Affairs (grant no. NE/L003279/1).

References

Abrahamsen MB, Browman HI, Fields DM, Skiftesvik AB (2010) The three-dimensional prey field of the northern krill, *Meganyctiphanes norvegica*, and the escape responses of their copepod prey. Mar Biol 157:1251–1258

Arístegui J, Duarte CM, Reche I, Gómez-Pinchetti JL (2014) Krill excretion boosts microbial activity in the Southern Ocean. PLoS One 9(2):e89391

Atkinson A, Sinclair JD (2000) Zonal distribution and seasonal vertical migration of copepod assemblages in the Scotia Sea. Polar Biol 23:46–58

Atkinson A, Snÿder R (1997) Krill-copepod interactions at South Georgia, I. Omnivory by *Euphausia superba*. Mar Ecol Prog Ser 160:63–76

Atkinson A, Whitehouse MJ (2001) Ammonium regeneration by Antarctic mesozooplankton: an allometric approach. Mar Biol 139:301–311

Atkinson A, Ward P, Hill A, Brierley AS, Cripps GC (1999) Krill-copepod interactions at South Georgia, Antarctica, II. *Euphausia superba* as a major control on copepod abundance. Mar Ecol Prog Ser 176:63–79

Atkinson A, Meyer B, Stübing D, Hagen W, Schmidt K, Bathmann UV (2002) Feeding and energy budgets of Antarctic krill *Euphausia superba* at the onset of winter – II. Juveniles and adults. Limnol Oceanogr 47:953–966

Atkinson A, Shreeve RS, Tarling GA, Hirst AG, Rothery P, Pond D, Korb R, Murphy EJ, Watkins JL (2006) Natural growth rates of Antarctic krill (*Euphausia superba*): II. Predictive models based on food, temperature, body length, sex, and maturity stage. Limnol Oceanogr 51:973–987

Atkinson A, Siegel V, Pakhomov EA, Rothery P, Loeb V, Ross RM, Quetin LB, Fretwell P, Schmidt K, Tarling GA, Murphy EJ, Fleming A (2008) Oceanic circumpolar habitats of Antarctic krill. Mar Ecol Prog Ser 362:1–23

Atkinson A, Schmidt K, Fielding S, Kawaguchi S, Geissler P (2012) Variable food absorption by Antarctic krill: relationships between diet, egestion rate and the composition and sinking rates of their fecal pellets. Deep-Sea Res II 59(60):147–158

Atkinson A, Hill SH, Barange M, Pakhomov EA, Raubenheimer D, Schmidt K, Simpson SJ, Reiss C (2014) Sardine cycles, krill declines and locust plagues: revisiting "wasp-waist" food webs. Trends Ecol Evol 29:309–316

Båmstedt U, Karlson K (1998) Euphausiid predation on copepods in coastal waters of the Northeast Atlantic. Mar Ecol Prog Ser 172:149–168

Barange M, Gibbons MJ, Carola M (1991) Diet and feeding of *Euphausia hanseni* and *Nematoscelis megalops* (Euphausiacea) in the northern Benguela Current: ecological significance of vertical space partitioning. Mar Ecol Prog Ser 73:173–181
Barkley E (1940) Nahrung und Filterapparat des Walkrebschens *Euphausia superba* Dana. Z Fisch Beih 1:65–156
Bathmann U, Fischer G, Müller PJ, Gerdes D (1991) Short-term variations in particulate matter sedimentation in KappNovegia, Weddell Sea, Antarctica: relation to water mass advection, ice cover, plankton biomass and feeding activity. Polar Biol 11:185–195
Beyer F (1992) *Meganyctiphanes norvegica* (Sars) (Euphausiacea). A voracious predator on *Calanus*, other copepods, and Ctenophores in Oslofjorden, Southern Norway. Sarsia 77:189–206
Boyd CM, Heyraud M, Boyd CN (1984) Feeding of Antarctic krill *Euphausia superba*. J Crustac Biol 4:123–141
Brierley AS et al (2002) Antarctic krill under sea ice: elevated abundance in a narrow band just south of ice edge. Science 295:1890–1892
Browman HI (2005) Applications of sensory biology in marine ecology and aquaculture. Mar Ecol Prog Ser 287:263–307
Cadée GC, González H, Schnack-Schiel SB (1992) Krill diet effects faecal string settling. Polar Biol 12:75–80
Clarke A, Morris DJ (1983) Towards an energy budget for krill: the physiology and biochemistry of *Euphausia superba*. Polar Biol 2:69–86
Clarke A, Tyler PA (2008) Adult krill feeding at abyssal depths. Curr Biol 18:282–285
Clarke A, Quetin LB, Ross RM (1988) Laboratory and field estimates of the rate of fecal pellet production by Antarctic krill, *Euphausia superba*. Mar Biol 98:557–563
Constable AJ et al (2014) Climate change and the Southern Ocean ecosystems I: how changes in physical habitats directly affect marine biota. Global Chang Biol 20:3004–3025
Cripps GC, Atkinson A (2000) Fatty acid composition as an indicator of carnivory in Antarctic krill, *Euphausia superba*. Can J Fish Aquat Sci 57:31–37
Cuzin-Roudy J, Labat JP (1992) Early summer distribution of Antarctic krill sexual development in the Scotia Weddell region: a multivariate approach. Polar Biol 12:65–74
Dalsgaard J, St John M, Kattner G, Müller-Navarra D, Hagen W (2003) Fatty acid trophic markers in the pelagic marine environment. Adv Mar Biol 46:227–340
Daly KL (1998) Physioecology of juvenile Antarctic Krill (Euphausia superba) during spring in ice-covered seas. In: Lizotte MP, Arrigo KR (eds) Antarctic Sea Ice: biological processes, interactions and variability. American Geophysical Union, Washington, DC, pp 183–198
Daly KL, Macaulay MC (1991) Influence of physical and biological mesoscale dynamics on the seasonal distribution and behaviour of *Euphausia superba* in the Antarctic marginal ice zone. Mar Ecol Prog Ser 79:37–66
Dortch Q (1990) The interaction between ammonium and nitrate uptake in phytoplankton. Mar Ecol Prog Ser 61:183–201
Ducklow HW et al (2013) West Antarctic Peninsula: an ice-dependent coastal marine ecosystem in transition. Oceanography 3:190–203
Everson I (2000) Krill: biology, ecology and fisheries. Blackwell Science, Oxford, 297 pp
Flores H, Atkinson A, Rebolledo E, Cirelli V, Cuzin-Roudy J, Fielding S, van Franeker JA, Groeneveld JJ, Haraldsson M, Kawaguchi S, Krafft BA, Lombana A, Marschoff E, Meyer B, Milinevsky G, Nicol S, Pakhomov EA, Vande Pute AP, Reiss C, Rombolá E, Schmidt K, Siegel V, Tarling GA, Teschke M, Tonkes H, Toullec J-Y, Trathan PN, Tremblay N, Werner R, Werner T (2012a) Krill and climate change. Mar Ecol Prog Ser 458:1–19
Flores H, van Franeker JA, Siegel V, Haraldsson M, Strass V, Meesters EH, Bathmann U, Wolff WJ (2012b) The association of Antarctic krill with the under-ice habitat. PLoS One 7(2): e31775
Fuchs HL, Franks PJS (2010) Plankton community properties determined by nutrients and size-selective feeding. Mar Ecol Prog Ser 413:1–15

Gibbons MJ, Pillar SC, Stuart V (1991) Selective carnivory by *Euphausia lucens*. Cont Shelf Res 11:625–640
Gigliotti JC, Davenport MP, Beamer SK, Tou JC, Jaczynski J (2011) Extraction and characterisation of lipids from Antarctic krill (*Euphausia superba*). Food Chem 125:1028–1036
Gleiber MR, Steinberg DK, Ducklow HW (2012) Time series of vertical flux of zooplankton fecal pellets on the continental shelf of the western Antarctic Peninsula. Mar Ecol Prog Ser 471:23–36
Gonzalez HE (1992) The distribution and abundance of krill faecal material and oval pellets in the Scotia and Weddell Seas (Antarctica) and their role in particle flux. Polar Biol 12:81–91
Gonzalez HE, Kurbjeweit F, Bathmann UV (1994) Occurence of cyclopoid copepods and faecal material in the Halley Bay region, Antarctica, during January–February 1981. Polar Biol 14:331–342
Granéli E, Granéli W, Mozzam Rabbani M, Daugbjerg N, Fransz G, Cuzin-Roudy J, Alder VA (1993) The influence of copepod and krill grazing on the species composition of phytoplankton communities from the Scotia-Weddell Sea. Polar Biol 13:201–213
Gutt J, Siegel V (1994) Benthopelagic aggregations of krill (*Euphausia superba*) on the deeper shelf of the Weddell Sea (Antarctic). Deep-Sea Res I 41:169–178
Guzmán O (1983) Distribution and abundance of Antarctic krill in the Bransfield Strait. Ber Polarforsch 4:169–190
Haberman KL, Ross RM, Quetin LB, Vernet M, Nevitt GA, Kozlowski W (2002) Grazing by Antarctic krill *Euphausia superba* on *Phaeocystis antarctica*: an immunochemical approach. Mar Ecol Prog Ser 241:139–149
Haberman KL, Ross RM, Quetin LB (2003) Diet of the Antarctic krill (*Euphausia superba* Dana): II. Selective grazing in mixed phytoplankton assemblages. J Exp Mar Biol Ecol 283:97–113
Hagen W, Kattner G (1998) Lipid metabolism of the Antarctic euphausiid *Thysanoessa macrura* and its ecological implications. Limnol Oceanogr 43:1894–1901
Hagen W, Kattner G, Terbrüggen van Vleet ES (2001) Lipid metabolism of the Antarctic krill (*Euphausia superba*) and its ecological implications. Mar Biol 139:95–104
Hamm CE, Merkel R, Springer O, Jurkovic P, Maier C, Prechel K, Smetacek V (2003) Architecture and material properties of diatom shells provide effective mechanical protection. Nature 421:841–843
Hamner WM, Hamner PP (1988) Biomechanics of filter feeding in the Antarctic krill *Euphausia superba* – review of past work and new observations. J Crustac Biol 8:149–163
Hamner WM, Hamner PP (2000) Behavior of Antarctic krill (*Euphausia superba*): schooling, foraging and antipredator behaviour. Can J Fish Aquat Sci 57:192–202
Hamner WM, Hamner PP, Strand SW, Gilmer RW (1983) Behavior of Antarctic krill, *Euphausia superba* – chemoreception, feeding, schooling and molting. Science 220:433–435
Hamner WM, Hamner PP, Obst BS, Carleton JH (1989) Field observations on the ontogeny of schooling of *Euphausia superba* furciliae and its relationship to ice in Antarctic waters. Limnol Oceanogr 34:451–456
Hansen B, Bjørnsen PK, Hansen PJ (1994) The size ratio between planktonic predators and their prey. Limnol Oceanogr 39:395–403
Haraldsson M, Siegel V (2014) Seasonal distribution and life history of *Thysanoessa macrura* (Euphausiacea, Crustacea) in high latitude waters of the Lazarev Sea, Antarctica. Mar Ecol Prog Ser 495:105–118
Hill SL, Lees K, Atkinson A, Murphy EJ (2012) A foodweb model to explore uncertainties in the South Georgia shelf pelagic ecosystem. Deep-Sea Res II 59(60):237–252
Holm-Hansen O, Huntley M (1984) Feeding requirements of krill in relation to food sources. J Crustac Biol 4:156–173
Hopkins TL (1985) The zooplankton community of Croker Passage, Antarctic Peninsula. Polar Biol 4:161–170
Hopkins TL (1987) Midwater food web in McMurdo Sound, Ross Sea, Antarctica. Mar Biol 96:93–106

Hopkins TL, Torres JJ (1989) Midwater food web in the vicinity of the marginal ice zone in the western Weddell Sea. Deep-Sea Res A 36:543–560

Hopkins TL, Ainley DG, Torres JJ, Lancraft TM (1993a) Trophic structure in open waters of the Marginal Ice Zone in the Scotia-Weddell Confluence region during spring. Polar Biol 13:389–397

Hopkins TL, Lanccraft TM, Torres JJ, Donnelly J (1993b) Community structure and trophic ecology of zooplankton in the Scotia Sea marginal ice-zone in winter (1988). Deep-Sea Res 1(40):81–105

Huntley M, Brinton E (1991) Mesoscale variation in growth and early development of *Euphausia superba* Dana in the western Bransfield Strait region. Deep-Sea Res A 38:1213–1240

Huntley ME, Zhou M (2004) Influence of animals on turbulence in the sea. Mar Ecol Prog Ser 273:65–79

Ikeda T, Dixon T (1982) Body shrinkage as a possible overwintering mechanism of the Antarctic krill, *Euphausia superba* Dana. J Exp Mar Biol Ecol 62:143–151

Ikeda T, Dixon P (1984) The influence of feeding on the metabolic activity of Antarctic krill (*Euphausia superba* Dana). Polar Biol 1:1–9

Jacques G, Panouse M (1991) Biomass and composition of size fractionated phytoplankton in the Weddell-Scotia confluence area. Polar Biol 11:315–328

Jarman S, Deagle B (2016) Genetics of Antarctic krill, *Euphausia superba*, In: Siegel V (ed) 'Biology and Ecology of Antarctic Krill'. Advances in polar biology. Springer, Cham, pp 247–278

John DD (1936) The southern species of *Euphausia*. Discov Rep 14:193–324

Ju SJ, Harvey HR (2004) Lipids as markers of nutritional condition and diet in the Antarctic krill *Euphausia superba* and *Euphausia crystallorophias* during austral winter. Deep-Sea Res II 51:2199–2214

Jumars PA (2000) Animal guts as chemical reactors: maximising absorption rates. Am Nat 155:527–543

Karlson K, Båmstedt U (1994) Planktivorous predation on copepods: evaluation of mandible remains in predator guts as a quantitative estimate of predation. Mar Ecol Prog Ser 108:79–89

Kato M, Segawa S, Tanoue E, Murano M (1982) Filtering and ingestion rates of the Antarctic krill *Euphausia superba* Dana. Trans Tokyo Univ Fish 5:167–175

Kattner G, Hagen W (1998) Lipid metabolism of the Antarctic euphausiid *Thysanoessa macrura* and its ecological implications. Limnol Oceanogr 43:1894–1901

Kawaguchi S (2016) Reproduction and larval development in Antarctic krill (*Euphausia superba*). In: Siegel V (ed) 'Biology and Ecology of Antarctic Krill'. Advances in polar biology. Springer, Cham, pp 225–246

Kawaguchi S, Nicol S (2007) Learning about krill from the fishery. Antarct Sci 19:219–230

Kawaguchi S, Takahashi Y (1996) Antarctic krill (*Euphausia superba*) eat salps. Polar Biol 16:479–481

Kawaguchi K, Ishikawa S, Matsuda O (1986) The overwintering strategy of Antarctic krill (*Euphausia superba* Dana) under the coastal fast ice off the Ongul islands in Lützow-Holm Bay, Antarctica. Mem NatI Inst Polar Res 44:67–85

Kawaguchi S, Ichii T, Naganobu M (1999) Green krill, the indicator of micro- and nano-size phytoplankton availability to krill. Polar Biol 22:133–136

Kawaguchi S, Ishida A, King R, Raymond B, Waller N, Constable A, Nicol S, Wakita M, Ishimatsu A (2013) Risk maps for Antarctic krill under projected Southern Ocean acidification. Nat Clim Chang 3(9):843–847, doi:10.1038/NCLIMATE1937

Kils U (1983) Swimming and feeding of Antarctic krill, *Euphausia superba* – some outstanding energetics and dynamics – some unique morphological details. Ber Polarforsch (Sonderh) 4:130–155

Kiørboe T (2011) How zooplankton feed: mechanisms, traits and trade-offs. Biol Rev 86:311–339

Kiørboe T, Hirst AG (2014) Shifts in mass scaling of respiration, feeding, and growth rates across life-form transitions in marine pelagic organisms. Am Nat 183:E118–E130

Kiørboe TF, Møhlenberg F, Hamburger K (1985) Bioenergetics of the planktonic copepod *Acartia tonsa*: relation between feeding, egg production and respiration, and composition of specific dynamic action. Mar Ecol Prog Ser 26:85–97

Kittel W, Ligowski R (1980) Algae found in the food of *Euphausia crystallorophias* (Crustacea). Pol Polar Res 1:129–137

Knox AG (1994) The biology of the Southern Ocean. In: Bliss L, Drewry DJ, Walton DWH, Williams PJ, Clarke A (eds) Studies in polar research. Cambridge University Press, Cambridge

Kolakowska A, Kolakowski E, Szczygielski M (1994) Winter season krill (*Euphausia superba* Dana) as a source of *n-3* polyunsaturated fatty acids. Nahrung 38:128–134

Kopczynska EE (1992) Dominance of microflagellates over diatoms in the Antarctic areas of deep vertical mixing and krill concentrations. J Plankton Res 14:1031–1054

Kottmeier ST, Sullivan CW (1987) Late winter primary production and bacterial production in sea ice and seawater west of the Antarctic Peninsula. Mar Ecol Prog Ser 36:287–298

Kreibich T, Saborowski R, Hagen W, Niehoff B (2008) Short-term variation of nutritive and metabolic parameters in *Temora longicornis* females (Crustacea, Copepoda) as a response to diet shift and starvation. Helgol Mar Res 62:241–249

Lancraft TM, Hopkins TL, Torres JJ, Donelly J (1991) Oceanic micronektonic/macrozooplanktonic community structure and feeding in ice-covered Antarctic waters during winter (AMERIEZ 1988). Polar Biol 11:157–167

Lasker R (1960) Utilization of organic carbon by a marine crustacean: analysis with carbon-14. Science 131:1098–1100

Lasker R (1966) Feeding, growth, respiration and carbon utilisation of a euphausiid crustacean. J Fish Res Board Can 23:1291–1317

Le Vay L, Jones DA, Puello-Cruz AC, Sangha RS, Ngamphongsai C (2001) Digestion in relation to feeding strategies exhibited by crustacean larvae. Comp Biochem Physiol A Mol Integr Physiol 128:623–630

Leakey RJG, Fenton N, Clarke A (1994) The annual cycle of ciliates in nearshore waters at Signy Island, Antarctica. J Plankton Res 16:841–856

Lehette P, Tovar-Sanchez A, Duarte CM, Hernández-León S (2012) Krill excretion and its effect on primary production. Mar Ecol Prog Ser 459:29–38

Lewis AG, Syvitski JPM (1980) The interaction of plankton and suspended sediment in fjords. Sediment Geol 36:81–92

Ligowski R (2000) Benthic feeding by krill, *Euphausia superba* Dana, in coastal waters off West Antarctica and in Admiralty Bay, South Shetland Islands. Polar Biol 23:619–625

Lindeque PK, Parry HE, Harmer RA, Somerfield PJ, Atkinson A (2013) Next generation sequencing reveals the hidden diversity of zooplankton assemblages. PLoS One 8(11):e81327

Litchman, Ohman MD, Kiørboe T (2013) Trait-based approaches to zooplankton communities. J Plankton Res 35:473–484

Lowe AT, Ross RM, Quetin LB, Vernet M, Fritsen CH (2012) Simulating larval Antarctic krill growth and condition factor during fall and winter in response to environmental variability. Mar Ecol Prog Ser 452:27–43

Mackey AP, Atkinson A, Hill SL, Ward P, Cunningham NJ, Johnston NM, Murphy EJ (2012) Antarctic macrozooplankton of the southwest Atlantic sector and Bellingshausen Sea: baseline historical distributions (*Discovery Investigations*, 1928–1935) related to temperature and food, with projections for subsequent ocean warming. Deep-Sea Res II 59/60:130–146

Manno C, Stowasser G, Enderlein P, Fielding S, Tarling GA (2015) The contribution of zooplankton faecal pellets to deep carbon transport in the Scotia Sea (Southern Ocean). Biogeosciences 12:1955–1965

Marschall HP (1988) The overwintering strategy of Antarctic krill under the pack ice of the Weddell Sea. Polar Biol 9:129–135

Martin DL, Ross RM, Quetin LB, Murray AE (2006) Molecular approach (PCR-DGGE) to diet analysis in young Antarctic krill *Euphausia superba*. Mar Ecol Prog Ser 319:155–165

Mauchline J (1989) Functional morphology and feeding of euphausiids. In: Felgehauer BE, Watling L, Thistle AB (eds) Functional morphology of feeding and grooming in crustacea. AA Balkema, Rotterdam, pp 173–184
Mauchline J, Fisher LR (1969) The biology of euphausiids. Adv Mar Biol 7:1–454
Mayzaud P, Farber-Lorda J, Corre MC (1985) Aspects of the nutritional metabolism of two Antarctic euphausiids: *Euphausia superba* and *Thysanoessa macrura*. In: Siegried WR, Condy PR, Laws RM (eds) Antarctic nutrient cycles and food webs. Springer, Berlin, pp 330–338
McClatchie S (1985) Feeding behaviour in *Meganyctiphanes norvegica* (M. Sars) (Crustacea: Euphausiacea). J Exp Mar Biol Ecol 86:271–284
McClatchie S, Boyd CM (1983) Morphological study of sieve efficiencies and mandibular surfaces in the Antarctic krill, *Euphausia superba*. Can J Fish Aquat Sci 40:955–967
McDonnell AP, Buesseler KO (2010) Variability in the average sinking velocity of marine particles. Limnol Oceanogr 55:2085–2096
McWhinnie MA, Denys CJ (1978) Biological studies of Antarctic krill, austral summer, 1977–1978. Antarct J US 13:133–135
Melnikov IA, Spiridonov VA (1996) Antarctic krill under perennial sea ice in the western Weddell Sea. Antarct Sci 8:323–329
Mendes CRB, Tavano VM, Leal MC, de Souza MS, Brotas V, Garcia CAE (2013) Shifts in the dominance between diatoms and cryptophytes during three late summers in the Bransfield Strait (Antarctic Peninsula). Polar Biol 36:537–547
Meyer B (2012) The overwintering of Antarctic krill *Euphausia superba*, from an ecophysiological perspective. Polar Biol 35:15–37
Meyer B, Oettl B (2005) Effects of short-term starvation on composition and metabolism of larval Antarctic krill *Euphausia superba*. Mar Ecol Prog Ser 292:263–270
Meyer B, Teschke M (2016) Physiology of *Euphausia superba*. In: Siegel V (ed) 'Biology and Ecology of Antarctic Krill'. Advances in polar biology. Springer, Cham, pp 145–174
Meyer B, Atkinson A, Stübing D, Oettl B, Hagen W, Bathmann UV (2002a) Feeding and energy budgets of Antarctic krill *Euphausia superba* at the onset of winter – I. Furcilia III larvae. Limnol Oceanogr 47:943–952
Meyer B, Saborowski R, Atkinson A, Buchholz F, Bathmann UV (2002b) Seasonal differences in citrate synthase and digestive enzyme activity in larval and postlarval Antarctic krill, *Euphausia superba*. Mar Biol 141:855–862
Meyer B, Atkinson A, Blume B, Bathmann UV (2003) Feeding and energy budgets of larval Antarctic krill *Euphausia superba* in summer. Mar Ecol Prog Ser 257:167–177
Meyer B, Fuentes V, Guerra C, Schmidt K, Atkinson A, Spahic S, Cisewski B, Freier U, Olarriaga A, Bathmann UV (2009) Physiology, growth and development of larval krill *Euphausia superba* in autumn and winter in the Lazarev Sea, Antarctica. Limnol Oceanogr 54:1595–1614
Miller DG, Hampton I (1989) Biology and ecology of the Antarctic krill (*Euphausia superba* Dana): a review. BIOMASS Sci Ser 9:1–166
Morris DJ, Priddle J (1984) Observations on the feeding and molting of the Antarctic krill, *Euphausia superba* Dana, in winter. Br Antarct Surv Bull 65:57–63
Morris DJ, Ricketts C (1984) Feeding of krill around South Georgia.1. A model of feeding – activity in relation to depth and time of day. Mar Ecol Prog Ser 16:1–7
Murphy EJ, Watkins JL, Trathan PN, Reid K, Meredith MP, Thorpe SE, Johnston NM, Clarke A, Tarling GA, Collins MA, Forcada J, Shreeve RS, Atkinson A, Korb R, Whitehouse MJ, Ward P, Rodhouse PG, Enderlein P, Hirst AG, Martin AR, Hill SL, Staniland IJ, Pond DW, Briggs DR, Cunningham NJ, Fleming AH (2007) Spatial and temporal operation of the Scotia Sea ecosystem: a review of large-scale links in a krill centered food web. Phil Trans R Soc B 362:113–148
Murphy EJ, Cavanagh RD, Hofmann EE, Hill SL, Constable AJ, Costa DP, Pinkerton MH, Johnston NM, Trathan PN, Klinck JM, Wolf-Gladrow DA, Daly KL, Maury O, Doney SC

(2012) Developing integrated models of Southern Ocean food webs: including ecological complexity, accounting for uncertainty and the importance of scale. Prog Oceanogr 102:74–92
Naito Y, Taniguchi A, Hamada E (1986) Some observations on swarms and mating behaviour of Antarctic krill (*Euphausia superba* Dana). Mem Natl Inst Polar Res 40:178–182
Nemoto T (1967) Feeding patterns of euphausiids and differentiations in their body characters. Info Bull Planktol Soc Jpn Commem Iss 61:157–174
Nicol S, Foster J, Kawaguchi S (2012) The fishery for Antarctic krill: recent developments. Fish Fish 13:30–40
Nikolioudakis N, Isari S, Pitta P, Somarakis S (2012) Diet of sardine *Sardina pilchardus*: an 'end to end' field study. Mar Ecol Prog Ser 453:173–188
Nordhausen W (1994) Winter abundance and distribution of *Euphausia superba*, *E crystallorophias*, and *Thysanoessa macrura* in Gerlache Strait and Crystal Sound, Antarctica. Mar Ecol Prog Ser 109:131–142
Nordhausen W, Huntley M, Lopez MDG (1992) RACER: carnivory by *Euphausia superba* during the Antarctic winter. Antarct J US 27(5):181–183
O'Brien DP (1987) Direct observations of the behavior of *Euphausia superba* and *Euphausia crystallorophias* (Crustacea: Euphausiacea) under pack ice during the Antractic spring of 1985. J Crustac Biol 7:437–448
O'Brien C, Virtue P, Kawaguchi S, Nichols PD (2011) Aspects of krill growth and condition during late winter-early spring off East Antarctica (110–130°E). Deep-Sea Res II 58:1211–1221
O'Rorke R, Lavery S, Chow S, Takeyama H, Tsai P, Beckley LE, Thompson PA, Waite AM, Jeffs AG (2012) Determining the diet of Western Rock Lobster (*Panulirus cygnus*) using high-thoughput DNA sequencing techniques. PLoS One 7(8):e42757
Ortega-Retuerta E, Frazer TK, Duarte CM, Ruiz-Halpern, Tovar-Snchez A, Arrieta JM, Reche I (2009) Bioregeneration of chromophoric dissolved organic matter by bacteria and krill in the Southern Ocean. Limnol Oceanogr 54:1941–1950
Pakhomov EA, Perissinotto R, Froneman PW, Miller DGM (1997) Energetics and feeding dynamics of *Euphausia superba* in the South Georgia region during the summer of 1994. J Plankton Res 19:399–423
Pakhomov EA, Froneman PW, Perissinotto R (2002) Salp/krill interactions in the Southern Ocean: spatial segregation and implications for the carbon flux. Deep-Sea Res II 49:1881–1907
Pakhomov EA, Atkinson A, Meyer B, Oettl B, Bathmann UV (2004) Daily rations and growth of larval krill *Euphausia superba* in the Eastern Bellingshausen Sea during austral autumn. Deep Sea Res II 51:2185–2198
Pakhomov EA, Dubischar CD, Strass V, Brichta M, Bathamnn UV (2006) The tunicate *Salpa thompsoni* ecology in the Southern Ocean. I. Distribution, biomass, demography and feeding ecophysiology. Mar Biol 149:609–623
Perissinotto R, Pakhomov EA (1996) Gut evacuation rates and pigment destruction in the Antarctic krill *Euphausia superba*. Mar Biol 125:47–54
Perissinotto R, Pakhomov EA, McQuaid CD, Froneman PW (1997) In situ grazing rates and daily ration of Antarctic krill *Euphausia superba* feeding on phytoplankton at the Antarctic Polar Front and the Marginal Ice Zone. Mar Ecol Prog Ser 160:77–91
Pinol J, San Andres V, Clare EL Mir G, Symondson WOC (2014) A pragmatic approach to the analysis of diets of generalist predators: the use of next-generation sequencing with no blocking probes. Mol Ecol Resour 14:18–26
Polito MJ, Reiss CS, Trivelpiece WZ, Patterson WP, Emslie SD (2013) Stable isotopes identify an ontogenetic niche expansion in Antarctic krill (*Euphausia superba*) from the South Shetland Islands, Antarctica. Mar Biol 160:1311–1323
Pond DW, Priddle J, Dargent JR, Watkins JL (1995) Laboratory studies of assimilation and egestion of algal lipid by Antarctic krill – methods and initial results. J Exp Mar Biol Ecol 187:253–268

Pond DW, Atkinson A, Shreeve RS, Tarling G, Ward P (2005) Diatom fatty acid biomarkers indicate recent growth rates in Antarctic krill. Limnol Oceanogr 50:732–736

Price HJ (1989) Swimming behavior of krill in response to algal patches – a mesocosm study. Limnol Oceanogr 34:649–659

Price HJ, Boyd KR, Boyd CM (1988) Omnivorous feeding behavior of the Antarctic krill *Euphausia superba*. Mar Biol 97:67–77

Priddle J, Watkins JL, Morris D, Ricketts C (1990) Variation of feeding by krill in swarms. J Plankton Res 12:1189–1205

Priddle J, Leakey RGJ, Symon CS, Whitehouse MJ, Robins D, Cripps G, Murphy EJ, Owens NJP (1995) Nutrient cycling by Antarctic marine microbial phytoplankton. Mar Ecol Prog Ser 116:181–198

Quetin LB, Ross RM (1985) Feeding by Antarctic krill, *Euphausia superba*: does size matter? In: Siegfried WR, Condy PR, Laws RM (eds) Antarctic nutrrient cycles and food webs. Springer, Berlin, pp 372–377

Quetin LB, Ross RM (1991) Behavioral and physiological characteristics of the Antarctic krill, *Euphausia superba*. Am Zool 31:49–63

Quetin LB, Ross RM, Clarke A (1994) Krill energetics: seasonal and environmental aspects of the physiology of *Euphausia superba*. In: El-Sayed SZ (ed) Southern Ocean ecology: the BIOMASS perspective. Cambridge University Press, Cambridge, pp 165–184

Reiss CS (2016) Age, growth, mortality, and recruitment of Antarctic Krill, *Euphausia superba*. In: Siegel V (ed) 'Biology and Ecology of Antarctic Krill'. Advances in polar biology. Springer, Cham, pp 101–144

Reiss CS, Walsh J, Goebel ME (2015) Winter preconditioning determines feeding ecology of *Euphausia superba* in the Antarctic Peninsula. Mar Ecol Prog Ser 519:89–101

Ritz DA, Hobday AJ, Montgomery JC, Ward AJW (2011) Social aggregation in the pelagic zone with special reference to fish and invertebrates. Adv Mar Biol 60:161–227

Ross RM, Quetin LB (2003) Working with living krill – the people and the places. Mar Freshw Behav Physiol 36:207–228

Ross RM, Quetin LB, Haberman KL (1998) Interannual and seasonal variability in the short-term grazing impact of *Euphausia superba* in nearshore and offshore waters west of the Antarctic Peninsula. J Mar Syst 17:261–273

Ross RM, Quetin LB, Baker KS, Vernet M, Smoth RC (2000) Growth limitation in young *Euphausia superba* under field conditions. Limnol Oceanogr 45:31–43

Ruck KE, Steinberg DK, Canuel EA (2014) Regional differences in quality of krill and fish as prey along the Western Antarctic Peninsula. Mar Ecol Prog Ser 509:39–55

Rudjakov JA (1996) Mesozooplankton biomass distribution in the Atlantic sector of the Southern Ocean. Antarct Sci 8:343–348

Ruiz-Halpern S, Duarte C, Tovar-Sanchez A, Pastor M, Horstkotte B, Lasternas S, Agusti S (2011) Antarctic krill as a source of dissolved organic carbon to the Antarctic ecosystem. Limnol Oceanogr 56:521–528

Saba GK, Schofield O, Torres JJ, Ombres EH, Steinberg DK (2012) Increased feeding and nutrient excretion of adult Antarctic krill, *Euphausia superba*, exposed to enhanced carbon dioxide (CO_2). PLoS One 7(12):e52224

Sabarowski R, Buchholz F (1999) A laboratory study on digestion processes in the Antarctic krill, *Euphausia superba*, with special regard to chitinolytic enzymes. Polar Biol 21:295–304

Saborowski R (2012) Related antipodes: a comparative study on digestive endopeptidases from northern krill and Antarctic krill (Euphausiacea). Polar Biol 35:1003–1012

Sailley SF, Ducklow HW, Moeller HV, Fraser WR, Schofield OM, Steinberg DK, Garzio LM, Doney SC (2013) Carbon fluxes and pelagic ecosystem dynamics near two western Antarctic Peninsula Adélie penguin colonies: an inverse model approach. Mar Ecol Prog Ser 492:253–272

Saiz E, Grifell K, Calbet A, Isari S (2014) Feeding rates and prey: predator size ratios of the nauplii and adult females of the marine cyclopoid copepod *Oithona davisae*. Limnol Oceanogr 59:2077–2088

Samyshev EZ (1991) Antarctic krill and the structure of the plankton community in its area of occurrence. Nauka Press, Moscow (in Russian)

Sands M, Nicol S, McMinn A (1998) Fluoride in Antarctic marine crustaceans. Mar Biol 132:591–598

Schmidt K (2010) Food and feeding in Northern krill (*Meganyctiphanes norvegica* Sars). Adv Mar Biol 57:127–171

Schmidt K, McClelland JW, Mente E, Montoya JP, Atkinson A, Voss M (2004) Trophic-level interpretation based on $\delta^{15}N$ values: implications of tissue-specific fractionation and amino acid composition. Mar Ecol Prog Ser 266:43–58

Schmidt K, Atkinson A, Petzke K-J, Voss M, Pond DW (2006) Protozoans as a food source for Antarctic krill, *Euphausia superba*: complementary insights from stomach content, fatty acids, and stable isotopes. Limnol Oceanogr 51:2409–2427

Schmidt K, Atkinson A, Steigenberger S, Fielding S, Lindsay MCM, Pond DW, Tarling GA, Klevjer TA, Allen CS, Nicol S, Achterberg EP (2011) Seabed foraging by Antarctic krill: implications for stock assessment, bentho-pelagic coupling and the vertical transfer of iron. Limnol Oceanogr 56:1411–1428

Schmidt K, Atkinson A, Venables HJ, Pond DW (2012) Early spawning of Antarctic krill in the Scotia Sea is fuelled by "superfluous feeding" on non-ice-associated phytoplankton blooms. Deep-Sea Res II 59(60):159–172

Schmidt K, Atkinson A, Pond DW, Ireland LC (2014) Feeding and overwintering of Antarctic krill across its major habitats: the role of sea ice cover, water depth, and phytoplankton abundance. Limnol Oceanogr 59:17–36

Schnack SB (1985a) Feeding by *Euphausia superba* and copepod species in response to varying concentrations of phytoplankton. In: Siegfried WR, Condy PR, Laws RM (eds) Antarctic nutrient cycles and food webs. Springer, Berlin, pp p311–p329

Schnack SB (1985b) A note on the sedimentation of particulate matter in Antarctic waters during summer. Meeresforschung 30:306–315

Schneppenheim R (1980) Concentration of fluoride in Antarctic animals. Meeresforsch Rep Mar Res 28:179–182

Siegel V, Watkins J (2016) Distribution, biomass and demography of Antarctic krill, *Euphausia superba*. In: Siegel V (ed) 'Biology and Ecology of Antarctic Krill'. Advances in polar biology. Springer, Cham, pp 21–100

Siegel V, Bergström B, Strömberg JO, Schalk PH (1990) Distribution, size, frequency and maturity stages of krill, *Euphausia superba*, in relation to sea-ice in the northern Weddell Sea. Polar Biol 10:549–557

Simpson SJ, Raubenheimer D (2012) The nature of nutrition: a unifying framework from animal adaptation to human obesity. Princeton University Press, Woodstock, 239 pp

Smetacek V, Veth C (1989) Introduction. In: Hempel I, Schalk PH Smetacek V (eds) The expedition Antarktis VII/3 (EPOS Leg 2) of RV "Polarstern" in 1988/89. Berichte zur Polarforschung, Alfred-Wegener-Institut für Polar- und Meeresforschung, Bremerhaven, 65:1–7

Smith CR, DeMaster DJ (2008) Preface and brief synthesis for the FOODBANCS volume. Deep-Sea Res II 55:2399–2403

Smith CS, Mincks S, DeMaster DJ (2006) A synthesis of bentho-pelagic coupling on the Antarctic shelf: food banks, ecosystem inertia and global climate change. Deep-Sea Res II 53:875–894

Spiridonov VA, Gruzov EN, Pushkin AF (1985) Investigations of schools of Antarctic *Euphausia superba* (Crustacea, Euphausiacea) under the ice. Zoologichesky Zh 11:1655–1660

Steinberg DK, Van Mooy BAS, Buesseler KO, Boyd PW, Kobari T, Karl DM (2008) Bacterial vs zooplankton control of sinking particle flux in the ocean's twilight zone. Limnol Oceanogr 53:1327–1338

Stowasser G, Atkinson A, McGill RAR, Phillips RA, Collins MA, Pond DW (2012) Food web dynamics in the Scotia Sea in summer: a stable isotope study. Deep-Sea Res II 59(60):208–221

Stretch JJ, Hamner PP, Hamner WM, Michel WC, Cook J, Sullivan CW (1988) Foraging behaviour of Antarctic krill *Euphausia superba* on sea ice microalgae. Mar Ecol Prog Ser 44:131–139

Stübing D, Hagen W (2003) Fatty acid biomarker ratios – suitable trophic indicators in Antarctic euphausiids? Polar Biol 26:774–782

Stübing D, Hagen W, Schmidt K (2003) On the use of lipid biomarkers in marine food webs: an experimental case study on the Antarctic krill, *Euphausia superba*. Limnol Oceanogr 48:1685–1700

Suh HL (1996) The gastric mill in euphausiid crustaceans: a comparison of eleven species. Hydrobiologia 321:235–244

Suh HL, Nemoto T (1987) Comparative morphology of filtering structure of five species of *Euphausia* (Euphausiacea, Crustacea) from the Antarctic Ocean. Proc NIPR Symp Polar Biol 1:72–83

Suh HL, Nemoto T (1988) Morphology of the gastric mill in ten species of euphausiids. Mar Biol 97:79–85

Suzuki H, Sasaki H, Fukuchi M (2003) Loss processes of sinking fecal pellets of zooplankton in the mesopelagic layers of the Antarctic Marginal Ice Zone. J Oceanogr 59:809–818

Tarling GA, Fielding S (2016) Swarming and behavior in Antarctic krill. In: Siegel V (ed) 'Biology and Ecology of Antarctic Krill'. Advances in polar biology. Springer, Cham, pp 279–320

Tarling GA, Stowasser G, Ward P, Poulton AJ, Zhou M, Venables HJ, McGill RAR, Murphy EJ (2012) Seasonal trophic structure of the Scotia Sea pelagic ecosystem considered through biomass spectra and stable isotope analysis. Deep-Sea Res II 59–60:222–236

Teschke M, Kawaguchi S, Meyer B (2007) Simulated light regimes affect feeding and metabolism of Antarctic krill, *Euphausia superba*. Limnol Oceanogr 52:1046–1054

Thomas DB, Fordyce RE, Gordon KC (2013) Evidence for a krill-rich diet from non-destructive analysis of penguin bone. J Avian Biol 44:203–207

Thor P, Wendt I (2010) Functional response of carbon absorption efficiency in the pelagic calanoid copepod *Acartia tonsa* Dana. Limnol Oceanogr 55:1779–1798

Töbe K, Meyer B, Fuentes V (2010) Detection of zooplankton items in the stomach and gut content of larval krill, *Euphausia superba*, using a molecular approach. Polar Biol 33:407–414

Torgersen T (2001) Visual predation by the euphausiid *Meganyctiphanes norvegica*. Mar Ecol Prog Ser 209:295–299

Torres JJ, Donnelly J, Hopkins TL, Lancraft TM, Aarset AV, Ainley DG (1994) Proximate composition and overwintering strategies of Antarctic micronektonic Crustacea. Mar Ecol Prog Ser 113:221–232

Tovar-Sanchez A, Duarte CM, Hernández-Leon, Sanudo-Wilhelmy SA (2007) Krill as a central node for iron cycling in the Southern Ocean. Geophys Res Lett 34:L11601. doi:10.1029/2006GL029096

Trathan PN, Hill SL (2016) The importance of krill predation in the Southern Ocean. In: Siegel V (ed) 'Biology and Ecology of Antarctic Krill'. Advances in polar biology. Springer, Cham, pp 321–350

Troedsson C, Simonelli P, Nagele V, Nejstgaard JC, Frischer ME (2009) Quantification of copepod gut content by differential length amplification quantitative PCR (dla-qPCR). Mar Biol 156:253–259

Trumble SJ, Kanatous SB (2012) Fatty acid use in diving mammals: more than merely fuel. Front Physiol 3(184):1–7. doi:10.3389/fphys.2012.00184

Ullrich B, Storch V, Marschall HP (1991) Microscopic anatomy, functional morphology, and ultrastructure of the stomach of *Euphausia superba* Dana (Crustacea, Euphausiacea). Polar Biol 11:203–211

Vaughan DG, Marshall GJ, Connolley WM, Parkinson CL, Mulvaney R, Hodgson DA, King JC, Pudsey CJ, Turner J (2003) Recent rapid regional climate warming on the Antarctic Peninsula. Clim Change 60:243–274

Venables H, Moore CM (2010) Phytoplankton and light limitation in the Southern Ocean: learning from high-nutrient, high chlorophyll areas. J Geophys Res 115:C02015

Virtue P, Nicholls PD, Nicol S, Hosie G (1996) Reproductive trade-off in male Antarctic krill, *Euphausia superba*. Mar Biol 126:521–527
von Bodungen B (1986) Phytoplankton growth and krill grazing during spring in the Bransfield Strait, Antarctica – implications from sediment trap collections. Polar Biol 6:153–160
Wakefield ED, Phillips RA, Belchier M (2012) Foraging black-browed albatrosses target waters overlaying moraine banks – a consequence of upward benthic-pelagic coupling? Antarct Sci 24:269–280
Wefer G, Fischer G, Füetter D, Gersonde R (1988) Seasonal particle flux in the Bransfield Strait, Antarctica. Deep-Sea Res A35:891–898
Whitehouse MJ, Atkinson A, Ward P, Korb RE, Rothery P, Fielding S (2009) Role of krill versus bottom-up factors in controlling phytoplankton biomass in the northern Antarctic waters of South Georgia. Mar Ecol Prog Ser 393:69–82
Whitehouse MJ, Atkinson A, Rees AP (2011) Close coupling between ammonium uptake by phytoplankton and excretion by Antarctic krill, *Euphausia superba*. Deep-Sea Res I 58:725–732
Whiteley M (2003) The influence of Antarctic krill (*Euphausia superba*) on carbon fluxes in the Southern Ocean. Honours Dissertation, The University of Western Australia, 87 pp
Wickham SA, Berninger U-G (2007) Krill larvae, copepods and the microbial food web: interactions during the Antarctic fall. Aquat Microb Ecol 46:1–13
Wirtz KW (2012) Who is eating whom? Morphology and feeding type determine the size relations between planktonic predators and their ideal prey. Mar Ecol Prog Ser 445:1–12
Wirtz KW (2013) How fast can plankton feed? Maximum ingestion rate scales with digestive surface area. J Plankton Res 35:33–48
Wirtz KW (2014) A biomechanical and optimality-based derivation of prey-size dependencies in planktonic prey selection and ingestion rates. Mar Ecol Prog Ser 507:81–94

Chapter 6
Reproduction and Larval Development in Antarctic Krill (*Euphausia superba*)

So Kawaguchi

Abstract An understanding of the reproduction and larval development of Antarctic krill is important since reproductive output is a key factor influencing their overall population size. Seasonal cycles of maturity in krill are known to be finely synchronized with seasonal cycles of food, sea-ice and the light regime in the Antarctic environment. This chapter will describe the progression of Antarctic krill development throughout their life cycle, including their maturation process, fecundity, spawning, and larval development, in relation to various environmental factors that are thought to govern these processes. The chapter will close with some remarks on possible effects of climate change on krill life history.

Keywords Maturation process • Light regime • Seasonal cycle • Sea ice • Climate change

6.1 Introduction

The reproductive output of Antarctic krill is a key factor influencing their overall population size. Seasonal cycles of reproduction and larval development are synchronized with seasonal cycles of food, sea-ice, and the light regime. Since the mid-1990s there has been a dramatic increase in the number of studies looking into the effects of environmental variability on reproductive biology and recruitment of Antarctic krill. A major breakthrough in krill husbandry techniques has meant the entire lifecycle can successfully occur in captivity, permitting experimental studies on all stages of krill under a range of controlled environments. The intention of this review is to build on an earlier review of krill reproduction (Ross and Quetin 2000) and to examine advances in our understanding of the maturation process and larval development in relation to the seasonal cycle of the environment as revealed in recent publications and research.

S. Kawaguchi (✉)
Australian Antarctic Division, 203 Channel Highway, Kingston, TAS 7050, Australia

Antarctic Climate and Ecosystems Cooperative Research Centre, 20 Castray Esplanade, Hobart, TAS 7000, Australia
e-mail: So.Kawaguchi@aad.gov.au

V. Siegel (ed.), *Biology and Ecology of Antarctic Krill*, Advances in Polar Ecology,
DOI 10.1007/978-3-319-29279-3_6

6.2 Maturation Process

6.2.1 *Sexual Differentiation*

The genital apparatus of Malacostraca crustaceans is known to differentiate during post-embryonic development. Transformation of the primordial gonia into oogonia or spermatogonia occurs at the last larval stage or during the juvenile stage, however external sex characteristics are not yet differentiated. The genital apparatus at this stage is still identical in both sexes and the gonads only contain gonia (Cuzin-Roudy 1987a, b; Charniaux-Cotton and Payen 1985).

Sexual differences in gonads first occur in subadult krill (>24 mm) where secondary sex characteristics have not yet developed. If gonad differentiation is proceeding along the female pathway, young oocytes (slightly larger sexual cells) begin to be observed at the periphery of the lobes. Subadults of the same size without young oocytes developing are likely to be males. The development of spermatocytes, spermatids, spermatozoa, and spermatophores occurs in subadults between 24 and 36 mm (Cuzin-Roudy 1987b). In earlier studies Bargmann (1937) anatomically documented rudiments (or early forms) of male and female reproductive systems through dissection in juveniles with body length of 18–20 mm, and even down to 10 mm size specimens in her 1945 report (Bargmann 1945). The occurrence of gonadal differentiation was shown to be highly variable between regions and environments.

The mechanism for determining the sex of krill is still unknown. In fact, krill do not have heteromorphic sex chromosomes (Thiriout-Quevereux et al. 1998) which raises the possibility that environmental factors could determine initial gender as in other animals (Cook 2002). Further, Malacostraca are the only crustaceans known to possess an androgenic gland; this gland regulates male sex differentiation, plays a crucial role for development and maintenance of male primary and secondary sex characters in a number of representative species of a wide array of Malacostraca (Charniaux-Cotton and Payen 1985; Ventura et al. 2011), including euphausiids (Cuzin-Roudy 1987b).

6.2.2 *Development of Secondary Sexual Characteristics*

Secondary sexual (external) characteristics develop progressively in the late subadult stage. Length at maturity is nominally 35 mm for females and 43 mm or larger for males, although variations exist among years (Siegel and Loeb 1994). Based on staging using external characteristics (Makarov and Denys 1981) immature males are defined as the period from the appearance of small bud on the anterior surface of the internal ramus of the first pleopods until the bud grows into the fully developed petasma. The petasma is the hand-like organ that is used in spermataphore transfer during mating. In adult males, the fully developed petasmae are clearly visible through the gills, and spermatophores can be present in the ejaculatory duct ready to be ejected. In sub-adult females the thelycum (the organ to which spermataphores are attached) is present but not fully established (sternal plate not greater than coxal plates) and is pale or colourless. In adult females, the thelycum is fully developed, red in colour, and clearly visible through the gills. Adult females are characterized as being from stage IIIA to IIIE depending on whether or not spermatophores are attached on the thelycum, and also on the size of ovary, the development of a swollen thorax, and presence of empty space in the internal cavity of swollen thorax due to release of eggs (Makarov and Denys 1981). See Chap. 1 (Siegel 2016) for more information on external maturity identification.

6.2.3 Gonad Maturation

Ovarian development and the definition of sexual development stages (SDS) in female Antarctic krill has been categorised into ten SDS phases, and are described in detail by Cuzin-Roudy (2000), Cuzin-Roudy and Amsler (1991), and reviewed in Ross and Quetin (2000). During the first year of life when krill are larvae or juveniles and have not yet developed their secondary sexual characteristics the primary oogonia (og1) multiply into secondary oogonia (og2) in the germinal zone of a small ovary (gametogenesis) (SDS 1). Once they attain secondary sexual characteristics the ovary then enlarges and og2 are transformed into young primary oocytes (yoc) (oogenesis) (SDS2). Yolk accumulation in the oocytes in krill then occurs during previtellogenesis (SDS 3–4) and vitellogenesis (SDS 5–6) (Cuzin-Roudy 2010). Previtellogenesis is a process starting from early spring to the end of reproductive season (Cuzin-Roudy 2000) to turn groups of yoc into type 1 oocyte (oc 1) through accumulation of glycoproteic yolk (Cuzin-Roudy 1993; Cuzin-Roudy and Amsler 1991). During vitellogenesis oc1 will then further develop into oc2, 3, and 4. When oocytes mature (SDS 7), the ready-to-spawn ovary appears opaque or even bluish (Ross and Quetin 1983). After spawning a batch of eggs, the size of the ovary reduces in size (oviposition phase SDS 8) but the ovary may still contain previtellogenic oocytes and may repeat vitellogenesis (SDS 5–7) and produce multiple batches of eggs until all the oocytes are exhausted (the end of the reproductive cycle). Once the ovary completes reproduction for the season it enters the ovarian reorganisation phase (reproductive rest) for the overwintering period (SDS 9–10) (Cuzin-Roudy and Amsler 1991; Makarov 1975) (Fig. 6.1).

Reproduction for krill is an energetically demanding process (Nicol et al. 1995; Ross and Quetin 1983). In the process described above, there are three plateau steps involved in the ovarian developmental process. These steps are (1) accumulation of adequate reserves in fat bodies to fuel previtellogenesis, (2) beginning of vitellogenesis which require accumulation of resources enough to complete

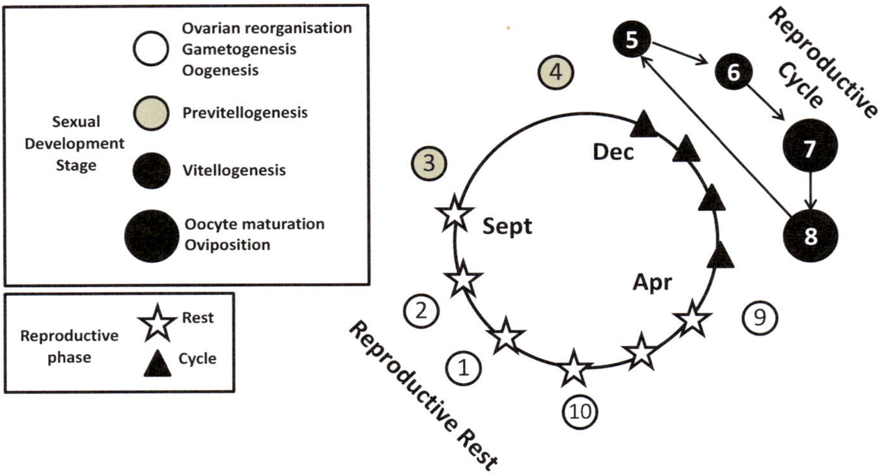

Fig. 6.1 Alternation of reproductive cycle (*black triangles* and *white stars*) for the Antarctic krill from west of the Antarctic Peninsula. Numbers in *circles* refer to sexual development stages (SDS) (Rearranged and redrawn from Fig. 6.5 of Ross and Quetin (2000))

vitellogenesis, and (3) just prior to release of last batch of embryos, preparation of the next set of oocytes for the following batch of eggs to be produced (Cuzin-Roudy and Amsler 1991; Quetin and Ross 2001).

The fat body, which extends loosely inside the ventral part of the thoracic cavity, is thought to play a major role in the synthesis and transformation of glycol- and lipoproteins into vitellins (glycolipoproteins), which form the yolk platelets of the mature oocytes. Vitellogenesis ceases when previtellogenesis is arrested for the spawning season and all the unreleased oocyctes from the developing brood undergo oosorption (Cuzin-Roudy 2000). The long previtellogenetic accumulation of glycoproteic yolk (Cuzin-Roudy and Amsler 1991) and the constitution of reserves of early summer may depend upon the late winter/early spring food supply from ice-algae and ice-edge blooms (Cuzin-Roudy and Labat 1992), and the vitellogenesis is primarily fuelled by summer phytoplankton production, and therefore not dependent on the energy resources carried over from the previous year (Cuzin-Roudy 2000; Hagen et al. 1996; Mayzaud et al. 1998). Continuous availability of food throughout the spawning season is critical for completion of the whole reproduction process (Quetin and Ross 2001).

For males, the spermatogonia in the testis produce spermatocysts through mitosis in testicular cysts, and they become spermatids through meiosis. These will then differentiate into non-flagellated spermatozoa. Spermatozoa then leave the testis and are packed into spermatophores and a pair of spermatophores is stored in the ampullae ready for release at mating (Cuzin-Roudy 2010). Males start to mature a few weeks earlier in the season than females (Siegel 1988, 2012; Brown et al. 2010).

6.3 Seasonal Maturity Cycle

A reproductive resting period (re-juvenation cycle) is one of the key features of euphausiids maturation (Cuzin-Roudy 2000; Ross and Quetin 2000). Once krill reach maturity for the first time, the sexual characteristics undergoes a cycle, annually oscillating between the fully mature and immature forms for both males and females for the rest of their life (Poleck and Denys 1982). In males the petasma which is fully developed in the reproductive season regress to a less developed form during non-reproductive season (Poleck and Denys 1982). Female maturity undergoes a cycle starting from a spent female still with fully mature red thelycum at the end of its reproductive season, regressing its thelycum to an immature form during reproductive resting period, then re-maturing to a fully developed red thelycum regardless of provision of constant food and light condition (Denys and McWhinnie 1982; Poleck and Denys 1982), including starvation and complete darkness (Thomas and Ikeda 1987), suggesting this is generally a free running cycle once krill are entrained in a cycle but this could also be manipulated under certain light conditions (see Sect. 6.3.1). There are also field observation recording females in mid-winter with clear in colour but fully mature thelycums in its form (e.g. Siegel 1989). This suggests they may not need to totally regress their thelycum during winter season if food condition is adequate. During the reproductive rest krill re-organise their ovary following a reproductive season (Cuzin-Roudy 2000; Makarov 1975; Ross and Quetin 2000). During the winter period it is unlikely that krill encounter feeding conditions sufficient to satisfy their reproductive needs, despite the use of multiple strategies to survive the winter, e.g. flexible feeding strategies such as switching their diet to zooplankton (Huntley et al. 1994), detritus (Kawaguchi et al. 1986), the use of body mass from reductions in size (Ikeda and Dixon 1982), and lowered metabolic rates (Meyer 2012; Quetin and Ross 2000; Brown et al. 2013).

6.3.1 Role of Light and Food

In polar regions the light regime is thought to be one of the important signals for the seasonal physiological cycle of the resident organisms. In many regions of the Southern Ocean dense winter sea-ice concentrations are thought to promote extensive growth of sea ice algal communities as well as an intense phytoplankton bloom following the sea ice retreat. The ice-mediated food supply fosters good feeding and early female gonadal development through early spring and early summer spawning (Siegel and Loeb 1995; Quetin et al. 2007). South Georgia is sea ice free year-round but the area exhibits one of the highest krill densities around the continent (Marr 1962), and large gravid female krill are frequently encountered around South Georgia (Tarling et al. 2007). Therefore existence of sea ice itself, ice algae, or ice edge bloom are not prerequisites for krill to successfully mature. More critically krill require access to appropriate amounts of food at the specific time in their seasonal maturity cycle in order to complete the reproductive cycle. For example female krill at SDS stage 3 will not complete the reproductive cycle that season and delay until the following year if they cannot accumulate appropriate amount of lipid during spring to summer (Quetin et al. 2007; Ross and Quetin 2000).

Recent developments of krill husbandry technique and sophisticated aquarium systems has contributed to closing the life cycle of Antarctic krill (Hirano and Matsuda 2003; Hirano et al. 2003) and to induce natural schooling behaviour in captivity (Kawaguchi et al. 2010a). This technical advancement allowed various experimental approaches to be taken to understand krill life history including reproduction. Laboratory experiments have also revealed that the seasonal cycle in krill physiology and behaviour is likely to be caused by an inherent annual rhythm, irrespective of ambient food levels (Ikeda 1987; Thomas and Ikeda 1987). This inherent cycle is primarily triggered by seasonal light regime that enables krill to adapt to the extreme seasonal changes in food availability (Meyer 2012; Teschke et al. 2007). Under a long-term controlled laboratory environment, summer and spring/autumn light conditions promoted higher feeding, faster growth and higher metabolism when compared to winter light conditions (Seear et al. 2009; Teschke et al. 2007) and accelerated the succession of maturity stages (Teschke et al. 2008). Maturation of krill commences shortly after an increase in photoperiod, demonstrating that maturity development is either directly or indirectly affected by differing simulated light regimes, which cue their overall seasonal physiological cycle (Hirano et al. 2003; Teschke et al. 2008). Timing of the krill maturity cycle is thus flexible to manipulation of light regime, and a period of darkness induces rapid regression and as a result the overall seasonal maturity cycle can be re-scheduled (Brown et al. 2011) (For additional information on photoperiod and physiology see Chap. 4 (Meyer and Teschke 2016), and Chap. 8 (Tarling and Fielding 2016) for behaviour. The series of experimental studies collectively demonstrate that various light regimes at certain points of the krill maturity cycle (e.g. the timing of darkness) could play an important role in synchronizing the krill maturity cycle to the environment (Brown et al. 2011). Overall, krill are adjusting their seasonal biological cycles with the cue of seasonal light regimes to cope with varying environmental conditions, and time their physiology ready for maturation at the point when the food (ice algae and phytoplankton bloom) in their habitat is most likely to be sufficient to complete previtellogenesis and vitellogenesis under the highly variable Antarctic environment (Kawaguchi et al. 2007a).

It is still unknown whether length of daylight, the intensity of the ambient light, or a combination of these factors are actually triggering the onset of maturity, and these factors could also be different for different phases of the maturity cycle (Meyer 2012). We still do not know how initiation of the first ovarian maturation for an individual can be cued for krill raised under constant food and darkness condition from egg (Ikeda 1987). Because the light regime varies with latitude and therefore specific light signal influencing the phenology of krill may be limited in latitudinal extent, the potential latitudinal range for krill to be successful may be restricted (Quetin et al. 2007).

Further field and laboratory research is required to understand the overall regulations of the physiological processes. Studies at the molecular levels are of particular importance in order to examine the sequence and functions related to the maturation process (or combination of them) and how they are expressed under various controlled environmental conditions (also see Chap. 4, Meyer and Teschke 2016, and Chap. 7, Jarman and Deagle 2016).

6.3.2 Maturity Cycle and the Environment

Initiation of krill reproduction requires considerable energy input at the right time and in the right amounts (Ross and Quetin 2000). Kawaguchi et al. (2007a) put forward a conceptual model for krill reproduction, which captures likely mechanisms of how the seasonality of the Antarctic environment may be linked to the endogenous physiological cycle of krill. This model highlighted the importance of food during winter including the amount of sea ice algae available during late winter/early spring in the lead up to the reproductive season. Availability of winter food may also influence how far regression of their maturity went and therefore could influence the timing of their re-maturation. Thus, winter food availability has important implications for the physiological readiness to commence vitellogenesis, coincident with the onset of sea ice retreat and the associated phytoplankton bloom (Fig. 6.2). Field surveys indicate that in October/November (spring) adult-sized females are in stage 3A (in previtellogenesis state with thelycum fully developed but not mated and ovary not mature, Ross and Quetin 2000) (e.g. Siegel 1988, 2012) and ready to start vitellogenesis and ovarian cycle if a pulse of food occurs though the start of the spring bloom. Primary production during summer is critical for completion of the reproductive cycle to mature the eggs and also to prolong the reproductive season by recycling their ovary. The summer period is longer in lower latitudes compared to higher latitudes. Thus there is a wider seasonal window for reproduction at lower latitudes and enables krill to sustain the ovarian cycle for prolonged period, potentially contributing to an overall high level of fecundity in more northern areas such as at South Georgia (Kawaguchi et al. 2007a). Increased temperature in more northern areas may also increase growth rates, assuming krill follow normal crustacean temperature–growth relation (Hartnoll 2001) and larger females produce more eggs per female biomass (Cuzin-Roudy 2000). Longer daylight periods further promote higher feeding, faster growth and metabolism (Teschke et al. 2007). Overall, the length of the primary production season dictates the length of the period during which ovarian cycles are sustained (Kawaguchi et al. 2007a).

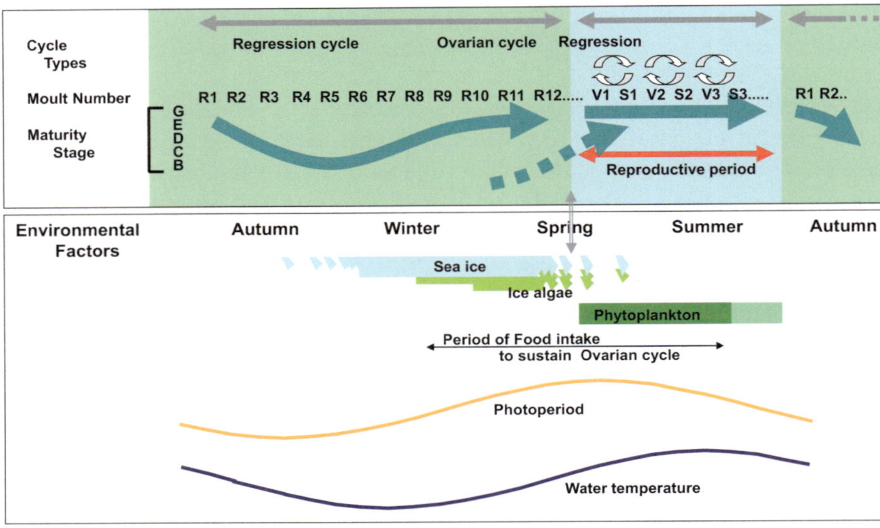

Fig. 6.2 Conceptual model of krill maturity cycle in relation to environmental factors. *R1*, *R2*... number of moulting event since start of resting period; *V* vitellogenic moult cycle, *S* spawning moult cycle; *Thick arrow* general progression of female maturity; *Thick dotted arrow* maturity progression of a first time spawner; *White circulating arrows* ovarian cycle (Reproduced from Fig. 6.4 of Kawaguchi et al. (2007a))

6.4 Sex Ratio

Krill mating behavior observed in the wild involves high speed pursuit of females by multiple males and rapid circular swimming actions (Kawaguchi et al. 2011a) which are likely to be energy demanding and require a heavy energetic investment in successful fertilization, and this places significant demands on their lipid reserves (Virtue et al. 1996). Female krill often carry several spermatophores (suggesting multiple pairing) and those spermatophores attached are lost through regular moulting therefore males continuously produce and mate throughout the reproductive season (Virtue et al. 1996). This energetic behaviour is thought to lead to mortality of large male krill skewing the sex ratio of larger krill towards females (Marr 1962; Virtue et al. 1996). This size-differentiated sex ratio can be observed in datasets derived from scientific surveys and from the krill fishery, and has been attributed to male krill growing faster and having a shorter lifespan than females (Kawaguchi et al. 2007b). If the longevity of males is considerably less than that of females then changes in sex ratio with time can significantly distort the population size structure in a population with only episodic recruitment. However, the skewed sex ratio often observed in krill swarms can also be partially explained by a combination of physical factors and ontogenetic behaviour. For example differential swimming abilities may also spatially separate krill at different sizes, stages and gender (Quetin and Ross 1984a; Tarling et al. 2009; Watkins et al. 1992) (Also read Chap. 3 (Reiss 2016) for different opinion on sex ratios).

6.5 Fecundity

Population fecundity can be derived from the abundance of females, the percentage of reproductive females, the number of eggs per spawning batch and the number of times a female recycles the ovary to produce eggs. In any 1 year, the percentage of the female population that is in the reproductive cycle in the western Antarctic Peninsula area ranged from 10 % to nearly 100 % (Quetin and Ross 2001). The areal extent of sea ice, the timing of sea-ice retreat, and associated phytoplankton bloom appear to be the important factors affecting the population fecundity. The conditions that provide adequate food at the appropriate time for the ovarian development are the most important for optimal population fecundity (Quetin and Ross 2001; Quetin et al. 2007).

Estimates of Antarctic krill fecundity in early studies assumed only one batch of eggs per year (Bargmann 1937; Naumov 1962; Jazdzewksi et al. 1978) but field evidence from later studies demonstrated that individual females could produce multiple batches of eggs in a single spawning season (Tarling et al. 2007 and references therein). This process of ovarian recycling is detailed in the earlier section in this Chapter (Sect. 6.2.3 (Gonad maturation)). The size of a single batch of eggs can be estimated by counting the number of fully mature oocytes since eggs are produced periodically through synchronous development of batches of oocytes (Cuzin-Roudy 2000). Although fecundity is highly variable between individuals, the size of a batch of eggs is significantly correlated with body size (Tarling et al. 2007). The body length-batch size relation also varies considerably between regions and years, but fecundity is generally higher in the Southwest Atlantic area (Antarctic Peninsula, Scotia Sea) than off East Antarctica (Brown 2005), probably due to difference in food availability between these areas. The concentration of food will have a large influence on the size of the egg batch that individuals can produce (Nicol et al. 1995). Further, even at the same location and season (mid- to late January) large difference in egg batch size can be observed between different years (years 2001 and 2003) (Brown 2005) (Fig. 6.3).

Female krill are estimated to spawn up to three batches of eggs or more during one spawning season (Cuzin-Roudy 2000; Tarling et al. 2007). Tarling et al. (2007) developed a semi-empirical model to predict the number of egg batches produced per season. This model predicted that 11 % of females produced only one batch per year, 60 % completed two, and 29 % completed three or more, and estimated that a South Georgian female krill would release on average 12,343 eggs per year assuming unlimited food. This is higher than the estimate of annual production of

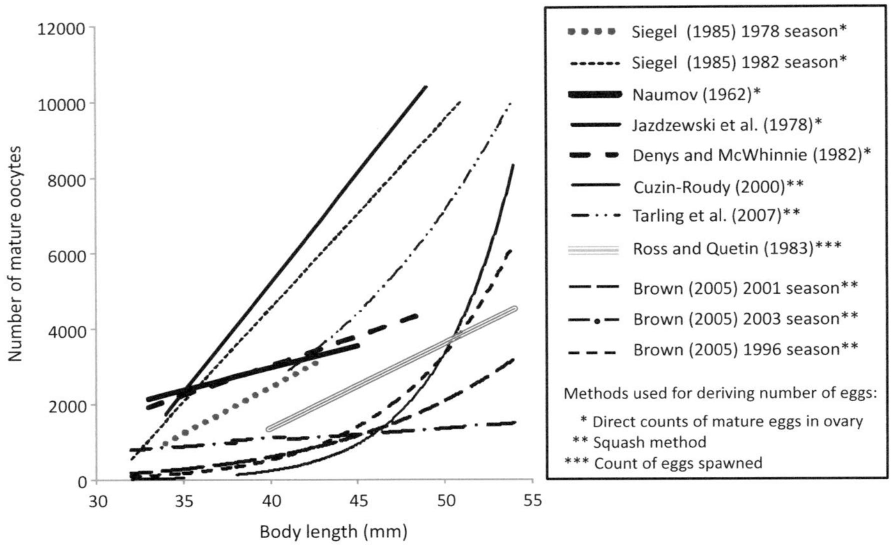

Fig. 6.3 Antarctic krill egg batch: body length relationships found by previous studies

7200 eggs based on lipid loss during spawning season (Clarke 1980) or the 7500 eggs estimated from the assumption that females produce only two batches of eggs per season (Cuzin-Roudy 2000). The effect of food limitation such as that caused by a delay in the phytoplankton bloom can be simulated by either delaying previtellogenesis by 20–40 days or delaying ovarian development by one additional moult cycle (total of three IMP rather than two IMP for one ovarian development cycle). Under this simulation, annual egg production was nearly halved, suggesting that episodic decrease in primary productivity observed in South Georgia could reduce the egg production for the season (Tarling et al. 2007).

6.6 Timing of Spawning

In much of Antarctica the driving force for boosting population maturity and spawning is the phytoplankton bloom that occurs following the seasonal sea-ice retreat. In ice covered waters, therefore, krill maturation progress from north to south (Cuzin-Roudy and Labat 1992; Kawaguchi et al. 2007a). Krill spawning around the Antarctic continent exhibits large regional differences in timing of the start and end of the spawning season, and the duration of the reproductive period (Fig. 6.4). The timing of spawning and its duration around the Antarctic continent has been analysed through examining adult maturity stages and the abundance of

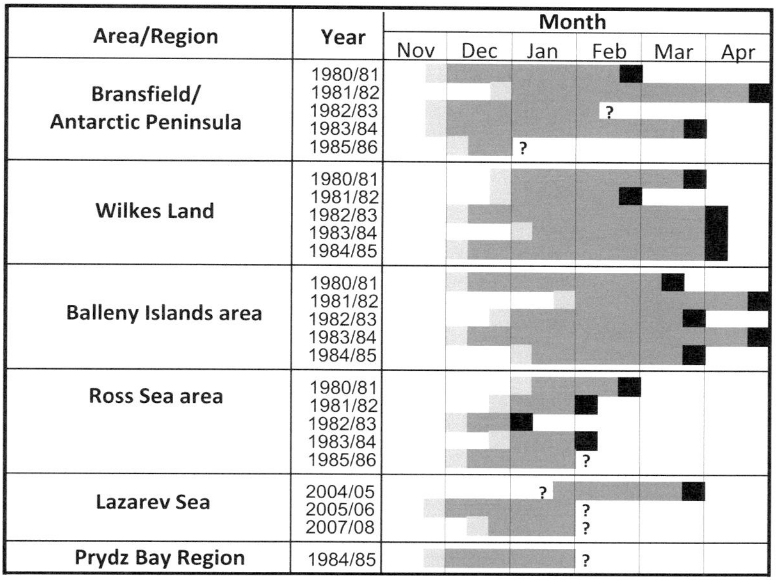

Fig. 6.4 Timing of Antarctic krill spawning around the continent. *Light grey*: onset of spawning, *Dark grey*: reproductive season, *Black*: end of spawning season, *?*: timing of the end of spawning season not derived in the original study (Data source: Bransfield and Antarctic Peninsula, Wilkes Land, Balleny Islands, and Ross Sea; Spiridonov 1995, Lazarev Sea; Siegel 2012; Prydz Bay region; Hosie 1991)

larval stages (Spiridonov 1995). Early starts to the spawning season were observed in low latitude areas, such as the Antarctic Peninsula region, due to sea-ice retreat occurring early in the spring. Delayed starts to the spawning season, with short spawning duration, were observed in high latitude areas, such as the Ross Sea, and coastal areas due to late sea-ice break-up resulting with delayed and short spring bloom which is not ideal for extensive reproduction. In the Lazarev Sea larval krill and mature adults have been observed early in the season indicating that active and successful spawning can occur in a high latitude region which was still under 60–70 % pack-ice cover (Siegel 2012). Early spawning in the Lazarev Sea appears to be facilitated by the opening up of polynyas which allows an early start of phytoplankton growth and this has been observed in other ice-covered areas (Siegel 2012; Arrigo and van Dijken 2003; Kawaguchi et al. 2010b; Schwarz et al. 2010). In the Prydz Bay region krill are also capable of spawning as early as the end of November as indicated by the presence of a few advanced developmental stages in January samples (Hosie 1991). Overall, there are various factors other than light regime that affect the timing, magnitude and duration of spawning, and, more studies on krill condition during late winter to early spring are essential to further understand the process of early onset of their reproduction events.

6.7 Early Development

6.7.1 Embryonic Development

Antarctic krill lay sinking embryos. Embryonic development of Antarctic krill is completed in 5–7 days (George 1984; Quetin and Ross 1984b). Within the first 2 days the embryo completes gastrulation; the body plan formed from three layers of germ cells and the different organ systems begin to differentiate. From the second day onwards the development of organ system accelerates. By the third to fourth day embryos enter limb bud stage, and in 5–7 days they hatch as nauplii (George 1984; Quetin and Ross 1984b; Jia et al. 2014).

The development rate increases with temperature (George 1984; Ross et al. 1988; Yoshida et al. 2004). Moderate pressure of 5–20 atm has also been reported to accelerate embryonic development (George and Stromberg 1985). There is an enormous range of values measured for the sinking rate (George and Stromberg 1985), but this was quite similar within a stage (Quetin and Ross 1984b; Ross and Quetin 1985). Pressure may have a significant effect on embryonic sinking and development but this is not yet fully quantified. Embryonic metabolism is only 3.45 % of the total metabolic demand of all the non-feeding stages, therefore, most of the embryonic lipid reserves can be passed onto the larvae for use during development to the first feeding stage (Quetin and Ross 1989). The development rate is slower in lower temperature and did not continue to develop at −1.5 °C (Ross et al. 1988), but another study showed successful embryonic development below this temperature down to −2.1 °C (Yoshida et al. 2004). At the higher range of the temperature embryonic development was inhibited at 4 °C (George and Stromberg 1985).

The fatty acid profiles in the diet of maternal krill are known to affect the fatty acid composition of embryos and in turn hatching success, indicating that the specific fatty acids are playing important roles during embryogenesis of krill (Yoshida et al. 2011).

6.7.2 Larval Growth and Development

Once they have hatched Antarctic krill larvae undergo a complex developmental processes with four phases consisting of a total of 12 stages (Fig. 6.5, Table 6.1) (Fraser 1936; Marr 1962; Kirkwood 1982; Jia et al. 2014). The developmental ascent as described by Marr (1962) is discussed in more detail in Chap. 2 (Siegel and Watkins 2016).

Survival during crustacean larval stages is expected to be lower than the survival rate of post larval stages. Although not yet estimated in the wild, in an aquarium study an average survival rate (the percentage of individuals surviving to juvenile stage from embryo) of 3 % (range 0–10 %) has been reported (Hirano et al. 2003).

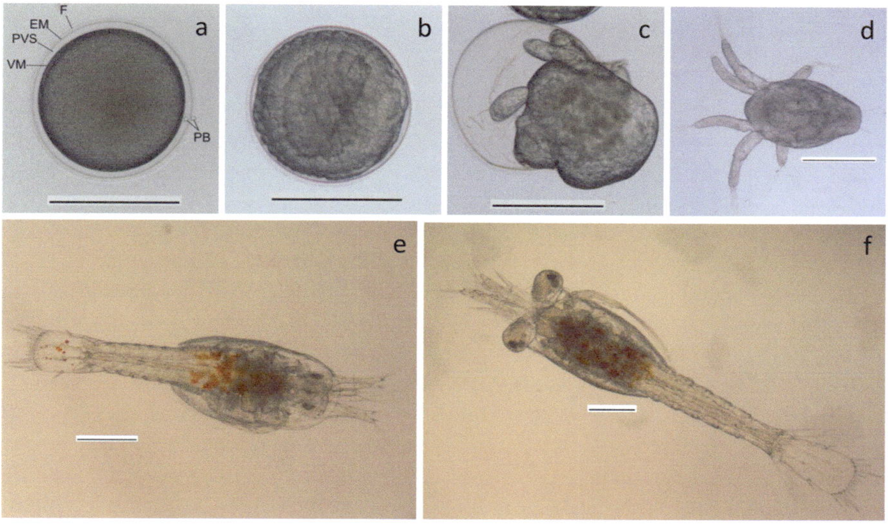

Fig. 6.5 Antarctic krill. (**a**) Newly spawned egg (*F* fertilisation jelly, *EM* embryo membrane, *PVS* perivitelline space, *VM* vitelline membrane, *PB* polar body), (**b**) late gastrula, (**c**) hatching larvae, (**d**) nauplius II, (**e**) calyptopis II, (**f**) furcilia I. Scale bars 500 μm (Source Jia et al. (2014))

Table 6.1 Morphological and behavioural characteristics of Antarctic krill larvae.

Stages	Key morphological characteristics	Behavioural characteristics
Nauplius I–II	Unsegmented body with 3 pairs of appendages. No eyes	No feeding
Metnauplius	Develops a carapace and abdomen folded under the carapace	
Calyptopis I–III	The first feeding phase with a distinct cephalothorax and unsegmented abdomen. Eyes and mouthparts with thoracic legs developed but still no abdominal appendages	First feeding phase
Furcilia I	Mobile eyes	Furcilia phase: mobile eyestalk, gradually take on the shape of the juvenile, involves the most complex morphological changes, especially in the Furcilia III stage and the development of the feeding basket which enable the late furcilia stage krill (stage IV-) to form schools and move the thoracic basked in the same way that adults feed on phytoplankton and ice algae
Furcilia II	Photophores start developing	
Furcilia III	Four photophores on the abdomen. All five pairs of pleopods have setae and fully developed	
Furcilia IV	Fully developed the feeding basket. Number of terminal spines on telson reduced from 7 to 5	
Furcilia V	Number of terminal spines on telson reduced from 5 to 3	
Furcilia VI	Number of terminal spines on telson reduced from 3 to 1	

Information on key morphological characteristics is based on Jia et al. (2014), and behavioural characteristics based on Hamner et al. (1989) and Jia et al. (2014)

There are two critical larval periods in the first year of a krill's life (Ross and Quetin 1989). The first is within the first few weeks after hatch when developmental rate is slow due to low temperature. This leads to shortage of maternal energy reserve before they turned into this first feeding stage (Ross et al. 1988). The second critical period is the first winter when phytoplankton in water column is scarce (Smith et al. 1996). This underscores the importance of warm circumpolar deep water to accelerate embryonic development to ensure their development into the feeding stage before they run out of the energy reserve inherited maternally.

The development and growth of larval krill show strong seasonal pattern from autumn and winter. The first feeding stage, Calyptopis I, is critical as the larvae have to find adequate food within 10–14 days (point-of-no-return). In the western Antarctic Peninsula area, larvae continue to develop during autumn and reach the furcilia V or VI stage in early winter. Development and growth are slow throughout winter, but increased again in late winter (Quetin and Ross 2003). Larval growth was strongly correlated with day length, suggesting an indirect effect of day length on either larval krill behaviour or primary production in the sea ice. Sea ice condition itself was a poor predictor of larval krill growth and development (Quetin and Ross 2003). Comparison of larval krill growth over winter between two winters (August 2001 and 2002) revealed significant nonlinear relationship between pigment content in larvae and their growth increments. Larvae shrunk or stayed the same size in 2001 when sea ice microbial community (SIMCO) biomass and Chla in ice cores were low whereas larvae showed positive growth in 2002 when SIMCO biomass was tenfolds higher. This suggests interannual variability in SIMCO and pigment in the sea ice habitat varied in its ability to enhance condition and growth in larval krill (Quetin et al. 2007).

In the Lazarev Sea, zero to negative larval growth rates have been recorded in winter with smaller body sizes and weights in winter than in autumn of equivalent larval stages (Meyer et al. 2009). Negative growths in larvae were also observed in field experiments conducted in the west Antarctic Peninsula region; −3.42 % per moult during light ice cover year (Ross and Quetin 1991), and average of −1.6 % per moult from field experiments covering area of approximately 160 km^2 in July to August 2001 (Ross et al. 2004). This may be a result of extremely limited food on the underside of pack ice and in the water column as illustrated by the study west of the Antarctic Peninsula (Quetin et al. 2007).

Larval krill do not have enough energy reserves as the adults do to grow and/or survive through the winter therefore their growth and condition factor depends on availability of ice algae (Quetin and Ross 2003), heterotrophic diet, as well as their own body protein for energy provision for their survival (Meyer et al. 2009).

The autumn phytoplankton dynamics has been highlighted as a major driver of krill development in an individual based model developed for the Western Antarctic Peninsula (Lowe et al. 2012). Other key factors indicated by the modelling process include sea ice advance, development of sea ice microbial communities, and the late winter increase in sea ice microbial community biomass. These environmental processes controlled food availability, and consequently, physiological condition of krill during their first winter.

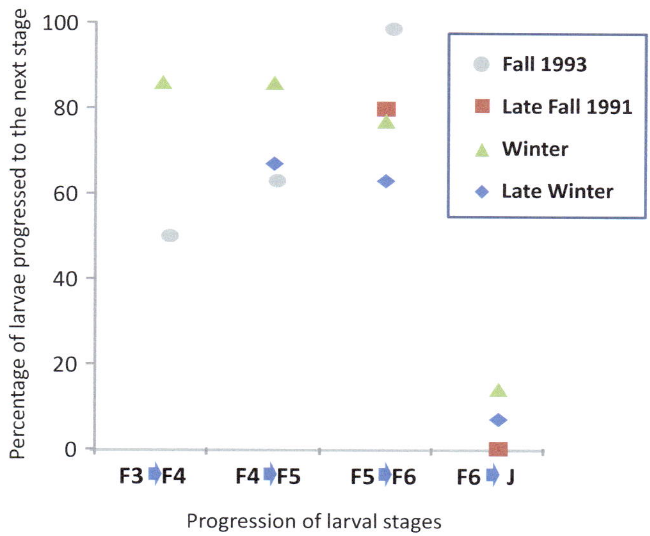

Fig. 6.6 Percentage of larvae advancing their stage to the next at a moulting event (Data source: Table 3 of Quetin and Ross 2003)

Stepwise advancement from one larval stage to the next does not always happen, especially in the late larval states. Regional variability in growth and development is often observed in the field and this is attributed to variability in the availability of food (Daly 2004; Fraser 1936; Brinton and Townsend 1984; Quetin et al. 2007; Brinton et al. 1986; Meyer et al. 2009). A very low proportion of furcilia VI stage larvae (less than 10 %) advance to the juvenile stage in winter (Quetin and Ross 2003) (Fig. 6.6). This developmental pathway is considered to be one of the strategies to time the larval development so that they metamorphose into the energy demanding juvenile stages coincident with the onset of spring bloom. This would enhance their chances of withstanding variable conditions and thus optimise the population survival (Brinton et al. 1986; Daly 2004; Feinberg et al. 2006; Jia et al. 2014; Meyer et al. 2009). Further study efforts are needed to enhance our understanding on the interaction between sea ice process and larval krill, especially in relation to their behaviour and access to food source.

6.8 Environmental and Climate Change Effects on Reproductive Success

Life history and population dynamics of Antarctic krill is likely to be impacted by the climate change attributed to increasing CO_2 in the atmosphere (Flores et al. 2012). Habitat range is likely to reduce because increasing seawater temperature in the northern limit of their habitat may become suboptimal for krill growth and development (Hill et al. 2013). Krill population dynamics is thought to be strongly linked to sea ice, and their reproductive output and recruitment success been related to extent, timing and the duration of winter sea ice cover. The underside structures of sea ice are generally thought to provide good feeding ground as well as nursery grounds for overwintering larvae. Extensive winter sea ice extent promotes strong spring phytoplankton blooms when retreating which are the key to fuel the adults' reproductive output for their summer spawning season (Quetin and Ross 2001). There are several mechanisms to explain the underlying correlation between reproductive success, larval survival and environmental timings in relation to climatic forcing and climate change.

SIMCO biomass accumulates over time in first year ice, and this accumulation rate slows down from March to June as the daily solar radiation declines (Fritsen et al. 2008). Therefore higher sea ice microbial community biomass is generally expected when ice advances earlier in the season (Quetin et al. 2007). Sea ice advancement in the west Antarctic Peninsula region is later now by more than 1 month than in 1980 (Stammerjohn et al. 2012), thus now sea ice advance in fall is happening further south with coincident lower solar energy input than the latitude where it used to occur. This means food availability and larval survival will decrease with the same timing at more southerly latitudes, and the krill population may be negatively impacted by this change (Quetin et al. 2007). A 200 km southward shift of the location of highest krill densities have been observed in Palmer-LTER area (Ross et al. 2014), but shift to the higher latitudes could be detrimental because light signals influencing krill phenology may be restricted by latitude (Quetin et al. 2007). A possible disjunct between the summer spawning position and winter nursery for larvae due to changes in seasonal sea ice dynamics may also affect the viability of current spawning regions (Quetin et al. 2007).

A similar southward shift of krill populations has been predicted by an individual-based model that explored the impact of photoperiod and under-ice food availability on the growth and demographics of krill (Groeneveld et al. 2015). This shift was attributed to reduction in the spatial sea ice extent, which is consistent with field observations. The model highlighted the importance for modelling krill population dynamics of the spatio-temporal dynamics of sea ice, especially for larval krill, together with temporal synchronization through internal biological clocks triggered by environmental factors (photoperiod and food) in adult krill (Groeneveld et al. 2015; Meyer 2012).

Large scale climatic forcing factors such as the Southern Annular Mode (SAM) and ENSO affect winter sea ice extent and water column stability, hence habitat condition for krill (Loeb et al. 2009; Saba et al. 2014; Steinberg et al. 2015). SAM is an index that describes the north–south movement of the westerly wind belt that circles Antarctica. In the Palmer-LTER area, in a negative SAM event, cold southerly winds move across the Antarctic Peninsula, winter ice extent and duration increase, and the number of windy days decreases. These conditions promote accumulation of ice algal biomass and ice-edge bloom and provide good habitat for adult krill reproductive output as well as survival of larvae in the first year, and vice versa (Saba et al. 2014).

Recent experimental studies revealed a sharp decline in embryonic development rates at increased CO_2 levels (Kawaguchi et al. 2011b), and suggested that the important krill habitats of the Weddell Sea and to the east are likely to become high-risk areas for krill recruitment within 100 years, and the entire Southern Ocean krill population could collapse by 2300 unless CO_2 emissions are mitigated (Kawaguchi et al. 2013). Hatching success is but one of the many population parameters that would be affected by ocean acidification. Krill respond to the elevated CO_2 levels with enhanced energetic requirements which may compromise their physiological performance (Saba et al. 2012). If elevated CO_2 concentrations persist, these physiological disturbances may retard their growth and reproduction, and may affect their behaviour (Kawaguchi et al. 2013).

The krill population is already experiencing a number of stresses due to a changing climate (Atkinson et al. 2004; Kawaguchi et al. 2009; Meredith and King 2005). Seawater temperature near the Antarctic Peninsula is rising, and sea ice cover declining. Ocean acidification is likely to put further pressure on the krill population. The multiple stressors are likely to impose synergistic, rather than additive impacts on krill population in a negative way (Flores et al. 2012).

6.9 Concluding Remarks

In this chapter I have highlighted the importance of the timing of the occurrence of seasonal events for reproductive output and successful larval survival. The reproductive strategy of Antarctic krill optimises its output in its highly variable environment. Earlier studies on reproduction and development highlighted the plasticity of krill biology, especially in their overwintering strategies, to explain this species' prosperity in the highly variable Southern Ocean. Recent studies have revealed that aspects of the physiological cycle of krill appear to be hard-wired to the natural light regime. The ongoing changes in the timing of important events such as sea ice advance and retreat and the phytoplankton bloom may therefore potentially cause a disjunct between krill's biological timings and the future environment. Krill have been shown to be a remarkably adaptable species so may well have some capacity to adjust to environmental changes but this possibility is yet to be investigated.

Krill is the fundamental component of the Southern Ocean ecosystem. Reproduction and early development are the key determinants of population size and it is important that we enhance our mechanistic understanding of these aspects of krill biology to improve the understanding and management of the krill-dominated ecosystem in a changing climate. Technological advancements including husbandry, modelling, analytical, and molecular techniques in recent years are allowing us to understand biological and environmental processes and their interaction even at the molecular levels. These tools will allow us to make considerable progress in improving our understanding of the reproductive processes of krill over the next decade.

References

Arrigo KR, van Dijken GL (2003) Phytoplankton dynamics within 37 Antarctic coastal polynya systems. J Geophys Res 108(C8):27

Atkinson A, Siegel V, Pakhomov E, Rothery P (2004) Long-term decline in krill stock and increase in salps within the Southern Ocean. Nature 432:100–103

Bargmann HE (1937) The reproductive system of *Euphausia superba*. Disc Rep 14:237–249

Bargmann HE (1945) The development and life history of adolescent krill *Euphausia superba*. Disc Rep 23:103–176

Brinton E, Townsend AW (1984) Regional relationships between development and growth in larvae of Antarctic krill, *Euphausia superba*, from field samples. J Crustacean Biol 4(Spec Issue 1):224–246

Brinton E, Huntley M, Townsend AW (1986) Larvae of *Euphausia superba* in the Scotia Sea and Bransfield Strait in March 1984 – development and abundance compared with 1981 larvae. Polar Biol 5:221–234

Brown M (2005) Reproduction of Antarctic krill (*Euphausia superba* Dana) in the Indian Ocean sector: with special reference to ovary maturation. Honours thesis, University of Tasmania

Brown M, Kawaguchi S, Candy S, Virtue P (2010) Temperature effects on the growth and maturation of Antarctic krill (*Euphausia superba*). Deep-Sea Res II 57:672–682

Brown M, Kawaguchi S, King R, Virtue P, Nicol S (2011) Flexible adaptation of the seasonal krill maturity cycle in the laboratory. J Plankton Res 33:821–826

Brown M, Kawaguchi S, Candy S, Yoshida T, Virtue P, Nicol S (2013) Long-term effect of photoperiod, temperature and feeding regimes on the respiration rates of Antarctic krill (*Euphausia superba*). Open Mar Sci 3:40–51

Charniaux-Cotton H, Payen G (1985) Sexual differentiation. In: Bliss DE, Mantel LH (eds) The biology of crustacean, vol 9. Academic, London, pp 217–299

Clarke A (1980) The biochemical composition of krill Euphausia superba (Dana), from South Georgia. J Exp Mar Biol Ecol 43:221–236

Cook JM (2002) Sex determination in invertebrates. In: Hardy ICW (ed) Sex ratios: concepts and research methods. Cambridge University Press, Cambridge, pp 178–194

Cuzin-Roudy J (1987a) Gonad history of the Antarctic krill *Euphausia superba* Dana during its breeding season. Polar Biol 7:237–244

Cuzin-Roudy J (1987b) Sexual differentiation in the Antarctic krill *Euphausia superba* Dana (Crustacea: Euphausiacea). J Crustacean Biol 7:518–524

Cuzin-Roudy J (1993) Reproductive strategies of the Mediterranean krill, *Meganyctiphanes norvegica* and the Antarctic krill, *Euphausia superba* (Crustacea: Euphausiacea). Invertebr Reprod Dev 23:105–114

Cuzin-Roudy J (2000) Seasonal reproduction, multiple spawning, and fecundity in northern krill, *Meganyctiphanes norvegica*, and Antarctic krill, *Euphausia superba*. Can J Fish Aquat Sci 57(Suppl 3):6–15

Cuzin-Roudy J (2010) Reproduction in Northern krill (*Meganyctiphanes norvegica* Sars). Adv Mar Biol 57:199–230

Cuzin-Roudy J, Amsler MO (1991) Ovarian development and sexual maturity staging in Antarctic krill, *Euphausia superba* Dana (Euphausiacea). J Crustracean Biol 11:236–249

Cuzin-Roudy J, Labat JP (1992) Early summer distribution of Antarctic krill sexual development in the Scotia-Weddell region: a multivariate approach. Polar Biol 12:65–74

Daly KL (2004) Overwintering growth and development of larval *Euphausia superba*: an interannual comparison under varying environmental conditions west of the Antarctic Peninsula. Deep-Sea Res II 51:2139–2168

Denys CJ, McWhinnie MA (1982) Fecundity and ovarian cycles of the Antarctic krill *Euphausia superba* (Crustacea, Euphausiacea). Can J Zool 60:2414–2423

Feinberg LR, Shaw CT, Peterson WT (2006) Larval development of *Euphausia pacifica* in the laboratory: variability in developmental pathways. Mar Ecol Prog Ser 316:127–137

Flores H, Atkinson A, Kawaguchi S, Pakhomov E, Quetin L, Ross R, Hill S, Reiss C, Siegel V, Tarling G (2012) Impact of climate change on Antarctic krill. Mar Ecol Prog Ser 458:1–19

Fraser FC (1936) On the development of the young stages of krill (*Euphausia superba*). Discov Rep 14:1–192

Fritsen CH, Memmott J, Stewart FJ (2008) Inter-annual sea-ice dynamics and micro-algal biomass in winter pack ice of Marguerite Bay, Antarctica. Deep-Sea Res II 55:2059–2067

George RY (1984) Ontogenetic adaptation in growth and respiration of *Euphausia superba* in relation to temperature and pressure. J Crustacean Biol 4(Spec Issue 1):252–262

George RY, Stromberg JO (1985) Development of eggs of Antarctic krill *Euphausia superba* in relation to pressure. Polar Biol 4:125–133

Groeneveld J, Johst K, Kawaguchi S, Meyer B, Teschke M, Grimm V (2015) How biological clocks and changing environmental conditions determine local population growth and species distribution in Antarctic krill (*Euphausia superba*): a conceptual model. Ecol Model 303:78–86

Hagen W, Van Vleet ESV, Kattner G (1996) Seasonal lipid storage as overwintering strategy of Antarctic krill. Mar Ecol Prog Ser 134:85–89

Hamner WM, Hamner PP, Obst BS, Carleton JH (1989) Field observation on the ontogeny of schooling of *Euphausia superba* furciliae and its relationship to ice in the Antarctic waters. Limnol Oceanogr 34:451–456

Hartnoll RG (2001) Growth in crustacean-twenty years on. Hydrobiologia 449:111–122

Hill SL, Phillips T, Atkinson A (2013) Potential climate change effects on the habitat of Antarctic Krill in the Weddell quadrant of the Southern Ocean. PlosOne 8, e72246

Hirano Y, Matsuda T (2003) Antarctic krill breeding facilities at Port of Nagoya public aquarium. Mar Fresh Behav Physiol 36:249–258

Hirano Y, Matsuda T, Kawaguchi S (2003) Breeding Antarctic krill in captivity. Mar Fresh Behav Physiol 36:259–269

Hosie GW (1991) Distribution and abundance of euphausiid larvae in the Prydz Bay region, Antarctica. Antarct Sci 3:167–180

Huntley M, Nordhausen W, Lopez MDG (1994) Elemental composition, metabolic activity and growth of Antarctic krill *Euphausia superba* during winter. Mar Ecol Prog Ser 107:23–40

Ikeda T (1987) Mature Antarctic krill (*Euphausia superba* Dana) grown from eggs in the laboratory. J Plankton Res 9:565–569

Ikeda T, Dixon T (1982) Body shrinkage as a possible overwintering mechanism of the Antarctic krill, *Euphausia superba* Dana. J Exp Mar Biol Ecol 62:143–151

Jarman S, Deagle B (2016) Genetics of Antarctic krill. In: Siegel V (ed) Biology and ecology of Antarctic krill, Advances in polar biology. Springer, Cham, pp 247–278

Jazdzewksi K, Dzik J, Porebski J, Rakusa-Suszczewski S, Witek Z, Wolnomiejski N (1978) Biological and populational studies on krill near South Shetland Islands, Scotia Sea and South Georgia in the summer 1976. Pol Arch Hydrobiol 25:607–631

Jia ZN, Virtue P, Swadling KM, Kawaguchi S (2014) A photographic documentation of the development of Antarctic krill (*Euphausia superba*) from egg to early juvenile. Polar Biol 37:165–179

Kawaguchi K, Ishikawa S, Matsuda O (1986) The overwintering strategy of Antarctic krill (*Euphausia superba* Dana) under the coastal fast ice of the Ongul Islands in Lützow-Holm Bay, Antarctica. Mem Nat Inst Polar Res Spec Iss 44:67–85

Kawaguchi S, Yoshida T, Finley L, Cramp P, Nicol S (2007a) The krill maturity cycle: a conceptual model of the seasonal cycle in Antarctic krill. Polar Biol 30:689–698
Kawaguchi S, Finley LA, Jarman S, Candy SG, Ross RM, Quetin LB, Siegel V, Trivelpiece W, Naganobu M, Nicol S (2007b) Male krill grow fast and die young. Mar Ecol Prog Ser 345:199–210
Kawaguchi S, Nicol S, Press AJ (2009) Direct effects of climate change on the Antarctic krill fishery. Fish Manage Ecol 16:424–427
Kawaguchi S, King RA, Meijers R, Osborn JE, Swadling KM, Ritz DA, Nicol S (2010a) An experimental aquarium for observing the schooling behaviour of Antarctic krill. Deep-Sea Res II 57:683–692
Kawaguchi S, Nicol S, Virtue P, Davenport SR, Casper R, Swadling KM, Hosie GW (2010b) Krill demography and large scale distribution in the Western Indian Ocean sector of the Southern Ocean (CCAMLR Division 58.4.2) in Austral summer of 2006. Deep-Sea Res II 57:934–947
Kawaguchi S, Kilpatrick R, Roberts L, King RA, Nicol S (2011a) Ocean-bottom krill sex. J Plankton Res 33:1134–1138
Kawaguchi S, Kurihara H, King R, Hale L, Berli T, Robinson JP, Ishida A, Wakita M, Virtue P, Nicol S, Ishimatsu A (2011b) Will krill fare well under Southern Ocean acidification? Biol Lett 7:288–291
Kawaguchi S, Ishida A, King R, Raymond B, Waller N, Constable A, Nicol S, Wakita M, Ishimatsu A (2013) Risk maps for Antarctic krill under projected Southern Ocean acidification. Nat Clim Chang 7:843–847
Kirkwood JM (1982) A guide to the Euphausiacea of the Southern Ocean. ANARE Res Notes 1:1–45
Loeb VJ, Hofmann EE, Klinck JM, Holm-Hansen O, White WB (2009) ENSO and variability of the Antarctic Peninsula pelagic marine ecosystem. Antarct Sci 21:135–148
Lowe AT, Ross RM, Quetin LB, Vernet M, Fritsen CH (2012) Simulating larval Antarctic krill growth and condition factor during fall and winter in response to environmental variability. Mar Ecol Prog Ser 452:27–43
Makarov RR (1975) A study of the second maturation of euphausiid (Eucarida, Euphausiacea) females (in Russian). Zoologicheskiy Zh 54(5):670–681
Makarov RR, Denys C (1981) Stages of sexual maturity of *Euphausia superba* Dana. BIOMASS Handb 11:11
Marr JWS (1962) The natural history and geography of the Antarctic krill (*Euphausia superba* Dana). Discov Rep 32:33–464
Mayzaud P, Alvessard E, Cuzin-Roudy J (1998) Changes in lipid composition of the Antarctic krill *Euphausia superba* in the Indian sector of the Antarctic Ocean: influence of geographical location, sexual maturity stages and distribution among organs. Mar Ecol Prog Ser 173:149–162
Meredith MP, King JC (2005) Rapid climate change in the ocean west of the Antarctic Peninsula during the second half of the 20th century. Geophys Res Lett 32, L19604
Meyer B (2012) The overwintering of Antarctic krill, *Euphausia superba*, from an ecophysiological perspective. Polar Biol 35:15–37
Meyer B, Teschke M (2016) Physiology of *Euphausia superba*. In: Siegel V (ed) Biology and ecology of Antarctic krill, Advances in polar biology. Springer, Cham, pp 145–174
Meyer B, Fuentes V, Guerra C, Schmidt K, Atkinson A, Spahic S, Cisewski B, Freier U, Olariaga A, Bathmann U (2009) Physiology, growth and development of larval krill *Euphausia superba* in autumn and winter in the Lazarev Sea, Antarctica. Limnol Oceanogr 54:1595–1614
Naumov AG (1962) Biological state of aggregations of *Euphausia superba* Dana found near the Balleny Islands. Inf Byull Sov Antarkt Eksped 4:277–280
Nicol S, de la Mare WK, Stolp M (1995) The energetic cost of egg production in Antarctic krill (*Euphausia superba* Dana). Antarct Sci 7:25–30
Poleck TP, Denys CJ (1982) Effect of temperature on the molting, growth and maturation of the Antarctic krill *Euphausia superba* (Crustacea: Euphausiacea) under laboratory conditions. Mar Biol 70:255–265

Quetin LB, Ross RM (1984a) School composition of the Antarctic krill *Euphausia superba* in the austral summer of 1982. J Crustacean Biol 4(Spec Issue 1):96–106
Quetin LB, Ross RM (1984b) Depth distribution of developing *Euphausia superba* embryos, predicted from sinking rates. Mar Biol 79:47–53
Quetin LB, Ross RM (1985) The effect of pressure on the sinking rates of the embryos of the Antarctic krill, *Euphausia superba*. Deep-Sea Res 32(7):799–807
Quetin LB, Ross RM (1989) Effects of oxygen, temperature and age on the metabolic rate of the embryos and early larval stage of the Antarctic krill *Euphausia superba* Dana. J Exp Mar Biol Ecol 133:103–127
Quetin LB, Ross RM (2001) Environmental variability and its impact on the reproductive cycle of Antarctic krill. Am Zool 41:74–89
Quetin LB, Ross RM (2003) Episodic recruitment in Antarctic krill *Euphausia superba* in the Palmer LTER study region. Mar Ecol Prog Ser 259:185–200
Quetin LB, Ross RM, Rristen CH, Vernet M (2007) Ecological responses of Antarctic krill to environmental variability: can we predict the future? Antarct Sci 19:253–266
Reiss CS (2016) Age, growth, mortality, and recruitment of Antarctic Krill, *Euphausia superba*. In: Siegel V (ed) Biology and ecology of Antarctic krill, Advances in polar biology. Springer, Cham, pp 101–144
Ross RM, Quetin LB (1983) Spawning frequency and fecundity of the Antarctic krill *Euphausia superba*. Mar Biol 77:201–205
Ross RM, Quetin LB (1989) Ecology and physiology of larval euphausiids, *Euphausia superba* (Euphausiacea). Mem Queensl Mus 31:321–333
Ross RM, Quetin LB (1991) Ecological physiology of larval euphausiids, Euphausia superba (Euphausiacea). Mem Queensl Mus 31:321–333
Ross RM, Quetin LB (2000) Reproduction in euphausiacea. In: Everson I (ed) Krill biology, ecology and fisheries, fish and aquatic resources, vol 6. Blackwell Science, London, pp 150–181
Ross RM, Quetin LB, Kirch E (1988) Effect of temperature on development times and survival of early larval stages of *Euphausia superba* Dana. J Exp Mar Biol Ecol 121:55–71
Ross RM, Quetin LB, Newberger T, Oakes SA (2004) Growth and behavior of larval krill (*Euphausia superba*) under the ice in late winter 2001 west of the Antarctic Peninsula. Deep-Sea Res II 51:2169–2184
Ross R, Quetin LB, Newberger T, Shaw CT, Jones JL, Oakes SA, Moore KJ (2014) Trends, cycles, interannual variability for three pelagic species west of the Antarctic Peninsula 1993–2008. Mar Ecol Prog Ser 515:11–32
Saba GK, Schofield O, Torres JJ, Ombres EH, Steinberg DK (2012) Increased feeding and nutrient excretion of adult Antarctic krill, *Euphausia superba*, exposed to enhanced carbon dioxide (CO_2). PLoS One 7, e52224
Saba GK, Fraser WR, Saba VS et al (2014) Winter and spring controls on the summer food web of the coastal West Antarctic Peninsula. Nat Commun 5:4318
Schwarz JN, Raymond B, Williams GD, Pasquer B, Marsland SJ, Gorton RJ (2010) Biophysical coupling in remotely-sensed wind stress, sea surface temperature, sea ice and chlorophyll concentrations in the South Indian Ocean. Deep-Sea Res II 57:701–722
Seear P, Tarling GA, Teschke M, Meyer B, Thorne MAS, Clark MS, Gaten E, Rosato E (2009) Effects of simulated light regimes on gene expression in Antarctic krill (*Euphausia superba* Dana). J Exp Mar Biol Ecol 381:57–64
Siegel V (1985) On the fecundity of Antarctic krill, *Euphausia superba* (Euphausiacea). Arch FischWiss 36:185–193
Siegel V (1986) Structure and composition of the Antarctic krill stock in the Bransfield Straight (Antarctic Peninsula) during the Second International BIOMASS Experiment (SIBEX). Arch Fisch 37:51–72
Siegel V (1988) A concept of seasonal variation of krill (*Euphausia superba*) distribution and abundance west of the Antarctic Peninsula. In: Sahrhage D (ed) Antarctic ocean and resources variability. Springer, Berlin, pp 219–230

Siegel V (1989) Winter and spring distribution and status of the krill stock in Antarctic Peninsula waters. Arch Fisch 39:45–72

Siegel (2012) Krill stocks in high latitudes of the Antarctic Lazarev Sea: seasonal and interannual variation in distribution, abundance and demography. Polar Biol 35:1151–1177

Siegel V (2016) Introducing Antarctic krill *Euphausia superba* Dana, 1850. In: Siegel V (ed) Biology and ecology of Antarctic krill, advances in Polar ecology. Springer, Cham, pp 1–20

Siegel V, Loeb V (1994) Length and age at maturity of Antarctic krill. Antarct Sci 6:479–482

Siegel V, Loeb V (1995) Recruitment of Antarctic krill *Euphausia superba* and possible causes for its variability. Mar Ecol Prog Ser 123:45–56

Siegel V, Watkins JL (2016) Distribution, biomass and demography of Antarctic krill, *Euphausia superba*. In: Siegel V (ed) Biology and ecology of Antarctic krill, advances in Polar ecology. Springer, Cham, pp 21–100

Smith RC, Dierssen H, Vernet M (1996) Phytoplankton biomass and productivity in the western Antarctic Peninsula region. In: Ross RM, Hofmann EE, Quetin LB (eds) Foundations for ecological research west of the Antarctic Peninsula. American Geophysical Union, Washington, DC, pp 333–356

Spiridonov VA (1995) Spatial and temporal variability in reproductive timing of Antarctic krill (*Euphausia superba* Dana). Polar Biol 15:161–174

Stammerjohn SS, Massom RA, Rind D, Martinson DG (2012) Regions of rapid sea ice change: an inter-hemispheric seasonal comparison. Geophys Res Lett 39, L06501. doi:10.1029/2012GL050874

Steinberg DK, Ruck KE, Gleiber MR, Garzio LM, Cope JS, Bernard KS, Stammerjohn SE, Schofield OME, Quetin LB, Ross RM (2015) Long-term (1993–2013) changes in macrozooplankton off the western Antarctic Peninsula. Deep-Sea Res I 101:54–70

Tarling GA, Fielding S (2016) Swarming and behaviour in Antarctic krill. In: Siegel V (ed) Biology and ecology of Antarctic krill, advances in Polar ecology. Springer, Cham, pp 279–320

Tarling GA, Cuzin-Roudy J, Thorpe SE, Shreeve RS RS, Ward P, Murphy EJ (2007) Recruitment of Antarctic krill *Euphausia superba* in the South Georgia region: adult fecundity and the fate of larvae. Mar Ecol Prog Ser 331:161–179

Tarling GA, Klevjer T, Fielding S, Watkins J, Atkinson A, Murphy E, Korb R, Whitehouse M, Leaper R (2009) Variability and predictability of Antarctic krill swarm structure. Deep-Sea Res I 56:1994–2012

Teschke M, Kawaguchi S, Meyer B (2007) Simulated light regimes affect feeding and metabolism of Antarctic krill, *Euphausia superba*. Limnol Oceanogr 52:1046–1054

Teschke M, Kawaguchi S, Meyer B (2008) Effects of simulated light regimes on maturity and body composition of Antarctic krill, *Euphausia superba*. Mar Biol 154:315–324

Thiriout-Quevereux C, Leiato A, Cuzin-Roudy J (1998) Chromosome diversity in Mediterranean and Antarctic euphausiid species (Euphausiacea). J Crustacean Biol 18:290–297

Thomas PG, Ikeda T (1987) Sexual regression, shrinkage, re-maturation and growth of spent female *Euphausia superba* in the laboratory. Mar Biol 95:357–363

Ventura T, Rosen O, Sagi A (2011) From the discovery of the crustacean androgenic gland to the insulin-like hormone in six decades. Gen Comp Endocrinol 173:381–388

Virtue P, Nichols PD, Nicol S, Hosie G (1996) Reproductive trade-off in male Antarctic krill, *Euphausia superba*. Mar Biol 126:521–527

Watkins JL, Buchholz F, Priddle J, Morris DJ, Ricketts C (1992) Variation in reproductive status of Antarctic krill swarms; evidence for a size related sorting mechanism? Mar Ecol Prog Ser 82:163–174

Yoshida T, Toda T, Hirano Y, Matsuda T, Kawaguchi S (2004) Effect of temperature on embryo development time and hatching success of the Antarctic krill *Euphausia superba* Dana in the laboratory. Mar Fresh Behav Physiol 37:137–145

Yoshida T, Virtue P, Kawaguchi S, Nichols PD (2011) Factors determining the hatching success of Antarctic krill *Euphausia superba* embryo: lipid and fatty acid composition. Mar Biol 158:2313–2325

Chapter 7
Genetics of Antarctic Krill

Simon N. Jarman and Bruce E. Deagle

Abstract From a genetic perspective, *Euphausia superba* (krill) can be described as a non-model organism with a large genome and a large population size. The population genetics of krill has been studied extensively and a consensus on population structure is now emerging. Some preliminary characterization of the krill genome has been accomplished, but the genome is too large to sequence with current technologies. The major genetic resources currently available for krill are transcriptome assemblies and DNA sequences for specific regions. Genetic technologies have advanced rapidly in the past decade and a range of new genetic approaches for studying krill population genetics, physiology, gene function and ecology are now available. Several krill gene expression studies in recent years have revealed genes involved in a range of physiological processes and highlight the potential of this approach for answering a range of questions in krill biology. The rapid expansion in genetic methods available for studying non-model organisms like krill means many new questions can be addressed with these approaches. This is an exciting time for geneticists and krill biologists considering these questions.

Keywords Gene • Genome • Phylogeny • Panmixia • Expression

7.1 Introduction

Krill has one of the largest population sizes among sexually reproducing animal species. The Southern Ocean is one of earth's most extensive continuous habitats and krill are found throughout most of its vast area at variable population densities (see Chap. 2, Siegel and Watkins 2016). For these reasons and because krill is

S.N. Jarman (✉)
CIBIO-InBIO, Centro de Investigação em Biodiversidade e Recursos Genéticos, Universidade do Porto, 4485-661 Vairão, Portugal

Australian Antarctic Division, 203 Channel Highway, Kingston, TAS 7050, Australia
e-mail: simon.jarman@gmail.com

B.E. Deagle
Australian Antarctic Division, 203 Channel Highway, Kingston, TAS 7050, Australia

V. Siegel (ed.), *Biology and Ecology of Antarctic Krill*, Advances in Polar Ecology,
DOI 10.1007/978-3-319-29279-3_7

commercially exploited and managed by a science-informed organization (Everson 2000) (see Chap. 11, Nicol and Foster 2016), population genetics has been one of the most thoroughly investigated aspects of krill genetics. Another reason for the early focus on krill population genetics was that techniques for investigating genetic polymorphism in enzymes in almost any animal were available in the 1970s (Ayala et al. 1975). Many other aspects of krill genetics could not be investigated easily until more recent technological advances enabled studies of features such as gene expression, gene function and genomics.

Many biological questions can be addressed with modern genetic technologies that extend far beyond the realm of classical genetic studies of inheritance or population genetics. Modern methods for studying many aspects of genetics can now be easily applied to non-model organisms like krill, where previously such detailed analyses were restricted to model organisms with well-characterised genetics like *Drosophila melanogaster* (Ekblom and Galindo 2011). High throughput DNA sequencing (HTS) has transformed analysis of the diversity and identity of nucleic acids since it began to be widely available in about 2008 (Metzker 2010). It has been applied to analysis of many aspects of krill biology such as the characterization of messenger RNA that gives researchers access to methods for studying the activity of large numbers of genes involved in many biological functions, which links genetics to physiology, growth and metabolism. Hybridisation methods such as microarrays, often developed from this transcriptomic data, have provided high-throughput options for analysis of activity of large numbers of genes in large numbers of samples. For more precise measurement of gene activity, quantitative polymerase chain reaction (qPCR) experiments are used to measure relative gene expression levels in small numbers of specific target genes. The applications of HTS extend to ecological questions such as species diversity and interactions (Ekblom and Galindo 2011; Bohmann et al. 2014). This approach can be used to identify microorganisms associated with krill and DNA sequencing and has been used to study krill diet as well as to identify krill in the diet of its predators. HTS is now even being applied to krill population genetics, bringing the oldest area of krill genetic research into the modern genomics era. The current range of genetic resources available for studying krill is summarised in Table 7.1.

7.2 Krill Population Genetics

Investigations of the population genetics of krill started over 40 years ago (Ayala et al. 1975) and this is still an area of active research (Candeias et al. 2014). In the first 15 years of study, genetic variation was measured by characterising enzyme variation (i.e. allozymes) using protein electrophoresis. This resulted in more than a dozen published studies (summarised in Fevolden 1988; Fevolden and Schneppenheim 1989). The high level of interest was spurred on in part by early indications that there was a genetic divide observed between samples on the east

Table 7.1 Genetic resources for krill. Sections are organised by the genetic resource type discussed in each chapter

Genetic resource type	Description	Reference	Accessions and URLs	Comments
Variable markers for population genetics (Sect. 7.2)	Mitochondrial NADH1	Bortolotto et al. (2011); Zane et al. (1998)	GenBank: JF766086- JF766209	Multiple polymorphisms in this region result in a high diversity of haplotypes useful for population genetics. A total of 129 haplotypes were recorded (Bortolotto et al. 2011)
	Mitochondrial COI	Goodall-Copestake et al. (2010)	GenBank: GQ305462–GQ305901	A portion of the mitochondrial cytochrome oxidase subunit I region was used to analyse genetic diversity among krill swarms in the same region. 440 haplotypes were identified from 504 samples
	Nuclear microsatellites	Candeias et al. (2014)	GenBank: KF648623-KF648632	Ten Polymorphic microsatellite loci were developed for future studies of krill population genetics
	Nuclear microsatellites	Bortolotto et al. (2011); De Pitta et al. (2008)	–	Three microsatellite loci were developed and applied to analyzing population genetic subdivision. Microsatellite sequences were derived from a previously published EST library (De Pitta et al. 2008)
	Nuclear RAD-sequences	Deagle et al. (2015)	Australian Antarctic Data Centre: http://dx.doi.org/10.4225/15/556FAB354BE19	A library of >1 billion Illumina reads of 90 bp from 148 krill. Reads are associated with *Sbf* I restriction sites in the krill genome (roughly 250,000 sites)
Markers for krill phylogenetics (Sect. 7.3)	Nuclear 18S/small subunit rDNA	Jenner et al. (2009)	GenBank: AY672801	A marker used in investigating relationships among Malacostracan orders
	Nuclear 28S/large subunit rDNA	Jarman et al. (2000b); Jarman (2001a, b)	GenBank: AF169700, AF169717, AF169734	Partial 28S rDNA sequences used for studying relationships among krill genera (Jarman 2001b) and between Euphausiaceae and other orders of Malacostraca (Jarman et al. 2000b)

(continued)

Table 7.1 (continued)

Genetic resource type	Description	Reference	Accessions and URLs	Comments
	Mitochondrial 16S/small subunit rDNA	Patarnello et al. (1996); Jarman et al. (2000a)	GenBank: M98483, EU583500, Z73805, JQ286349	A useful marker for inferring relationships among krill species. Also useful for krill species identification
	Mitochondrial COI	Jarman et al. (2000a)	AB084378	A useful marker for inferring relationships among krill species. Also used for krill species identification and population genetics
Mitochondrial genome sequences (Sect. 7.4.4)	Nearly complete mitochondrial DNA sequence	Shen et al. (2010); Johansson et al. (2012a, b)	GenBank: DQ201509	Most of the krill mitochondrial genome. A small section is hard to PCR amplify or sequence, so a complete sequence is not yet available. A large region has been sequenced in five different individuals to identify variable regions (Johansson et al. 2012a)
	Database of mitochondrial sequences	–	http://gobase.bcm.umontreal.ca/	493 sequences for Euphausia superba mitochondrial DNA
Transcriptome assemblies (Sect. 7.5.2)	Sanger sequence transcriptome	De Pitta et al. (2008)	http://paduaresearch.cab.unipd.it/1177/	1,017 contigs in total representing four tissue types
	Sanger sequence transcriptome	Seear et al. (2010)	dbEST: 68794898–68797989	302 contigs, 1624 singletons produced from head-derived RNA
			GenBank: GW421184-GW424275	
	Roche 454 transcriptome	Clark et al. (2011)	NCBI Short Read Archive: SRA023520	22,177 contigs of mean size 492 bp, ranging from 137 to 8515 bp
	Roche 454 transcriptome	Meyer et al. (2015)	EBI (Study PRJEB6147, Accession range HACF01000001-HACF01058581	57,343 contigs of mean size 691 bp, ranging from 300 to 11,127 bp. A 'master transcriptome' incorporating data from (Clark et al. 2011; De Pittà et al. 2013) as well as two novel cDNA libraries generated from whole krill and from krill heads

Suppressive subtractive hybridization library (Sect. 7.5.2)	Sanger sequences	Seear et al. (2009)	dbEST: 60125367-60126646	A library of 1,279 est sequences derived from krill heads
			GenBank: FL688135-FL689414	
Microarrays (Sect. 7.5.2)	All available transcript microarray	De Pittà et al. (2013)	Agilent eArray: 034463	32,217 oligonucleotide probe microarray
			GEO Accession: GPL16296	
	Moult cycle microarray	Seear et al. (2010)	ArrayExpress accession: A-MEXP-1800	10,176 clone-probe microarray featuring 4,792 unique clones
	Seasonal gene microarray	Seear et al. (2012)	ArrayExpress accession: A-MEXP-2146	4792 clone-probe microarray and 864 Sanger sequences
			GenBank: JK623235-623956	
			dbEST: JK623235-JK623956	
Markers for krill identification (Sect. 7.6.1)	Mitochondrial 16S gene/small subunit rDNA	Deagle et al. (2007)	GenBank: DQ201509,	Regions of the mitochondrial 16S rDNA gene are very useful for identifying krill DNA in mixed pools of DNA often found in environmental samples such as that purified from penguin faecal material. PCR primers designed for this region can be conserved across many taxa Deagle et al. 2014)
	Mitochondrial COI 'DNA barcoding' region	Bucklin et al. (2007)	GenBank: AF177182,AB084378	A portion of the mitochondrial cytochrome oxidase subunit I is amplifiable from krill with the standard PCR primer set used by the consortium for the barcoding of life. For samples of pure suspected krill tissue, this is a good approach, but if mixed with other biological material, then alternative methods should be considered (Deagle et al. 2014)
	Nuclear 18S/small subunit rDNA	Jarman et al. (2013)	GenBank: AY672801	Regions of this gene are PCR amplifiable from most organisms with 'universal' PCR primer sets. This makes this gene especially useful for identifying krill in the diet of other animals

(continued)

Table 7.1 (continued)

Genetic resource type	Description	Reference	Accessions and URLs	Comments
	Nuclear 28S/large subunit rDNA expansion segment	Jarman et al. (2002)	GenBank: AF169717	A region of the nuclear large subunit rDNA has an expansion segment that is unique to krill. PCR primers for this region allow amplification of krill DNA even in the presence of large amounts of DNA from other sources. Useful for identification of krill species in environmental samples
Markers for krill food identification (Sect. 7.6.2)	Nuclear 18S/small subunit rDNA	Martin et al. (2006); Passmore et al. (2006)	Taxonomically verified, single entry 18S sequences: http://www.arb-silva.de/	Several short regions of the small subunit rDNA can be used to identify a range of eukaryotic food species
	Nuclear 28S/large subunit rDNA	Vestheim and Jarman (2008)	Taxonomically verified, single entry 28S sequences: http://www.arb-silva.de/	A short region of the nuclear large subunit rDNA can be used to identify many eukaryotic food species consumed by krill, including animals, eukaryotic microbes. A non-amplifying 'blocking primer' was used to suppress amplification of krill DNA in the PCR reaction
	Mitochondrial COI regions	Töbe et al. (2010)	GenBank: FJ960447, AF33279, AF536518, FJ960446, AF177182	qPCR tests for four species of copepod were developed to study the incidence of predation on copepods by larval and adult krill

Genetic resource types are subdivided by published studies. The associated references; accession numbers for accessing data in public databases; or URLs are given and west of the Antarctic Peninsula (Fevolden and Ayala 1981). Studies initially focused on a regional scale (Schneppenheim and MacDonald 1984; MacDonald et al. 1986) but as the data accumulated, comparisons across the entire range were made by combining data sets (see Fevolden 1988; Fevolden and Schneppenheim 1989). The combined protein electrophoresis dataset include data from several thousand individual krill collected in more than 60 sampling events. The subdivision originally observed east and west of the Antarctic Peninsula were not found in subsequent sampling (Schneppenheim and MacDonald 1984) and were dismissed as temporal anomalies potentially resulting from selection acting on the enzyme markers. Some other apparent differences in allele frequencies between regions were attributed to the use of different electrophoretic methods (Fevolden and Schneppenheim 1989). The overall conclusions were that "the genetic data obtained to date have substantiated the hypothesis of a single genetically homogeneous breeding population of krill in Antarctic waters" (Fevolden and Schneppenheim 1989).

There was a period of relative inactivity in the study of krill population genetics before new methods to directly study DNA variation began to be applied (studies summarised in Fig. 7.1). Once again an early study indicated that there was small but significant genetic differentiation between two sampling sites (Zane et al. 1998), this time between samples collected in the Weddell Sea versus those collected near South Georgia. This finding was based on sequences from a fragment of the mtDNA ND1 gene. A subsequent study sequenced the same mtDNA fragment in samples from a wider distribution of collection sites, and also looked at variation in three microsatellite markers (Bortolotto et al. 2011). In this second study the only mtDNA differentiation observed was between two samples collected in different years at South Georgia; the level of divergence here was small, similar to that originally observed between Weddell Sea and South Georgia. In a combined analysis of mtDNA data from the two studies there is no significant differentiation between any sites after correction for multiple comparisons (Bortolotto et al. 2011). The microsatellite dataset also indicated the vast majority of variation (>99 %) was due to within location differences, rather than being partitioned between sample sites (Bortolotto et al. 2011).

The most recent examination of circum-continental genetic variation in krill applied a genotyping by sequencing approach (Deagle et al. 2015). In this approach specific parts of the genome (restriction-site associated DNA; RAD markers) are characterized in many individuals using HTS (Davey et al. 2011). Despite collecting over 90 billion base pairs of data in 148 krill samples, the reliability of the resultant genotype calls was problematic because of the highly repetitive nature of the krill genome (see Sect. 7.3.1). The presence of multiple copies of genetic regions makes it difficult to determine if variation seen in a recovered sequence is due to allelic variation (i.e. variation in a marker at a single locus) or variation between copies at separate loci. To circumvent this issue the data was analysed using counts of sequences (rather than the derived genotypes) from several thousand genomic markers. Using this dataset individual krill could be easily identified in replicate samples, but the analysis failed to detect any structuring between samples collected from five sites distributed across the species range. Deagle et al. (2015) also sequenced two mtDNA markers (ND1 and COI) in their samples. The mtDNA results

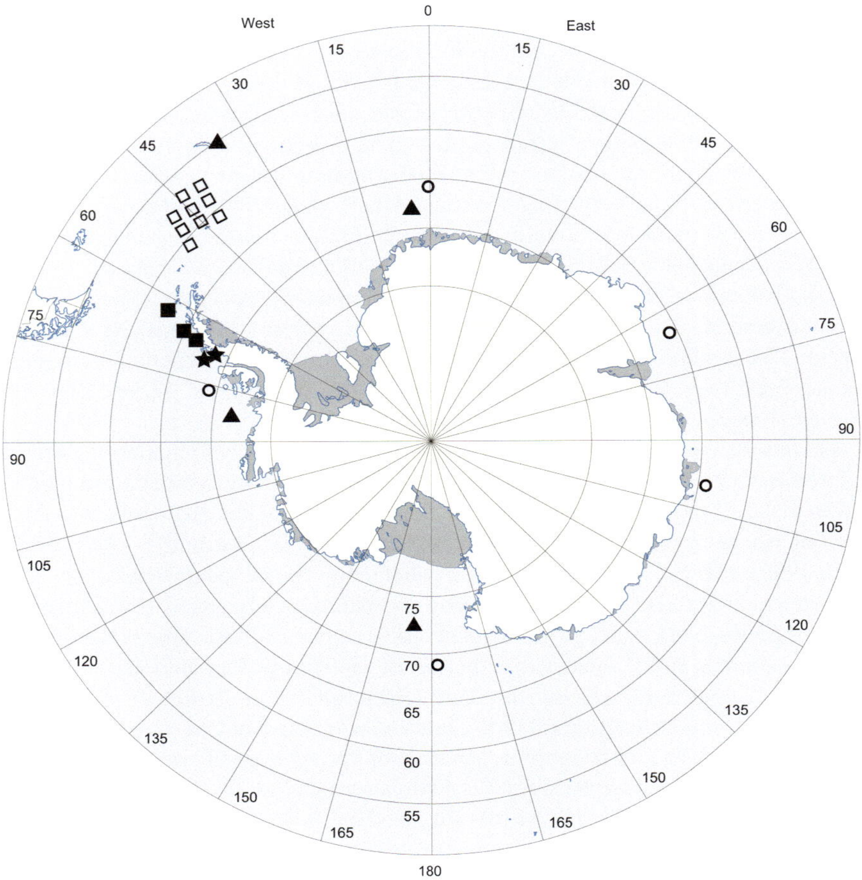

Study	mtDNA	Microsatellites	Nuclear SNPs
▲ Zane et al., 1998	√		
◇ Goodall-Copestake et al., 2010	√		
★ Batta-Lona et al., 2011	√		
▲■ Bortolotto et al., 2011	√	√	
○ Deagle et al., 2015	√		√

Fig. 7.1 Locations of krill sampling sites for population genetics studies using DNA markers. The types of DNA markers used in the five studies are indicated. The study of Batta-Lona et al. (2011) analysed multiple samples collected at a small spatial scale in the two areas indicated and samples were also collected in 2 years. The sample set analysed by Bortolotto et al. (2011) incorporated the same samples analysed in Zane et al. (1998), samples from multiple years collected near South Georgia, as well as samples from three new sites indicated by squares near the western Antarctic Peninsula

were remarkably congruent with the dataset produce by Bortolotto et al. (2011). Despite analysing samples from very different geographical areas, the frequency of the more common haplotypes ND1 was almost identical in both studies. Once again, no structuring between sites was observed.

There have also been two krill mtDNA population genetics studies that have looked at sequence variation on a relatively small spatial scale. The first focused on year to year variation in sites of the Western Antarctic Peninsula region (Batta-Lona et al. 2011). The study concluded that there was small but significant genetic heterogeneity between sampling events and this reflected multiple sources of recruitment in the region. The second looked at variation between krill sampled from different swarms in the Scotia Sea (Goodall-Copestake et al. 2010). Here, no significant swarm-related genetic structuring was observed, but the large amount of sequence data collected confirmed high genetic diversity within swarms.

The high genetic diversity of krill relative to other species with unstructured populations is likely a reflection of its exceptional abundance (Goodall-Copestake et al. 2012). Large populations tend to have more genetic variability, and measures of genetic diversity at specific loci can be used to estimate 'effective population size' which can be defined as the number of individuals that contribute genes equally to the next generation in an idealised population (Wright 1938). The effective population size of female krill based on mitochondrial markers range from a few hundred thousand to a few hundred million (Goodall-Copestake et al. 2010; Zane et al. 1998). Although these estimates have very wide confidence intervals, it is clear that it is a large number of individuals, but still far smaller than the census size of the female krill population (Zane et al. 1998). Genetic diversity can be measured as the number of genetic variants observed in a sample, or as the genetic distance between the variants. By comparing these measures past demography of a species can be estimated. In the case of krill, this type of analysis consistently indicates there has been a recent population expansion. Goodall-Copestake et al. (2010) have estimated that the population size doubled in the last 100,000 years, although an alternative explanation that selection on mtDNA is driving the result cannot be completely discounted. There is also supporting evidence from a small number of microsatellite markers for recent increase in population size (Bortolotto et al. 2011). This demonstrates that the population expansion inferred from mtDNA is not due to a selective sweep. It would be interesting to investigate this issue more completely with nuclear sequence markers.

Given the high degree of interest in the potential for population structure in Antarctic krill the scope of DNA-based studies carried out to date are still relatively limited. This reflects the difficultly in collecting comparable samples from appropriate sites around the Southern Ocean. But it also reflects the difficulty in developing markers for the species. In most species microsatellite repeats probably provide the most high resolution picture of population structure (Selkoe and Toonen 2006). However, these markers have proven very difficult to apply in krill because most krill microsatellites are extremely long and tend to be highly polymorphic (i.e. often only a few individuals share alleles of a given marker in sample sizes commonly used for population genetics studies) (Bortolotto et al. 2011; Jarman 2001a). However, with extensive development effort, shorter and less polymorphic microsatellites have been isolated and these are proving to be useful for population genetic studies (Bortolotto et al. 2011; Candeias et al. 2014). Where genetic differentiation exists among different populations of an organism, large sample sizes taken from each population make it likely that allele frequency differences will be detected (Waples 1998). Although most krill population genetic studies have been limited by the number of markers, many have had large sample sizes and this should enable genetic differentiation to be identified (see assessment of statistical power in Bortolotto et al. (2011)). Despite this, there is no convincing evidence for temporally stable genetic differences among krill from any Antarctic regions.

The overarching explanation for the lack of genetic differentiation is simply that there is ongoing migration (i.e. gene flow) between regions of the Southern Ocean where high densities of krill are found. However, this only tells part of the story because it does not clarify the level of population connectivity required to maintain this genetic homogeneity. When gene flow is restricted, genetic differentiation of neutral markers develops among populations through random genetic drift – a change in allele frequencies due to chance sampling error between successive generations (Slatkin 1987). The importance of genetic drift is strongly related to the effective population size; the larger the effective population, the smaller the effect of genetic drift on allele frequencies (Charlesworth 2009). An interesting corollary of this is that the migration rate required to homogenise gene frequency between regions is independent of population size (i.e. in larger populations relatively fewer individuals need to be exchanged to overcome the effect of drift). Given the large effective population size of krill, the speed at which differentiation occurs will be slow, and can be overcome by exchange of only a miniscule proportion of the population (i.e. migration of five individuals per generation between sampled regions might make detection of structure difficult using currently available datasets (Bortolotto et al. 2011)). Therefore, while all the current population genetic evidence indicates there is genetic connectivity among krill around Antarctica, the level of migration between regions may still be very low. This means population connectivity may not be high enough to allow krill in separate regions to be managed as a single stock (Bortolotto et al. 2011; Deagle et al. 2015). It also means different regions may still need to be considered to be semi-independent when modelling future demographic scenarios.

While genetic drift is limited in large populations, and this prevents divergence, the opposite is true for the influence of natural selection. In large populations different selection forces acting on sub-population are likely to result in genetic changes even when the selection is quite weak. Essentially, the lack of genetic drift magnifies genetic changes that occur due to selection (Charlesworth 2009). Therefore, the primary factor that could potentially lead to local genetic differentiation among krill populations may be natural selection. The main population genetics studies that used markers that could be subject to selection are the allozyme studies summarised above. None of the DNA-based studies to date has investigated allele frequency differences in protein coding genes. A potentially interesting avenue for future investigations of krill population genetics would be to identify different krill habitats that may impose different selection pressures on krill residing there and to focus on markers with frequencies that could be subject to the likely selection pressure.

7.3 Euphausiaceae Phylogenetics

No fossils of any species of euphausiid have ever been identified, so the evolution of the ancestors of existing euphausiid species can only be inferred by phylogenetic analysis (Jarman 2001a). Phylogenetic analysis was originally developed for application to morphological features that could be considered to be shared among a group of species (Hennig 1966; Felsenstein 2004) such as the pre-anal spine that is present in all Euphausiacea (Spiridonov 1992). Current taxonomic classifications of the crustacean class Malacostraca, which contains the order Euphausiaceae, are still based on morphological analyses. Since the 1980s, DNA sequencing has provided a huge number of markers for phylogenetic analysis. A major advantage of DNA sequence markers is that DNA regions that evolve at a rate that is appropriate for a particular phylogenetic question can be selected (Felsenstein 2004). This has been particularly important for analysis of the relationship of krill to other crustaceans because the divergence of these groups is estimated to have occurred >495 million years ago (MYA) by both DNA-based phylogenetics and from fossil evidence (Fig. 7.2) of the closest relatives of krill (Jarman 2001b). Morphological characteristics used for phylogenetic analyses often have a higher potential for homoplasy (reversion from a 'derived' form to a 'primitive' form) over very long timescales than slowly-evolving DNA sequences. Genes that encode fundamental components of the cellular machinery and are known to have a very slow evolutionary rate can be used to study divergence among very ancient lineages where homoplasy in morphological features can result in inaccurate phylogenies (Felsenstein 2004).

7.3.1 The Relationship of Euphausiacea to Other Crustacea

The degree of relatedness of other Euphausiacea to other crustacean groups and to other animals is useful information for many types of genetic analyses as krill is a 'non-model' organism. This means that its genome is not sequenced, there are no genetic maps for it and it has not yet had extensive characterization of gene expression in different tissues and stages. Many of the newer molecular methods for analyzing genetics of non-model organisms depend at least partly on comparing sequences of DNA, RNA or proteins identified in the non-model organism to the closest identifiable relatives that are 'model' organisms so that inferences can be made about the function of the molecule based on its primary sequence. Krill are particularly poorly served by this approach because they are quite distantly related to any model organism (Fig. 7.2).

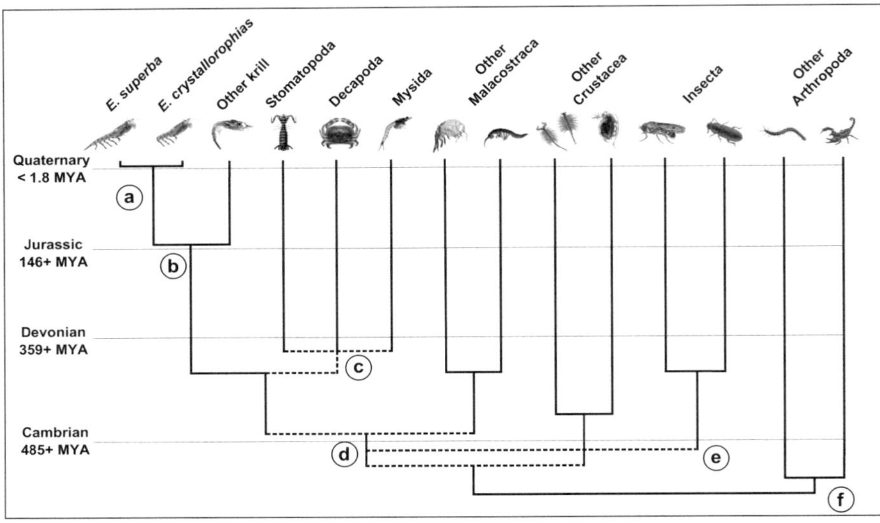

Fig. 7.2 The relationship of krill to other Arthropoda. (**a**) Krill is probably most closely related to *Euphausia crystallorophias* and the divergence between the species is estimated to have happened since the start of the Quaternary era 1.8 MYA (Patarnello et al. 1996; Jarman et al. 2000a). (**b**) The extant genera within the Euphausiidae are predicted from molecular evidence to share a common ancestor that existed since the Jurassic era 146 MYA (Jarman 2001b). (**c**) The relationship between Euphausiaceae and other orders of Malacostraca is equivocal, but fossil evidence for other orders of Malacostraca being established indicates that the divergence likely took place after the Devonian era 359 MYA (Jarman 2001b; Johansson et al. 2012b; Rolfe 1981). (**d**) The Malacostraca are supported as a monophyletic group by morphological and molecular evidence (Schram 1981). (**e**) The relationship between Insecta and Crustacea is not well understood. (**f**) Crustacea and Insecta are supported as sister taxa by molecular and morphological evidence and more distantly related to other Arthropoda such as Myriopoda and Arachnida (Meusemann et al. 2010)

The classic model organism *Drosophila melanogaster*, the vinegar fly, is the best characterised arthropod and relationships between krill messenger RNAs (mRNAs) and *Drosophila* mRNAs have been used to assign putative functions to genes. However, *Drosophila* is quite distantly related to krill and along with other dipteran insects has some strange genomic features such as lack of normal chromosome recombination in males (Chen et al. 1998) and very low levels of DNA methylation as a gene expression control mechanism (Takayama et al. 2014) so in some cases comparisons of krill genetics with *Drosophila* genetics may be confounded by *Drosophila*'s unusual biology. A range of other insect genomes such as several *Aedes* mosquitos and the flour beetle *Tribolium castaneum* have been sequenced and characterised. The silk moth *Bombyx mori* is probably the second best characterised model organism that is relevant to krill research. It is another good reference for comparisons of krill DNA, RNA or protein sequences. Among Crustacea, there are still no completed genome projects for any Malacostraca. There is a complete sequence for *Daphnia pulex*, but this belongs to the class

Branchiopoda and it is not necessarily more closely related to krill than insects are (Fig. 7.2). Insects are thought to have a common origin of about 479 MYA (Misof et al. 2014) and it is thought that Branchiopoda are more closely-related to Insecta than to Malacostraca (Meusemann et al. 2010). This research is based on a 'phylogenomic' approach where large numbers of mRNA transcripts are sequenced and their protein products compared in a massive phylogenetic analysis of markers from across the whole genome. This approach has not yet been applied to most orders in the class Malacostraca, including krill, but is an excellent example of the new information that modern genomic technologies and HTS can generate in comparison methods based on sequencing single molecules, which were unable to resolve this issue.

Identification of the closest relatives of Euphausiaceae is currently of primary interest for taxonomic classification and for understanding crustacean evolution. Historically, taxonomies based on morphological characters either placed Euphausiacea with Mysida in the 'Schizopoda,' Boas 1883, a taxon that is no longer used or with Decapoda in the 'Eucarida' Calman 1904, which is almost universally used today (Schram 1981). This classification is not, however, supported by current molecular evidence. A partial phylogeny of orders of Malacostraca and tests of each of these classifications as a phylogenetic hypothesis using nuclear large subunit rDNA sequences suggested that krill were more closely related to Mysida than to Decapoda (Jarman et al. 2000b). However, this analysis is based on only one gene and did not include all orders within the Malacostraca. More recent molecular evidence suggest that Stomatopoda may actually be very closely related to Euphausiaceae based on phylogenetic analysis of all protein coding genes in the mitochondria (Podsiadlowski and Bartolomaeus 2005) and both nuclear 18S and 28S sequences (Jenner et al. 2009). However, a different phylogenetic analysis based on all mitochondrial protein coding genes supported Euphausiacea being most closely-related to Decapoda (Johansson et al. 2012b). There has not yet been a thorough investigation of malacostracan phylogeny with enough taxonomic sampling and genetic markers to resolve the relationships between Euphausiaceae and other orders. This question was bound to be difficult because of the enormous diversity of Malacostraca, which are an old lineage, thought to date to the Cambrian >485 MYA (Rolfe 1981) in which many divergence events may have occurred close to each other in time. The recent advent of phylogenomic approaches (Misof et al. 2014) that have been successful in resolving similar issues in insect evolution suggest that this approach may in future resolve malacostracan phylogeny if combined with appropriate taxonomic sampling.

7.3.2 The Relationship of Krill to Other Euphausiaceae

Krill is most closely related to *Euphausia crystallorophias* based on mitochondrial small subunit rDNA (16S) sequences (Patarnello et al. 1996) and combined mitochondrial 16S and COI sequences of some members of the genus *Euphausia*

(Jarman et al. 2000a). *E. crystallorophias* is more abundant in neritic waters over the Antarctic continental shelf, where its range overlaps with krill, which peaks in abundance over deeper water (see Chap. 2, Siegel and Watkins 2016). A study that included 20 species of the 31 members of the genus Euphausia found weak bootstrap support (58 %) for *Euphausia lamelligera* being more closely related to krill than *E. crystallorophias* based on maximum likelihood inference of COI sequence phylogeny (Bucklin et al. 2007). However, this level of support is very low and as *E. lamelligera* is from the equatorial east Pacific, it is unlikely to be more closely related to krill than *E. crystallorophias* given the timescales in which these speciations are thought to have taken place in *Euphausia* (Patarnello et al. 1996).

None of these studies include the full range of 31 species in the genus *Euphausia* and as the genus may be paraphyletic, it is still possible that other closer relationships to krill exist that have not yet been identified. These studies are also all based only on mitochondrial DNA sequences, which are inherited as one non-recombining molecule, even where multiple regions are used e.g. Jarman et al. (2000a). The species phylogenies being inferred may not be congruent with the gene phylogeny estimated for mitochondrial DNA, so until there are multi-gene phylogenies containing a more complete taxonomic range of Euphausiaceae, the relationships as currently estimated are a working model. However, it is still reasonable to consider *E. crystallorophias* a close relative of krill and this is relevant for genetic studies in that it provides a good target for comparative studies. For example, the transcriptome (see Sect. 7.4) and peptidome of *E. crystallorophias* have been analysed and this will provide a useful comparison for studying gene function in krill (Toullec et al. 2013). Comparisons of some proteins and gene activities have recently been made between the two presumed sister species (Cascella et al. 2015). Another practical reason for identification of unique DNA sequences for related species of *Euphausia* is for the accurate identification of krill species and differentiation among them in a variety of circumstances where whole animals cannot be identified by morphology (see Sect. 7.6.1).

The relationship of the genus *Euphausia* to other krill genera has been examined by adult morphology (Casanova 1984), larval morphology (Maas and Waloszek 2001) and from a limited range of species by nuclear large subunit ribosomal rDNA analysis (Jarman 2001b). Based on shared larval morphological features, *Euphausia* is a sister group of the genus *Pseudeuphausia* and this grouping is highly derived within Euphausiaceae (Maas and Waloszek 2001). Phylogenetic analysis of adult morphological features suggests, however, that *Euphausia* is a basal genus along with *Tessarabranchion* and *Thysanopoda* (Casanova 1984). Molecular analyses suggest that *Euphausia* was a sister taxon to *Nyctiphanes* (Jarman 2001b). In conclusion, there is not currently a clear picture of how the genera of Euphausiaceae are related.

7.4 Krill Genomics

Genomes are inherently complicated and consequently difficult to study. Even the genomes of genetic model organisms such as *Drosophila melanogaster* that have been under investigation since the early 1900s are still under investigation and are producing novel results e.g. Takayama et al. (2014). The study of genomes involves a diverse range of genetic methods such as different types of karyotyping, fluorescent flow cytometry and HTS applied to different sub-regions of the genome. The genome of krill is currently poorly characterized, but we do have some knowledge of it and the potential to gather more is increasing constantly.

7.4.1 Genome Size and Features

The size of the krill genome has been estimated by DNA-specific fluorescent staining measured with both flow cytometry and images of Fuelgen stained nuclei at 48.53 ± 0.78 picograms per cell (47.46 ± 0.76 gigabases) (Jeffery 2011). This is quite a large animal genome, although not exceptional and several other malacostracan Crustacea have similarly large genomes. The largest recorded genome size for a crustacean is 64.62 ± 2.85 pg for the arctic amphipod *Ampelisca macrocephala* (Rees et al. 2007). The mean genome size in the 288 species of Crustacea that had been recorded in 2014 was close to 3 pg – similar to the size of the human genome (Gregory 2005). Most of the large genome sizes recorded for Crustacea to date are from Malacostraca and many have been recorded from polar species, which was suggested as a possible link between organisms living in cold environments and large genome sizes, but the reasons for this are not well understood (Jeffery 2011). The largest genome to be completely sequenced so far is for the loblolly pine, *Pinus taeda* and is 22 gigabases in size (Neale et al. 2014). The krill genome is more than twice this size and this makes sequencing it unfeasible with commonly used HTS approaches currently available. Newer long-read HTS technologies combined with lower costs for current short-read HTS data generation may soon make large genomes like this possible to sequence completely.

The overall base composition of the krill genome is richer in adenine-thymine base pairs than for most metazoans. This feature is normally expressed as GC% and krill have a GC% estimated at 32 % by thermal denaturation and sequencing of a very limited random portion of the krill genome (Jarman et al. 1999). This means that adjacent thymidines would be expected to be more common in the krill genome if bases are randomly distributed overall, which would make krill DNA more susceptible to DNA damage than DNA of other organisms. This may be the case, but krill have also been shown to have high levels of DNA repair mechanisms for dealing with damage from UV-B (Ban et al. 2007).

Very little is known about epigenetic features of the krill genome. DNA methylation has been examined at a very crude level by digestion of total krill DNA with the methylation sensitive restriction endonuclease *Hpa* II and a methylation-insensitive endonuclease *Msp* I that recognises the same sequence. After overdigestion with both enzymes, smears of the resulting fragments on low percentage agarose gels were identical, as seen for *Drosophila melanogaster* methylation negative control DNA, but not for *Mus musculus* positive control DNA, which showed a distinctly greater trend towards larger restriction fragments in the *Hpa* II digest where DNA methylation had protected some sites from cleavage (Jarman 2001a). It should be noted, however, that small amounts of DNA methylation have since been identified in the *D.melanogaster* genome, which was previously thought to not have any (Takayama et al. 2014). Modern methods for analysis of DNA methylation have not been applied to the krill genome, so it is still unknown whether krill use this epigenetic mechanism. No other types of epigenetic mechanisms for regulating gene expression or epigenetic modes of inheritance have been investigated in krill.

7.4.2 Chromosomes

Krill are apparently diploids with all chromosomes being bivalent and having centromeric regions based on large-scale chromosome morphology as studied by light microscopy. The chromosome complement has been reported as either 2n = 32 (Yabu and Kawamura 1984) or 2n = 34 (Thiriot-Quiévreux et al. 1998; Van Ngan et al. 1989). Chromosome aberrations were recorded in 2 of 21 examined individuals (Yabu and Kawamura 1984) which is an interesting finding that has not been investigated further. Non-diploid amplification patterns have also been observed for some microsatellite loci (Bortolotto et al. 2011), although whether this is the result of partial ploidy greater than 2n, or simply duplicated microsatellite loci is unknown. There are apparently no sex-specific chromosomes based on the gross morphology investigated in these studies, although it is possible that more detailed staining methods could identify sex-related chromosomes or regions if they do exist. It is also possible that there is partial polyploidy that cannot be identified with karyotypic methods applied to krill so far.

7.4.3 Repetitive DNA

The krill genome contains substantial quantities of repetitive DNA. This has been demonstrated by a genomic sequencing survey of a portion of the krill genome that produced ~250,000 'RAD' markers associated with *Sbf* I restriction sites (Deagle et al. 2015). If these markers were all present as single copies in the genome, it would be expected that most of the sequences would be represented once in a data subset with 75,000 sequences (i.e. if each had an equal chance of being picked, few

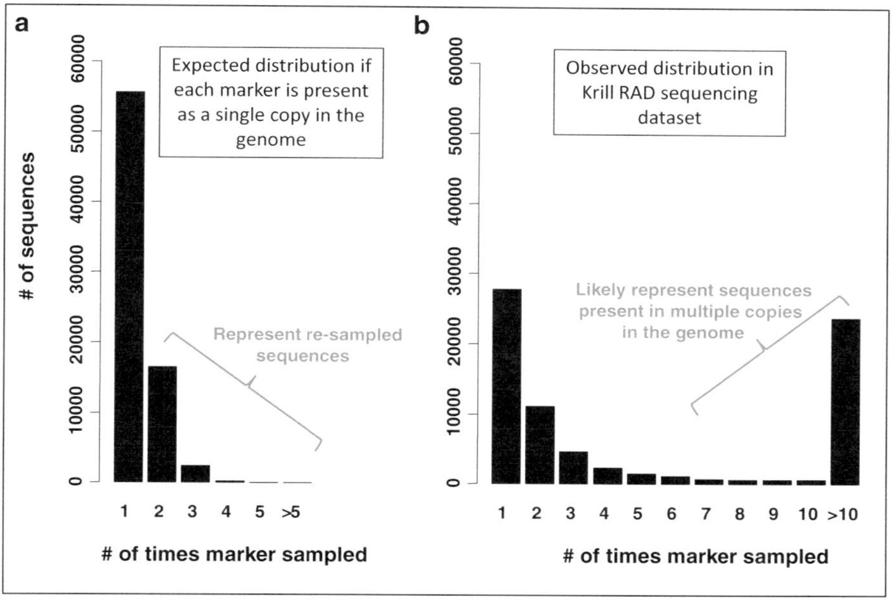

Fig. 7.3 Illustration of the repetitive nature of the krill genome. Barplots show the numbers of times closely related RAD sequences are recovered (**a**) in simulated data where each sequence is unique (**b**) in a krill RAD sequence data (Deagle et al. 2015). The over-representation of many sequences in the krill data indicates they are present in multiple copies in the krill genome. The number of RAD markers in the krill genome is roughly 250,000 and in both plots 75,000 sequences were sampled. The clusters group sequences with >85 % similarity

would be picked more than once with this low coverage; Fig. 7.3a). However, when we sampled 75,000 sequences from the >10 million sequences recovered from an individual krill it was found that about a third of the sequences are highly over-represented (Fig. 7.3b). These are likely derived from DNA present in multiple copies throughout the genome. Similarly, in their testing of microsatellite markers (Bortolotto et al. 2011) found that many loci had more than two alleles per individual, indicating these markers were in replicated genomic regions. All large genomes tend to have large amounts of repetitive DNA (Treangen and Salzberg 2012), so this characteristic is not unique to the krill genome. However, this feature can complicate large scale analysis of genomic DNA sequences.

Another low coverage genome sequencing dataset has been collected using 454 pyrosequencing of random DNA fragments (Leese et al. 2012). In this study, ~50,000 sequences of ~250 bp length were generated. This produced 12 million bp of krill DNA sequence, which represents 0.02 % of the genome. This dataset was analysed for the presence of short repetitive regions. Microsatellites, which are tandem repeats of two to five base pairs, were found in about 1 % of the reads, which is consistent with the incidence in other species of arthropods. Longer minisatellites repeats (unit size of 7–50 bp) formed a relatively small part of the reads (0.35 %) compared to other species. A search of the 'expressed sequence tag' (EST) database of 1,170 sequences derived from messenger RNA (De Pitta et al. 2008) identified 68 sequences (0.058 %) that contained microsatellites (Bortolotto et al. 2011). This lower incidence of microsatellites in transcribed regions is consistent with the pattern in other animal genomes (Treangen and Salzberg 2012).

7.4.4 Mitochondrial Genome

The krill mitochondrial genome is more than 15,498 nucleotides long and contains a short region surrounding the replication origin that is difficult to PCR amplify and sequence, so the exact length remains unknown (Shen et al. 2010; Machida et al. 2004). The GC% is 32.2 %, which is very close to the overall nuclear GC% (Machida et al. 2004). A complete mitochondrial genome sequence has been determined for *Euphausia pacifica* and this is 16,898 nucleotides (Shen et al. 2011) so it is likely that the krill mitochondrial DNA is of a similar size. The gene complement of the krill mitochondrial genome is the same as found in other arthropods (Machida et al. 2004; Shen et al. 2010). Rearrangements of the order and orientation of mitochondrial genes are moderately common, however, and krill has a gene arrangement that is apparently unique among crustacean mitochondrial genomes sequenced to date (see Table 7.1). The most common arrangement of mitochondrial genes in Crustacea is termed the 'pancrustacean ground pattern' and is found in most crustacean species, examples being *Macrobranchium rosenbergii* and *Palinurus japonicas*. Krill has five differences from this pattern in position and orientation of four tRNA genes (trnL1, trnL2, trnW, and trnI) and a duplication of the trnN tRNA gene (Shen et al. 2011). Variation in the mitochondrial genome has been assessed for small sections in population genetics studies (see Sect. 7.1). A 13,000 bp region of the mitochondrial genome was sequenced in five individuals collected near the west side of the Antarctic Peninsula. This study found that the *cox2*, *nad5*, and *nad6* regions are the most variable portion of the krill mitochondrial genome, suggesting that these regions would be useful markers for population genetics studies (Johansson et al. 2012a).

Complete mitochondrial genome sequences have been produced for *Euphausia pacifica* in two studies (Johansson et al. 2012b; Shen et al. 2011). The gene order reported for it differs from krill by having a duplication of the trnN tRNA gene and a translocation of the trnI tRNA gene (Shen et al. 2011). A nearly complete mitochondrial genome sequence was also produced for the krill *Thysanoessa raschii* and a similar gene order was found to that of the two *Euphausia* mitochondrial genomes (Johansson et al. 2012b).

7.5 Krill Gene Expression

7.5.1 *Quantification of Total RNA*

The earliest examinations of gene expression in krill was analysis of total RNA content compared to total DNA content as a proxy marker for growth rate (Ikeda 1989). As a proportion of total RNA, ribosomal RNA forms the largest component of about 70 % and this varies with growth rate even when different sets of messenger RNAs are being expressed, which form approximately 10 % of the total RNA in a cell. DNA quantity does not change in response to any physiological condition, so the ratio between total RNA and total DNA primarily measures overall levels of gene translation and has become a commonly used proxy for rate of growth in many species (Buckley 1984). This method has been used to measure krill growth rate since then (Shin et al. 2003) but it is more common now to apply Instantaneous Growth Rate measurement (see Chap. 3, Reiss 2016).

7.5.2 *Krill Messenger RNA Studies*

Analyses of gene expression on a broad scale with HTS are now relatively common in both model and non-model organisms. This approach is frequently used now to help understand the genes involved in a variety of physiological and developmental processes. This process involves purifying messenger RNA, converting it to its DNA equivalent (complementary DNA or cDNA) and sequencing the DNA version using HTS or Sanger sequencing in older studies. Several groups have produced transcriptomes for different krill tissues and life stages under different physiological conditions (Table 7.1). The first study to systematically analyse krill gene transcripts produced tissue-specific libraries from the head, abdomen, photophores and thoracopods (De Pitta et al. 2008). Sanger sequencing was used to produce 1770 high quality ‘expressed sequence tags’ (ESTs). Putative functions were assigned to approximately 30 % of these transcripts based on known functions in insects, other crustaceans and in a few instances from vertebrates. The seemingly large proportion of transctipts that were not identifiable relates to the problem of krill being so distantly related to any genetic model organism (see Sect. 7.2.1).

Krill experience a range of very different light regimes over the course of each year (see Chap. 4, Meyer and Teschke 2016) and the genetic control of biorhythms and metabolic processes in response to these conditions is one of the processes to be most extensively analysed at the gene expression level. The effects of experimentally simulated winter and summer light regimes on gene expression in krill was studied by subtractive hybridisation experiments with pooled summer-type krill cDNA being the ‘tester’ and winter-like cDNA the ‘driver’ (Seear et al. 2009). Eleven of the genes most differentially expressed between the winter and summer-like light conditions were further analysed by qPCR on multiple individual krill.

This study identified a range of genes that were down-regulated in the winter-like light conditions that are involved in processes such as cuticle degradation/formation and food metabolism, which were consistent with expectations for krill adopting a 'quiescent' physiological state in the winter. Changes in gene expression between krill captured from the wild in summer and in winter were examined with microarray experiments validated by qPCR (Seear et al. 2012). Samples were collected in winter from two locations, one near the Antarctic Peninsula and one near South Georgia. A sample was collected in summer from the Antarctic Peninsula. Gene expression from several whole krill was compared among these samples using a microarray with 6,712 transcripts represented on it. This study produced results broadly consistent with those of the earlier experimental study, but with a wider range of transcripts being identified with this technology. A broad-scale analysis of oscillatory gene expression linked to light/dark cycles identified 609 transcripts with sinusoidal expression patters (De Pittà et al. 2013). This study used microarrays based on normalised cDNA library sequences generated by 454 pyrosequencing. Once again, only a small proportion (~35 %) of the transcripts were able to be assigned function, but this is actually common in studies of non-model invertebrates and the authors point out that this compares favourably to similar studies of bivalves and corals.

Changes in expression of genes involved in the moult cycle of krill were investigated using microarrays to measure relative levels of 960 targets (Seear et al. 2010). A novel cDNA library was generated for this study and sequenced with Sanger sequencing. A microarray based on a transcriptome assembly that generated 302 contigs was produced. Twenty six different genes that change expression at different phases of the moult cycle were identified by purifying RNA from heads of krill sampled at five points of the cycle. Differential expression was determined from differential strength of cDNA binding to the microarray. These differentially expressed genes included several transcripts that could clearly be associated with cuticle formation or dissolution such as from trypsin and collagen genes. This study is a good example of the insights into krill biology that can be gained by studying gene expression associated with a specific physiological phenomenon. A later study of genes expressed in krill stress responses identified 22,177 contigs from a transcriptome assembly of 943,817 pyrosequencing reads (Clark et al. 2011). This study focused on characterising the classic 'stress proteins' HSP70, HSP90, ferritin and GST and was the first study to utilise HTS in analysis of krill genes. Contigs were assembled from the 454 sequencing data and identified by BLAST sequence similarity searches of GenBank, which identified 25 % of the sequences by homology with sequences from other Arthropods. A range of peptide stress hormones, including HSPs were identified from a transcriptomic study of *Euphausia crystallorophias* and these are likely to be highly similar to orthologous transcripts in krill (Toullec et al. 2013).

Transcriptome quality is dependent on many factors, but numbers of sequences and their length and quality are the fundamental considerations. Bioinformatic approaches for combining transcriptomic data generated in different experiments are available and under continued development. A recent transcriptome assembly derived from multiple data sets has been created to provide a resource with greater coverage than either original data set (Meyer et al. 2015). The genetic resources mentioned above (see Table 7.1) provide a good foundation for producing transcriptomes for future experiments and in the future it is hoped that greater sequencing depth combined with a well-annotated genome from a more closely-related arthropod will provide better results from these experimental approaches.

7.5.3 Krill Functional Genetics

Functional genetic studies aim to characterise the protein sequence encoded by a gene to determine its function. The first step in this process for individual genes is often to generate a phylogeny relating the protein to those already characterised in other organisms to infer likely function. For broader-scale transcriptomic studies, a simpler version of this approach is applied where similarity scores are used to assign putative transcripts to 'gene ontology' groups (e.g. De Pittà et al. (2013)). Some inferences about the function of a protein can also be made from its primary sequence, which allows likely features like degree of hydrophobicity of different regions of the protein to be predicted. Where a complete DNA sequence of a gene is available it is also possible to clone the gene into another organism. For krill, this has been done in a unicellular eukaryote *Pichia pastoris* in order to study the function of the native protein and to create mutants of it with different features (Benjamin et al. 2001). For confirmation of the function of any protein, introducing it into a mutant strain that lacks it and demonstrating restoration of the lost function in that species is a powerful demonstration of the protein's function. This has not yet been done with any krill genes, but the genetic model organisms *Drosophila melanogaster* would be the best available experimental system for this.

There are currently 891 GenBank accessions for krill coding regions. Several genetic systems have been characterised in krill. Some of these such as the 'heat shock' proteins and antioxidant enzymes that are involved in responses to environmental stress have been investigated in transcriptomic studies and some phylogenetic characterisation of these has been done (Clark et al. 2011). The specific thermal tolerance and expression dynamics characteristics of five HSP70 genes have been compared between krill and *Euphausia crystallorophias* (Cascella et al. 2015). Biorhythm genes are one of the most intensively studied groups of krill genes because of their fundamental roles in krill biology, as already mentioned (Teschke et al. 2011; Mazzotta et al. 2010) and several of these transcriptomic studies include some functional or phylogenetic characterisation of specific genes (see also Sect. 3.2.2). The first biorhythm gene isolated from krill was a *cryptochrome* orthologue of the insect *cry2* family (GenBank FM200054) (Mazzotta et al. 2010). Quantitative PCR measurement of the expression of *cry2* in krill kept under experimentally controlled light and dark conditions demonstrated oscillatory rhythms linked to light-dark cycles (Teschke et al. 2011). *Cry2* was the first 'canonical' biorhythm gene to be identified in krill and the oscillatory expression pattern was shown to be linked to expression of several metabolic enzymes by qPCR and enzyme assays as well as oxygen consumption. Opsins are light-sensitive proteins found in the retina and are part of the photoreception system of most animals. Two full-length krill opsin genes were identified from a transcriptomic study and phylogenetic analysis used to demonstrate that they are probable orthologues of type 1 opsins in other crustaceans (De Pittà et al. 2013).

There has also been some characterization of krill functional genetics because of the interest in krill products for 'bioceutical' or biotechnological applications. One example is research on krill proteases, which are cold-active and have uses as cleaning products (Benjamin et al. 2001; Lee et al. 2006). The potential for mixes of krill-derived enzymes to be used in biotechnological applications such as wound and infection treatment (Anheller et al. 1989) and general proteolysis (Osnes and Mohr 1985; Chen et al. 1978) and as cleaning agents was initially identified from the activities demonstrated by enzymes purified directly from krill. The properties of single krill enzymes such as xylanases have been explored by purification of the enzyme from total krill extract (Turkiewicz et al. 2000) or by isolating the genes encoding them and expressing the protein in culturable microorganisms (Lee et al. 2006). A cold-active collagen degrading enzyme termed 'euphauserase' was cloned and expressed in the yeast *Pichia pastoris* in wild type and mutant forms and trypsin expressed in a similar way from a fish gene used to ensure an active tertiary form. Mutagenesis of the plasmid containing the euphauserase gene allowed selection for mutant versions with higher thermal tolerance that could have wider biotechnological application as this allows room temperature activity of this highly active serine protease (Benjamin et al. 2001).

7.6 Krill DNA Metabarcoding and Metagenomics

Modern genetic methods for analysing the diversity and identity of nucleic acids purified from environmental samples have transformed our understanding of the natural world (Shokralla et al. 2012). DNA based analyses of environmental DNA have been applied to analysis of several aspects of krill biology such as identifying krill in the gut contents or faeces of krill predators; identifying the organisms that krill feed upon; and identifying the microorganisms that interact with krill as parasites, symbionts or commensal partners.

7.6.1 Genetic Detection and Identification of Krill Material

DNA sequences unique to krill are a useful tool for detecting and identifying krill material in a range of situations. Krill are probably the most significant food source for higher predators in the Southern Ocean (see Chap. 9, Trathan and Hill 2016). Krill-specific nuclear DNA sequences have been used to identify different krill species in the diet of whales and penguins, where DNA from animals consumed by the whale or penguin can be purified from faecal material to identify the species that the predators were feeding on (Jarman et al. 2002). In that study, a region of the nuclear large subunit ribosomal RNA gene (rDNA) that has an expansion segment unique to Euphausiaceae but variable in sequence among different Euphausiaceae species was used (see Table 7.1 for details). The nuclear small subunit rDNA of Antarctic krill has also been sequenced (Table 7.1) and has also been used as a dietary biomarker to identify krill in Adelie penguin diet (Jarman et al. 2013). Mitochondrial DNA regions are especially useful for discriminating among animal species and two regions of the mitochondria in particular have been used for this. The mitochondrial 16S rDNA (Table 7.1) is appropriate for identifying krill species in samples with mixed templates such as those purified from faecal samples of penguins because it contains conserved regions that make good PCR primer binding sites in a range of krill and other taxa, which is an important consideration in dietary studies (Deagle et al. 2014). For samples of krill tissue that are physically isolated from other biological material, the standard 'DNA barcoding' region favoured by the consortium for the barcoding of life (http://www.barcodeoflife.org/), the mitochondrial COI region can be used (Table 7.1). This approach is convenient in laboratories that commonly follow the barcoding of life standard protocols.

7.6.2 Genetic Identification of Krill Food

Identification of krill food species by morphological features of items found in the digestive gland is the standard approach for analysing krill diets (see Chap. 5, Schmidt and Atkinson 2016). This method is very good for identifying species that have hard parts that survive digestion such as the silica skeletons of diatoms and dinoflagellates. Diatoms are a particularly significant part of the diet of krill and microscopic identification of their frustules is considered a quantitative method that generally results in species-level identifications. This approach is therefore likely to remain the standard krill dietary analysis method. However, many groups of organisms consumed by krill are not so easily identified by morphology. Unicellular organisms such as silicoflagellates leave very little that can be identified after consumption (Passmore et al. 2006) and metazoans such as copepods are often damaged too extensively to be recognised (Martin et al. 2006; Töbe et al. 2010).

Identification of food consumed by krill can also be accomplished with DNA purified from the digestive gland or from krill faeces. This approach is not yet common, but it has some advantages for detection and identification of the many potential food species that do not leave recognisable hard parts. There are two broad approaches for DNA-based dietary analysis. One approach is to develop species-specific tests that recognise DNA sequences unique to a food species of interest. An example of this is the development of species-specific copepod PCR assays that were used to measure incidence of predation on copepods by krill (Töbe et al. 2010). The other main approach is to use 'universal' PCR primers that bind to short regions of DNA with conserved sequences in a range of organisms to be identified. This approach has been used with a short region of the nuclear large subunit ribosomal RNA gene (18S rDNA). DNA purified from faecal pellets was amplified with PCR primers specific to ends of a ~240 bp region of the 18S region. The resulting mix of amplicons was then separated using denaturing gradient gel electrophoresis (DGGE), which causes DNA to migrate to a point on a gel defined by its sequence. Bands were then identified by extracting them from the gel, Sanger sequencing them and comparing them to known 18S sequences in GenBank (Martin et al. 2006). This study successfully identified a range of copepods, diatoms, dinoflagellates and other unicellular eukaryotes such as Cercozoa, Cryptomonadaceae and Prasinophyceae that would not normally be identified by microscopic analysis of stomach contents or faeces. Another dietary analysis study that used a different, shorter region of the 18S (~140 bp) used DNA purified from the digestive gland as a substrate for PCR and instead of separating bands on a DGGE gel, cloned fragments into bacteria to isolate individual molecules from bacterial colonies, followed by Sanger sequencing and comparisons to known sequences to identify them (Passmore et al. 2006). This sort of approach has a slight advantage in that counting of sequences will produce a semi-quantitative, although highly biased, measure of quantity of material from each food species. This is useful for measuring change in relative biomass between samples, although it cannot provide an absolute estimate of relative biomass in single samples. All of the methods based on universal PCR primer sets have the problem that they will not only amplify DNA from species of food consumed by the krill, but also from krill symbionts and generally a lot from the krill itself. A method for reducing the PCR amplification of krill DNA in DNA-based diet analyses was developed for a region of the nuclear 28S rDNA gene that can be amplified with universal primers and a 'blocking primer' that reduces the quantity of krill DNA amplified in these PCRs (Vestheim and Jarman 2008). The universal diet analysis approaches are all ideal for use in combination with HTS now rather than the DGGE or clone library and sequencing approaches used in these older studies. It is expected that more studies of krill diet using this approach will appear soon.

7.6.3 *Identification of Krill Microbiota*

Most microorganisms cannot be cultured and many cannot be adequately identified microscopically, which has led to extensive use of DNA markers for their identification. PCR amplification of DNA markers followed by HTS is now the standard approach for identifying Bacteria and Archaea (Klindworth et al. 2013), viruses (Radford et al. 2012) and the most prevalent strategy for identifying eukaryotic microbes (Bik et al. 2012). The microbiota of krill remains largely uncharacterized by any method and there are no studies have yet attempted to apply HTS to identifying the broad-scale krill microbiota. Krill are host to numerous parasites and suffer from several known diseases and it is likely that HTS approaches will contribute to understanding these in the future (see Chap. 10, Gómez-Gutiérrez and Morales-Ávila 2016).

Some studies have used DNA markers to identify single species of microbe associated with krill. A cold adapted bacterium, *Psychrobacter proteolyticus* that secretes a metalloprotease active in cold conditions and potentially useful as a cold-water cleaning enzyme was isolated from a krill stomach and identified by its nuclear large subunit ribosomal DNA (16S) sequence (GenBank NR_028918) (Denner et al. 2001). Ciliates of the genus *Ephelota* are common ectoparasites of krill as both 'feeding forms' found predominately on the abdomen and legs and sessile cysts found commonly in the filtering basket. A 1380 bp fragment of the 18S gene was PCR amplified and sequenced from krill collected in the Antarctic Peninsula region (Stankovic et al. 2002). The sequence (GenBank AF515610) could be used for diagnostic PCR in recognizing *Ephelota* infestation in krill.

7.7 Conclusion and Outstanding Questions

Genetic research on krill has changed markedly in the past decade. Population genetics, some karyotyping studies and a small number of phylogenetic studies accounted for almost all krill genetics research that had been published by 2005. The main type of genetic study that has developed since then is analysis of gene expression. Because gene expression is the point at which information stored as DNA becomes part of an organism's function, almost all biological processes are controlled by it at some level. Several groups have used HTS, microarrays and qPCR to study a diverse range of physiological processes such as moulting, control of biorhythms and stress responses. The genetic resources developed for these studies are summarised in Table 7.1.

Many questions remain unanswered or not addressed at all in krill genetics and in areas where genetic methods are applicable. Some questions that we hope krill genetic research will cover in the near future are:

1. Do krill from different environments show different frequencies of alleles subject to differential selection? Selectively neutral markers have not revealed consistent population subdivision. Markers under selection may do so.
2. How can the phylogenetics of krill and the Malacostraca in general be resolved? Species sampling is the main limitation for inferring species relationships. Phylogenomic approaches appear capable of resolving ancient evolutionary events.
3. What genes are involved in regulating the complex, multi-stage development of krill? How are these genes regulated?
4. What epigenetic mechanisms do krill use to regulate gene expression? Do krill use DNA methylation, histone acetylation etc.? What is their microRNA complement like and what processes do krill microRNAs regulate?

5. How is the sex of krill determined? Is there a primary genetic mechanism for this or an environmental one, or mixtures of both? Can krill change sex? Transcripts putatively involved in female reproductive maturation have been identified (Seear et al. 2012) but very little else is known.
6. What can we infer about kinship in krill? Multiple matings occur and a sperm plug is produced (see Chap. 6, Kawaguchi 2016). How variable is the number of males who contribute to each female's brood of eggs?
7. What species of microbe inhabit different parts of the krill body? Modern molecular methods enable detection and identification of the many microorganisms that interact with krill, but there has been very little application of these methodologies to date.

Acknowledgements We thank Andrea Polanowski, Cassandra Faux, Léonie Suter, Mathias Teschke, Volker Siegel and Lorenzo Zane for their helpful comments. The authors were funded by Australian Antarctic Science project 4015.

References

Anheller J-E, Hellgren L, Karlstam B, Vincent J (1989) Biochemical and biological profile of a new enzyme preparation from Antarctic krill (*E. superba*) suitable for debridement of ulcerative lesions. Arch Dermatol Res 281(2):105–110

Ayala F, Valentine JW, Zumwalt G (1975) An electrophoretic study of the Antarctic zooplankter *Euphausia superba*. Limnol Oceanogr 20(4):635–640

Ban S, Ohi N, Leong S, Takahashi K, Riser C, Taguchi S (2007) Effect of solar ultraviolet radiation on survival of krill larvae and copepods in Antarctic Ocean. Polar Biol 30(10):1295–1302

Batta-Lona PG, Bucklin A, Wiebe PH, Patarnello T, Copley NJ (2011) Population genetic variation of the Southern Ocean krill, *Euphausia superba*, in the Western Antarctic Peninsula region based on mitochondrial single nucleotide polymorphisms (SNPs). Deep Sea Res Part II 58(13–16):1652–1661

Benjamin DC, Kristjánsdóttir S, Gudmundsdottir A (2001) Increasing the thermal stability of euphauserase. Eur J Biochem 268(1):127–131

Bik HM, Porazinska DL, Creer S, Caporaso JG, Knight R, Thomas WK (2012) Sequencing our way towards understanding global eukaryotic biodiversity. Trends Ecol Evol 27(4):233–243

Bohmann K, Evans A, Gilbert MTP, Carvalho GR, Creer S, Knapp M, Yu DW, de Bruyn M (2014) Environmental DNA for wildlife biology and biodiversity monitoring. Trends Ecol Evol 29(6):358–367

Bortolotto E, Bucklin A, Mezzavilla M, Zane L, Patarnello T (2011) Gone with the currents: lack of genetic differentiation at the circum-continental scale in the Antarctic krill *Euphausia superba*. BMC Genet 12:32

Buckley LJ (1984) RNA-DNA ratio: an index of larval fish growth in the sea. Mar Biol 80 (3):291–298

Bucklin A, Wiebe PH, Smolenack SB, Copley NJ, Beaudet JG, Bonner KG, Färber-Lorda J, Pierson JJ (2007) MtCOI sequences for euphausiids (Euphausiacea, Crustacea): DNA barcodes for species identification. J Plankton Res 29(6):483–493

Candeias R, Teixeira S, Duarte CM, Pearson GA (2014) Characterization of polymorphic microsatellite loci in the Antarctic krill *Euphausia superba*. BMC Res Notes 7(1):73

Casanova B (1984) Phylogénie des euphausiacés (Crustacés eucarides). Bull Mus Natl Hist Nat Sect A Zool Biol Écol Anim 6(4):1077–1089

Cascella K, Jollivet D, Papot C, Léger N, Corre E, Ravaux J, Clark MS, Toullec J-Y (2015) Diversification, evolution and sub-functionalization of 70 kDa heat shock proteins in two sister species of Antarctic krill: differences in thermal habitats, responses and implications under climate change. PLoS ONE 10(4):e0121642

Charlesworth B (2009) Effective population size and patterns of molecular evolution and variation. Nat Rev Genet 10(3):195–205

Chen CS, Yan TR, Chen HY (1978) Purification and properties of trypsin-like enzymes and a carboxypeptidase A from *Euphausia superba*. J Food Biochem 2(4):349–366

Chen B, Chu T, Harms E, Gergen JP, Strickland S (1998) Mapping of *Drosophila* mutations using site-specific male recombination. Genetics 149(1):157–163

Clark MS, Thorne MAS, Toullec J-Y, Meng Y, Guan LL, Peck LS, Moore S (2011) Antarctic krill 454 pyrosequencing reveals chaperone and stress transcriptome. PLoS ONE 6(1):e15919

Davey JW, Hohenlohe PA, Etter PD, Boone JQ, Catchen JM, Blaxter ML (2011) Genome-wide genetic marker discovery and genotyping using next-generation sequencing. Nat Rev Genet 12:499–510

De Pitta C, Bertolucci C, Mazzotta GM, Bernante F, Rizzo G, De Nardi B, Pallavicini A, Lanfranchi G, Costa R (2008) Systematic sequencing of mRNA from the Antarctic krill (*Euphausia superba*) and first tissue specific transcriptional signature. BMC Genomics 9:45

De Pittà C, Biscontin A, Albiero A, Sales G, Millino C, Mazzotta GM, Bertolucci C, Costa R (2013) The Antarctic krill *Euphausia superba* shows diurnal cycles of transcription under natural conditions. PLoS ONE 8(7):e68652

Deagle BE, Jarman SN, Coissac E, Pompanon F, Taberlet P (2014) DNA metabarcoding and the cytochrome c oxidase subunit I marker: not a perfect match. Biol Lett 10(9):20140562

Deagle BE, Faux C, Kawaguchi S, Meyer B,SNJ (2015) Antarctic krill population genomics: apparent panmixia, but genome complexity and large population size muddies the water. Mol Ecol 24(19):4943–4959

Denner EBM, Mark B, Mark B, Busse H-J, Busse H-J, Turkiewicz M, Lubitz W (2001) *Psychrobacter proteolyticus* sp. nov., a psychrotrophic, halotolerant bacterium isolated from the Antarctic krill *Euphausia superba* Dana, excreting xcreting a cold-adapted metalloprotease. Syst Appl Microbiol 24(1):44–53

Ekblom R, Galindo J (2011) Applications of next generation sequencing in molecular ecology of non-model organisms. Heredity 107(1):1–15

Everson I (2000) Krill: biology, ecology and fisheries. Blackwell Science, Oxford

Felsenstein J (2004) Inferring phylogenies. Sinauer Associates, Massachusetts, pp 123–146

Fevolden SE (1988) Biochemical genetics and population structure of *Euphausia superba*. Comp Biochem Phys B 90(3):507–513

Fevolden SE, Ayala FJ (1981) Enzyme polymorphism in Antarctic krill (Euphausiacea); genetic variation between populations and species. Sarsia 66(3):167–181

Fevolden SE, Schneppenheim R (1989) Genetic homogeneity of krill (*Euphausia superba* Dana) in the Southern Ocean. Polar Biol 9(8):533–539

Gómez-Gutiérrez J, Morales-Ávila JR (2016) Parasites and diseases. In: Siegel V (ed) The biology and ecology of Antarctic krill, *Euphausia superba* Dana, 1850. Springer, Cham, pp 351–386

Goodall-Copestake W, Perez-Espona S, Clark M, Murphy E, Seear P, Tarling G (2010) Swarms of diversity at the gene cox1 in Antarctic krill. Heredity 104(5):513–518

Goodall-Copestake W, Tarling G, Murphy E (2012) On the comparison of population-level estimates of haplotype and nucleotide diversity: a case study using the gene cox1 in animals. Heredity 109(1):50–56

Gregory (2005) Animal genome size database. http://www.genomesize.com/

Hennig W (1966) Phylogenetic systematics. University of Illinois Press, Urbana, IL, USA

Ikeda T (1989) RNA content of Antarctic krill (*Euphausia superba* Dana), an estimator of natural growth rate. Proc NIPR Symp Polar Biol 1989:26–33

Jarman SN (2001a) DNA evolution in krill. PhD thesis, University of Tasmania

Jarman SN (2001b) The evolutionary history of krill inferred from nuclear large subunit rDNA sequence analysis. Biol J Linn Soc 73(2):199–212
Jarman S, Elliott N, Nicol S, McMinn A, Newman S (1999) The base composition of the krill genome and its potential susceptibility to damage by UV-B. Antarct Sci 11(01):23–26
Jarman SN, Elliott NG, Nicol S, McMinn A (2000a) Molecular phylogenetics of circumglobal *Euphausia* species (Euphausiacea: Crustacea). Can J Fish Aquat Sci 57(S3):51–58
Jarman SN, Nicol S, Elliott NG, McMinn A (2000b) 28S rDNA evolution in the Eumalacostraca and the phylogenetic position of krill. Mol Phylogenet Evol 17(1):26–36
Jarman SN, Gales NJ, Tierney M, Gill PC, Elliott NG (2002) A DNA-based method for identification of krill species and its application to analysing the diet of marine vertebrate predators. Mol Ecol 11(12):2679–2690
Jarman SN, McInnes JC, Faux C, Polanowski AM, Marthick J, Deagle BE, Southwell C, Emmerson L (2013) Adélie penguin population diet monitoring by analysis of food DNA in scats. PLoS ONE 8(12):e0082227
Jeffery NW (2011) The first genome size estimates for six species of krill (Malacostraca, Euphausiidae): large genomes at the north and south poles. Polar Biol 35(6):959–962
Jenner RA, Dhubhghaill CN, Ferla MP, Wills MA (2009) Eumalacostracan phylogeny and total evidence: limitations of the usual suspects. BMC Evol Biol 9(1):21
Johansson M, Duda E, Sremba A, Banks M, Peterson W (2012a) Assessing population-level variation in the mitochondrial genome of *Euphausia superba* using 454 next-generation sequencing. Mol Biol Rep 39(5):5755–5760
Johansson M, Sremba A, Feinberg L, Banks M, Peterson W (2012b) The mitochondrial genomes of *Euphausia pacifica* and *Thysanoessa raschii* sequenced using 454 next-generation sequencing, with a phylogenetic analysis of their position in the Malacostracan family tree. Mol Biol Rep 39(9):9009–9021
Kawaguchi S (2016) Reproduction and larval development in Antarctic krill (*Euphausia superba*). In: Siegel V (ed) The biology and ecology of Antarctic krill, *Euphausia superba* Dana, 1850. Springer, Cham, pp 225–246
Klindworth A, Pruesse E, Schweer T, Peplies J, Quast C, Horn M, Glöckner FO (2013) Evaluation of general 16S ribosomal RNA gene PCR primers for classical and next-generation sequencing-based diversity studies. Nucleic Acids Res 41(1):e1
Lee C, Kibblewhite-Accinelli R, Wagschal K, Robertson G, Wong DS (2006) Cloning and characterization of a cold-active xylanase enzyme from an environmental DNA library. Extremophiles 10(4):295–300
Leese F, Brand P, Rozenberg A, Mayer C, Agrawal S, Dambach J, Dietz L, Doemel JS, Goodall-Copstake WP, Held C, Jackson JA, Lampert KP, Linse K, Macher JN, Nolzen J, Raupach MJ, Rivera NT, Schubart CD, Striewski S, Tollrian R, Sands CJ (2012) Exploring pandora's box: potential and pitfalls of low coverage genome surveys for evolutionary biology. PLoS ONE 7(11):e49202
Maas A, Waloszek D (2001) Larval development of *Euphausia superba* Dana, 1852 and a phylogenetic analysis of the Euphausiacea. Hydrobiologia 448(1–3):143–169
MacDonald CM, Williams R, Adams M (1986) Genetic variation and population structure of krill (*Euphausia superba* Dana) from the Prydz Bay region of Antarctic waters. Polar Biol 6(4):233–236
Machida R, Miya M, Yamauchi M, Nishida M, Nishida S (2004) Organization of the mitochondrial genome of Antarctic krill Euphausia superba (Crustacea: Malacostraca). Mar Biotechnol 6(3):238–250
Martin DL, Ross RM, Quetin LB, Murray AE (2006) Molecular approach (PCR-DGGE) to diet analysis in young Antarctic krill *Euphausia superba*. Mar Ecol Prog Ser 319:155–165
Mazzotta GM, De Pittà C, Benna C, Tosatto SC, Lanfranchi G, Bertolucci C, Costa R (2010) A *cry* from the krill. Chronobiol Int 27(3):425–445
Metzker ML (2010) Sequencing technologies—the next generation. Nat Rev Genet 11(1):31–46

Meusemann K, von Reumont BM, Simon S, Roeding F, Strauss S, Kück P, Ebersberger I, Walzl M, Pass G, Breuers S (2010) A phylogenomic approach to resolve the arthropod tree of life. Mol Biol Evol 27(11):2451–2464

Meyer B, Teschke M (2016) Physiology of *Euphausia superba*. In: Siegel V (ed) The biology and ecology of Antarctic krill, *Euphausia superba* Dana, 1850. Springer, Cham, pp 145–174

Meyer B, Martini P, Biscontin A, De Pitta C, Roumaldi C, Teschke M, Frichenhaus S, Harmns L, Frier U, Jarman S, Kawaguchi S (2015) Pyroseqencing and de novo assembly of the Antarctic krill (*Euphausia superba*) transcriptome to study the adaptability of krill to climate induced environmental changes. Mol Ecol Res 15(6):1460–1471

Misof B, Liu S, Meusemann K, Peters RS, Donath A, Mayer C, Frandsen PB, Ware J, Flouri T, Beutel RG, Niehuis O, Petersen M, Izquierdo-Carrasco F, Wappler T, Rust J, Aberer AJ, Aspöck U, Aspöck H, Bartel D, Blanke A, Berger S, Böhm A, Buckley TR, Calcott B, Chen J, Friedrich F, Fukui M, Fujita M, Greve C, Grobe P, Gu S, Huang Y, Jermiin LS, Kawahara AY, Krogmann L, Kubiak M, Lanfear R, Letsch H, Li Y, Li Z, Li J, Lu H, Machida R, Mashimo Y, Kapli P, McKenna DD, Meng G, Nakagaki Y, Navarrete-Heredia JL, Ott M, Ou Y, Pass G, Podsiadlowski L, Pohl H, von Reumont BM, Schütte K, Sekiya K, Shimizu S, Slipinski A, Stamatakis A, Song W, Su X, Szucsich NU, Tan M, Tan X, Tang M, Tang J, Timelthaler G, Tomizuka S, Trautwein M, Tong X, Uchifune T, Walzl MG, Wiegmann BM, Wilbrandt J, Wipfler B, Wong TKF, Wu Q, Wu G, Xie Y, Yang S, Yang Q, Yeates DK, Yoshizawa K, Zhang Q, Zhang R, Zhang W, Zhang Y, Zhao J, Zhou C, Zhou L, Ziesmann T, Zou S, Li Y, Xu X, Zhang Y, Yang H, Wang J, Wang J, Kjer KM, Zhou X (2014) Phylogenomics resolves the timing and pattern of insect evolution. Science 346(6210):763–767

Neale DB, Wegrzyn JL, Stevens KA, Zimin AV, Puiu D, Crepeau MW, Cardeno C, Koriabine M, Holtz-Morris AE, Liechty JD, Martinez-Garcia PJ, Vasquez-Gross HA, Lin BY, Zieve JJ, Dougherty WM, Fuentes-Soriano S, Wu LS, Gilbert D, Marcais G, Roberts M, Holt C, Yandell M, Davis JM, Smith KE, Dean JF, Lorenz WW, Whetten RW, Sederoff R, Wheeler N, McGuire PE, Main D, Loopstra CA, Mockaitis K, deJong PJ, Yorke JA, Salzberg SL, Langley CH (2014) Decoding the massive genome of loblolly pine using haploid DNA and novel assembly strategies. Genome Biol 15(3):R59

Nicol S, Foster J (2016) The fishery for Antarctic krill – its current status and management regime. In: Siegel V (ed) The biology and ecology of Antarctic krill, *Euphausia superba* Dana, 1850. Springer, Cham, pp 387–422

Osnes KK, Mohr V (1985) Peptide hydrolases of Antartic krill, *Euphausia superba*. Comp Biochem Phys B 82(4):599–606

Passmore AJ, Jarman SN, Swadling KM, Kawaguchi S, McMinn A, Nicol S (2006) DNA as a dietary biomarker in Antarctic krill, *Euphausia superba*. Mar Biotechnol 8(6):686–696

Patarnello T, Bargelloni L, Varotto V, Battaglia B (1996) Krill evolution and the Antarctic ocean currents: evidence of vicariant speciation as inferred by molecular data. Mar Biol 126(4):603–608

Podsiadlowski L, Bartolomaeus T (2005) Organization of the mitochondrial genome of mantis shrimp *Pseudosquilla ciliata* (Crustacea: Stomatopoda). Mar Biotechnol 7(6):618–624

Radford AD, Chapman D, Dixon L, Chantrey J, Darby AC, Hall N (2012) Application of next-generation sequencing technologies in virology. J Gen Virol 93(9):1853–1868

Rees DJ, Dufresne F, Glémet H, Belzile C (2007) Amphipod genome sizes: first estimates for Arctic species reveal genomic giants. Genome 50(2):151–158

Reiss CS (2016) Age, growth, mortality, and recruitment of Antarctic krill, *Euphausia superba*. In: Siegel V (ed) The biology and ecology of Antarctic krill, *Euphausia superba* Dana, 1850. Springer, Cham, pp 101–144

Rolfe WI (1981) Phyllocarida and the origin of the Malacostraca. Geobios 14(1):17–26

Schmidt K, Atkinson A (2016) Feeding and food processing in Antarctic krill (Euphausia superba Dana). In: Siegel V (ed) The biology and ecology of Antarctic krill, *Euphausia superba* Dana, 1850. Springer, Cham, pp 175–224

Schneppenheim R, MacDonald C (1984) Genetic variation and population structure of krill (*Euphausia superba*) in the Atlantic sector of Antarctic waters and off the Antarctic Peninsula. Polar Biol 3(1):19–28

Schram FR (1981) On the classification of Eumalacostraca. J Crust Biol 1(1):1–10
Seear P, Tarling GA, Teschke M, Meyer B, Thorne MAS, Clark MS, Gaten E, Rosato E (2009) Effects of simulated light regimes on gene expression in Antarctic krill (*Euphausia superba* Dana). J Exp Mar Biol Ecol 381(1):57–64
Seear PJ, Tarling GA, Burns G, Goodall-Copestake WP, Gaten E, Ozkaya O, Rosato E (2010) Differential gene expression during the moult cycle of Antarctic krill (*Euphausia superba*). BMC Genomics 11(1):582
Seear PJ, Goodall-Copestake WP, Fleming AH, Rosato E, Tarling GA (2012) Seasonal and spatial influences on gene expression in Antarctic krill *Euphausia superba*. Mar Ecol Prog Ser 467:61–75
Selkoe KA, Toonen RJ (2006) Microsatellites for ecologists: a practical guide to using and evaluating microsatellite markers. Ecol Lett 9(5):615–629
Shen X, Wang H, Ren J, Tian M, Wang M (2010) The mitochondrial genome of *Euphausia superba* (Prydz Bay) (Crustacea: Malacostraca: Euphausiacea) reveals a novel gene arrangement and potential molecular markers. Mol Biol Rep 37(2):771–784
Shen X, Wang H, Wang M, Liu B (2011) The complete mitochondrial genome sequence of *Euphausia pacifica* (Malacostraca: Euphausiacea) reveals a novel gene order and unusual tandem repeats. Genome 54(11):911–922
Shin H-C, Nicol S, King RA (2003) Nucleic acid content as a potential growth rate estimator of Antarctic krill; results from field-caught krill from the Indian sector of the southern ocean. Mar Freshw Behav Phys 36(4):295–305
Shokralla S, Spall JL, Gibson JF, Hajibabaei M (2012) Next-generation sequencing technologies for environmental DNA research. Mol Ecol 21(8):1794–1805
Siegel V, Watkins W (2016) Distribution, biomass and demography of Antarctic krill, *Euphausia superba*. In: Siegel V (ed) The biology and ecology of antarctic krill, *Euphausia superba* Dana, 1850. Springer, Cham, pp 21–100
Slatkin M (1987) Gene flow and the geographic structure of natural populations. Science 236 (4803):787–792
Spiridonov VA (1992) The preanal plate and spine of Euphausiacea (Crustacea, Eucarida), their variability and homology (with particular reference to the austral species of the genus *Euphausia* Dana). Russ J Arthropodan Res 1(4):3–15
Stankovic A, Borsuk P, Koper M, Weglenski P (2002) Studies on *Ephelota* sp., an epizoic suctorian found on Antarctic krill. Polar Biol 25(11):827–832
Takayama S, Dhahbi J, Roberts A, Mao G, Heo S-J, Pachter L, Martin D, Boffelli D (2014) Genome methylation in *D. melanogaster* is found at specific short motifs and is independent of DNMT2 activity. Genome Res 24(5):821–830
Teschke M, Wendt S, Kawaguchi S, Kramer A, Meyer B (2011) A circadian clock in antarctic krill: an endogenous timing system governs metabolic output rhythms in the euphausid species *Euphausia superba*. PLoS ONE 6(10):e26090
Thiriot-Quiévreux C, Leitão A, Cuzin-Roudy J (1998) Chromosome diversity in Mediterranean and Antarctic Euphausiid species (Euphausiacea). J Crust Biol 18(2):290–297
Töbe K, Meyer B, Fuentes V (2010) Detection of zooplankton items in the stomach and gut content of larval krill, *Euphausia superba*, using a molecular approach. Polar Biol 33(3):407–414
Toullec J-Y, Corre E, Bernay B, Thorne MAS, Cascella K, Ollivaux C, Henry J, Clark MS (2013) Transcriptome and peptidome characterisation of the main neuropeptides and peptidic hormones of a euphausiid: the ice krill, *Euphausia crystallorophias*. PLoS ONE 8(8):e71609
Trathan PN, Hill SL (2016) The importance of krill predation in the Southern Ocean. In: Siegel V (ed) The biology and ecology of Antarctic krill, *Euphausia superba* Dana, 1850. Springer, Cham, pp 321–350
Treangen TJ, Salzberg SL (2012) Repetitive DNA and next-generation sequencing: computational challenges and solutions. Nat Rev Genet 13(1):36–46

Turkiewicz M, Kalinowska H, Zielińska M, Bielecki S (2000) Purification and characterization of two endo-1,4-β-xylanases from Antarctic krill, *Euphausia superba* Dana. Comp Biochem Phys B 127(3):325–335

Van Ngan P, Gomes V, Suzuki H, Passos MJACR (1989) Preliminary study on chromosomes of Antarctic krill, *Euphausia superba* Dana. Polar Biol 10(2):149–150

Vestheim H, Jarman SN (2008) Blocking primers to enhance PCR amplification of rare sequences in mixed samples – a case study on prey DNA in Antarctic krill stomachs. Front Zool 5(1):5–12

Waples RS (1998) Separating the wheat from the chaff: patterns of genetic differentiation in high gene flow species. J Hered 89(5):438–450

Wright S (1938) Size of population and breeding structure in relation to evolution. Science 87(2263):430–431

Yabu H, Kawamura A (1984) Chromosomes of *Euphausia superba* Dana. Bull Plank Soc Jpn 31:61–63

Zane L, Ostellari L, Maccatrozzo L, Bargelloni L, Battaglia B, Patarnello T (1998) Molecular evidence for genetic subdivision of Antarctic krill (*Euphausia superba* Dana) populations. Proc R Soc Lond B Biol 265(1413):2387–2391

Chapter 8
Swarming and Behaviour in Antarctic Krill

Geraint A. Tarling and Sophie Fielding

Abstract The behavioural ecology of Antarctic krill is dominated by their tendency to swarm. They form amongst the largest monospecific aggregations of biomass in the animal kingdom, with some swarms measuring up to 100 km^2 and containing 2 million tonnes of krill. Swarms come in a multitude of shapes and sizes, and a greater understanding of the functional attributes of different swarm types is starting to emerge. This chapter will consider the spectrum of krill-swarms and -schools that have been described and some of the latest approaches taken to understand their shape and formation. The fundamental needs to avoid predation, feed, mate and spawn have often been attributed to being a major influence on swarming and we will examine these behaviours and their wider impacts. This chapter also considers how krill position themselves in the water column, altering their depth over diel and seasonal cycles, with further levels of modification depending on the environmental context. The ability of krill to migrate large distances is a major ecological feature of the Southern Ocean ecosystem, affecting the productivity of both planktonic and upper-trophic level communities, and we consider how such migrations are driven at the level of the swarm. New technologies are emerging that are providing previously unreported krill behaviours and we assess the future potential of these technologies to develop an even deeper appreciation of krill ethology. Also, we assess what impact predicted changes to the Southern Ocean environment will have on krill behavioural traits.

Keywords Schooling • Aggregation • Swimming • Migration • Superswarm

8.1 General Introduction

Krill arguably have the largest biomass of any single metazoan species (Atkinson et al. 2009), and their behavioural flexibility and capacity to swarm is likely to contribute to their success. There have been a number of reviews of krill behaviour, social aggregation and swarming, some aimed specifically at Antarctic krill (Miller

---

G.A. Tarling (✉) • S. Fielding
British Antarctic Survey, Natural Environment Research Council, High Cross, Madingley Rd, Cambridge CB3 0ET, UK
e-mail: gant@bas.ac.uk; sof@bas.ac.uk

V. Siegel (ed.), *Biology and Ecology of Antarctic Krill*, Advances in Polar Ecology,
DOI 10.1007/978-3-319-29279-3_8

and Hampton 1989; Siegel and Kalinowski 1994), others, on a wider range of euphausiid species (Watkins 2000) and further reviews have covered all socially aggregating invertebrates, where Antarctic krill have been specifically mentioned (Ritz 1994; Ritz et al. 2011). It is not our intention to repeat the information within these reviews, although many of the general themes that they cover remain pertinent to the way that behaviour and swarming is studied in Antarctic krill today. Our main focus has been on work published since the most recent review of euphausiid behaviour by Watkins (2000), although seminal works before this time are still mentioned. We searched for all published works from 2000 onwards using the terms 'Antarctic' and 'krill' and 'swarming, schooling or behaviour (behavior)' in 'Google-scholar' and 'Web of Science' search engines. Further literature was identified through cross-referencing and forward citation from those articles identified. The structure of this review reflects the balance of the literature identified through this means, which we believe indicates the present direction of research into behaviour and swarming in Antarctic krill.

We subdivided this body of work into five sections based on what we considered to be discrete areas of research within this field: field-based observations; laboratory-based observations; models of behaviour and swarming; temporal and spatial patterns of aggregations; and ways ahead for future study on krill behaviour. In common with a number of previous krill reviews, there is a need to define important terms at the outset, since many are used interchangeably throughout the literature, obscuring the intended meaning. We follow the definitions of Ritz (1994) where 'aggregations' will be used as a general term describing a group of conspecific individuals where there may or may not be a level of mutual attraction; 'swarms' will be taken to mean a discrete integrated social group with members that are regularly spaced but not polarized; 'schools' contain members assumed to be polarized and moving in the same direction; 'shoals' will be used when referring to larger groupings containing subregions where members are either swarming or schooling.

The main body of cited literature deals specifically with Antarctic krill. However, in several instances, it was necessary to refer to important insights gained from other euphausiids, because they reflected patterns that are likely to exist also in Antarctic krill but have yet to be studied or discerned. A thorough overview across all euphausiid species is warranted, but is beyond the limits of the present work. Antarctic krill share a large number of traits with many other euphausiid species, but also embody an extremity in terms of biomass dominance and prolific swarming capabilities.

8.2 Field-Based Observations

8.2.1 Introduction

Swarming is a fundamental part of the life of *Euphausia superba*, yet we still know very little about what drives the considerable variability in swarm shape, size and biomass. In this section, we describe how technological improvements in field-surveying methods and analysis over the past 15 years have advanced our ability to study swarm diversity. This has not only allowed levels of variability in shape and structure to be documented comprehensively but has opened up new ways in which to investigate the drivers of swarming patterns. Furthermore, we consider the new ways in which the movement of swarms, and the individuals within them, have been studied and resolved.

8.2.2 Swarm Types and Their Properties

Initial studies into swarm structures were mainly concerned with describing variability in swarm size, shape and packing-density but, in more recent times, there has been greater focus on why such variability exists and whether it is possible to predict it.

Krill have been observed in small compact aggregations (10–100 m long, 2–20 m high; Kalinowski and Witek 1985), extensive layers (41 km long; Watkins and Murray 1998), formations that are dispersed throughout the water column (Everson 1982) and superswarms covering several km horizontally and >100 m vertically (Mathisen and Macaulay 1983; Macaulay et al. 1984; Schulenberger et al. 1984; Higginbottom and Hosie 1989; Lawson et al. 2008; Tarling et al. 2009; Nowacek et al. 2011; Fielding et al. 2012). Various classification schemes have been developed in response to the need to summarise this large degree of variability in swarm structure (Mauchline 1980; Kalinowski and Witek 1985; Murphy et al. 1988; Miller and Hampton 1989; Siegel and Kalinowski 1994; Tarling et al. 2009). That developed by Kalinowski and Witek (1985), and since modified by Miller and Hampton (1989) and Siegel and Kalinowski (1994), first distinguishes between scales of <1 and 1–100 km, referring to features at the latter scale as "concentrations", and subdividing features at the former scale into "cohesive forms", "irregular forms", "layers" and "superpatches". The scheme has an intuitive appeal but does rely on a degree of subjectivity when dividing up the wide spectrum of shapes, sizes and densities of swarms resolved by echosounders.

One of the most significant breakthroughs since 2000 has been the inclusion of school detection algorithms (e.g. SHAPE Shoal Analysis and Patch Estimation System; Coetzee 2000) in the development and increasing use of processing software (e.g. Echoview (Myriax) and LSSS (Marec)). These techniques are now capable of rapid, automatic identification of krill swarms and subsequent measurement of multiple parameters describing shape, density and distribution across large-scale acoustic transects. This has allowed a new wave of swarm analysis, in which subjective interpretations have been replaced by objective, statistical methods to identify clusters of swarms based on similarities across a large number of different parameters. For instance, Tarling et al. (2009) analyzed the area, length, height, depth, packing concentration and inter-swarm distance of 4525 positively identified krill swarms across the Scotia Sea. Hierarchical cluster analysis revealed two major types of swarm, one type in which swarms were generally small (<50 m long) and not very tightly packed (<10 ind. m^{-3}) and another, which were an order of magnitude larger in size and had packing concentrations that were ten times greater than the smaller swarm type. This multivariate approach to swarm classification has since been used by Cox et al. (2011a) and Krafft et al. (2012, Table 8.1).

The approach allows the physical properties of swarms to be related to potential forcing factors in their environment. For instance, Tarling et al. (2009) found that, at a mesoscale level within the Scotia Sea, highest explanatory power towards swarm type distribution was achieved by a combination of four variables, fluorescence, PAR, krill maturity and krill body length. Cox et al. (2011a) found a strong relationship between the distribution of swarm types and the on-shelf to off-shelf gradient, with the high krill-density swarm type being more common in the on-shelf environment. Krafft et al. (2012), who mainly surveyed open-ocean regions, found a relationship between packing density and remoteness to other swarms, with isolated swarms being more densely packed. They proposed these isolated swarms were actively migrating, potentially in search of new food resources.

A further development in recent decades has been in the increasing use of multibeam echosounders, capable of resolving swarms in three dimensions (Korneliussen et al. 2009; Cox et al. 2009, Fig. 8.1). Using a multi-beam echosounder system (MBES, 200 kHz SM20 Kongsberg Mesotech Ltd Canada), Cox et al. (2010) resolved complex 3-D shapes in spatially segregated swarms. Brierley and Cox (2010) used such data to afford new insights into swarm structure. They found that the 3-D shape, as described by the surface area:volume ratio (or roughness) was distributed narrowly about a mean of 3.3 m^{-1} in a shape they described as a multifaceted lozenge. This swarm-shape property was also found to be common to other shoaling species, such as sardine and anchovies, across geographically and oceanographically diverse locations.

8.2.3 Swarm Movement

Antarctic krill appear to be able to influence their distribution at large oceanic scales (Atkinson et al. 2008), which suggests that active swimming by krill affects large scale patterns of movement. Individual swimming speeds have been measured at between 15 and 30 cm s^{-1} (Kils 1981; Murphy et al. 2011). Nevertheless, it is net movement of swarms relative to the background flow that will ultimately determine larger scale distribution patterns.

One approach to measuring swarm movement is to track swarms over time. Clarke and Morris (1983) managed to track and continuously sample a krill swarm over 6 days, but such dedicated time to a single purpose is rarely afforded to science research vessels. Krill fishery vessels, on the other hand, deliberately track krill and Kanda et al. (1982) documented two cases of active migration of swarms of 560 m and 1089 m diameter aboard such vessels. Both aggregations moved consistently southwards for more than 2 weeks with average velocities of 14 cm s^{-1} and with a maximum speed of 90 cm s^{-1}. Despite the potential for this approach, further measurements of this sort are yet to be made.

Another approach than can reveal longer term, larger scale movements of swarms is to track the net movement of predators that prey on krill swarms. For instance, Fauchald and Tveraa (2006) examined the movement patterns of the Antarctic petrel (*Thalassoica antarctica*), whose diet is mainly focussed on Antarctic krill. They reported an active and massive migration of Antarctic krill prey in an upstream direction over a period of 3 months.

The approach taken by Tarling and Thorpe (2014) was to measure instantaneous rates of movement over large numbers of swarms ($n > 4000$). The study combined the use of echosounders, nets and an acoustic Doppler current profiler. Measurements were obtained of not only total (or gross) instantaneous swarm movement but also of movement relative to background flow, through comparing flow measurements made inside the swarm to those immediately outside. The study succeeded in resolving differences in both the magnitude and direction of swarm movement relative to background flow across large areas of the Scotia Sea. The degree to which movement deviated from the background flow depended on a number of factors, including the size of the swarm, the vicinity of the ice edge, salinity and fluorescence. In particular, the direction of swarm movement deviated from the background flow most when fluorescence levels were high, which is likely to be a response to retaining a favourable feeding environment. The effects of drag on swarm movement were also identified, with swarms becoming increasingly slower than the background flow with increasing current velocity. The study has the drawback that it only considered movement patterns at the instant the swarm was observed and did not involve subsequent tracking to determine longer-term trajectories of movement. Nevertheless, the fact that the measurements were made over so many swarms over such a large area does give scope to generalize about patterns of swarm movement.

A remaining question is the relationship between best estimates of swimming speed of individuals and that of swarms. In the natural environment, both individuals and swarms have to contend with one of the world's strongest ocean currents, the Antarctic Circumpolar Current, with average flow rates of between 10 and 20 cm s^{-1} (Tarling and Thorpe 2014). Theoretically, krill have the individual swimming capability to oppose these flows but whether they are capable of doing so when in swarm formation remains unanswered. Although Tarling and Thorpe (2014) showed that swarm trajectories deviated from background flows, the estimates fell short of making absolute estimates of net swarm velocity because the influence of other acoustic scatterers could not be completely eliminated. A combination of tracking and instantaneous rate movement estimates is necessary in order to resolve this issue.

8.2.4 Migration and Distribution in the Vertical Realm

The traditional view is that adult Antarctic krill is an epipelagic species, residing within the top 250 m of the water column (Marr 1962). Recently this tenet has been

Table 8.1 Krill swarm categorisation using multivariate approaches

Authors	Southern Ocean region (dates)	On-shelf/ off-shelf	Acoustic system	Number of swarms analysed	Number swarm types identified	Mean or median* dimensions of largest swarm type	Mean or median* dimensions of densest swarm type	% biomass in largest swarm type (% frequency)	% biomass in densest swarm (% frequency)
Tarling et al. (2009)	Mainly Scotia Sea (Jan–Feb 2003)	On and off-shelf	Single (split)-beam echosouder (38 kHz and 120 kHz SIMRAD EK60)	4,525	5	Length: 1174 m	Length: 104 m	55 % (4 %)	43 % (37 %)
						Height: 29 m	Height: 10 m		
						Depth: 50 m	Depth: 45 m		
						Interswarm distance: 450 m	Interswarm distance: 560 m		
						Density: 22 ind. m^{-3}	Density: 65 ind. m^{-3}		
Cox et al. (2010)	Antarctic Peninsula (Feb 2006)	On-shelf	Multibeamechosounder (200 kHz SM20 Kongsberg Mesotech)	1,006	3	Length: 276 m	Length: 82 m	32 % (16 %)	68 % (43 %)
						Height: 13 m	Height: 12 m		
						Depth: 55 m	Depth: 58 m		
						Interswarm distance: 38 m	Interswarm distance: 48 m		
						Density: −51.89 dB re 1 m^{-1}	Density: −51.65 dB re 1 m^{-1}		

Cox et al. (2011a)	South Georgia (austral summer 1997, 1998, 1999)	On-shelf	Single (split)-beam echosouder (38 kHz and 120 kHz SIMRAD EK500)	2,990	3	Length: 1,608 m	Length: 166 m	26 % (28 %)	74 % (28 %)
						Height: 13 m	Height: 13 m		
						Depth: 87 m	Depth: 66 m		
						Interswarm distance: not given	Interswarm distance: not given		
						Density: 2 g m^{-3}	Density: 94 g m^{-3}		
Krafft et al. (2012)	Mainly 10°W–10°E (Jan–Feb 2008)	Off-shelf	Single (split)-beam echosouder (38 kHz and 120 kHz SIMRAD EK60)	4,791	4	Length: 108 m*	Length: 40 m*	19 % (18 %)	78 % (30 %)
						Height: 18 m	Height: 9 m		
						Depth: 57	Depth: 52		
						Interswarm distance:104 m	Interswarm distance:493 m		
						Density: 27 ind m^{-3}	Density: 226 ind m^{-3}		

* denotes median rather than mean value

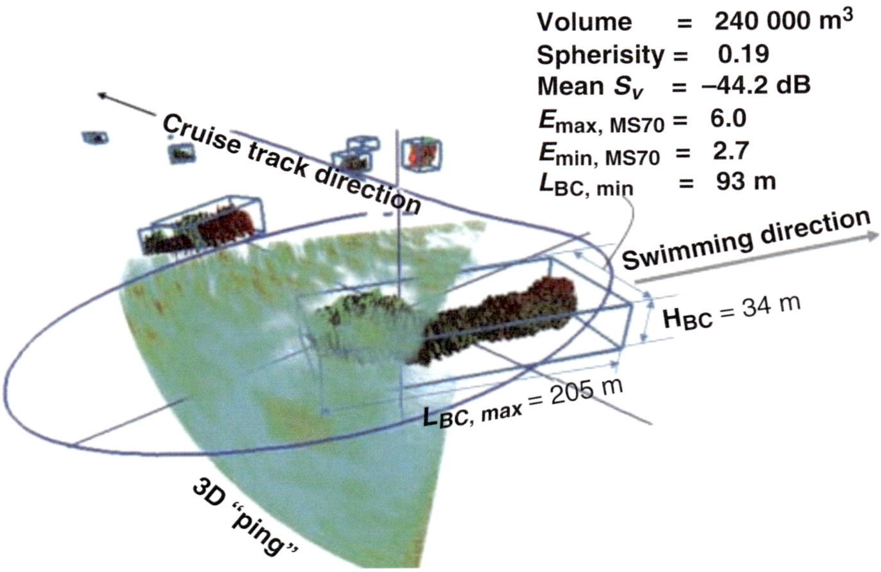

Fig. 8.1 A three-dimensional visualization of Antarctic krill schools, the cruise track (*blue curve*), and one MS70 ping partly covering a single school (Korneliussen et al. 2009). The schools are detected with several pings. The bounding *boxes* are the outer extent of the three-dimensional schools. The quantitative measures are only for the marked school (Reproduced with permission)

challenged by new video observations of adult krill actively feeding at abyssal depths (Clarke and Tyler 2008) and mating in deep water (Kawaguchi et al. 2011). Brierley (2008) proposed that these deep dwelling krill could be members of a distinct and permanent deep sea population or, more likely, Antarctic krill could undertake deep sea foraging migrations. Schmidt et al. (2011) compiled observations from over 30 studies to show that Antarctic krill frequented benthic habitats year round, in both shelf and oceanic waters, with krill feeding on the seabed. Using net and acoustic data, they show that, even in summer, typically 2–20 % of krill resided below 200 m and, on some occasions, this could be as large as 75 % of the population (Fig. 8.2). Lithogenic particles and benthic diatoms were found in the guts of some krill, suggesting that there is a dynamic exchange of individuals between the surface and the seabed. Schmidt et al. (2011) proposed that individuals are most likely to make the migrations to the 'benthic food bank' when feeding conditions in the surface layers are sub-optimal.

The ability to migrate large vertical distances and exploit deep food sources may also enable krill to survive unsuitable surface water temperatures. For instance, during a year of anomalously high surface water temperatures around South Georgia (2009), there were very few krill swarms in the surface layers (<250 m, Fielding et al. 2014). This was accompanied by low foraging success in air-breathing predators (Hill et al. 2009). Yet the starry skate, *Amblyraja georgiana*,

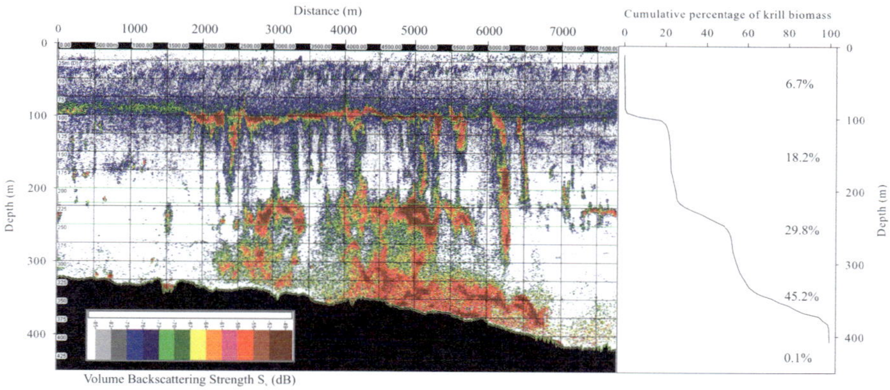

Fig. 8.2 Volume backscattering strength (S_v, dB re 1 m^{-1}) at 120 kHz showing *Euphausia superba* aggregations at the seabed and in midwater layers (NE of South Georgia, 54° 7.5′S; 35° 25.6′W, Schmidt et al. 2011). Krill density was estimated from the 120 kHz data using a variable $\Delta S_{v120-38}$ dB window identification technique and the Stochastic Distorted-Wave Born Approximation acoustic model (CCAMLR 2010) (Reproduced with permission)

was found to have feasted on large Antarctic krill at the seabed in the same area (Main and Collins 2011). This evidence suggests that Antarctic krill use their versatile vertical range to avoid adverse surface temperatures.

Antarctic krill commonly exhibit diurnal vertical migration, moving towards the surface at night, where they change from swarms to more dispersed formations (Demer and Hewitt 1995; Ross et al. 1996; Godlewska 1996; Watkins and Murray 1998; Zhou and Dorland 2004; Siegel 2012). However, considerable variation from this pattern has also been described, with swarms remaining coherent throughout the day and night (Watkins et al. 1992), changing depth and density (Klevjer et al. 2010) and ceasing diel vertical migration behaviour (Hernandez-Leon et al. 2001). Ritz (1994) proposed that social aggregation serves the same purpose as diel vertical migration, and the fact that krill adopts both strategies may contribute to our confused view of their vertical migration behaviour. The efficacy of a dual strategy was illustrated in a modelling study by Sainmont et al. (2013) who showed it to be optimal in minimising predation risk.

Tarling and Johnson (2006) proposed that, rather than undertaking only one single migration per 24 h, Antarctic krill may make as many as three vertical migrations per night. They showed that krill with moderately full stomachs adopt a parachute mode, associated with descent through the water column, significantly more often than those with an empty stomach. These observations provide support for the hunger/satiation hypothesis, where organisms make short forays into surface waters containing high food concentrations, and then return to intermediate depths to reduce predation risk whilst they digest it (Swadling 2006). Through this mechanism, krill could be indirectly responsible for transferring as much as 6 % more carbon to deep ocean sediments than previously assumed by global models

(Tarling and Johnson 2006). These studies identify the need to resolve individual behaviour within aggregations, as well as the vertical movement of the aggregations themselves.

8.3 Laboratory-Based Measurements

8.3.1 Introduction

Aquarium studies can be used to understand krill biology in a controlled situation, allowing experimental method to be applied, under the caveat that it is an imperfect representation of how individuals will behave in their natural environment (Kawaguchi et al. 2010). O'Brien (1989) and Strand and Hamner (1990) were the first to induce schooling behaviour in krill in a laboratory setting, the former aboard a research vessel, the latter at Palmer Station in the Antarctic Peninsula. Nevertheless, such behaviour has been hard to replicate outside of Antarctica until the development of the Australian Antarctic Division krill incubation facility in Kingston, Tasmania. The facility, which contains four separate chilled sea water recirculating systems up to 8000 L each and is restocked each year with fresh Antarctic krill, has reinvigorated behavioural investigations. Combined with work in smaller aquariums elsewhere, as well as field incubations, large advances have been made over the last 15 years in investigating swimming behaviour, conspecific communication and stereotypic responses to predatory threats.

8.3.2 Mechanism of Swimming

Determination of the pattern of swimming has been a major recent focus in Antarctic krill research. Krill provide propulsion through the continuous beating of five pairs of swimming appendages (pleopods) located on the abdomen. The power stroke of these swimming legs is performed in a metachronal fashion while the recovery is synchronous. The metachronal wave travels in the same direction as the animal (adlocomotory system), with the posterior pair of swimming legs initiating the power stroke, followed by its adjacent leg pair after a phase-lag. The sequence continues until all legs have stroked. After the anterior-most pair has stroked, the posterior pair, having already performed the recovery stroke, begins a new power stroke.

Alben et al. (2010) showed that a metachronal stroke (as opposed to a synchronous or intermediate stroke) gave the largest average body speed for shrimp-like crustaceans. Metachronal swimming in these types of organisms is a drag based mechanism, in which the swimming appendages serve as paddles (Murphy et al. 2011). In order to generate forward propulsion, the drag of the appendage during the power stroke must be larger than during the recovery stroke. This is achieved through varying the surface area of the appendage during the stroke cycle.

In krill, the limb flexes during the recovery stroke, decreasing the drag. The distal limb is biramous and these two segments are actively spread laterally during the power stroke to increase drag, and folded together during the recovery stroke to decrease it.

Murphy et al. (2011) examined free swimming Antarctic krill as a model species and identified three distinct behavioural swimming gaits: hovering, fast-forward swimming and upside-down swimming. Quantification of these gaits via high speed video image analysis revealed that, in order to increase swimming speed, krill firstly increased stroke amplitude and, secondly, increase beat frequency. The beat frequency was 3.0 Hz in the hovering gait, 5.7 Hz in the swimming gait and 3.7 Hz during upside down swimming. Of the three gaits, fast forward swimming was found to have the largest stroke amplitude. The hovering gait was investigated further by Murphy et al. (2013) using high resolution volumetric flow data. In this gait, water was pushed downwards in a jetlike flow at a force sufficient to counteract the negative buoyancy of the body. Although the paddling system used by krill is of lower efficiency compared to other propulsion mechanisms, this study illustrated its advantages, particularly its capacity to generate higher thrust forces which can provide superior manoeuvrability.

Johnson and Tarling (2008) considered how body size and physiological state altered the pattern of swimming and level of propulsion in tethered Antarctic krill. They found beat rate decreased and beat strength increased with increasing size. Adult females had faster but weaker pleopod beats than males of equivalent size, while individuals that were about to moult, or had recently moulted, had significantly lower swimming capacities than their intermoult counterparts. Both the rate and the strength of the pleopod beat were significantly lower in satiated compared to unsatiated individuals. These state-related differences in swimming capacities may provide a means of sorting within and between swarms, and explain field observations, such as synchronicity in moult stage and sex-bias. Dispersion of swarms during feeding may also be explained by the influence of satiation on swimming performance.

8.3.3 *Cost of Swimming*

In gregarious mysids, Ritz (2000) found that weight-specific oxygen uptake was reduced by about seven times when they formed cohesive aggregations compared to when they were in uncohesive small groups. It was posited that the mysids exploited updrafts generated by their neighbours to reduce the energy required to counteract sinking. To test this principle within Antarctic krill, Swadling et al. (2005) measured both respiration rate and pleopod beat rate of krill within a closed recirculating system capable of generating current speeds up to a maximum of 17 cm s^{-1}. Different size groups (50, 100, 300) were released into the system and, in turn, exposed to six different current speeds. The relationship between respiration and current speed was not linear over the six speeds. Respiration rate increased linearly from 1.8 mg O_2 g_D h^{-1} at 3 cm s^{-1} to 8.0 mg O_2 g_D h^{-1} at 17 cm s^{-1}. However, contrary to expectations, the size of the group tested did not have a significant effect on pleopod beating rates or oxygen consumption. Swadling et al. (2005) were not able to induce swarming behaviour in any of the experimental groups, so were unable to validate the benefits of formation swimming. Interestingly, maximum cost of transport did not occur at the highest current speed but at 5 cm s^{-1} possibly because krill swarm more horizontally at higher current speeds, which probably decreased the cost of swimming. This is in line with Kils (1981) and Torres et al. (1994), who both predicted that metabolic rate stabilizes or even decreases towards the maximum swimming speed of krill, at around 40 cm s^{-1}. The overall cost of transport was estimated to be around 73 % of total daily metabolic expenditure, at least during early summer. This compares to an estimate of 60 % of total daily metabolic expenditure by Kils (1981), which he calculated through measuring the cost of hovering and assuming that forward propulsion at speeds of up to 25 cm s^{-1} could be attained at no extra cost through changing body angle.

Murphy et al. (2013) took a different approach to estimating the cost of swimming. A high-speed tomographic particle image velocimetry system was used to measure flow around the pleopods and in the wake. The flow measurements were combined with actuator disk theory to estimate the energy required for hovering. By the time muscular efficiency and hydromechanical energy losses were taken into account, these values (8.4×10^{-6}–1.19×10^{-5} W) were similar to those made by Swadling et al. (2005) when converted into the same units, reinforcing the view that the continuous-swimming lifestyle of Antarctic krill comes at a high metabolic cost.

8.3.4 Communication Within Swarms

The coordinated way in which individuals position themselves and move within swarms demands a rapid and pervasive mechanism of communication. Although vision and smell are likely to play a part in this coordination, the major focus since 2000 has been on the role of hydrodynamic cues.

Patria and Wiese (2004) performed a study on tethered Northern krill in which they stimulated the antennules with water oscillations of between 3 and 10 Hz, as a simulation of the vibrations that would typically be generated by the beating of swimming appendages of both Northern krill (*Meganyctiphanes norvegica*) and Antarctic krill. The water oscillations entrained the pleopod beat to a particular frequency, so long as the frequency was within the limited range to which the antennules were sensitive. Patria and Wiese (2004) concluded that adjacent krill in school formation remain coordinated with each other through beating their pleopods at the same frequency. As a consequence, a disturbance at the school edge could rapidly propagate through the school as individuals react to the altered hydrodynamic signals of others around them.

Although studies on tethered specimens can reveal a great deal about the neurological and physiological attributes of swimming performance, they are compromised with respect to visualizing flow fields around individuals and within groups, since they impart an unbalanced force on the fluid. Catton et al. (2011) used infrared particle image velocimetry measurements to analyse the flow field of free-swimming solitary specimens and coordinated groups of 3 to 6 *E. superba*. They found hydrodynamic disturbance to be observable above background level at nearest neighbour distances (NNDs) of four body lengths. NNDs in schools have been measured to be around 2–3 body lengths (Kawaguchi et al. 2010), meaning that hydrodynamic signals are viable cues for conspecific communication. Deterministic models of swarm organization based on hydrodynamic cues are considered further in Sect. 8.4.5.

8.3.5 Schooling Behaviour in Captivity

The first reported inducement of schooling of Antarctic krill in captivity was by Strand and Hamner (1990). Krill kept in a circular experimental tank began to school as soon as contours were removed from the visual field using indirect lighting of a uniform white surround. A black standpipe in the centre of the tank had a positive effect on schooling, driving all the krill to the periphery of the tank, presumably an evasive response with the level of avoidance proportional to the size of the object. A predator model 10 cm in diameter and 20 cm long stalled the school and, if moved towards the school, reversed it and chased it in the opposite direction.

Kawaguchi et al. (2010) reported on how Antarctic krill, within the Australian Antarctic Division krill incubation facility, were induced to school repeatedly over a period of a year. Similar to Strand and Hamner (1990), krill aggregated more tightly when kept within a white featureless background and responded as a group to dark nearby objects but not to less distinct objects (Fig. 8.3). Furthermore, simulated diel cycles had a profound effect on swarming behaviour, with tight schools forming during daytime but not at night. No schooling behaviour was observed when maintained at densities of 300 individuals within a 1170 L tank. Doubling this density, however, resulted in them swimming in loose groups, while ribbon-like schools occurred at high densities (9000 individuals in 1860 L or 4839 ind. m^{-3}).

The response of these krill to the introduction of a dense phytoplankton bloom appeared to go through a number of discrete phases. Firstly, the krill formed rapidly swimming schools which circled the food source. On encountering it, the schools broke up in a spiral formation and individuals started to filter feed. As the phytoplankton patch dispersed in the tank, krill distribution became random. Following this (and after a period of 15–30 min from the introduction of the phytoplankton), most of the krill started swimming in schools, at moderate to high velocities, while continuously filter feeding.

Analysis of NNDs in polarized schools with stereo cameras revealed minimal distances of ~150 mm (approximately three body lengths of separation) while NNDs were more variable in non-schooling krill. These NNDs are larger than the two body lengths measured by Catton et al. (2011), although it is unclear whether there was any size differences between the krill used in the two studies. Individuals swarm at around 20 cm s^{-1} within schools, but around 10 cm s^{-1} when not schooling.

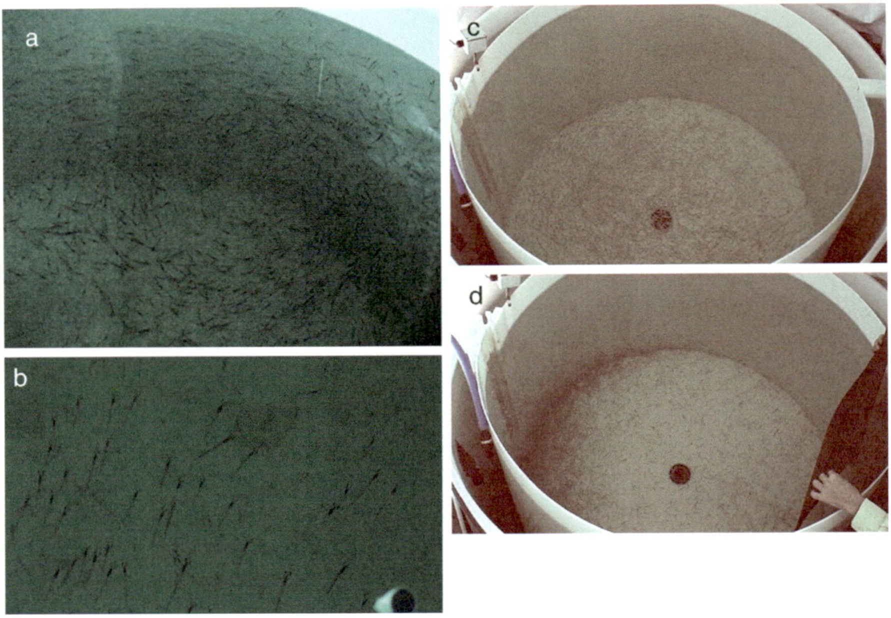

Fig. 8.3 Schooling behaviour induced in a laboratory: (**a**) A tight active circular school; (**b**) a close up of schooling krill; (**c**) random distribution before escape response; (**d**) escape response from a large black object (Kawaguchi et al. 2010, reproduced with permission)

A further interesting observation was that the schools swam in a figure of eight formation within the tanks when swimming in polarized groups. Such a formation was first described by Marr (1962) when observing schooling krill from a jetty in South Georgia. Therefore, despite the constraint of captivity, incubated krill are still able to display stereotypical behaviour patterns, which support the validity of other behavioural measurements.

8.4 Models of Behaviour and Swarming in Krill

8.4.1 Introduction

Behavioural modelling is an important tool for investigating swarm processes (Hofmann et al. 2004) and has also been one of the most rapidly growing research areas in krill science over the past 15 years. A wide range of modelling approaches have been taken over this time. We have partitioned these into generic model types (Table 8.2), although numerous models incorporate features that transcend our attempt at categorization. All the models described in this section have the ultimate aim of identifying how krill swarm, with a number also addressing the question as to why.

Table 8.2 Modelling approaches used to examine swarming and vertical migration behaviour in euphausiid species, particularly Antarctic krill

Authors	Model type	Parameters predicted	Input parameters and outcomes
Rule-based mechanistic models			
Hofmann et al. (2004)	2-D Lagrangian including density dependence	Swarm size, packing density	*Key parameters*: Food availability, detection distance
			Model outcomes: Food availability found to be a major driver in swarming
Zhou and Huntley (1996)	Reaction-diffusion	Swarm size, packing density	*Key parameter*: Constant of animal attraction derived from individual krill mass
			Model outcomes: Simulation of observed swarm structure
Grunbaum (1998)	Mechanistic rule-based social taxis	Time to detect and move along an environmental gradient	*Key parameters*: Detection distance, attractive-repulsive mechanism, alignment rule
			Model outcome: Identifies an optimum level of schooling dependent on spatio-temporal scales of environmental variation
Verdy and Flierl (2008)	Reaction-diffusion with an evolutionary social taxis element	Tendency to grouping or non-grouping behaviour	*Key parameters*: Density-dependent reproductive function, phytoplankton dynamics, predation pressure
			Model outcome: Swarming evolves when mixing occurs rapidly enough for resources to remain available to the clustered organisms
Individual trade-off models			
Alonzo and Mangel (2001)	Dynamic state-variable	Distribution between inshore and offshore habitats, feeding behaviour	*Parameters:* Temperature, food availability, predation risk
			Model outcomes: Shrinking predicted, even under positive energy budgets
Burrows and Tarling (2004)	Genetic Algorithm including density dependence	Depth, dispersion	*Parameters*: Light levels (predation risk), food distribution
			Model outcomes: Dispersion but not mean depth found to be density dependent
Cresswell et al. (2009)	State-dependent life history model	Depth, swarm density	*Parameters*: Food availability, predation risk
			Model outcomes: Vicinity of predator colonies drove krill deeper when food was low.

(continued)

Table 8.2 (continued)

Authors	Model type	Parameters predicted	Input parameters and outcomes
			Swarm density predicted to change little across the region
Willis (2007) and Willis (2014)	Dynamic state-variable. Willis (2014) includes nutrient recycling	Depth, abundance	*Parameters*: Predation risk from whales, food availability
			Model outcomes: Krill became shallower and more productive when whales were abundant
Predator-prey models			
Brierley and Cox (2010)	Risk-benefit trade-off	Swarm density, swarm shape	*Key parameters*: Oxygen availability, predation risk
			Model outcomes: Optimal swarm surface area to volume ratio of 3.3 m^{-1}
Olson et al. (2013)	Predator-prey evolution	Selective pressure to swarm	*Key parameters:* Sensory input (vision) and internal state, swarm-position based mortality
			Model outcomes: Whales exert an opposite selective pressure to penguins and seals, the latter being dominant in the evolution of krill swarms
Alonzo et al. (2003)	Dynamic optimisation	Habitat selection in penguins and krill	*Key parameters*: Size dependent vertical migration, penguin foraging rule, on-shelf-offshelf gradients in environmental conditions
			Model outcomes: prediction of non-linear effects of penguin-krill interactions matching observations
Cresswell et al. (2009)	Stochastic dynamic programming	Directional swimming, swarming density, depth	*Key parameters*: Food distribution, predator distribution
			Model outcomes: Krill retain favourable position at shelf-edge through increased turning
Deterministic behavioural models			
Godlewska (1996)	Statistical model	Vertical migration depth range and periodicity	*Key parameters*: Depth-discrete net catches and acoustic records
			Model outcomes: Vertical migration has a periodicity of 12 and 24 h, both depth range and periodicity are affected by food availability and individual body size and maturity

(continued)

Table 8.2 (continued)

Authors	Model type	Parameters predicted	Input parameters and outcomes
Wiese (1996)	Schematic model	Positioning of individuals within krill schools	*Key parameters*: Water disturbance pattern behind forerunning krill
			Model outcomes: Optimal positioning relative to polarised neighbours to maximise efficiency of swimming stroke
Catton et al. (2011)	Flow-field time-series	Distance of hydrodynamic disturbance from individual krill	*Key parameters*: Inter-individual distance within schools, pattern of flow field generated behind free-swimming krill
			Model outcomes: Hydrodynamic disturbances are viable cues for conspecific communication within schools
Murphy et al. (2013)	Flows modelled by actuator disk theory	Energy requirement for hovering	*Key parameters*: Flow speeds around swimming-legs and in the wake of individual krill
			Model outcomes: Drag-based paddling was found to be both an efficient and highly manoeuvrable form of locomotion

8.4.2 *Rule-Based Mechanistic Models*

One approach to investigating the processes controlling swarming and schooling in krill is to consider the consequences of mechanistic rules of attraction and dispersion. The study of Hofmann et al. (2004) presents a good type example of the simulation of krill swarming through a rule-based mechanistic approach. They developed a two-dimensional Lagrangian particle model with the location of particles (representing individual krill) being determined at each time step by random diffusion, foraging activity and movement induced by the presence of neighbours (which can be sensed within a certain range). There was also a foraging rule implemented in which krill turn more frequently in areas of high food concentration. The model included a density-dependent term, meaning that individuals were repulsed from each other when the density was high, and more attracted to each other when it was low. Individuals were also removed from the model by an indiscriminate predation mortality function. The model showed that the formation of swarms can occur rapidly as a product of the way in which individuals forage. High food conditions were found to result in the formation of only small, low density swarms because of the need for rapid turning and slowing to sample the environment. The largest swarms, with the highest density of animals, occurred when food was low. The model also predicted that swarms were less likely to form when the ability to sense neighbouring individuals was reduced. The study succeeded in simulating a number of traits that have been observed in the field. For example, Hamner et al. (1983) proposed that their observations of krill feeding conformed to a pattern of area-intensive searching followed by rapid feeding, which is reflected in model simulations where swarms search in larger groups and break up into smaller groups when a food source is located. Krill found under-ice do not commonly occur within large swarms (O'Brien 1987; Frazer et al. 1997), which the present model would predict to be the result of the impedance that ice causes in sensing neighbouring krill.

Other models which have taken a rule-based mechanism approach to simulating krill swarming patterns include Morin et al. (1988) and Azzali et al. (1999) which are both based on reaction-diffusion equations in which there is a non-random force of attraction between individuals counteracting physical diffusive forces. Zhou and Huntley (1996) also modeled the effect of attractive forces on the structure of krill swarms through the development of what they termed 'the biocontinuum theory of attraction', where the attractive force in animal aggregations is proportional to the inverse of the distance squared. They applied their model to acoustic measurements made on an Antarctic krill aggregation, in which abundance was found to increase monotonically towards the centre, as predicted by their model.

Grunbaum (1998) used a mechanistic rule-based approach to examine the utility of schools as a means of achieving efficient movement along preferred environmental gradients. The model contained three basic rules: (1) decreasing probability of detection or responsiveness to neighbours at large separation distances, (2) a social response that includes some sort of switch from attractive to repulsive interactions with neighbours mediated by either separation distance or local density of animals, and (3) a tendency to align with neighbours. Results of simulations showed that schooling individuals moved in an up-gradient direction more rapidly than asocial searchers. The study illustrated that averaging decisions among a large number of individuals at one instant can substitute in part for averaging many decisions by a single individual over space and time, so that each member of the group may arrive at the right decision more quickly and with greater accuracy than it would in isolation.

8.4.3 Individual Trade-Off Models

Adopting a swarm formation can be viewed as a trade-off between a number of benefits and costs. In terms of the benefits of swarming, there is the reduction in predation risk and the location of new resources, while, in terms of costs, resources must be shared, oxygen levels may be depleted and the risk of infection from parasites and diseases increased. Some, if not all, of these benefits and costs have been used in models to predict swarm structure, particularly in considering how optimal patterns of behaviour and swarming may alter as a result of changes in both external factors and the internal condition of individuals.

Cresswell et al. (2009) provides a type-example of such a trade-off model. This study predicted how Antarctic krill responded to changes in food availability (using satellite derived Chl-a as a proxy) and predation risk (in terms of distance from the nearest predator colony) as they travelled along predetermined trajectories through the Scotia Sea. Individual krill could respond to these external changes in two ways, firstly, through altering their vertical position in the water column and, secondly, through altering distance to their nearest neighbour i.e. swarm packing density. Individuals migrating deeper and packing closer experienced lower predation risk but at the cost of lowered feeding success. By contrast, individuals moving closer to the surface and becoming more dispersed achieved a greater reward from feeding but at a greater risk. The model was run for three different size classes of krill to consider the influence of life-stage on the trade-off. A state dynamic optimization framework was used to predict the optimal response to each set of conditions (Mangel and Clark 1988). The model predicted that, in the vicinity of predator colonies, krill would adopt deeper swarms, particularly when food availability was low. Optimal depth was closer to the surface in smaller krill in food scarce situations, as a result of them being more able to afford the cost of reduced food intake. Swarm packing densities were not predicted to be very responsive to environmental conditions. Model predictions compare favourably to acoustic survey observations which show krill occur deeper in predator-rich shelf environments (e.g. the Antarctic Peninsula) compared to the open ocean, where predator densities are lower (e.g. Klevjer et al. (2010; Sect. 8.5.3.2). However, the lack of responsiveness in swarm packing concentrations in the model does not match observations (e.g. Tarling et al. 2009), which probably reflects that the costs and benefits of this trait were not fully captured within the model formulation.

Alonzo and Mangel (2001) used state dynamic optimization modelling to determine optimal trade-offs in krill growth and behaviour. Physiological and environmental parameters were integrated into a model used to predict krill size and distribution between habitats (inshore, offshore-deep and offshore-shallow). In the presence of predation risk when feeding, model krill experience a trade-off between growth and survival and shift their feeding behaviour and habitat distribution accordingly. The model predicted that krill will shrink when experiencing extreme temperatures or food deprivation because the energy budgets become negative. However, more unexpectedly, it also predicted that shrinkage may be an optimal choice when the energy budget was positive, since larger krill can be seen from a greater distance than smaller krill, and the according risk from visual predators is decreased through shrinking. Whether krill have the capacity to shrink in direct response to increased visual predation threat awaits empirical testing.

Willis (2007, 2014) used a state dynamic optimization model to examine the effect of whale predation on krill vertical migration behaviour, particularly to determine whether krill productivity was affected by the impact of whaling. The models were run for two scenarios, one where whales were present and the other where they were absent. In the former, predation risk during daytime was greater in the deep than in the surface while, in the latter, predation risk during daytime was greater in the surface than in the deep. In Willis (2014), there was also a whale-induced enhancement of nutrient levels, which increased food productivity levels at the surface, so increasing the level of reward. The models predicted that krill would spend a greater time in the surface layers when whales were present, which increased krill growth rates. This results in krill population abundance being enhanced by the presence of whales, despite their increased predatory impact.

Burrows and Tarling (2004) used a genetic algorithm modelling framework to examine the factors that contribute to the dispersion of krill around a mean depth. Specifically, they considered if density-dependence could drive different members of the population to choose different depths on a dynamic basis. The model predicted diel vertical migration behaviour based on internal state (i.e. energy reserves), risk of predation and location of conspecifics, with the parameterization being taken from studies on Northern krill in the Clyde Sea area. In being dynamic, the modelling framework allowed for optimal policies to be responsive to changing circumstances through time. For example, the depth of maximum food availability may become less optimal if it is also occupied by many other conspecifics, making alternative depths preferable. The model found that increasing density dependence in feeding success did not affect the mean depth chosen at night, but did increase the spread of the population.

8.4.4 Predator-Prey Models

Predator-prey models, in which both the prey and the predator respond to each other, have yet to be extensively used in predicting swarm behavioural patterns. However, the few examples that have been published provide useful insights into the further complexities that influence behavioural and swarming patterns. Alonzo et al. (2003) simulated a spatial ecological game between penguin predators and krill prey. The model predicted habitat selection in penguins and krill during the period of land-based reproduction for penguins. The model penguins were allowed to use one of two optimal foraging strategies: (1) maximization of food intake or (2) minimization of foraging effort. Krill could move between inshore or offshore habitats and could migrate between surface and deep strata when offshore. The model found size-dependent diel vertical migration in krill had one of the strongest effects on penguin foraging patterns. However, it was also apparent that the model was capable of predicting non-linear effects that help explain reported relationships between krill abundance and penguin reproductive success (Boyd and Murray 2001).

Olson et al. (2013) considered the evolution of swarming behaviour as a response to attacks from predators. The study examined the selfish herd hypothesis (Hamilton 1971) which states that concentrated groups arise because prey selfishly attempt to place their conspecifics between themselves and the predator, thus causing an endless cycle of movement towards the centre of the group. Within this context, there are attack modes that select for and against swarming behaviour in their prey. Blue whales (*Balaenoptera musculus*) are known to dive into the densest areas in swarms of krill, consuming hundreds of thousands of krill in the middle of a swarm in a single attack. Such a mode of attack (called a high-density area attack) clearly selects against swarming behaviour because it targets the prey that swarm the most. The study therefore questions why krill evolved swarming behaviour at all. The proposed explanation is that krill swarms are also fed on by smaller species (seals and penguins) that consistently attack krill on the outside of swarms. In effect, krill swarms experience two forms of attack simultaneously, high-density area attacks from whales and outside attacks from penguins and seals. Thus it is possible that the outside attacks outweigh the selection pressure to disperse from the high-density area attacks. Simulations showed that swarming behaviour evolved in krill only when the high-density area attacks were too infrequent relative to outside attacks.

Brierley and Cox (2010) also used a selfish-herd approach in designing a model to explain why swarms of krill (and fish) have a relatively invariant surface area to volume ratio (R). Because individuals favour an interior position to minimize their risk of predation, predation pressure drives swarms to adopt spherical shapes. However, countering this is the limitation of oxygen availability within the shoal interior. The authors calculated a metric they defined as a benefit factor (B) which was a function of the number of animals, the number at the edge of the swarm, and the time to oxygen depletion. The value of B was calculated for a number of different geometric shapes, varying in R. The model predicted that the maximal B was attained when R was ~3.3 m^{-1}, which agrees with their empirical analyses of wild krill swarms. It was concluded that swarms with R ~ 3.3 m^{-1} achieve an optimal trade-off between predator avoidance and oxygen acquisition.

8.4.5 Deterministic Behavioural Models

These types of models describe observed swarming patterns in a statistical manner. Although not predictive, they provide a useful perspective on how krill behave and organize themselves. Two examples where this approach has proved valuable are in the consideration of vertical migration and internal swarm structure.

Vertical migration in Antarctic krill appears to be highly variable across a wide range of temporal and spatial scales. From the analysis of a large body of data, Godlewska (1996) revealed several universal patterns. Firstly, that the majority of vertical migrations were within the range of 0–100 m, with the maximum depth being dependent on a number of biotic and abiotic factors. Secondly, migrations occur at a cyclicity of either 12 h or 24 h. When food is scarce, migrations are shallower and have a 12 h cycle, but revert to a 24 h deeper migration cycle when food is abundant. Individual size was also seen to have an influence, with adults more likely to exhibit a 24 h migration cycle over large amplitudes, while juveniles have a greater tendency to have 12 h cycles and shallower migration amplitudes.

Within krill swarms, Wiese (1996) considered how propulsion jets are generated by individuals during swimming and determined the optimum organization within polarized schools to take advantage of these jets. Turbulent fringes of neighbouring jets meet at some distance behind and below each rank of krill and, at these meeting points, the jet flow either becomes stationary or moves in narrow circles. This provides each swimmeret with a solid abutment, which is more efficient than if producing a powerstroke within the fast moving water of the jet from the forerunner. Wiese (1996) concluded that individuals within krill swarms should organize themselves in a staggered formation if they aim to optimize swimming efficiency. Catton et al. (2011) used infrared particle image velocimetry measurements to analyse the flow field of coordinated groups of three to six *E. superba* and found that individuals avoided the region of the strongest flow generated by their neighbours, as predicted by Wiese (1996).

8.5 Temporal and Spatial Patterns in Swarming

8.5.1 Introduction

Murphy et al. (1988) introduced the concept of scale in characterizing the range of interactions that influence krill behaviour and swarm structure. Abiotic and biotic influences were regarded as forming hierarchies of different scales, with each scale being relevant to a different type of process within krill biology. For instance, turbulent mixing processes, which act over minutes to an hour and up to tens of metres in space were considered to be most relevant to individual-level responses. Turbulent diffusion (hundreds to thousands of metres, days to weeks) operates at scales likely to have most impact on swarm structures. Features such as fronts, eddies and areas of upwelling can lead to concentrations of swarms at scales of tens to hundreds of kilometers over periods of weeks to months. This hierarchical view of how krill interact with their environment still underlies how modern-day krill research is carried out, from small-scale process studies, to mesoscale surveys, to analyses of large-scale patterns of aggregation. The link between behaviour and scale (both temporal and spatial) therefore remains key to understanding behaviour and swarming in krill.

8.5.2 Temporal Scales

8.5.2.1 Swarm Fidelity

Marr (1962) posited that the similarity in size- and age-structure in swarms is a result of them being comprised of cohorts of shared origin. The alternative is that the pattern is a passive product of the behaviour and the swimming capabilities of swarm members. The issue was tackled by Goodall-Copestake et al. (2010) who determined the genetic diversity of nine spatially separated Antarctic krill swarms, through sequencing the population marker (barcode) cox1. Analysis of molecular variance did not show any significant genetic structure, with the genetic variance between individuals within swarms being as great at the variance between swarms. The implication is that swarms do not constitute discrete genetic units and that there is mixing between swarms on both small and large scales.

Hamner et al. (1983) provided an evocative example of passive processes causing sorting within swarms. Divers manipulated a small school with a bimodal size distribution to swim faster but not panic which resulted in larger faster-swimming krill moving to the front of the school and elongating it until it pinched in half. At a much larger scale, Farber-Lorda et al. (2009) used a Differentiation Index (DI) to separate out krill according to morphometric parameters over a sampling grid that was 20° in longitude and 5° in latitude. They found that small males were separate from larger males and argued that this reflected the differing swimming capacities of the different DI stages.

Section 8.3.2 discusses how the different swimming capabilities of individuals in different moult stages and of different sexes can lead to passive sorting within and between swarms (Johnson and Tarling 2008). Spawning may also be a further means of such passive sorting. Tarling et al. (2009) recorded the process of females extruding their eggs from their thelycum. The eggs were released in small batches as part of a cyclic pattern in which the swimming legs varied in beat rate. The entire batch of eggs was released over a period of up to 10 h. The change in swimming behaviour that was adopted by the spawning females is likely to impact their overall swimming performance and lead to dissociation from non-spawning conspecifics. Such behaviour may lead to the spawning swarms reported by Mauchline (1980) and Hamner (1984a, b).

8.5.2.2 Diel Periodicity

Most diel cycles are coordinated by the light cycle. Onsrud and Kaartvedt (1998) reported that euphausiids can see in a light field as low as 10^{-6} μ Einstein $cm^{-2}\,s^{-1}$, which would enable krill to detect daylight to depths of 250 m in typical Southern Ocean conditions (Zhou and Dorland 2004), well below their typical vertical migration range of around 100 m (Godlewska 1996). Nevertheless, the difficulty of maintaining diel periodicity in high latitude environments is that the day/night cycle seasonally shifts from extremes of almost constant darkness during winter to continual daylight during summer. Against this background, krill must coordinate metabolic and physiological processes alongside those of feeding and digestion, which cannot all occur simultaneously. It has also been hypothesized that krill must disperse at regular periods in order to flush out waste products, parasites and pathogens and replenish oxygen concentrations (Kawaguchi et al. 2010). For these reasons, it has been inferred that krill must operate an endogenous clock (see also Chap. 4, Meyer and Teschke 2016).

Gaten et al. (2008) carried out behavioural experiments in which Antarctic krill were maintained in activity-monitors and exposed to either constant darkness or a light-dark cycle. It was found that krill has complex rhythms made of two circadian components, one shorter than 24 h and one longer than 24 h. Also noteworthy was the observation that krill did not display a robust 24 h rhythm even under a light-dark cycle, which suggests that light may not be a dominant coordinating factor (*Zeitgeber*).

Both Mazzotta et al. (2010) and Teschke et al. (2011) investigated the molecular aspects of endogenous clocks in Antarctic krill. Mazzotta et al. (2010) found krill to contain the clock-gene cryptochrome (*cry2*), a cardinal component of the clock-work machinery in several organisms. The expression of mRNA of this gene in krill captured in their natural environment at different times of day was found to show periodicity, with peak levels occurring at around 06:00. Nevertheless, this periodicity was not mirrored in the corresponding protein. Teschke et al. (2011) incubated krill under different light regimes and measured the expression of *cry2*. Gene expression was found to be highly rhythmic both in a light-dark cycle and in constant darkness, with the rhymicity under constant darkness having a period of around 18 h. The authors speculated that the circadian system of krill has to delay for several hours every day to keep a stable phase of entrainment in the light-dark environment. This may be an adaptation to allow krill to entrain to the wide range of light regimes that they experience at high latitudes.

8.5.2.3 Seasonality

Individual response At the individual level, Antarctic krill has been reported to alter activity levels according to time of year. Reduced metabolic rates during winter have been observed in Antarctic krill in a number of different areas (Morris and Priddle 1984; Quetin and Ross 1991; Atkinson et al. 2002; Meyer et al. 2002; Cullen et al. 2003). Nevertheless, it remains unclear whether a seasonal change in metabolic rate simply reflects a change in ingestion rate, or alternatively, a more fundamental switch to a different physiological state, cued by external or internal factors (Meyer et al. 2002; Meyer 2012). Evidence for the former is presented by Ikeda and Dixon (1982) who showed krill starved for 24 h lowered their oxygen uptake to 63 % of that of wild specimens. In support of the latter, Atkinson and Snyder (1997) and Atkinson et al. (2002) found that autumn krill did not raise their relatively low feeding rate when placed in food-saturated conditions, which indicated that they were in a state of quiescence (see also Chap. 4, Meyer and Teschke 2016).

Switching to a different physiological state between seasons suggests that a trigger mechanism exists in order to time the switch. Meyer et al. (2010) proposed that entry and exit from overwintering quiescence is triggered by the photoperiod or light level (see also Chap. 4, Meyer and Teschke 2016). In incubated Antarctic krill, it has been shown than the light regime triggers the onset of reproductive maturation (Hirano et al. 2003; Teschke et al. 2008) and alters the rate of oxygen consumption and feeding (Teschke et al. 2007). Seear et al. (2009) showed that gene expression patterns may change within 7 days of being exposed to an altered light regime.

In wild caught Antarctic krill, gene expression studies by Seear et al. (2010) showed that there may be considerable flexibility in seasonal activity levels of krill according to location. At the Antarctic Peninsula, there was evidence of greater activity in summer compared to winter. Genes involved in feeding and digestion, respiration, motor activity, immunity and egg-production were up-regulated (more highly expressed) in summer (see also Chap. 7, Jarman and Deagle 2016). When comparing overwintering krill from the Peninsula (60°S) to those at South Georgia (54°S), the more northerly krill exhibited up-regulation in feeding and digestion, and immunity, indicating that they were more actively feeding. Nevertheless, there was no differential expression between the two locations in respiration, motor activity or vitellogenesis, indicating that individual activity levels were similar. It was concluded that, despite differences in feeding, both populations of overwintering krill were in a similarly quiescent state.

Swarm response Siegel and Kalinowski (1994) proposed that there was little evidence of swarming during the overwintering period and that much of the krill biomass that lives beneath the ice is dispersed. However, a number of studies have since reported extensive swarming during the winter period (Lascara et al. 1999; Taki et al. 2005; Lawson et al. 2008). Lascara et al. (1999) observed a cyclical change in dimensions and density of Antarctic krill aggregations with the smallest, densest aggregations being found in the summer, while the tendency towards autumn and then winter was for aggregations to become larger and less dense. Lawson et al. (2008) considered the distribution of swarms over two consecutive winters and found a large difference between years with biomass levels being two orders of magnitude greater in the winter of 2002 compared to that in winter 2001. From using an acoustic device towed at depth (BIOMAPER-II), they showed that the majority of krill biomass in winter occurred below a depth of 100 m, much deeper than at other times of year.

Fielding et al. (2012) compared the dimensions and densities of krill between seasons and found krill density was similar in spring (27.7 g m^{-2}) and summer (27.2 g m^{-2}) but lower (2.2 g m^{-2}) in autumn. During spring, the high mean density resulted from a small number of swarms containing a large number of small krill while, in summer, the high mean krill density resulted from a large number of swarms containing a low number of large krill. In autumn, the low density resulted from a small number of swarms containing a small number of large krill. Krill swarms tended to become deeper over the course of the season, being located at a mean depth of 58 m in spring, 94 m in summer and 123 in autumn. Similarly, in an analysis of Japanese fishery data by Taki et al. (2005), it was found that the average trawling depth tended to be shallower during austral summer and early autumn, deepened gradually from autumn and reached maximum depth in winter, before shifting to shallower depths again in early spring. Siegel (2012) did not consider depth changes between seasons, but did report a large drop in krill abundance between 0 and 200 m at night compared to day during autumn and winter in the Lazarev Sea. No such diel shift in abundance was found during summer in this region.

Flores et al. (2012) investigated the seasonal association of Antarctic krill with the underside of sea-ice. In winter, Antarctic krill were absent from the ice-water interface by day but were found there in high densities at night, indicating that the association with the underside of sea ice may be limited to the dark hours in winter. However, during the summer, the opposite pattern prevailed. Therefore, seasonal switches may not only be evident at the individual level but also at the level of the aggregation.

8.5.3 Spatial Scales

8.5.3.1 Horizontal Patchiness

One of the earliest approaches to determining horizontal patchiness in krill swarms was carried out by Weber et al. (1986). They applied power spectral analysis to continuous horizontal records of surface seawater temperature, surface in vivo fluorescence and acoustically derived estimates of integrated krill density. Over the range of 4–20 km, variability in phytoplankton biomass was largely determined by physical processes. By contrast, there was little difference in the level of horizontal variability in krill biomass between large and small scales, which implies that physical processes are of lesser importance to determining scales of patchiness in krill distributions compared to other plankton. The study was important in demonstrating social behaviour as being the main driver of variability in small-scale density distributions of Antarctic krill, and was the basis of a number of subsequent studies that modelled swarming behaviour in krill (see 8.4).

Folt and Burns (1999) considered the main behavioural mechanisms that can result in zooplankton patchiness as being diel vertical migration, predator avoidance, food finding and mating behaviour and each of these has been considered in other sections of the present review and previous reviews of Antarctic krill behaviour. A further addition to this list is the influence of ontogenetic stage because the ratio of the costs to benefits of swarming change as individuals grow in size and mature, as does their ability to maintain tighter, less random aggregations through improved swimming capabilities. Decima et al. (2010) demonstrated this concept in other species of euphausiids. They measured patchiness using a count based statistic applied to an 11 year time series of euphausiid densities in the California Current System, including size-specific enumerations. The most common pattern was a U-shaped curve, showing elevated patchiness in the smallest size classes, a rapid decrease in patchiness in intermediate sizes and increased patchiness in adults at the onset of reproductive maturity. The initial descending limb of the curve appears to be caused by turbulent diffusion while the later ascending limb was consistent with the onset of predator-induced aggregation behaviour.

The influence of predation on patchiness in Antarctic krill was investigated by Reid et al. (2000) who examined the level of association between baleen whales and krill under a number of spatial scales in survey regions close to South Georgia. They found a positive relationship between whale abundance and mean krill density at the largest spatial scale (80 × 100 km) but, at progressively smaller scales, the relationship weakened due mainly to the increase in frequency of areas of high krill density where whales were not recorded. In particular, whales were absent from inshore areas that had high krill mean densities. This observation was similar to that reported by Greene et al. (1994) when comparing the distribution of chinstrap penguins to Antarctic krill distributions within a small scale grid near the South Shetland Islands. They found that penguin surface distributions did overlap with areas of moderate krill abundance but never overlapped with areas of highest krill abundance. It was suggested that the penguins did not hunt where krill were most strongly aggregated since coordinated schools are effective at avoiding predation, making them a comparatively unrewarding target on which to forage (see also Chap. 9, Trathan and Hill 2016).

Brierley and Cox (2015) provided a different perspective on horizontal patchiness in Antarctic krill through comparing the number and size of schools between years with different overall stock-biomass levels. They found that, in years of low stock-biomass levels, school sizes remain approximately the same but the overall number of schools declines. In other words, individuals were distributed among fewer schools and not smaller schools. It implies that inter-swarm distances increase in low biomass-stock years, increasing the scale of patchiness. The pattern may be further influenced by interannual fluctuations in body-size distributions given that both Tarling et al. (2009) and Fielding et al. (2012) found there to be a strong relationship between swarm size and individual body size, with many of the largest swarms being comprised of young, small krill. In years of strong recruitment and high stock-biomass, the proportion of smaller krill increase (Fielding et al. 2014), so increasing the frequency of larger swarms. Therefore, in good recruitment years, average swarm size is likely to increase. Brierley and Cox (2015) did not cover years of strong recruitment since krill biomass was in a low phase during the time of their study (Fielding et al. 2014). The effect of high biomass levels and recruitment on horizontal patchiness in Antarctic krill is a subject worthy further investigation.

8.5.3.2 Influence of Habitat

The interactions of krill with both sea-ice and the sea-bed have been prominent in krill studies over the past 15 years. Both habitats provide food resources and, in the case of sea-ice, a place of refuge from predators. The influence of these habitats on krill behaviour and swarming is profound, although responses vary according to geographic region. With regards sea-ice, we refer here specifically to those aspects which affect swarming and behaviour, while a wider consideration of krill interactions with sea-ice is covered in Chap. 2 (Siegel and Watkins 2016).

Sea-ice habitat Brierley and Watkins (2000) compared swarm characteristics in areas within and outside of ice cover in the Bellingshausen, Weddell and Scotia seas. They found no significant differences between horizontal and vertical extents of swarms or inter-swarm distance in ice-covered or open waters. This differs from Zhou and Dorland (2004) at the Antarctic Peninsula, who found smaller-size aggregations with a higher abundance in the ice free region and larger-size aggregations with a lower abundance in the ice covered region.

Brierley and Watkins (2000) did find a difference between ice and non-ice habitats in swarm mid-depths, being generally shallower under ice than in open water in the Weddell and Scotia Seas. However, the opposite pattern prevailed in the Bellingshausen Sea. This was explained as being a result of the different types of ice between regions, with the Weddell and Scotia Sea containing multiyear ice, more likely to contain food and be able to provide refuge than the first year ice present in the Bellingshausen Sea. The older ice was also thicker and may attenuate more light leading to shallower distributions.

Beyond the edge of the sea-ice, Brierley et al. (2002) found a concentrated band of Antarctic krill 1–13 km in from the ice-edge through observations made by the autonomous underwater vehicle *Autosub-2*. Krill densities were significantly higher under the sea-ice than in adjacent open water. The under-ice swarms were particularly associated with the presence of keels and other topographic features, which is consistent with scuba-diver observations (Siegel et al. 1990). Within ice distributions were also investigated by Flores et al. (2012) using a Surface and Under Ice Trawl (SUIT) which sampled the 0–2 m surface layer under ice. They found a year-round association of Antarctic krill with the under-ice habitat, hundreds of km into the ice-covered area of the Lazarev Sea. Within-ice locations may represent a trade-off between productive ice-edges and refuges from air-breathing predators unable to dive through the ice (Brierley et al. 2002).

Bathymetric gradients The effect of the offshelf-onshelf gradient on swarming behaviour was clearly illustrated by Klevjer et al. (2010) from acoustic surveys carried out in the Scotia Sea. They found cross-sectional areas of swarms were significantly larger inshore with a mean value of 120 m^2 in a zone 0–50 km from shore, compared to <80 m^2 further offshore. The packing concentrations of krill within inshore swarms was also significantly greater, with an average density of 95 ind. m^{-3} compared to between 24 and 31 ind. m^{-3} elsewhere. A significantly greater number of large, biomass-rich swarms occurred in the 0–50 km zone compared to all other zones. Similarly, Lawson et al. (2008) found highest krill biomass was located in regions close to land in surveys at the Antarctic Peninsula carried out in autumn and winter, while surveys in the same region during summer by Cox et al. (2011a) reported mean densities of swarms to be higher on-shelf than off-shelf. Cox et al. (2011a) proposed that the higher densities were a response to greater levels of predation risk, particularly in reducing detectability to visual predators, given that mortality from land-based predators increases closer to shore (Murphy and Reid 2001). In East Antarctica, Pauly et al. (2000) found that krill aggregations were most frequently encountered on the shelf break region at the convergence of the summer ice-edge, the

1000 m isobath and the Antarctic slope current. They surmised that the association of krill swarms with continental shelf and shelf-edge environments was a response to the presence of semi-permanent phytoplankton blooms at these sites (Korb et al. 2004; Pollard et al. 2007, 2009).

8.5.3.3 Super-Sized Swarms

The term 'superswarm' was originally applied to Antarctic krill by Cram et al. (1979) when describing his results from a krill survey aboard the *SAS Protea*. Superswarms were described as being "swarms of large horizontal dimension (of the order of thousands of metres or greater) and perhaps more important, very high densities". Although the term has since been used to describe swarms of various sizes, it is manly applied to aggregations that stretch for several kilometers in length, tens of metres in depth and contain concentrated rather than dispersed densities of krill (Lascara et al. 1999; Lawson et al. 2008; Tarling et al. 2009; Nowacek et al. 2011; Fielding et al. 2012).

The subject of superswarms has been reviewed several times (Miller and Hampton 1989; Siegel and Kalinowski 1994; Watkins 2000), describing the many locations in which swarms have been reported and proposed mechanisms leading to their formation. Here, we will describe two significant studies that have advanced our understanding of superswarms since these reviews.

In a multidisciplinary survey of the Scotia Sea, Tarling et al. (2009) considered the factors that could predict the occurrence of superswarms. The study found that superswarms made up less than 1 % of all swarms but accounted for more than 56 % of all biomass within the survey region. Analyses showed that superswarms occurred more frequently in regions where surface fluorescence levels were low, which may reflect a need to swarm in order to search efficiently for further phytoplankton patches. Superswarms were also most likely to be found in regions dominated by smaller and immature krill. Larger krill have greater energetic demands than smaller krill, and the lack of superswarms containing large krill may be a result of their need to break-up into smaller feeding swarms more frequently.

Zhou and Dorland (2004) considered what may limit the maximum size of superswarms of Antarctic krill. In their study, they compared horizontal scales of Antarctic krill aggregations with horizontal gradients of currents. Zhou and Dorland (2004) estimated a parameter termed the critical horizontal scale beyond which the swimming capacity of krill will not be able to counteract the shear currents acting to disperse the aggregation. This scale is determined by the internal Rossby Radius, which is around 5 km in the Southern Ocean (Huntley and Niiler 1995). In their study area, they estimated the critical horizontal scale to be in the order of 4 km, assuming krill swimming velocities of 15 cm s^{-1}. This scale was comparable to the horizontal scales of krill aggregations in their study region (Laubeuf fjord, Marguirite Bay) of between 1 and 6 km. Therefore, shear currents at scales acting on the outer limits of superswarms most likely limit their maximum size.

8.6 Field-Based Studies of Krill Behaviour into the Future

The challenge for understanding social behaviour in pelagic organisms such as Antarctic krill is in developing observing systems that can resolve (1) an individual's behaviour within a group and the cues that affect these, (2) multiple aggregations over large spatial scales simultaneously, repeatedly and the cues that cause them and (3) aggregations in difficult-to-sample regions such as near surface, near sea-bed, under ice or at depth. Here, we review recently developed observing systems that can advance our ability to address these challenges.

8.6.1 Multibeam Echosounders (MBES)

Initially developed for seafloor mapping, MBES now enable us to image a synoptic slice of the water column (Mayer 2006) and are attractive for behavioural studies by simultaneously ensonifying whole aggregations of fish or krill as well as discriminating and tracking the position of individual targets within the swath (Mayer 2006; Colbo et al. 2014). A key strength of MBES is the ability to rapidly encounter hundreds of schools and identify school shape parameters (e.g. length, width, height, surface area, volume and ratios thereof) and composition parameters (e.g. number and position of nuclei, number of vacuoles). This can be achieved whilst external parameters are measured (e.g. depth, distance to thermocline or fluorescence maximum, height above bottom, predator presence/absence), with the ability to gain a better understanding of the drivers of individual and group behaviour (Axelsen et al. 2001; Mayer et al. 2002; Gerlotto et al. 2006; Benoit-Bird 2009; Colbo et al. 2014). MBES systems have been used by De Robertis et al. (2003), Cox et al. (2009, 2010, 2011b), and Brierley and Cox (2010, 2015) and have already provided unique insights into krill swarm structure and individual behaviour.

8.6.2 Autonomous- and Remotely Operated Underwater Vehicles (AUVs and ROVs)

AUVs are now becoming routinely deployed to undertake physical and biological measurements. There are two main types, buoyancy-driven gliders, and propeller vehicles (Griffiths 2003; Rudnick et al. 2004; Johnson et al. 2009). These AUVs

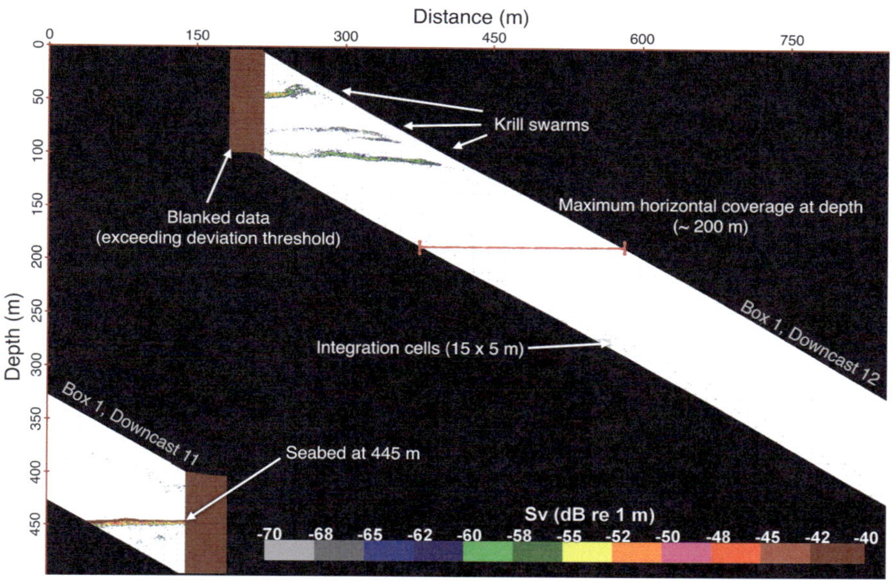

Fig. 8.4 Georeferenced and depth-corrected glider-derived volume backscattering strength (S_V, dB re 1 m^{-1}) at 120 kHz showing krill swarms, seabed, blanked data and undulating coverage (Guihen et al. 2014, reproduced with permission)

profile the oceans typically to at least 1000 m, although new developments have increased this capacity to 6000 m (Furlong et al. 2012; Osse and Eriksen 2007). A key strength of the AUVs is to make observations in hard-to-access areas and for temporal periods beyond the scope of expensive research vessels. Brierley et al. (2002) used the propelled vehicle Autosub-2 to make continuous high-resolution acoustic measurements of krill under ice, reaching 27 km beyond the ice edge (Sect. 8.5.3.2). Autosub-2 was also used to investigate behavioural responses of krill to survey vessels (Brierley et al. 2003).

More recently, Schofield et al. (2013) and Guihen et al. (2014) have used simple echosounders mounted on underwater gliders to observe krill layers and swarms at the Antarctic Peninsula (Fig. 8.4), with the latter presenting observations to depths of 1000 m. Gliders can be deployed for several months, travelling at speeds of 0.25 m s^{-1}, and are ideal platforms to quantify krill behaviour and aggregations at the locations and scales at which predators operate (Kohut et al. 2014). A fleet of gliders equipped with acoustic instruments could enable synoptic observations over large scales (Greene et al. 2014).

ROVs, tethered to a ship, have a more restricted range than AUVs, but can be piloted to target specific features of interest identified in real-time (Amon et al. 2013). They have a relatively large payload capacity and, powered from the ship, can carry high specification sonar and camera systems. Using an ROV, Clarke and Tyler (2008) observed live gravid female krill feeding on phytoplankton at depths up to 3500 m, which have challenged our view of krill as an epipelagic species (Brierley 2008).

8.6.3 Camera Systems

Modern digital technology has provided a step-change in our ability to image the sea optically: with increased performance in illumination and recording, the availability of more sophisticated processing methods, miniaturisation of technology and a decrease in the cost of the equipment (Jaffe 2013). The increasing use of towed and lowered camera systems to observe krill at the seabed was reviewed by Schmidt et al. (2011) and provided evidence of the importance of the seabed as an alternative habitat for krill.

Recently developed animal-borne miniaturised camera/video technology have presented a unique way of studying prey distribution and predator-prey interactions, and these cameras have been deployed on seals and penguins (Fuiman et al. 2002; Hooker et al. 2002, 2008; Watanabe and Takahashi 2013). Takahashi et al. (2008) observed krill swarms in both the pelagic and benthic realms. They showed that penguins swam just above the sea floor and foraged on krill in a downward direction, so using the sea-floor to trap the krill. Watanabe and Takahashi (2013) also used penguins as platforms to observe krill swarms under sea ice.

8.6.4 Long-Ranging Systems

The ability to detect and measure krill swarms instantaneously over large spatial areas remains one of the ultimate challenges in krill science. Two emerging technologies could help achieve this goal. LIDARs (Light detection and ranging) have been used to detect and describe the density and shape of fish schools and zooplankton (Squire and Krumboltz 1981; Churnside and Thorne 2005; Carrera et al. 2006). LIDAR is an optical remote sensing technology that detects and ranges on targets in the surface 100 m of the water column (Liu 1990). It can be either airborne, collecting large amounts of data over a large area, or shipborne, collecting the same large volume of data but over a smaller area. Churnside and Thorne (2005) and Churnside (2007) showed that zooplankton layers and relative abundance could be successfully determined using LIDAR. Ocean Acoustic Waveguide Remote Sensing (OAWRS) offers a method for instantaneously imaging and monitoring marine populations over continental shelf-scale areas (Makris et al. 2006). Jagannathan et al. (2009) considered it possible to resolve Antarctic krill with this method, assuming that krill swarms are very dense ($>10^3$ ind. m^{-3}) and typically located in the surface 100 m of the water column.

8.7 Concluding Remarks

This review has focused on the work carried out on krill swarming over the past 15 years. Across these many investigative approaches, probably the only consistent thing to emerge is the high degree of variability in both swarm structures themselves and the behavioural responses that influence them. Antarctic krill exist in a highly heterogeneous environment, with rapidly advancing and retreating ice-shelves, variable bathymetry, patchily distributed food and dynamic advective forces. Their distributional range covers almost 30° of latitude (Atkinson et al. 2008), entailing strong gradients in light- and temperature fields. Temporally, they are part of an ecosystem that has been severely perturbed by past exploitation of their major predators (Laws 1977), in regions undergoing some of the most rapid increases in water temperature in the modern ocean (Meredith and King 2005; Whitehouse et al. 2008). Therefore, variability and change is a defining aspect of the environment in which Antarctic krill exist.

Yet, the phenomenal success of Antarctic krill, as one of the most biomass-rich species on Earth, may not be despite of this variability, but because of it. The heterogenous environment of the Southern Ocean provides numerous opportunities for organisms capable of exploiting them. The variable behavioural responses of Antarctic krill give it the flexibility to take advantage of these opportunities. Krill behavioural modes and swarm structures change according not only to aspects programmed within their own biology, namely larval development, moulting, mating and spawning, but also to external aspects, such as movement into and out of hot-spots of predator activity, food patches and ice-shelves. The ways in which Antarctic krill deal with these challenges allows them to succeed like no other pelagic organism in the Southern Ocean.

Over the past 15 years, we have advanced our ability to detect patterns and trends within the flexible behavioural repertoire of Antarctic krill. The challenge for the coming decades is to increase our ability to predict how the behaviour of this species will respond to future changes to the Southern Ocean.

Acknowledgements We are grateful to R. Kornelliusson, K. Schmidt, D. Guihen and S. Kawaguchi for their permission to reproduce their figures within this review. GAT and SF were supported by the Ecosystems programme at the British Antarctic Survey, Natural Environment Research Council.

References

Alben S, Spears K, Garth S, Murphy D, Yen J (2010) Coordination of multiple appendages in drag-based swimming. J R Soc Interface 7:1545–1557, rsif20100171

Alonzo SH, Mangel M (2001) Survival strategies and growth of krill: avoiding predators in space and time. Mar Ecol Prog Ser 209:203–217

Alonzo SH, Switzer PV, Mangel M (2003) Ecological games in space and time: the distribution and abundance of Antarctic krill and penguins. Ecology 84:1598–1607

Amon DJ, Glover AG, Wiklund H, Marsh L, Linse K, Rogers AD, Copley JT (2013) The discovery of natural whale fall in the Antarctic deep sea. Deep-Sea Res II 92:87–96

Atkinson A, Snyder R (1997) Krill-copepod interactions at South Georgia, Antarctica, I. Omnivory by *Euphausia superba*. Mar Ecol Prog Ser 160:63–76

Atkinson A, Meyer B, Stubing D, Hagen W, Schmidt K, Bathmann UV (2002) Feeding and energy budgets of Antarctic krill *Euphausia superba* at the onset of winter – II. Juveniles and adults. Limnol Oceanogr 47:953–966

Atkinson A, Siegel V, Pakhomov EA, Rothery P, Loeb V, Ross RM, Quetin LB, Schmidt K, Fretwell P, Murphy EJ, Tarling GA, Fleming AH (2008) Oceanic circumpolar habitats of Antarctic krill. Mar Ecol Prog Ser 362:1–23

Atkinson A, Siegel V, Pakhomov EA, Jessopp MJ, Loeb V (2009) A re-appraisal of the total biomass and annual production of Antarctic krill. Deep-Sea Res I 56:727–740

Axelsen BE, Anker-Nilssen T, Fossum P, Kvamme C, Nøttestad L (2001) Pretty patterns but a simple strategy: predator-prey interactions between juvenile herring and Atlantic puffins observed with multibeam sonar. Can J Zool 79:1586

Azzali M, Kalinowski J (1999) Spatial and temporal distribution of krill *Euphausia superba* biomass in the Ross Sea (1989–1990 and 1994). In: Faranda F, Guglielmo L, Ianora A (eds) Ross Sea ecology. Springer, Berlin, pp 433–455

Azzali M, Kalinowski J, Lanciani G, Cosimi G (1999) Characteristic properties and dynamic aspects of krill swarms from the Ross Sea. In: Faranda F, Guglielmo L, Ianora A (eds) Ross Sea ecology. Springer, Berlin

Benoit-Bird KJ (2009) Dynamic 3-dimensional structure of thin zooplankton layers is impacted by foraging fish. Mar Ecol Prog Ser 396:61–76

Boyd IL, Murray AWA (2001) Monitoring a marine ecosystem using responses of upper trophic level predators. J Anim Ecol 70:747–760

Brierley AS (2008) Antarctic ecosystem: are deep krill ecological outliers or portents of a paradigm shift. Curr Biol 18:R252–R254

Brierley AS, Cox MJ (2010) Shapes of krill swarms and fish schools emerge as aggregation members avoid predators and access oxygen. Curr Biol 20:1758–1762

Brierley AS, Cox MJ (2015) Fewer but not smaller schools in declining fish and krill populations. Curr Biol 25:75–79

Brierley AS, Watkins JL (2000) Effects of sea ice cover on the swarming behaviour of Antarctic krill, *Euphausia superba*. Can J Fish Aquat Sci 57:24–30

Brierley AS, Fernandes PG, Brandon MA, Armstromg F, Millard NW, McPhail SD, Stevenson P, Pebody M, Perrett J, Squires M, Bone DG, Griffiths G (2002) Antarctic krill under sea ice: elevated abundance in a narrow band just south of ice edge. Science 295:1890–1892

Brierley AS, Fernandes PG, Brandon MA, Armstrong F, Millard NW, McPhail SD, Stevenson P, Pebody M, Perrett J, Squires M, Bone DG, Griffiths G (2003) An investigation of avoidance by Antarctic krill of RRS *James Clark Ross* using the *Autosub-2* autonomous underwater vehicle. Fish Res 60:569–576

Burrows MT, Tarling GA (2004) Effects of density dependence on diel vertical migration of populations of Northern krill: a genetic algorithm model. Mar Ecol Prog Ser 277:209–220

Carrera P, Churnside JH, Boyra G, Marques V, Scalabrin C, Uriarte A (2006) Comparison of airborne lidar with echosounders: a case study in the coastal Atlantic waters of Southern Europe. ICES J Mar Sci 63:1736–1750

Catton KB, Webster DR, Kawaguchi S, Yen J (2011) The hydrodynamic disturbances of two species of krill: implications for aggregation structure. J Exp Biol 214:1845–1856

CCAMLR (2010) Report of the fifth meeting of the Subgroup on Acoustic Survey and Analysis methods. SC-CCAMLR-XXIX/6, CCAMLR, Hobart, Tasmania

Churnside JH (2007) Lidar detection of plankton in the ocean, Geoscience and Remote Sensing Symposium. IEEE, Barcelona

Churnside JH, Thorne RE (2005) Comparison of lidar measurements with 420 kHz echosounder measurements of zooplankton. Appl Optics 44:5504–5511

Clarke A, Morris DJ (1983) Towards an energy budget for krill: the physiology and biochemistry of *Euphausia superba* Dana. Polar Biol 2:69–86

Clarke A, Tyler PA (2008) Adult Antarctic krill feeding at abyssal depths. Curr Biol 18:282–285

Coetzee J (2000) Use of a shoal analysis and patch estimation system (SHAPES) to characterize sardine schools. Aquat Living Resour 13:1–10

Colbo K, Ross T, Brown C, Weber T (2014) A review of oceanographic applications of water column data from multibeam echosounders. Estuar Coast Shelf Sci 145:41–56

Cox MJ, Warren JD, Demer DA, Cutter GR, Brierley AS (2009) Multibeam observations of Antarctic krill provide new insight to krill-predator interactions. Mar Ecol Prog Ser 378:199–209

Cox MJ, Warren JD, Demer DA, Cutter GR, Brierley AS (2010) Three-dimensional observations of swarms of Antarctic krill (*Euphausia superba*) made using a multi-beam echosounder. Deep-Sea Res II 57:508–518

Cox MJ, Watkins JL, Reid K, Brierley AS (2011a) Spatial and temporal variability in the structure of aggregations of Antarctic krill (*Euphausia superba*) around South Georgia, 1997–1999. ICES J Mar Sci 68:489–498

Cox MJ, Borchers DL, Demer DA, Cutter GR, Brierley AS (2011b) Estimating the density of Antarctic krill (*Euphausia superba*) from multibeam echosounder observations using distance sampling methods. J R Stat Soc: Ser C (Appl Stat) 50:301–316

Cram DJ, Agenbag JJ, Hampton I, Robertson AA (1979) SAS Protea cruise, 1978: the general results of the acoustics and remote sensing study, with recommendations for estimating the abundance of krill (*Euphausia superba* Dana). S Afr J Antarct Res 9:3–14

Cresswell K, Tarling G, Thorpe S, Burrows M, Wiedemann J, Mangel M (2009) Diel vertical migration of Antarctic krill (*Euphausia superba*) is flexible during advection across the Scotia Sea. J Plankton Res 31:1265–1281

Cullen M, Kaufmann RS, Lowery MS (2003) Seasonal variation in biochemical indicators of physiological status in *Euphausia superba* from Port Foster, Deception Island, Antarctica. Deep-Sea Res II 50:1787–1798

De Robertis A, Schell C, Jaffe JS (2003) Acoustic observations of the swimming behaviour of the euphausiid *Euphausia pacifica* Hansen. ICES J Mar Sci 60:885–898

Décima M, Ohman MD, Robertis AD (2010) Body size dependence of euphausiid spatial patchiness. Limnol Oceanogr 55:777–788

Demer DA, Hewitt RP (1995) Bias in acoustic biomass estimates of *Euphausia superba* due to diel vertical migration. Deep-Sea Res I 42:455–475

Everson I (1982) Diurnal variations in mean volume backscattering strength of an Antarctic krill (*Euphausia superba*) patch. J Plankton Res 4:155–162

Färber-Lorda J, Beier E, Mayzaud P (2009) Morphological and biochemical differentiation in Antarctic krill. J Mar Syst 78:525–535

Fauchald P, Tveraa T (2006) Hierarchical patch dynamics and animal movement pattern. Oecologia 149:383–395

Fielding S, Watkins J, Collins M, Enderlein P, Venables H (2012) Acoustic determination of the distribution of fish and krill across the Scotia Sea in spring 2006, summer 2008 and autumn 2009. Deep-Sea Res II 59–60:173–188

Fielding S, Watkins JLW, Trathan PN, Enderlein P, Waluda CM, Stowasser G, Tarling GA, Murphy EJ (2014) Interannual variability in Antarctic krill (*Euphausia superba*) density at South Georgia, Southern Ocean: 1997–2013. ICES J Mar Sci 71:2578–2588

Flores H, Van Franeker JA, Siegel V, Haraldsson M, Strass V, Meesters EH, Bathmann U, Wolff WJ (2012) The association of Antarctic krill *Euphausia superba* with the under-ice habitat. PLoS One 7:e31775

Folt CL, Burns CW (1999) Biological drivers of zooplankton patchiness. Trends Ecol Evol 14:300–305

Frazer TK, Quetin LB, Ross RM (1997) Abundance and distribution of larval krill, *Euphausia superba*, associated with annual sea ice in winter. In: Battaglia B, Valencia J, Walton DWH

(eds) Antarctic communities: species, structure and survival. Cambridge University Press, Cambridge, pp 107–111
Fuiman LA, Davis RW, Williams TM (2002) Behaviour of midwater fishes under the Antarctic ice: observations by a predator. Mar Biol 140:815–822
Furlong ME, Paxton D, Stevenson P, Pebody M, McPhail SD, Perrett J (2012) Autosub long range: a long range deep diving AUV for ocean monitoring, Autonomous Underwater Vehicles (AUV), 2012 IEEE/OES. IEEE, Southampton
Gaten E, Tarling G, Dowse H, Kyriacou C, Rosato E (2008) Is vertical migration in Antarctic krill (*Euphausia superba*) influenced by an underlying circadian rhythm? J Genet 87:473–483
Gerlotto F, Bertrand S, Bez N, Gutierrez M (2006) Waves of agitation inside anchovy schools observed with multibeam sonar: a way to transmit information in response to predation. ICES J Mar Sci 63:1405–1417
Godlewska M (1996) Vertical migrations of krill (*Euphausia superba* Dana). Pol Arch Hydrobiol 43:9–63
Goodall-Copestake WP, Perez-Espona S, Clark MS, Murphy EJ, Seear PJ, Tarling GA (2010) Swarms of diversity at the gene cox1 in Antarctic krill. Heredity 104:513–518
Greene CH, Wiebe P, Zamon J (1994) Acoustic visualization of patch dynamics in oceanic ecosystems. Oceanography 7:4–12
Greene CH, Meyer-Gutbrod EL, McGarry LP, Hufnagle LC, Chu D, McClatchie S, Packer A, Jung J-B, Acker T, Dorn H, Pelkie C (2014) A wave glider approach to fisheries acoustics: transforming how we monitor the nation's commercial fisheries in the 21st century. Oceanography 27:168–174
Griffiths G (ed) (2003) Technology and applications of autonomous underwater vehicles, 2nd edn. Taylor and Francis Group, New York
Grunbaum D (1998) Schooling as a strategy for taxis in a noisy environment. Evol Ecol 12:503–522
Guihen D, Fielding S, Murphy EJ, Heywood KJ, Griffiths G (2014) An assessment of the use of ocean gliders to undertake acoustic measurements of zooplankton: the distribution and density of Antarctic krill (*Euphausia superba*) in the Weddell Sea. Limnol Oceanogr Methods 12:373–389
Hamilton WD (1971) Geometry for the selfish herd. J Theor Biol 31:295–311
Hamner WM (1984a) Behavior of Antarctic krill *Euphausia superba*: chemoreception, feeding, schooling and molting. Science 220:433–435
Hamner WM (1984b) Aspects of schooling of *Euphausia superba*. J Crustac Biol 4(Special Issue):67–74
Hamner WM, Hamner PP, Strand MP, Gilmer RW (1983) Behaviour of Antarctic krill (*Euphausia superba*): chemoreception, feeding, schooling and moulting. Science 220:433–435
Hernandez-Leon S, Portillo-Hahnefeld A, Almeida C, Becognee P, Moreno I (2001) Diel feeding behaviour of krill in the Gerlache strait, Antarctica. Mar Ecol Prog Ser 223:235–242
Higginbottom IR, Hosie GW (1989) Biomass and population structure of a large aggregation of krill near Prydz Bay, Antarctica. Mar Ecol Prog Ser 58:197–203
Hill SL, Belchier M, Collins MA, Fielding S, Murphy EJ, Trathan PN, Venables HJ, Waluda CM (2009) Multiple indicators suggest a strong ecosystem anomaly at South Georgia in 2009. CCAMLR WG-EMM-09/23. p 16
Hirano Y, Matsuda T, Kawaguchi S (2003) Breeding Antarctic krill in captivity. Mar Freshw Behav Physiol 36:259–269
Hofmann EE, Haskell AE, Klinck JM, Lascara CM (2004) Lagrangian modelling studies of Antarctic krill (*Euphausia superba*) swarm formation. ICES J Mar Sci 61:617–631
Hooker SK, Boyd IL, Jessop M, Cox O, Blackwell J, Boveng PL, Bengtson JL (2002) Monitoring the prey-field of marine predators: combining digital imaging with datalogging tags. Mar Mamm Sci 18:680–697
Hooker SK, Heaslip SG, Matthiopoulos J, Cox O, Boyd IL (2008) Data sampling options for animal-borne video cameras: considerations based on deployments with Antarctic fur seals. Mar Technol Soc J 42:65–75

Huntley ME, Niiler PP (1995) Physical control of population dynamics in the Southern-Ocean. ICES J Mar Sci 52:457–468
Ikeda T, Dixon P (1982) Body shrinkage as a possible over-wintering mechanism of the Antarctic Krill, *Euphausia superba* Dana. J Exp Mar Biol Ecol 62:143–151
Jaffe JS (2013) A contemporary perspective on underwater optical imaging, OCEANS-Bergen, 2013 MTS/IEEE. Scripps Inst. of Oceanogr., U.C. San Diego, La Jolla
Jagannathan S, Bertsatos I, Symonds DT, Chen T, Nia HT, Jain AD, Andrews M, Gong Z, Nero RW, Ngor L, Jech M, Godø OR, Lee S, Ratilal P, Makris NC (2009) Ocean acoustic waveguide remote sensing (OAWRS) of marine ecosystems. Mar Ecol Prog Ser 395:137–160
Jarman S, Deagle BE (2016) Genetics of Antarctic krill. In: Siegel V (ed) The biology and ecology of Antarctic krill, *Euphausia superba* Dana, 1850. Springer, Cham, pp 247–278
Johnson ML, Tarling GA (2008) Influence of individual state on swimming capacity and behaviour of Antarctic krill *Euphausia superba*. Mar Ecol Prog Ser 366:99–110
Johnson KS, Berelson WM, Boss ES, Chase Z, Claustre H, Emerson SR, Gruber N, Kortzinger A, Perry MJ, Riser SC (2009) Observing biogeochemical cycles at global scales with profiling floats and gliders prospects for a global array. Oceanography 22:216–225
Kalinowski J, Witek Z (1985) Scheme for classifying Antarctic krill. Biomass Handb Ser 27:1–12
Kanda K, Takagi K, Seki Y (1982) Movement of the larger swarms of Antartic krill *Euphausia superba* population off enderby land during 1976–1977 season. J Tokyo Univ Fish 68:25–42
Kawaguchi S, King R, Meijers R, Osborn JE, Swadling KM, Ritz DA, Nicol S (2010) An experimental aquarium for observing the schooling behaviour of Antarctic krill (*Euphausia superba*). Deep-Sea Res II 57:683–692
Kawaguchi S, Kilpatrick R, Roberts L, King RA, Nicol S (2011) Ocean-bottom krill sex. J Plankton Res 33:1134–1138
Kils U (1981) Swimming behaviour, swimming performance and energy balance of Antarctic krill *Euphausia superba*. BIOMASS Sci Ser 3:1–121
Klevjer TA, Tarling GA, Fielding S (2010) Swarm characteristics of Antarctic krill *Euphausia superba* relative to the proximity of land during summer in the Scotia Sea. Mar Ecol Prog Ser 409:157–170
Kohut J, Bernard K, Fraser W, Oliver MJ, Statscewich H, Winsor P, Miles T (2014) Studying the impact of local oceanographic processes on Adelie penguin foraging ecology. Mar Technol Soc J 48:25–34
Korb RE, Whitehouse MJ, Ward P (2004) SeaWiFS in the southern ocean: spatial and temporal variability in phytoplankton biomass around South Georgia. Deep-Sea Res II 51:99–116
Korneliussen RJ, Heggelund Y, Eliassen IK, Øye OK, Knutsen T, Dalen J (2009) Combining multibeam-sonar and multifrequency echosounder data: examples of the analysis and imaging of large euphausiid schools. ICES J Mar Sci 66:991–997
Krafft BA, Skaret G, Knutsen T, Melle W, Klevjer T, Soiland H (2012) Antarctic krill swarm characteristics in the Southeast Atlantic sector of the Southern Ocean. Mar Ecol Prog Ser 465:69–83
Lascara CM, Hofmann EE, Ross RM, Quetin LB (1999) Seasonal variability in the distribution of Antarctic krill, *Euphausia superba*, west of the Antarctic Peninsula. Deep-Sea Res I 46:951–984
Laws RM (1977) Seals and whales of the Southern Ocean. Phil Trans R Soc Lond Ser B 279:81–96
Lawson GL, Wiebe PH, Ashjian CJ, Stanton TK (2008) Euphausiid distribution along the Western Antarctic Peninsula – part B: distribution of euphausiid aggregations and biomass, and associations with environmental features. Deep-Sea Res II 55:432–454
Liu Z-S (1990) Estimate of maximum penetration depth of LIDAR in coastal water of the China Sea. Proc SPIE – Int Soc Opt Eng 1302:655–661
Macaulay MC, English TS, Mathisen OA (1984) Acoustic characterization of swarms of Antarctic krill (*Euphausia superba*) from Elephant Island and Bransfield Strait. J Crustac Biol 4(Special Issue):16–44

Main CE, Collins MA (2011) Diet of the Antarctic starry skate *Amblyraja georgiana* (Rajidae, Chondrichthyes) at South Georgia (Southern Ocean). Polar Biol 34:389–396

Makris NC, Ratilal P, Symonds DT, Jagannathan S, Lee S, Nero RW (2006) Fish population and behaviour revealed by instantaneous continental shelf scale imaging. Science 311:660–663

Mangel M, Clark CW (1988) Dynamic modelling in behavioral ecology. Princeton University Press, Princeton, p 308

Marr JWS (1962) The natural history and geography of the Antarctic krill. Discov Rep 32:33–464

Mathisen OA, Macaulay MC (1983) The morphological features of a super swarm of krill *Euphausia superba*. Mem Natl Inst Polar Res Spec Issue 28(Special Issue):153–164

Mauchline J (1980) Studies on patches of krill *Euphausia superba* Dana. Biomass Handb Ser 6:36

Mayer L (2006) Frontiers in seafloor mapping and visualization. Mar Geophys Res 27:7–17

Mayer L, Li Y, Melvin G (2002) 3D visualization for pelagic fisheries research and assessment. ICES J Mar Sci 59:216–225

Mazzotta GM, De Pittà C, Benna C, Tosatto SC, Lanfranchi G, Bertolucci C, Costa R (2010) A cry from the krill. Chronobiol Int 27:425–445

Meredith MP, King JC (2005) Rapid climate change in the ocean west of the Antarctic Peninsula during the second half of the 20th century. Geophys Res Lett 32:L19604

Meyer B (2012) The overwintering of Antarctic krill, *Euphausia superba*, from an ecophysiological perspective. Polar Biol 35:15–37

Meyer B, Teschke M (2016) Physiology of *Euphausia superba*. In: Siegel V (ed) The biology and ecology of Antarctic krill, *Euphausia superba* Dana, 1850. Springer, Cham, pp 145–174

Meyer B, Atkinson A, Stubing D, Oettl B, Hagen W, Bathmann UV (2002) Feeding and energy budgets of Antarctic krill *Euphausia superba* at the onset of winter – I. Furcilia III larvae. Limnol Oceanogr 47:943–952

Meyer B, Auerswald L, Siegel V, Spahic S, Pape C, Fach BA, Teschke M, Lopata AL, Fuentes V (2010) Seasonal variation in body composition, metabolic activity, feeding, and growth of adult krill *Euphausia superba* in the Lazarev Sea. Mar Ecol Prog Ser 398:1–18

Miller DGM, Hampton I (1989) Biology and ecology of the Antarctic krill, vol 9. SCOR, Cambridge, pp 1–166

Morin A, Okubo A, Kawasaki K (1988) Acoustic data analysis and models of krill spatial distribution. Scientific Committee for the Conservation of Antarctic Marine Living Resources, Selected Scientific Papers part I. CCAMLR, Hobart, Tasmania, pp 311–329

Morris DJ, Priddle J (1984) Observations on the feeding and moulting of the Antarctic krill, *Euphausia superba*. Br Antarct Surv Bull 65:57–63

Murphy EJ, Reid K (2001) Modelling Southern Ocean krill population dynamics: biological processes generating fluctuations in the South Georgia ecosystem. Mar Ecol Prog Ser 217:175–189

Murphy E, Morris DJ, Watkins JL, Priddle J (1988) Scales of interaction between Antarctic krill and the environment. In: Sahrhage D (ed) Antarctic ocean and resources variability. Springer, Berlin, pp 120–130

Murphy D, Webster D, Kawaguchi S, King R, Yen J (2011) Metachronal swimming in Antarctic krill: gait kinematics and system design. Mar Biol 158:2541–2554

Murphy DW, Webster DR, Yen J (2013) The hydrodynamics of hovering in Antarctic krill. Limnol Oceanogr: Fluids Environ 3:240–255

Nowacek DP, Friedlaender AS, Halpin PN, Hazen EL, Johnston DW, Read AJ, Espinasse B, Zhou M, Zhu YW (2011) Super-aggregations of krill and humpback whales in Wilhelmina Bay, Antarctic Peninsula. PLoS One 6:e19173

O'Brien DP (1987) Direct observations of the behavior of *Euphausia superba* and *Euphausia crystallorophias* (Crustacea, Euphausiacea) under pack ice during the Antarctic spring of 1985. J Crustac Biol 7:437–448

O'Brien DP (1989) Analysis of the internal arrangement of individuals within crustacean aggregations (Euphausiacea, Mysidacea). J Exp Mar Biol Ecol 128:1–30

Olson RS, Knoester DB, Adami C (2013) Evolution of swarming behavior is shaped by how predators attack. arXiv preprint arXiv:13106012
Onsrud MSR, Kaartvedt S (1998) Diel vertical migration of the krill *Meganyctiphanes norvegica* in relation to physical environment, food and predators. Mar Ecol Prog Ser 171:209–219
Osse TJ, Eriksen CC (2007) The Deepglider: a full ocean depth glider for oceanographic research. Oceans 2007. IEEE
Patria MP, Wiese K (2004) Swimming in formation in krill (Euphausiacea), a hypothesis: dynamics of the flow field, properties of antennular sensor systems and a sensory-motor link. J Plankton Res 26:1315–1325
Pauly T, Nicol S, Higginbottom I, Hosie G, Kitchener J (2000) Distribution and abundance of Antarctic krill (*Euphausia superba*) off East Antarctica (80–150°E) during the austral summer of 1995/1996. Deep-Sea Res II 47:2465–2488
Pollard R, Sanders R, Lucas M, Statham P (2007) The CROZet natural iron bloom and EXport experiment (CROZEX). Deep-Sea Res II 54:1905–1914
Pollard RT, Salter I, Sanders RJ, Lucas MI, Moore CM, Mills RA, Statham PJ, Allen JT, Baker AR, Bakker DCE, Charette MA, Fielding S, Fones GR, French M, Hickman AE, Holland RJ, Hughes JA, Jickells TD, Lampitt RS, Morris PJ, Nedelec FH, Nielsdottir M, Planquette H, Popova EE, Poulton AJ, Read JF, Seeyave S, Smith T, Stinchcombe M, Taylor S, Thomalla S, Venables HJ, Williamson R, Zubkov MV (2009) Southern ocean deep-water carbon export enhanced by natural iron fertilization. Nature 457:577–U581
Quetin LB, Ross RM (1991) Behavioural and physiological characteristics of Antarctic krill *Euphausia superba* Dana. Am Zool 31:49–63
Reid K, Brierley AS, Nevitt GA (2000) An initial examination of relationships between the distribution of whales and Antarctic krill *Euphausia superba* at South Georgia. J Cetac Res Manage 2:143–149
Ritz DA (1994) Social aggregation in pelagic invertebrates. Adv Mar Biol 30:155–216
Ritz DA (2000) Is social aggregation in aquatic crustaceans a strategy to conserve energy. Can J Fish Aquat Sci 57:59–67
Ritz DA, Hobday AJ, Montgomery JC, Ward AJW (2011) Social aggregation in the pelagic zone with special reference to fish and invertebrates. In: Lesser M (ed) Advances in marine biology, vol 60. Academic, London, pp 161–227
Ross RM, Quetin LB, Lascara CM (1996) Distribution of Antarctic krill and dominant zooplankton west of the Antarctic Peninsula. In: Ross RM, Hofmann EE, Quetin CM (eds) Foundations for ecological research west of the Antarctic Peninsula. AGU Antarctic Research Series American Geophysical Union, Washington, DC, pp 199–217
Rudnick DL, Davis RE, Eriksen CC, Fratantoni DM, Perry MJ (2004) Underwater gliders for ocean research. Mar Technol Soc J 38:73–84
Sainmont J, Thygesen UH, Visser AW (2013) Diel vertical migration arising in a habitat selection game. Theor Ecol 6:241–251
Schmidt K, Atkinson A, Steigenberger S, Fielding S, Lindsay M, Pond D, Tarling G, Klevjer T, Allen C, Nicol S, Achterberg E (2011) Seabed foraging by Antarctic krill: Implications for stock assessment, bentho-pelagic coupling, and the vertical transfer of iron. Limnol Oceanogr 56:1411–1428
Schofield O, Glenn S, Moline M (2013) The robot ocean network. Am Sci 101:434–441
Schulenberger E, Wormuth JH, Loeb VJ (1984) A large swarm of *Euphausia superba*: overview of patch structure and composition. J Crustac Biol 4(Special Issue):75–95
Seear P, Tarling GA, Teschke M, Meyer B, Thorne MAS, Clark MS, Gaten E, Rosato E (2009) Effects of simulated light regimes on gene expression in Antarctic krill (*Euphausia superba* Dana). J Exp Mar Biol Ecol 381:57–64
Seear PJ, Tarling GA, Burns G, Goodall-Copestake WP, Gaten E, Özkaya Ö, Rosata E (2010) Differential gene expression during the moult cycle of Antarctic krill (*Euphausia superba*). BMC Genomics 11:582

Seear PJ, Goodall-Copestake WP, Fleming AH, Rosato E, Tarling GA (2012) Seasonal and spatial influences on gene expression in Antarctic krill *Euphausia superba*. Mar Ecol Prog Ser 467:61

Siegel V (2012) Krill stocks in high latitudes of the Antarctic Lazarev Sea: seasonal and interannual variation in distribution, abundance and demography. Polar Biol 35:1151–1177

Siegel V, Kalinowski J (1994) Krill demography and small-scale processes: a review. In: El-Sayed SZ (ed) Southern ocean ecology: the BIOMASS perspective. Cambridge University Press, Cambridge, pp 145–164

Siegel V, Watkins JL (2016) Distribution, biomass and demography of Antarctic krill, *Euphausia superba*. In: Siegel V (ed) The biology and ecology of Antarctic krill, *Euphausia superba* Dana, 1850. Springer, Cham, pp 21–100

Siegel V, Bergström B, Strömberg J, Schalk P (1990) Distribution, size frequencies and maturity stages of krill, *Euphausia superba*, in relation to sea-ice in the northern Weddell Sea. Polar Biol 10:549–557

Squire JL, Krumboltz H (1981) Profiling pelagic fish schools using airborne optical lasers and other remote sensing techniques. Mar Technol Soc J 15:27–31

Strand SW, Hamner WM (1990) Schooling behavior of Antarctic krill (*Euphausia superba*) in laboratory aquariums – reactions to chemical and visual stimuli. Mar Biol 106:355–359

Swadling KM (2006) Krill migration: up and down all night. Curr Biol 16:R173–R175

Swadling KM, Ritz DA, Nicol S, Osborn JE, Gurney LJ (2005) Respiration rate and cost of swimming for Antarctic krill, *Euphausia superba*, in large groups in the laboratory. Mar Biol 146:1169–1175

Takahashi A, Kokubun N, Mori Y, Shin HC (2008) Krill-feeding behaviour of gentoo penguins as shown by animal-borne camera loggers. Polar Biol 31:1291–1294

Taki K, Hayashi T, Naganobu M (2005) Characteristics of seasonal variation in diurnal vertical migration and aggregation of Antarctic krill (*Euphausia superba*) in the Scotia Sea, using Japanese fishery data. CCAMLR Sci 12:163–172

Tarling GA, Johnson ML (2006) Satiation gives krill that sinking feeling. Curr Biol 16:83–84

Tarling GA, Thorpe SE (2014) Instantaneous movement of krill swarms in the Antarctic circumpolar current. Limnol Oceanogr 59:872–886

Tarling GA, Klevjer T, Fielding S, Watkins JL, Atkinson A, Murphy E, Korb R, Whitehouse M, Leaper R (2009) Variability and predictability of Antarctic krill swarm structure. Deep-Sea Res I 56:1994–2012

Teschke M, Kawaguchi S, Meyer B (2007) Simulated light regimes affect feeding and metabolism of Antarctic krill, *Euphausia superba*. Limnol Oceanogr 52:1046–1054

Teschke M, Kawaguchi S, Meyer B (2008) Effects of simulated light regimes on maturity and body composition of Antarctic krill, *Euphausia superba*. Mar Biol 154:315–324

Teschke M, Wendt S, Kawaguchi S, Kramer A, Meyer B (2011) A circadian clock in Antarctic krill: an endogenous timing system governs metabolic output rhythms in the euphausid species *Euphausia superba*. PLoS One 6:e26090

Torres JJ, Donnelly J, Hopkins TL, Lancraft TM, Aarset AV, Ainley DG (1994) Proximate composition and overwintering strategies of Antarctic micronektonic crustacea. Mar Ecol Prog Ser 113:221–232

Trathan PN, Hill SL (2016) The importance of krill predation in the Southern Ocean. In: Siegel V (ed) The biology and ecology of Antarctic krill, *Euphausia superba* Dana, 1850. Springer, Cham, pp 321–350

Verdy A, Flierl G (2008) Evolution and social behavior in krill. Deep-Sea Res II 55:472–484

Watanabe YY, Takahashi A (2013) Linking animal-borne video to accelerometers reveals prey capture variability. Proc Natl Acad Sci 110:2199–2204

Watkins JL (2000) Aggregation and vertical migration. In: Everson I (ed) Krill biology, ecology and fisheries. Blackwell Science, Oxford, pp 80–102

Watkins JL, Murray AWA (1998) Layers of Antarctic krill, *Euphausia superba*: are they just long krill swarms? Mar Biol 131:237–247

Watkins JL, Buchholz F, Priddle J, Morris DJ, Ricketts C (1992) Variation in reproductive status of Antarctic krill swarms; evidence for a size-related sorting mechanism? Mar Ecol Prog Ser 82:163–174
Weber LH, El-Sayed SZ, Hampton I (1986) The variance spectra of phytoplankton, krill and water temperature in the Antarctic ocean, south of Africa. Deep-Sea Res 33:1327–1343
Whitehouse MJ, Meredith MP, Rothery P, Atkinson A, Ward P, Korb RE (2008) Rapid warming of the ocean around South Georgia, Southern Ocean, during the 20th century: forcings, characteristics and implications for lower trophic levels. Deep-Sea Res I 55:1218–1228
Wiese K (1996) Sensory capacities of euphausiids in the context of schooling. Mar Freshw Behav Physiol 28:183–194
Willis J (2007) Could whales have maintained a high abundance of krill? Evol Ecol Res 9:651
Willis J (2014) Whales maintained a high abundance of krill; both are ecosystem engineers in the Southern Ocean. Mar Ecol Prog Ser 513:51–69
Zhou M, Dorland RD (2004) Aggregation and vertical migration behaviour of *Euphausia superba*. Deep-Sea Res II 51:2119–2137
Zhou M, Huntley ME (1996) The principle of biological attraction, demonstrated by the bio-continuum theory of zooplankton patch dynamics. J Mar Res 54:1017–1037

Chapter 9
The Importance of Krill Predation in the Southern Ocean

Philip N. Trathan and Simeon L. Hill

Abstract Antarctic krill is a major prey species for a diverse array of Southern Ocean predators. The amount of krill that predators consume, and how this changes over space and time, is a key issue in understanding both regional and circumpolar aspects of the Southern Ocean food-web. We assess current knowledge of consumption by the various predator groups, and the ecological processes through which krill and its predators influence each other. Knowledge has improved greatly over recent decades and has revealed a high level of complexity in the processes that govern krill consumption. We focus on the Antarctic Peninsula and Scotia Sea region where both krill and its consumers occur in significant concentrations and where an updated estimate of krill consumption by the main vertebrate groups is 55 million tonnes per year. Research has mainly focused on mammalian and avian predators of post-larval krill, particularly penguins. Potentially important consumer groups like fish, cephalopods and carnivorous zooplankton remain poorly understood, as does consumption of the early life stages of krill. As a consequence of these knowledge gaps and the variability that arises from complexity, a reliable seasonally, spatially or taxonomically resolved description of krill consumption remains elusive. One of the key motivations for attempting to estimate krill consumption is to understand how changes in krill availability impact predator populations. Such understanding is an important requirement for ecosystem based management of the Antarctic krill fishery. We therefore propose that integrated study areas in strategic fishing locations should be developed to directly assess the response of krill predators to changes in krill availability.

Keywords Krill consumption • Top-down control • Krill surplus hypothesis • Ecosystem perturbation • Ecosystem recovery

---

P.N. Trathan (✉) • S.L. Hill
British Antarctic Survey, Natural Environment Research Council, High Cross, Madingley Road, Cambridge CB3 0ET, UK
e-mail: pnt@bas.ac.uk; sih@bas.ac.uk

V. Siegel (ed.), *Biology and Ecology of Antarctic Krill*, Advances in Polar Ecology,
DOI 10.1007/978-3-319-29279-3_9

9.1 Introduction

Antarctic krill, *Euphausia superba*, is a widespread and abundant species in the Southern Ocean and most of its life history stages are important prey for a wide range of organisms (Miller and Hampton 1989). These predators include species of commercial interest (fish and squid) and conservation concern (marine mammals and seabirds) as well as an array of zooplankton and benthic invertebrates. Krill also indirectly supports predators that feed on other krill consumers. Consistent with its key role in many predators' diets, krill apparently exerts bottom-up control on predator variables such as reproductive output and, ultimately, population size (Murphy 1995; Reid et al. 2005; Atkinson et al. 2014; Forcada and Trathan 2009). Conversely, the idea that predators exert top-down controls on the abundance and distribution of krill has been influential in the study of Southern Ocean ecology (Laws 1977; Mori and Butterworth 2006; Hill et al. 2006). Because of its circumpolar distribution, krill connects many Southern Ocean food-webs, while highly migratory predators such as baleen whales and albatrosses extend the influence of these trophic interactions into other oceans. Thus, to fully comprehend the ecological importance of krill requires quantification of trophic interactions and energy flows involving all life history stages at a range of spatial and temporal scales.

Knowledge about krill consumption is key to many of the critical questions about Southern Ocean food-webs, including:

(i) How do these food-webs function both temporally and spatially, at the regional and circumpolar scale,
(ii) How do the population processes and reproductive output of predator species respond to natural variation in the availability of krill,
(iii) What is the natural mortality of krill and how does it affect krill population dynamics,
(iv) What are the implications of altered krill availability in food-webs which might result from rapid, regional changes in climate, and
(v) What are the implications of krill harvesting for Southern Ocean food-webs?

Despite its clear importance as a central process in Southern Ocean food-webs, there is considerable uncertainty about many aspects of krill consumption, which even extends to identifying which predators are the major consumers (Atkinson et al. 2012). The reasons for these uncertainties include sparse data on many predator species and significant inter-annual variability in krill population size, which is connected to bottom-up environmental variation (Atkinson et al. 2004; Trathan et al. 2006, 2007, Murphy et al. 2007; Steinberg et al. 2015), and which may interact with top-down predation controls operating over multi-decadal scales.

The study of interactions between krill and its predators is a major part of Southern Ocean ecology, which has produced a substantial literature and increased understanding of the many influences on krill consumption. One consequence of these insights is a growing appreciation of the complex role that krill plays in

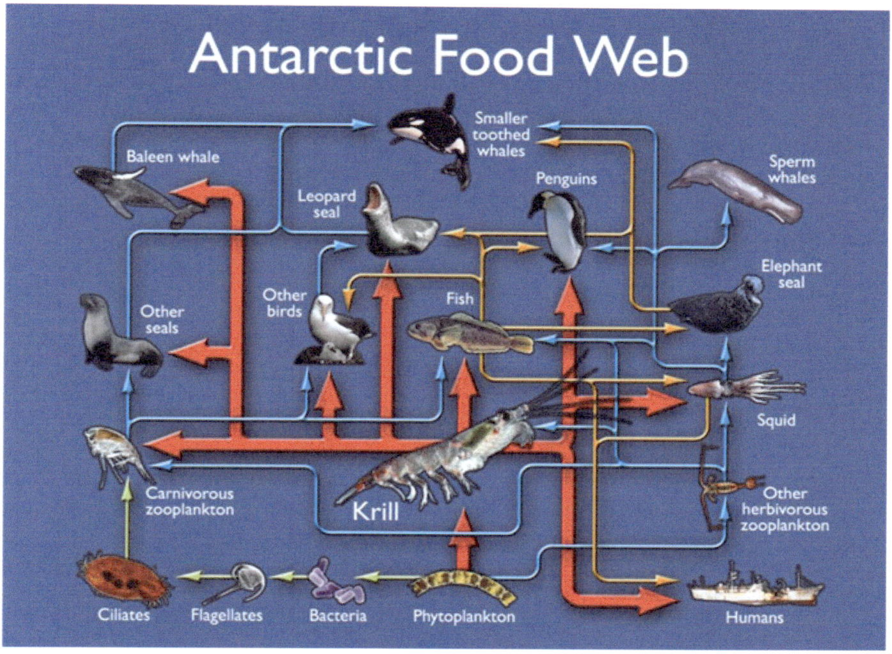

Fig. 9.1 The major predators of Antarctic krill (Source: www.discoveringantarctica.org.uk). Individual predators in the figure are representative of a broad range of individual species; e.g. Penguins might variously include Adélie, chinstrap, gentoo, emperor and macaroni

Southern Ocean food-webs. For example, recent studies have shown that krill feed on the benthos year round throughout Antarctica, with an estimated ~20 % of the stock deeper than 200 m at any one time (Schmidt et al. 2011), although others (see Chap. 2, Siegel and Watkins 2016) suggest that the percentage below 200 m may only be around 5 %. Further, Clarke and Tyler (2008) have shown that krill habitat also extends to abyssal depths (including to 3000 m and even 3500 m). These observations revise our understanding of the depth distribution and ecology of Antarctic krill, suggesting that krill connects food-webs not just horizontally, between regions, but vertically, between depth zones.

In this chapter we consider the consumption of krill by its many predators (Fig. 9.1), and discuss aspects of predator-prey ecology that contribute to our understanding of consumption. We focus mainly on the Antarctic Peninsula and Scotia Sea region where there is a greater biomass of krill (Atkinson et al. 2009) and probably also of krill consumers; further, information from this region also extends back in time for a number of decades given the interest of early integrated ecological studies such as the *Discovery Investigations* (e.g. Marr 1962; Mackintosh 1974). We present information about consumption by various taxonomic groups, and highlight penguins as one of the most studied groups of krill predators, but emphasise that important gaps still exist even for this group. We recognise that there are regional differences in the diets of many predators, which have a strong dependency on krill where it is regionally abundant, but have other dominant prey elsewhere (e.g. Staniland et al. 2010). This is one of many complexities that mean that accurately describing the many trophic interactions, particularly those related to the early life stages of krill, remains challenging. We highlight a number of major uncertainties that remain to be addressed if we are to properly understand the role of krill in the Antarctic marine ecosystem. We also consider a major application of this understanding, in the ecosystem-based management of Antarctic krill fisheries, and suggest a way of gaining the necessary insight into the relevant predator-prey dynamics despite uncertainties in consumption estimates.

9.2 Historical Perspective

Estimates of krill consumption by marine predators date back at least to Marr (1962), who estimated the annual consumption of krill by large baleen whales during the mid 1930s. Later estimates by Everson (1977) built on calculations of krill consumption by seals (Laws 1977), birds (Prevost 1981) and whales (e.g. Mackintosh 1974; Laws 1977), but recognised that there were no consumption estimates for other important krill-eating groups, including fish and squid. The importance of krill to predators in the Southern Ocean and the potential for commercial krill fishing to affect these predators became central to the establishment of the ecosystem-based management approach adopted by the Commission for the Conservation of Antarctic Marine Living Resources (CCAMLR) (Everson 2000).

Estimating consumption generally requires information about diet composition, feeding rate, and population size. Information on each of these elements has increased considerably since CCAMLR first met in 1982. For example, Croxall et al. (1985a) compiled the available information to estimate consumption of krill by seabirds in the Scotia Sea, and identified macaroni penguins *Eudyptes chrysolophus* breeding at South Georgia as a major krill consumer. Boyd (2002) arrived at a similar consumption estimate for macaroni penguins and provided insight into some of the uncertainties in the estimate. Further studies (Green et al. 2001, 2002) suggested that macaroni penguin metabolic rates are higher than those assumed by Boyd (2002) but that population sizes are now considerably smaller, as a result of a decline since the late 1970s (Trathan et al. 2012). However, consumption estimates remain uncertain because of a lack of information about pre-breeders and about winter diets of all demographic classes. Knowledge about krill consumption by macaroni penguins has therefore accumulated over time but the ecosystem has also changed within this same time interval, and uncertainties in consumption estimates therefore remain. This is despite the fact that macaroni penguins are relatively easy to observe compared to fully marine predator groups such as cephalopods, fish, whales and planktonic invertebrates, where the complexities are more difficult to overcome.

Table 9.1 Available estimates of Antarctic krill consumption (million tonnes $year^{-1}$) by various predator groups at the circumpolar and Scotia Seascales

Predator group	Circumpolar						Circumpolar		Scotia sea[m]
							Min	Max	
Whales	34[a]	43[b]	85[c]	3–120[d]			3	120	2
Minke	8	20	75		36[e]	140[f]			
Seals	129[a]	64[b]		24–73[d]			24	129	8
Crabeater	128	63							1
Antarctic fur	0	0							7
Leopard	1	1							0
Seabirds	15–20[g]						15	20	8
Penguins									5
Fish	18–20[h]	*21–36*[i]	40–50[j]	*28–61*[k]			18	61	37
Mesopelagic		20–35		*5–32*					25
Cephaplopods	30–56[l]						30	56	
Total							**90**	**387**	**55**

Values in italics are not full circumpolar estimates (see notes). Shaded rows show selected subgroups of the group above
[a]Bengston (1984)
[b]Laws (1977)
[c]Everson (2000)
[d]Mori and Butterworth (2006)
[e]Armstrong and Siegfried (1991)
[f]Ichii and Kato (1991)
[g]Everson (1977)
[h]Lubimova and Shust (1980)
[i]Kock (1992). The demersal fish component of the consumption estimate is for the Scotia Arc only
[j]Hureau (1994)
[k]Kock et al. (2012). The mesopelagic fish component of the consumption estimate is for the Atlantic sector only
[l]Everson (1984)
[m]Hill et al. (2007). We have updated the consumption estimate to include new information about pack ice seals (Forcada et al. 2012), Antarctic fur seals (Trathan et al. 2012) and macaroni penguins (Trathan et al. 2012) and to include flying seabirds (Croxall et al. 1985a)

Nonetheless, the literature contains circumpolar krill consumption estimates for many predator groups including seals and whales (Laws 1977; Bengston 1984; Armstrong and Siegfried 1991; Ichii and Kato 1991; Mori and Butterworth 2006), seabirds (Everson 1977; Croxall et al. 1984, 1985a), cephalopods (Everson 1984) and fish (Lubimova and Shust 1980; Kock 1992; Hureau 1994). Various authors, including Everson (1977), Miller and Hampton (1989), Barrera-Oro (2002), and Mori and Butterworth (2006), have drawn on these sources to estimate total circumpolar krill consumption in the post-whaling era. These estimates are in the range of 90–387 million tonnes per year (Table 9.1), which overlaps with the range of krill gross post-larval production estimates (342–536 million tonnes per year) presented in Atkinson et al. (2009). For comparison, Hill et al. (2007) compiled estimates of krill consumption in the Scotia Sea region (FAO Subareas 48.1, 48.2 and 48.3), which we have here updated to include two key missing groups: flying seabirds (Croxall et al. 1985a) and pack ice seals (Forcada et al. 2012), and revised estimates for macaroni penguins and fur seals (Trathan et al. 2012). This suggests that krill consumption in the Scotia Sea region is at least 55 million tonnes per year.

These consumption estimates give an indication of the importance of krill to higher trophic levels, but they also highlight some of the uncertainties that affect our understanding of the Southern Ocean food-web. For example, there is a difference of almost 300 million tonnes year^{-1} between the minimum and maximum circumpolar consumption estimates. This difference is greater than the uncertain estimate of late twentieth century circumpolar krill biomass (~215 million tonnes: Atkinson et al. 2009, updated in Hill 2013). The available circumpolar estimates suggest that the dominant predator group could be any of crabeater seals, minke whales, cephalopods, or fish, whereas comparison with historical Scotia Sea estimates reveals an apparent underestimate of the importance of Antarctic fur seals at the circumpolar scale.

With the exception of estimates of consumption by mesopelagic fish, circumpolar consumption estimates are generally concerned with predation on post-larval krill, and so omit the significant consumption of eggs and larvae (see below). This omission reflects gaps in the available information, which also mean that the level of information varies between predator groups. Much research has continued to focus mainly on birds and seals and, even in these groups, the level of information varies widely between locations and taxa. For example, in 2007, CCAMLR, which is responsible for managing the krill fishery and its effects on dependent and related species, established a specialist sub-group to address the need for information about predator abundance as part of the requirement to estimate predator demand for krill (SC-CAMLR 2007). In 2008, the sub-group identified short and medium term aims which concerned only birds and seals, with an acknowledgement that future work should also include fish (SC-CCAMLR 2008). At present there is no equivalent strategic focus on cephalopods, other invertebrates, or even whales. This restricted focus reflects the limited resources available to the sub-group, and therefore the size of the challenge implied by the ambition to estimate predator demand for krill.

In the last decade, several authors have developed food-web models which provide holistic representations of consumption and production throughout the food-web (Cornejo-Donoso and Antezana 2008; Pinkerton et al. 2010; Hill et al. 2012; Ballerini et al. 2014; Gurney et al. 2014). Such models provide a useful synthesis of available information, which often includes new consumption estimates for predators which have not been studied directly. They are also useful for exploring uncertainties in consumption estimates (Hill et al. 2012). The current generation of models focus on relatively small geographical areas but it is technically possible to use this approach to produce synoptic estimates of krill consumption at the regional scale (e.g. for the Scotia Sea) or for the whole Southern Ocean. Such models, and the plausible regional estimates of krill consumption they produce, have an important role to play in addressing the critical questions about Southern Ocean food-webs outlined above. However, the ability to constrain and validate such models is limited by the available data and knowledge. In data poor situations, models can be particularly valuable for identifying critical knowledge gaps, such as the assumptions which most strongly influence conclusions about krill consumption (Hill et al. 2012; Hill and Matthews 2013; Southwell et al. 2015).

9.3 Regional Perspectives

Atkinson et al. (2008) highlight that, in contrast to many other species of zooplankton, the distribution of krill is concentrated between 0 and 90°W, with more than 70 % of the krill stock found in the Atlantic sector. These high levels of abundance are also partly why the commercial krill fishery now only operates in these waters. The concentration of krill in the Atlantic sector also means that there are large numbers of krill predators in this region. Consequently, many studies of predator-prey dynamics have been undertaken in the Atlantic sector, particularly within the Scotia Sea. In addition to the long history of research, this part of the krill-based ecosystem is also the focus of current attempts by CCAMLR to produce krill consumption estimates for use in management of the krill fishery.

In addition to the Atlantic sector, studies of predator-prey interactions are conducted in various other Southern Ocean regions, with important foci in the Ross Sea and in East Antarctica. These complementary studies have provided important advances in understanding about predator-prey dynamics as well as identifying regional contrasts and similarities. For example, studies from East Antarctica by Deagle et al. (2007, 2008) have shown that DNA analysis of faecal material from macaroni penguins accords well with dietary studies based on other analysis techniques. Similarly, Jarman et al. (2013) used DNA analysis of faecal material from Adélie penguins *Pygoscelis adelie* to show spatial and temporal changes in diet from 12 populations and that prey diversity was greater than previously thought. Jarman et al. (2013) showed that krill, fish, copepods and amphipods were the most important prey, which accords with other studies based on hard part remains or stable isotope analysis. However, their DNA analysis also showed that a substantial proportion of Adélie penguin diet comprised gelatinous groups such as jellyfish and comb jellies, and a range of other prey not previously identified in the diet of these penguins.

9.4 Problems in Relating Consumption to Production and Standing Stock

Croxall et al. (1985a) considered various ecological interactions between krill and its predators, highlighting some of the difficulties that exist when attempting to make comparisons between the amounts of krill consumed by seabirds and marine mammals and estimates of the standing stock of krill available in the water, or at least as estimated by acoustic or net-based surveys. They emphasised that important sources of error result from a number of factors. (i) The flux of krill, as water (and krill) is transported across a region. For example, they noted that the standing stock of krill in the Scotia Sea is likely to be replenished by upstream sources including from both the Weddell Sea and the Bellingshausen Sea, making balancing ecological budgets extremely difficult. While this conundrum remains valid today, flux is not the only process contributing to the turnover of krill. Atkinson et al. (2009) estimated that the total circumpolar gross post-larval production was about 2.5–4 times circumpolar biomass. This might overestimate the production available to predators as it does not include the effects of mortality but it suggests that such production might exceed the standing stock. Thus, the actual quantity of krill available to predators might be a complex interaction between standing stock, production, and flux. (ii) The diel vertical migration of krill continually alters the distribution of the standing stock. Individual krill migrate through the water column, moving to the surface at night, and returning to deeper depths through the day. This migration not only affects foraging seabirds and seals which are restricted in their dive depth range, but also affects any acoustic estimates of krill biomass which might be used when attempting to understand whether biomass is sufficient to meet consumption demand. This is because acoustic surveys generally cover a limited depth range, omitting the surface layer and deeper reaches of the water column. (iii) The patchy nature of krill aggregations also leads to difficulties in surveying all potential habitats with adequate survey effort when generating estimates of standing stock. With constraints on research budgets, this remains a constant challenge. For example, the last large scale krill survey in the Scotia Sea was 16 years ago in 2000 (Trathan et al. 2001; Hewitt et al. 2004). (iv) Finally, Croxall et al. (1985a) noted that estimates of standing stock are generally restricted in their temporal coverage whereas it is now apparent that krill density can vary considerably between seasons (e.g. Saunders et al. 2007). As predator consumption varies between seasons, extrapolation of temporally limited data for either predators or prey can compound uncertainties.

Standing stock estimates provide an indicator, but not a definitive measure of krill availability to predators. Caution is therefore necessary in comparing consumption to standing stock. Alternative comparisons to production are useful for assessing impacts over short time scales (e.g. Shreeve et al. 2009). Model-based studies generally include an explicit representation of turnover (such as the production to biomass ratio in mass-balanced food-web models e.g. Hill et al. 2012), but the difficulties with estimating these parameters suggested by Croxall et al. (1985a) still apply.

9.5 Consumption of Krill Eggs and Larvae

Adult female krill are reproductively active during summer, showing a comparatively high investment in the ovary, which may reach up to 46 % of the total wet weight of krill at some locations in some years (Tarling et al. 2007). A semi-empirical model used by Tarling et al. (2007) predicted that 11 % of females complete one spawning episode year^{-1}, 60 % complete two, and 29 % complete three or more. The number of spawning events may vary spatially and temporally, but on average, krill release many thousands of eggs each year. For example, at South Georgia the average female krill releases 12,343 eggs year^{-1}. Tarling et al. (2007) show that where eggs are unable to complete the descent–ascent developmental cycle because bathymetry is too shallow (i.e. on-shelf) eggs will sink to the sea floor; eggs that sink to the seabed are then vulnerable to predation by the benthos. In deeper water (i.e. off-shelf), eggs may return to the surface either as a metanauplii or as the first calyptopis stage. Tarling et al. (2007) concluded that predation on larvae is a major cause of local recruitment failure. Observed concentrations of larvae can be very high; for example, in the eastern Bellingshausen Sea, Pakhomov et al. (2004) found mean concentrations of almost 9000 larvae m^{-2} and maximum concentrations in excess of 30,000 larvae m^{-2}. At the Antarctic Peninsula, Brinton and Townsend (1984) found 45,000 larvae m^{-2}. In contrast, Siegel et al. (2013) reported much higher larval numbers along the western Antarctic Peninsula, with average values of 65,000 m^{-2} and with many stations having numbers in excess of 200,000 m^{-2}. Even higher levels have been reported in the Scotia Sea where a maximum of 1.8 10^6 larvae m^{-2} were recorded during 1981 (Siegel 2005). Such high larval numbers (see Chap. 2, Siegel and Watkins 2016) indicate the considerable predation potential.

In the maritime Sub-Antarctic, for example at South Georgia, some of the principle predators of krill larvae are Myctophid fish (Pakhomov et al. 1996), the hyperiid amphipod *Themisto gaudichaudii* (Pakhomov and Perissinotto 1996) and chaetognaths (Øresland 1990). An individual Myctophid fish can consume approximately 30 larvae day^{-1}, while an individual *Themisto gaudichaudii* might predate about 3 larvae day^{-1} and an individual chaetognath may eat around 0.5 larvae day^{-1} (Tarling et al. 2007). When the relative abundance of these predators in the South Georgia region is considered, estimates indicate that they have the potential to consume approximately 500 larvae m^{-2} day^{-1}. A succession of such predators feeding on a patch of larvae would therefore diminish numbers quite rapidly (Tarling et al. 2007).

Predation on eggs and larvae almost certainly occurs elsewhere, including in Antarctic waters. Understanding exactly where and when eggs and larvae are produced is therefore important when considering potential predation levels. Hofmann and Hüsrevoğlu (2003) showed that, as successful completion of the descent–ascent cycle is determined by both bathymetry and water mass, there are likely to be significant regional differences in the production of larvae. Predation on eggs and larvae is thus likely to be intimately linked to krill biogeography.

The nutritional value of eggs and larvae lost to predation, either pelagically, or within the benthos is potentially enormous, but remains largely unquantified. The consumption is probably highly seasonal, as eggs and larvae are available mainly during late summer and autumn, though multiple spawning episodes will prolong this period. Estimates of development times, from egg to post larva, vary but are of the order of 40–120 days (Ikeda 1984; Ross and Quetin 1982, 1983; Brinton and Townsend 1984), meaning that species that predate krill eggs and larvae must either cease feeding or find alternate nutritive sources at other times of year.

9.6 Consumption of Krill by Fish and Squid

There are a number of species of fish that eat Antarctic krill, including species from two (largely) demersal families, the Nototheniidae and Channichthyidae (order Perciformes), as well as species from the pelagic family Myctophidae. Life history information is scarce for many of these krill-eating fish, so considerable uncertainty remains about the importance of krill in their diets.

As a group, the Perciform fish occupy a number of different habitats (which are even more diverse if different life stages are also considered), therefore the diets of these fish are also highly variable. For some Perciform fish, krill may be only one of many different prey items; such species include *Notothenia rossii*, *Pseudochaenichthys georgianus* and *Chionodraco* spp. (Kock 1992). In contrast krill is probably the main prey item for others, including for *Champsocephalus gunnari*, *Lepidonotothen larseni*, *Chaennodraco wilsoni* and *Pleuragramma antarctica* (Kock 1992). Fish are clearly the most important predators in many other marine food-webs (Scheffer et al. 2005). However, the balance of predators in the Southern Ocean has been affected by serial over-exploitation of seals, whales, and demersal fish. The many effects of this include an apparent reduction in demersal fish populations to levels far below those which were present in the 1960s (Myers and Worm 2003). Thus, it is possible that these species were a more important group of krill consumers before they were depleted.

Antarctic krill occur in the diets of many Myctophid fish species, including *Electrona carlsbergi*, *Electrona antarctica*, *Krefftichthys anderssoni*, *Gymnoscopelus braueri*, *Gymnoscopelus nicholsi* and *Gymnoscopelus opisthopterus* (Pusch et al. 2004; Shreeve et al. 2009; Saunders et al. 2014). For *E. antarctica* and *Gymnoscopelus* spp., they are probably the dominant prey species (Kock 1992). In addition to the Myctophids, other abundant mesopelagic fish, including *Notolepis coatsi*, *Notolepis annulata* and *Bathylagus antarcticus* predominantly feed on krill (Kock 1992).

Rates of daily food intake vary considerably during the ontogenetic development of different fish species and they also vary seasonally (Kock 1992). For example, Kock (1992) reports that *N. rossii* shows reduced feeding in winter compared with summer, possibly by as much as a factor of 2 or 3. Feeding in some fish species also decreases or even stops prior to and during spawning; for example *N. rossii* and *C. gunnari* often have empty stomachs during this period, with as many as 90 % of fish fasting.

Hill et al. (2007) suggest that estimates of krill consumption per unit biomass are an order of magnitude lower for fish than for penguins and seals. However, Hill et al. (2007) also suggest that Myctophid fish are possibly the main consumers of krill in the Southern Ocean, though estimates of Myctophid abundance need to be re-evaluated to reduce existing levels of uncertainty. Collins et al. (2012) suggested that Myctophid biomass in the Scotia Sea is around 4.5 million tonnes with zooplankton consumption around 25 million tonnes year^{-1}, although they did not estimate the krill fraction of this consumption.

The potential importance of Antarctic krill in the diets of squid has been suggested (e.g. Miller and Hampton 1989), but is not well established (Collins and Rodhouse 2006). Serological methods have shown the presence of krill in the diet of *Psychroteuthis glacialis*, *Mastigoteuthis psychrophila*, *Moroteuthis knipovitchi*, *Moroteuthis robsoni*, *Slosarczykovia circumantarctica* and *Martialia hyadesi* (Kear 1992), though there are generally too few dietary data to determine whether krill is an important prey item for other squid species (Collins and Rodhouse 2006). Krill are certainly not a major prey item for Sub-Antarctic species such as *Martialia hyadesi* or *Moroteuthis ingens*, the only squid species for which there have been detailed dietary studies (Collins and Rodhouse 2006). That said, krill have been found to be an important dietary component for those squid species taken as by-catch in the Japanese krill fishery (Nemoto et al. 1985, 1988). These squid were all small and it is possible that krill might only be important during certain early developmental phases. Filippova and Yukhov (1979) suggest that squid that live near the ocean surface probably feed almost exclusively on crustaceans (krill, hyperiid amphipods and mysids), while larger species that inhabit the mesopelagic and bathypelagic zones probably feed on crustaceans during early life.

Thus, the importance of krill consumption by squid remains difficult to quantify; many pelagic squid species are relatively short lived but exhibit very rapid growth rates. Given that krill occur in the diets of some species, and are probably important for many more during different developmental periods, consumption of krill by squid certainly should not be ignored (Miller and Hampton 1989).

9.7 Consumption of Krill by Seabirds

One of the most comprehensive regional assessments of Antarctic krill consumption by seabirds is that of Croxall et al. (1985a). The focus of this work was the Scotia Sea which potentially includes 26–28 % of the circumpolar krill stock (Atkinson et al. 2008). The Scotia Sea is a highly productive area that is important for a broad guild of seabirds that feed on krill. Thus, though the work of Croxall et al. (1985a) is now some three decades old, their assessment remains one of the key pieces of work on krill consumption, especially during summer. These authors show that krill is the most important dietary item for most seabirds in the Scotia Sea, comprising some 82 % of their diet; by comparison, copepods, squid, fish and amphipods comprise only 8 %, 4 %, 3 % and 2 % respectively. Croxall et al. (1985a) report that 21 species of seabirds predate krill in this region, with the major consumers being macaroni penguins, chinstrap penguins *Pygoscelis antarctica*, and Antarctic prions *Pachyptila desolata* which together account for over 50 % of all krill consumed by seabirds.

Techniques to better estimate the population sizes of some krill predators, particularly surface breeding species such as penguins, are now becoming available with the advent of high resolution satellite imagery (e.g. Fretwell and Trathan 2009; Fretwell et al. 2012). These methods still need to be refined, but these new approaches look promising (e.g. Schwaller et al. 2013; LaRue et al. 2014) and will hopefully provide in the future better population estimates and hence better krill consumption estimates.

In 1985 Croxall and co-workers noted that there were important gaps in our understanding of food-web connections related to krill. These gaps still largely remain today in 2016. Croxall et al. (1985a) highlighted the need for better estimates of many predator population sizes, including for macaroni and chinstrap penguins, and small petrel species. Only macaroni penguin population estimates have been revised in the intervening period (Trathan et al. 2012), although this revision does not include estimates of pre-breeders which may comprise a sizable proportion of the population. Other factors which are still important to re-evaluate include better data on dietary composition and the energy available from krill, especially in winter, and better information about bio-energetics and the energy cost of certain activities. Gorman et al. (2014) have shown that such issues are complex, highlighting previously unknown trophic pathways and the need to consider gender and sex-specific foraging niches, particularly in relation to environmental variability and potential changes in future habitat.

CCAMLR has established an ecosystem monitoring programme focusing on krill-eating seals and seabirds, especially penguins, to better understand whether changes in the ecosystem might be brought about by fishery impacts or by natural variability.

9.8 Consumption of Krill by Marine Mammals

Croxall et al. (1985a) also provide estimates of krill consumption by a number of marine mammals. As with seabirds, krill consumption estimates for most mammalian species depend upon reliable estimates of their population size, yet these still remain uncertain. For example, revised population estimates exist for crabeater seal *Lobodon carcinophagus* close to the Antarctic Peninsula (Forcada et al. 2012) and Antarctic fur seal *Arctocephalus gazella* at South Georgia (Boyd 1993; Boyd 2002); however, populations of these major krill consumers are known to be changing (e.g. Forcada and Hofmann 2014).

Recently, it has been inferred that elephant seals *Mirounga leonina* feed upon krill (Walters et al. 2014) which constitutes an abundant, easily accessible source of prey in water masses used by sub-yearling seals. Walters et al. (2014) used isotopic assessment and concurrent tracking of seals to successfully identify ontogenetic shifts in broad-scale foraging habitat use and diet in these animals. Although elephant seal population sizes in the Scotia Sea are reasonably well known (Boyd et al. 1996), the consumption of krill by juvenile elephant seals has not previously been considered.

The importance of krill in the diets of various species of baleen whale has long been recognised, and krill consumption by cetaceans has been studied over many years, including by the *Discovery Investigations* (e.g. Mackintosh 1974). Blue whales *Balaenoptera musculus*, fin whales *Balaenoptera physalus*, sei whales *Balaenoptera borealis*, minke whales *Balaenoptera bonaerensis*, humpback whales *Megaptera novaengliae* and southern right whales *Eubalaena australis*, all consume krill, though they all also predate other species including other crustaceans and sometimes fish. In some areas, particularly the Ross Sea, crystal krill *Euphausia crystallorophias* may become a dominant prey item. The baleen whales which feed in the Southern Ocean breed at lower latitudes. Each species arrives in the Southern Ocean at slightly different times of year and preferentially feeds in different habitats (Lockyer 1981a). For example minke whales are species of the pack ice (Friedlaender et al. 2014), while blue whales are species of the ice edge, fin whales are species of more open water and humpback whales are more coastal (Ropert-Coudert et al. 2014).

Reilly et al. (2004) reviewed techniques used to estimate krill consumption by baleen whales together with estimates of whale abundance. They concluded that consumption estimates are still very much less than they would be if baleen whale populations were to recover to pre-exploitation levels.

Baleen whales historically consumed substantial amounts of krill, not just in coastal areas such as at South Georgia, but across the entire Southern Ocean. The relationship between whale catch distribution and various physical properties of the Southern Ocean has been noted before (e.g. de la Mare 1997, 2009, Ackley et al. 2003). This distribution also mirrors the circumpolar distribution of krill (Atkinson et al. 2004, 2008), reflecting the broadest scale over which krill predators operate.

9.9 The Krill Surplus Hypothesis

Understanding variability in patterns of distribution, abundance of species and structure of communities, together with the consequences for species diversity has been a central motivation for ecologists for many decades (Hutchinson 1959); notwithstanding, ecological knowledge about some of the processes that generate these patterns remains elusive. Perturbations to natural systems cause change, and observing these changes as they occur over various timescales can provide insight into the underlying processes. To quote Richard Feynman (Feynman et al. 1963), "We do not know what the rules of the game are; all we are allowed to do is to watch the playing. Of course, if we watch long enough, we may eventually catch on to a few of the rules".

Some systems are particularly difficult to understand and ecological patterns only emerge after decades of observation. The Southern Ocean is one ecosystem where understanding has been slow to emerge (Hill et al. 2006). This is not simply because the Southern Ocean is remote and logistically difficult to access; it is also because it is more complex than commonly held and has been perturbed over decades of commercial exploitation. However, in one sense, the perturbation caused by the removal of seals and whales through decades of unregulated exploitation can be viewed as a major *experiment* which can provide an opportunity for us to gain insight into ecological processes. Two contrasting ideas about the effects of this perturbation are that the removal of the great whales led to a possible "krill surplus" (Laws 1977; Ballance et al. 2006) or that it led to a possible reduction in krill (Nicol et al. 2010), (see below).

Early commercial exploitation of Antarctic fur seals removed a considerable amount of krill-dependent biomass from the Southern Ocean, possibly many tens of thousands of tonnes (Mori and Butterworth 2006). Although these figures are vague, it is certain that the removal of so many seals meant that impacts on krill, their main prey species, would have been substantial (Emslie and Patterson 2007). Later, as commercial interests in the Southern Ocean shifted, the initial biomass of whales, estimated to have been 45.6 million tonnes, was reduced to just 8.2 million tonnes (Laws 1977), again with major impacts upon their main prey, krill. Further impacts on krill will also have resulted from the commercial fisheries which removed hundreds of thousands of tonnes of krill dependant finfish (Kock 1992). Commercial harvesting for finfish in the 1960s is known to have profoundly altered fish communities in the Sub-Antarctic, many of which fed upon krill (Kock 1992).

The subsequent annual release of prey resulting from just the removal of whales alone was thought to be enormous; approximately 147.0 million tonnes of krill, 7.1 million tonnes of squid and 3.3 million tonnes of fish (Laws 1977). It was recognised that such substantial changes in the marine ecosystem would have resulted in a major ecosystem response across a variety of trophic levels (Sladen 1964; Laws 1977; Ainley et al. 2007). The additional release of krill arising from the removal of seals and fish, and its consequences for other predator populations, has not been adequately assessed, nor have the implications of the fact that some of these fish were alternative prey for other krill predators. Any such assessment would probably now be confounded by the serial nature of commercial removals, which would have facilitated different ecosystem responses, including in some of these harvested components.

Sladen (1964) first recognised that increases in populations of both chinstrap penguins and Adélie penguins in the Scotia Sea were likely to be a direct result of the declines in the baleen whale stocks. Documented increases in other species, including gentoo penguins *Pygoscelis papua*, and recovering Antarctic fur seal populations (Payne 1977), were also thought to be due to the much more abundant food supplies.

The Southern Ocean marine system has therefore shown clear signs of ecosystem change in the past century, following the massive perturbation caused by the removal of fur seals and the great whales. However, how such changes will act out into the future and how long the ecosystem will take to recover is still unknown (e.g. Murphy 1995).

9.10 Recovery of the Seal and Whale Stocks

Other changes in the Antarctic marine ecosystem may complicate the recovery of seal and whale populations. Mostly importantly, Antarctic krill is now thought to have undergone more than a twofold decrease in abundance, at least in the Scotia

Sea, between the early 1970s and late 1980s (Atkinson et al. 2004, 2008, 2014). Abundance estimates for the 1990s and 2000s indicate significant inter-annual variability but no trend (Atkinson et al. 2004, 2008, 2014). Atkinson et al. (2004) based their abundance estimates on samples taken from research nets. More recently, Fielding et al. (2014) have shown that density estimates based on acoustic surveys indicate that, at least at South Georgia, no further decline has been detectable since the mid 1990s. However, any change in krill abundance is highly likely to have had a major impact upon those avian and mammalian consumers, such as penguins, seals and baleen whales that depend upon krill (Croxall et al. 1984). Other changes in the zooplankton community have also been reported, but these may be more subtle. Thus, studies (Ward et al. 2008) showed that in some regions zooplankton abundance values were very much lower in the early part of the twentieth century than in recent years; however, their analyses did not reveal any systematic differences in species composition across years. The changes in zooplankton communities are not fully understood; however, changes in sea ice (Parkinson 2002, 2004) and water temperature (Whitehouse et al. 2008) have been implicated (Atkinson et al. 2004).

Populations of Antarctic fur seals, harvested almost to extinction in the eighteenth and nineteenth centuries, have nearly now recovered (Barlow et al. 2002). Indeed, populations at South Georgia, the main breeding centre for the species, have increased from a few tens of individuals in the 1950s to over 1.5 million by the early 1990s (Boyd 1993). Boyd (1993) reports that the population growth rate decreased from the exceptionally high level of 16.8 % over the period 1955–1971 (Payne 1977) to 9.8 % in the period 1977–1990. This is further supported by recent work that suggests that density dependent effects may be becoming important (Reid and Forcada 2005); data from long-term monitoring sites on Bird Island (see Forcada et al. 2005; Forcada and Hoffman 2014) indicate that the number of pups born annually at this site has actually decreased over the period 1984/1985–2003/2004.

Thus, fur seal population numbers may be approaching their upper limit in parts of their range at South Georgia. Consumption of krill by fur seals, based on existing energetic analyses (Boyd 2002), suggests that if populations were as high as four million animals (BAS unpublished data), then krill consumption would probably be approaching 6.80 million tonnes year^{-1}.

Although exploitation of whales in the Southern Ocean ceased with the moratorium on commercial whaling in 1986, comprehensive data are not yet available to describe the recovery of stocks which forage in Antarctic waters. Nevertheless, it is possible to deduce some of the likely changes in whale numbers using knowledge of the changes in their breeding population size based on counts in their calving areas. Two stocks that are known to be increasing are the humpback and southern right whales that calve in the coastal waters off Brazil and Argentina respectively (Zerbini et al. 2004; Leaper et al. 2006). These stocks feed around South Georgia during the summer months and both are thought to feed on krill when in the Scotia Sea (Tormosov et al. 1998; Reilly et al. 2004).

Thus, populations of marine mammals (seals and whales) are showing clear signs of recovery from historical exploitation, and as such, the hypothesised "krill surplus" (Laws 1977) might be expected to decrease. This will have important consequences for populations of other krill-dependant predators, such as penguins, that may have expanded as a consequence of increased food availability following the exploitation of seals and whales.

9.11 Guilds of Krill Predators

Antarctic krill aggregate in a variety of forms, including loose, diffuse layers and dense, tightly packed swarms (Miller and Hampton 1989; see Chap. 8, Tarling and Fielding 2016). As such there is a broad guild of species that prey upon krill and which, depending upon krill aggregation state, potentially benefit from increased feeding opportunities in ways that are not always easy to model or predict. Thus, the relative success of different predator species in a given situation is likely to depend on the interaction between the form of the aggregation and species-specific foraging strategies.

Minke whales apparently specialize in swallow feeding under ice flows (Friedlaender et al. 2014), which may be made easier by their head shape and jaw structure (Lockyer 1981a, b). Other baleen whales that feed by swallowing are the blue, fin and humpback whales. In contrast, right whales are classed as skimmers (Lockyer 1981a, b). Swallow feeders rely upon concentrated plankton swarms, whilst skimmers swim with open mouths sifting the plankton from the water.

Other krill-eating species also have preferred foraging strategies, with different predators either surface seizing or plunge diving to find krill. The feeding strategies of pursuit divers, such as seals and penguins, have been studied extensively, respectively showing analogies with swallowing and skimming whales. For example, fur seals are known to feed in bouts (Boyd 1996) taking multiple krill during each dive within a bout. Most diving behaviour occurs at night, and most dives are shallow (Croxall et al. 1985b; Boyd and Croxall 1992). Mouth opening events occur mainly during the bottom phase of a dive (Iwata et al. 2012). Boyd (1996) has shown that the time taken for a fur seal to locate a new patch after leaving an old one is an indication of the distance between patches and that the intervals between bouts changes between years. These changes suggest that the structure and/or the spatial distribution of krill swarms varies between years and that fur seals must adapt their foraging behaviour accordingly. In contrast, penguins feed mainly during the day when they dive repeatedly with foraging dives relatively constant over the course of the foraging trip (e.g. Croxall et al. 1993). Penguins feed during periods when they undulate, or change depth rapidly, also generally during the bottom phase of a dive (Takahashi et al. 2004). During the course of these undulations the number of underwater beak-opening events increases, suggesting that the relative feeding intensity of each dive can be represented by depth-profile data. Underwater beak-opening patterns of krill-feeding penguin species are comparable with similar data from fish- and squid-feeding penguins (Takahashi et al. 2004).

The relationship between a particular krill-eating species and variation in the availability of krill is influenced by many factors including (i) the factors that determine prey availability such as density, aggregation and location, (ii) how the species selects and interacts with its prey, including in the presence and absence of other krill-eating predators, (iii) how the species' demography and reproductive output interact with its diet, and, (iv) the species', demographic processes, including their rates of maturation and onset of sexual maturity (e.g. Lockyer 1972, 1974; Laws 1977).

While knowledge of these issues is increasing for some species Atkinson et al. (2012) note that inferences about krill populations derived from predator performance and diet can be prone to bias, nonlinearity and noise. A major source of bias is the fact that predators are not random samplers of the wider krill population. Factors such as the vertical and horizontal distribution of krill in relation to the foraging range and diving depth of predators, the density of swarms and the presence of competitors or natural enemies of the predators may all influence this (Croxall et al. 1985b, 1988; Reid et al. 1996; Trathan et al. 2012). Predators also select krill according to size, sex, etc. (Hill et al. 1996). Such interactions (including multiple predator interactions) are important to consider in ecosystem models (Hill et al. 2006).

A major nonlinearity arises from the predator's functional response, i.e. the way consumption rate changes with prey availability (Holling 1959). Various authors (e.g. Boyd and Murray 2001; Reid et al. 2005; Cury et al. 2011) present evidence for asymptotic functional responses in some krill predators. However, Waluda et al. (2012) suggest that the functional response for macaroni penguins is sigmoidal. This type of response is associated with abrupt switching from one prey type to another, which raises the additional complication that the functional response is modified by the availability of alternative prey (Hill et al. 2005).

9.12 Seasonality and Temporal Considerations

The consumption of krill, whether by other zooplankton, fish, squid, seabirds or marine mammals, varies seasonally. Each predator has a natural cycle that governs its life-history processes; this not only includes its ontogenetic development from juvenile to adult, but also any recurrent cycle as it reproduces. At any different life stage an individual predator may feed upon different parts of the krill stock and have varying degrees of dependence upon krill.

Long-lived vertebrates potentially exemplify the complexity of these cycles. For example, sub-yearling elephant seals may feed upon krill, but once they mature, they feed upon other prey. Other predators, such as seals or whales may feed upon krill while they are in the Antarctic, but not whilst they are outside the Antarctic.

Any or all of a predator's weight gain, body condition, propensity to breed, or reproductive output may vary in response to prey availability. Such variation in life-history parameters might not be immediately apparent, but may be lagged by days, weeks or months. Where reproductive output is affected, changes in population might not be apparent until cohorts recruit months or years into the future. Thus, understanding how predator-prey interactions impact predator populations can be extremely complex.

9.13 Feed-Backs from Predators to Krill

Predators have the potential to alter not only the density, but also the structure of their prey field; for example, humpback whales feeding on krill may use streams of bubbles, so called bubble-netting, to cause krill to aggregate (e.g. Jurasz and Jurasz 1979). Diving predators may cause swarms to move deeper in the water column, possibly outside the reach of other species of diving predator, whilst demersal fish may cause krill to move off the seabed to mid water depths. Such predator-prey interactions have been suggested over many years.

A different sort of interaction between whales and krill has also been postulated more recently (e.g. Smetacek 2008; Nicol et al. 2010). Iron is generally thought to be the limiting micronutrient for phytoplankton in the Southern Ocean, which when released and re-mineralised following grazing by krill and other herbivores, cycles back through the marine system. An alternate major mechanism for recycling iron could also have been defecation by baleen whales after they have consumed krill. Nicol et al. (2010) show that whale faecal iron concentration is approximately ten million times that of Antarctic seawater, suggesting that whales could be an important source of the micronutrient. Nicol et al. (2010) also calculate that the krill population currently contains approximately 24 % of the total iron in the surface waters of the Southern Ocean, thus acting as a long-term reservoir of iron. Pre-exploitation populations of whales must also have stored larger quantities of iron and recycled more iron in surface waters, enhancing overall ocean productivity through a positive feedback loop. Nicol et al. (2010) therefore speculate that allowing the baleen whales to recover could actually increase Southern Ocean productivity through making enhanced iron levels in surface waters available to phytoplankton.

Coupled with the ongoing debate about the relative impacts on krill of historical harvesting versus climate change (Trathan and Reid 2009; Trathan et al. 2012), uncertainty about whether krill predators actually enhance the abundance of their prey suggests a wide range of possible scenarios for how the Southern Ocean ecosystems might develop in the future. This highlights the uncertainty in any attempt to project the future state of the system and emphasises the challenges in managing activities that might influence this state.

9.14 Why Do We Need to Know About Krill Consumption?: A Re-evaluation

At the start of this chapter we highlighted a number of reasons why so much emphasis has been placed on determining the levels of krill consumed by its various predators. Ultimately all of these reasons are about understanding ecosystem change, and managing human activities (particularly fishing) that potentially drive such change. Important progress has been made on quantifying some aspects of the general krill consumption problem (e.g. macaroni penguin population estimates, macaroni penguin winter distribution, etc.), but many aspects still remain unresolved (e.g. squid population estimates, Myctophid fish population estimates). Progress towards a comprehensive understanding of krill consumption and how this fits into Southern Ocean food-webs has therefore been relatively modest since Everson (1977) first addressed the issue.

The motivating questions outlined above all remain vitally important, particularly as we try to project how the Southern Ocean ecosystem might respond to ongoing climate change or react to increased levels of commercial krill harvesting. Such issues might become increasingly important as human populations continue to increase and governments seek to sustainably develop any remaining under-exploited sources of marine protein (such as krill). Therefore, we need realistic ways to characterise how krill is connected to its predators in the marine food-web.

CCAMLR established its ecosystem monitoring programme (CEMP) in the mid-1980s, with the aspiration of using the programme in managing the commercial fishery for Antarctic krill (SC-CAMLR 1985). CEMP monitors land-based krill predators and aims to understand whether changes in the ecosystem might be brought about by fishery impacts. At present, the CEMP has almost 30 years of monitoring data but is not well integrated into management. These data, analysed in conjunction with krill indices, have provided insights into the relationships between krill and its avian and mammalian predators which demonstrate, *inter alia*, that (i) significant natural variability in the physical marine ecosystem, which alters ecological relationships involving krill (Trathan et al. 2003; Murphy et al. 2007), also affects krill predators such that some show very low levels of reproductive output in years of low krill availability (Reid et al. 2005; Trathan et al. 2006), whilst others might switch to alternative prey and still produce offspring (Waluda et al. 2012); and (ii) climate change may be having impacts on both on krill (Atkinson et al. 2004) and its predators (Forcada and Hofmann 2014). These issues provide important context for any proposed management system that relies on CEMP data.

There are additional challenges which remain to be addressed, including that (i) CEMP sites are few and far between, especially compared to the wide geographic distribution of the krill fishery (e.g. Murphy et al. 1997); (ii) the functional responses of the monitored species to differing levels of krill abundance are not well understood (e.g. Boyd and Murray 2001; Barlow et al. 2002); and (iii) the monitored species are unlikely to be the major consumers of krill and the

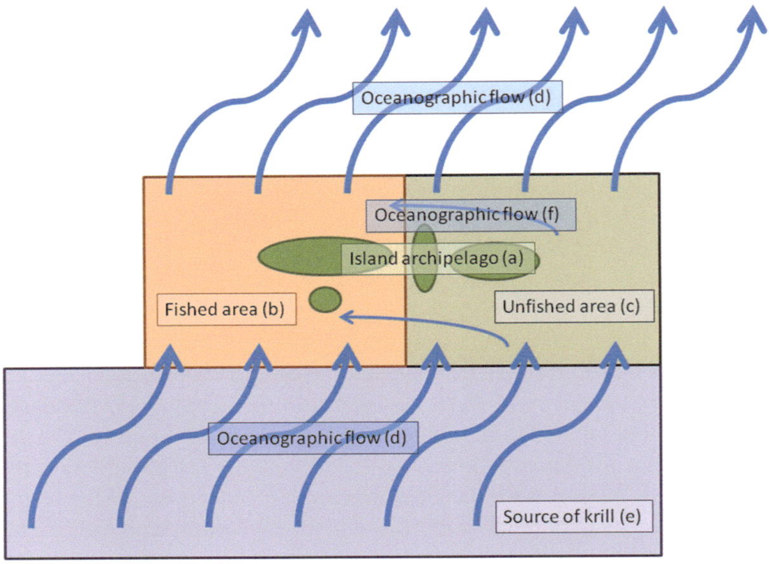

Fig. 9.2 Potential experimental design of an integrated study region including a reference area; (**a**) archipelago of islands; (**b**) area where fishing for krill is allowed; (**c**) reference area where no krill fishing is allowed; (**d**) regional oceanographic flow; (**e**) regional source of krill; (**f**) local oceanographic flow (See text for explanation)

relationships between these species and the state of the wider group of krill predators is not well understood.

To some extent, CEMP offers a practical framework for assessing the ecosystem effects of krill fishing through monitoring accessible predators. However the utility of such information could be significantly improved by studying more directly how predator populations respond to local changes in krill.

This approach would require regular estimates of krill availability (distribution and abundance) and of the abundance of representative predators. The local scale might be more tractable than quantifying the seasonal demands of the whole guild of krill predators at a regional or circumpolar scale. However such local study areas would need to be understood in the regional context, particularly in terms of krill flux and seasonal variations in predator foraging range and diet. Such an approach might also usefully include reference areas (Fig. 9.2) where the confounding impacts of climate change and harvesting might be more easily disentangled.

An integrated study area (Fig. 9.2) should take into account both the regional and local oceanographic flow and flux of krill at varying scales (e.g. Murphy et al. 1998; Pinõnes et al. 2011; 2013; Dinniman and Klink 2004) when identifying the location of reference areas, which should then be located upstream of any harvesting impacts. This is vital if the confounding impacts of fishing and climate change are to be resolved. Harvesting in reference areas would need to be prohibited, or stringently regulated. Harvesting in fished areas should be subject to a catch limit based on acoustic surveys to estimate regional krill biomass and assess whether depletion could be detected in the fished area over the course of the fishing season. This might necessitate a regional acoustic survey each year, followed by a series of local surveys in the area of fishing.

Many krill-eating species, particularly zooplankton, squid and fish are much more difficult to monitor than land-based predators. Therefore, like CEMP, an integrated study might usefully focus on a restricted set of ecosystem indicator species. These might include land-based predators such as penguins, and pelagic predators such as baleen whales. These species should be monitored across the region, including in both fished areas and reference areas. The problem remains that indicator species might have complex relationships with the state of the wider group of krill predators. This can be addressed to some extent by including contrasting species, but insights into the wider state of the ecosystem gained, for example, through monitoring predator diets, would be an important objective.

Modelling work has shown there is little evidence that krill predators maintain similar ecological niches across years (Friedlaender et al. 2011). Different species have unique physiological constraints and foraging behaviours which may have a stronger influence on ecological niche than proximity to prey for some species (penguins and pack-ice seals). In contrast, fully marine predators (cetaceans), occupy ecological niches closer to prey. Nonetheless, Friedlaender et al. (2011) also suggest that the amount of overlap between modelled niches is relatively small, even for species with similar energetic requirements. Thus, in a rapidly changing environment, the relationships between monitored predators, their prey, and the wider ecosystem are likely to change as well. Detailed information about the local ecosystem context might therefore be essential in interpreting data from indicator species.

Integrated study areas are not a new concept and have been advocated by CCAMLR in the past. Indeed, integrated studies currently exist e.g. the Palmer LTER (Ducklow et al. 2007), the US AMLR grid (Trivelpiece et al. 2011) and the South Georgia Core Box (Murphy et al. 2007; Trathan et al. 2014). These studies are generally focussed on ecological understanding. We suggest implementing a similar scientifically robust framework in priority areas for fisheries management; any such efforts would almost certainly require scientific contributions from a number of research groups as well as from the krill-fishing industry.

The response of predators to prey availability has been a key ecological issue for many decades. In seeking to understand how krill predators respond to krill, integrated studies have the potential to greatly improve our understanding. For example, the Palmer LTER has revealed some of the ecological complexity found within predator-prey interactions. There, recent work has demonstrated that the foraging patterns of penguins, including foraging distance, relate to tidal phase (Bernard and Steinberg 2013). Moreover, Chapman et al. (2010) have shown that penguin chicks that fledge and survive to recruit into the breeding population are generally heavier at fledging than those that do not survive to breed. Spatial differences in prey nutritive value, as well as temporal differences, have also been shown to have important implications for predators. Ruck et al. (2014) showed that across all sexes and maturity stages, krill in the southern part of the Palmer study region had higher total lipid content than those in the north. This regional variability in prey quality could affect the ability of apex predators to meet their energetic demands. These examples demonstrate the complexity of predator-prey interactions involving krill; other integrated study areas have revealed similar levels of complexity. Without such sustained long-term ecological research many such predator-prey complexities would be difficult to elucidate.

9.15 Conclusion

Antarctic krill is rightly characterised as a central component of Southern Ocean food-webs and a key food source for a diverse array of predators in a wide range of marine habitats. It is true that simple three-step food chains (phytoplankton – krill – predators) are an important feature of these food-webs. However, this simplistic representation belies the complexity which decades of study into predation on krill has revealed. The size of krill changes by two orders of magnitude as it grows from egg to adult. It is found throughout the water column from surface layers to abyssal depths and at most latitudes south of the Polar Front. It has a highly heterogeneous distribution at a variety of spatial scales, from dense swarms at the 10 m scale to the concentration of circumpolar biomass in a few productive regions such as the Scotia Sea. These distribution patterns are constantly changing as a result of krill behaviour, flux and the influence of predators. The predators themselves are rarely obligate krill feeders, and so their diets vary ontogenetically, regionally and seasonally with the availability of krill and alternative prey. They employ a wide range of foraging strategies and travel different distances to feed, from the trans-ocean migrations of baleen whales to the relative immobility of some benthic organisms. Superimposed on this trophic complexity are the effects of environmental variability and change, which can cause major spatial and temporal shifts in productivity and habitat quality, affecting the population dynamics of predators and prey alike. Moreover, the historical overexploitation of seals, whales and demersal fish has perturbed these dynamics, which continue to change as a result.

With this level of complexity, and the vastly different levels of information available on different krill predators, it is hardly surprising that although understanding of the trophic role of krill continues to grow, attempts to assess krill consumption at the circumpolar or regional scale yield crude and incomplete estimates. The key motivation behind such estimates is to understand how the food-web operates and how it responds to variability in the availability of krill. Climate change and krill fishing are potential drivers of this variability which require monitoring and, in the case of fishing, managing. Integrated study areas, designed to control the spatial patterns of fishing and monitor both the krill stock and krill predators offer a potentially more tractable means of understanding the food-web response to changes in the availability of krill.

Acknowledgement We thank the many colleagues at BAS, within the CCAMLR community, and more widely for their valuable help and discussion over many years.

References

Ackley S, Wadhams P, Comiso JC, Worby AP (2003) Decadal decrease of Antarctic sea ice extent inferred from whaling records revisited on the basis of historical and modern sea ice records. Polar Res 22:19–25

Ainley DG, Ballard G, Ackley S, Blight LK, Eastman JT, Emslie SD, Lescroël A, Olmastroni S, Townsend SE, Tynan CT, Wilson P, Woehler E (2007) Paradigm lost, or is top-down forcing no longer significant in the Antarctic marine ecosystem? Antarct Sci 19:283–290

Armstrong AJ, Siegfried WR (1991) Consumption of Antarctic krill by minke whales. Antarct Sci 3:13–18

Atkinson A, Hill SL, Barange M, Pakhomov EA, Raubenheimer D, Schmidt K, Simpson SJ, Reiss C (2014) Sardine cycles, krill declines, and locust plagues: revisiting 'wasp-waist' food webs. Trends Ecol Evol 29(6):309–316

Atkinson A, Siegel V, Pakhomov EA, Rothery P (2004) Long term decline in krill stock and increase in salps within the Southern Ocean. Nature 432:100–103

Atkinson A, Siegel V, Pakhomov EA, Rothery P, Loeb V, Ross RM, Quetin LB, Schmidt K, Fretwell P, Murphy EJ, Tarling GA, Fleming AH (2008) Oceanic circumpolar habitats of Antarctic krill. Mar Ecol Prog Ser 362:1–23

Atkinson A, Siegel V, Pakhomov EA, Jessopp MJ, Loeb V (2009) A re-appraisal of the total biomass and annual production of Antarctic krill. Deep-Sea Res Pt I 56:727–740

Atkinson A, Nicol S, Kawaguchi S, Pakhomov E, Quetin L, Ross R, Hill S, Reiss C, Siegel V (2012) Fitting *Euphausia superba* into Southern Ocean food-web models: a review of data sources and their limitations. CCAMLR Sci 19:219–245

Ballance LT, Pitman RL, Hewitt RP, Sineff DB, Trivelpiece WZ, Clapham PJ, Brownell RL Jr (2006) The removal of large whales from the Southern Ocean: evidence for long-term ecosystem effects? In: Estes JA, Demaster DP, Doak DF, Williams TE, Brownell RL Jr (eds) Whales, whaling, and ocean ecosystems. University of California Press, Berkeley, pp 215–230

Ballerini T, Hofmann EE, Ainley DG, Daly K, Marrari M, Ribic CA, Smith WO Jr, Steele JH (2014) Productivity and linkages of the food web of the southern region of the western Antarctic Peninsula continental shelf. Prog Oceanogr 122:10–29

Barlow KE, Boyd IL, Croxall JP, Reid K, Staniland IJ, Brierley AS (2002) Are penguins and seals in competition for Antarctic krill at South Georgia? Mar Biol 140:205–213

Barrera-Oro E (2002) The role of fish in the Antarctic marine food web: differences between inshore and offshore waters in the southern Scotia Arc and west Antarctic Peninsula. Antarct Sci 14:293–309

Bengston JL (1984) Review of Antarctic marine fauna. In: Selected papers presented to the Scientific Committee of CCAMLR, 1982–1984, Part 1. CCAMLR, Hobart, pp 1–126

Bernard KS, Steinberg DK (2013) Krill biomass and aggregation structure in relation to tidal cycle in a penguin foraging region off the Western Antarctic Peninsula. ICES J Mar Sci 70:834–849

Boyd IL (1993) Pup production and distribution of breeding Antarctic fur seals (*Arctocephalus gazella*) at South Georgia. Antarct Sci 5:17–24

Boyd IL (1996) Temporal scales of foraging in a marine predator. Ecology 77:426–434

Boyd IL (2002) Estimating food consumption of marine predators: Antarctic fur seals and macaroni penguins. J Appl Ecol 39:103–119

Boyd IL, Croxall JP (1992) Diving behavior of lactating Antarctic fur seals. Can J Zool 70:919–928

Boyd IL, Murray AWA (2001) Monitoring a marine ecosystem using responses of upper trophic level predators. J Anim Ecol 70:747–760

Boyd IL, Walker TR, Poncet J (1996) Status of southern elephant seals at South Georgia. Antarct Sci 8:237–244

Brinton E, Townsend AW (1984) Regional relationships between development and growth in larvae of Antarctic krill, *Euphausia superba*, from field samples. J Crustac Biol 4(Special Issue):224–246

Chapman EW, Hofmann EE, Patterson DL, Fraser WR (2010) The effects of variability in Antarctic krill (*Euphausia superba*) spawning behavior and sex/maturity stage distribution on Adélie penguin (*Pygoscelis adeliae*) chick growth: a modelling study. Deep-Sea Res Pt II 57:543–558

Clarke A, Tyler PA (2008) Adult Antarctic krill feeding at abyssal depths. Curr Biol 18:282–285

Collins MA, Rodhouse PGK (2006) Southern ocean cephalopods. Adv Mar Biol 50:191–265

Collins MA, Stowasser G, Fielding S, Shreeve R, Xavier JC, Venables HJ, Enderlein P, Cherel Y, Van de Putte A (2012) Latitudinal and bathymetric patterns in the distribution and abundance of mesopelagic fish in the Scotia Sea. Deep-Sea Res Pt II 59–60:189–198

Cornejo-Donoso J, Antezana T (2008) Preliminary trophic model of the Antarctic Peninsula ecosystem (Subarea CCAMLR 48.1). Ecol Model 218:1–17

Croxall JP, Ricketts C, Prince PA (1984) Impact of seabirds on marine resources, especially krill, of South Georgia waters. In: Whittow GC, Rahn H (eds) Seabird energetics. Plenum Press, New York, pp 285–318

Croxall JP, Prince PA, Ricketts C (1985a) Relationships between prey life cycles and the extent, nature and timing of seal and seabird predation in the Scotia Sea. In: Siegfried WR, Condy PR, Laws RM (eds) Antarctic nutrient cycles and food webs. Springer, Berlin, pp 516–533

Croxall JP, Everson I, Kooyman GL, Ricketts C, Davis RW (1985b) Fur seal diving behaviour in relation to vertical distribution of krill. J Anim Ecol 54:1–8

Croxall JP, Davis RW, O'Connell MJ (1988) Diving patterns in relation to diet of gentoo and macaroni penguins at South Georgia. Condor 90:157–167

Croxall JP, Briggs DR, Kato A, Naito Y, Watanuki Y, Williams TD (1993) Diving pattern and performance in the macaroni penguin *Eudyptes chrysolophus*. J Zool 230:31–47

Cury PM, Boyd IL, Bonhommeau S, Anker-Nilssen T, Crawford RJM, Furness RW, Mills JA, Murphy EJ, Österblom H, Paleczny M, Piatt JF, Roux J-P, Shannon L, Sydeman WJ (2011) Global seabird response to forage fish depletion – one-third for the birds. Science 334:1703–1706

de la Mare WK (1997) Abrupt mid-twentieth-century decline in Antarctic sea ice extent from whaling records. Nature 389:57–60

de la Mare WK (2009) Whaling records and changes in Antarctic sea ice: consistency with historical records. Polar Rec 38:355–360

Deagle BE, Gales NJ, Evans K, Jarman SN, Robinson S, Trebilco R, Hindell MA (2007) Studying seabird diet through genetic analysis of faeces: a case study on macaroni penguins (*Eudyptes chrysolophus*). PLoS One 2:e831

Deagle BE, Gales NJ, Hindell MA (2008) Variability in foraging behaviour of chick-rearing macaroni penguins *Eudyptes chrysolophus* and its relation to diet. Mar Ecol Prog Ser 359:295–309

Dinniman MS, Klink JM (2004) A model study of circulation and cross-shelf exchange on the west Antarctic Peninsula continental shelf. Deep-Sea Res Pt II 51:2003–2022

Ducklow HW, Baker K, Martinson DG, Quetin LB, Ross RM, Smith RC, Stammerjohn SE, Vernet M, Fraser W (2007) Marine ecosystems: the West Antarctic Peninsula. Philos Trans R Soc B 362:67–94

Emslie SD, Patterson WP (2007) Abrupt recent shift in δ 13C and δ 15 N values in Adélie penguin eggshell in Antarctica. Proc Natl Acad Sci U S A 104:11666–11669

Everson I (1977) The living resources of the Southern Ocean. Southern Ocean fisheries survey programme. FAO Report GLO/SO/77/1, 156 pp

Everson I (1984) Marine interactions. In: Laws RM (ed) Antarctic ecology, vol 2. Academic, London, pp 783–819

Everson I (2000) Role of krill in marine food webs: the Southern Ocean. In: Everson I (ed) Krill: biology, ecology and fisheries. Blackwell Science, Oxford, pp 194–201

Feynman RP, Leighton RB, Sands M (1963) The Feynman lectures on physics. California Institute of Technology, California

Fielding S, Watkins JL, Trathan PN, Enderlein P, Waluda CM, Stowasser G, Tarling GA, Murphy EJ (2014) Interannual variability in Antarctic krill (*Euphausia superba*) density at South Georgia, Southern Ocean: 1997–2013. ICES J Mar Sci 71:2578–2588

Filippova JA, Yukhov VL (1979) Specific composition and distribution of cephalopod molluscs in meso- and bathypelagic Antarctic waters. Antarktika Doklady Komissi 18:175–187

Forcada J, Hoffman JI (2014) Climate change selects for heterozygosity in a declining fur seal population. Nature 511:462–465

Forcada J, Trathan PN (2009) Penguin responses to climate change in the Southern Ocean. Glob Chang Biol 15:1618–1630

Forcada J, Trathan PN, Reid K, Murphy EJ (2005) The effects of global climate variability in pup production of Antarctic fur seals. Ecology 86:2408–2417

Forcada J, Trathan PN, Boveng PL, Boyd IL, Burns JM, Costa DP, Fedak M, Rogers TL, Southwell CJ (2012) Responses of Antarctic pack-ice seals to environmental change and increasing krill fishing. Biol Conserv 149:40–50

Fretwell PT, Trathan PN (2009) Penguins from space: faecal stains reveal the location of emperor penguin colonies. Global Ecol Biogeogr 18:543–552

Fretwell PT, LaRue MA, Morin P, Kooyman GL, Wienecke B, Ratcliffe N, Fox AJ, Fleming AH, Porter C, Trathan PN (2012) An emperor penguin population estimate: the first global, synoptic survey of a species from space. PLoS One 7:e33751

Friedlaender AS, Johnston DW, Fraser WR, Burns J, Halpin PN, Costa DP (2011) Ecological niche modeling of sympatric krill predators around Marguerite Bay, Western Antarctic Peninsula. Deep-Sea Res Pt II 58:1729–1740

Friedlaender AS, Goldbogen JA, Nowacek DP, Read AJ, Johnston D, Gales N (2014) Feeding rates and under-ice foraging strategies of the smallest lunge filter feeder, the Antarctic minke whale (*Balaenoptera bonaerensis*). J Exp Biol 217:2851–2854

Gorman KB, Williams TD, Fraser WR (2014) Ecological sexual dimorphism and environmental variability within a community of Antarctic penguins (Genus *Pygoscelis*). PLoS One 9:e90081

Green JA, Butler PJ, Woakes AJ, Boyd IL, Holder RL (2001) Heart rate and rate of oxygen consumption of exercising macaroni penguins. J Exp Biol 204:673–684

Green JA, Butler PJ, Woakes AJ, Boyd IL (2002) Energy requirements of female macaroni penguins breeding at South Georgia. Funct Ecol 16:671–681

Gurney LJ, Pakhomov EA, Christensen V (2014) An ecosystem model of the Prince Edward Island archipelago. Ecol Model 294:117–136

Hewitt RP, Watkins JL, Naganobu M, Sushin V, Brierley AS, Demer D, Kasatkina S, Takao Y, Goss C, Malyshko A, Brandon MA, Kawaguchi S, Siegel V, Trathan PN, Emery J, Everson I, Miller DG (2004) Biomass of Antarctic krill in the Scotia Sea in January/February 2000 and its use in revising an estimate of precautionary yield. Deep-Sea Res Part II 51:1215–1236

Hill SL (2013) Prospects for a sustainable increase in the availability of long chain omega 3s: lessons from the Antarctic krill fishery. In: De Meester F, Watson RR, Zibadi S (eds) Omega 6/3 fatty acids functions, sustainability strategies and perspectives. Humana Press, New York, pp 267–296

Hill SL, Matthews J (2013) The sensitivity of multiple output statistics to input parameters in a krill–predator–fishery ecosystem dynamics model. CCAMLR Sci 20:97–118

Hill HJ, Trathan PN, Croxall JP, Watkins JL (1996) A comparison of Antarctic krill, *Euphausia superba*, caught by nets and by macaroni penguins, *Eudyptes chrysolophus*: evidence for selection? Mar Ecol Prog Ser 140:1–11

Hill SL, Reid K, North AW (2005) Recruitment of mackerel icefish (*Champsocephalus gunnari*) at South Georgia indicated by predator diets and its relationship with sea surface temperature. Can J Fish Aquat Sci 62:2530–2537

Hill SL, Murphy EJ, Reid K, Trathan PN, Constable AJ (2006) Modelling Southern Ocean ecosystems: krill, the food-web, and the impacts of harvesting. Biol Rev 81:581–608

Hill SL, Reid K, Thorpe SE, Hinke J, Watters GM (2007) A compilation of parameters for ecosystem dynamics models of the Scotia Sea – Antarctic Peninsula region. CCAMLR Sci 14:1–25
Hill SL, Keeble K, Atkinson A, Murphy EJ (2012) A foodweb model to explore uncertainties in the South Georgia shelf pelagic ecosystem. Deep-Sea Res Pt II 59–60:237–252
Hofmann EE, Hüsrevoğlu YS (2003) A circumpolar modelling study of habitat control of Antarctic krill (*Euphausia superba*) reproductive success. Deep-Sea Res Pt II 50:3121–3142
Holling CS (1959) The components of predation as revealed by a study of small-mammal predation of the European pine sawfly. Can Entomol 91:293–320
Hureau JC (1994) The significance of fish in the marine Antarctic ecosystems. Polar Biol 14:307–313
Hutchinson GE (1959) Why are there so many kinds of animals? Am Nat 93:145–159
Ichii T, Kato H (1991) Food and daily food consumption of southern minke whales in the Antarctic. Polar Biol 11:479–487
Ikeda T (1984) Development of the larvae of the antarctic krill (*Euphausia superba* Dana) observed in the laboratory. J Exp Mar Biol Ecol 75:107–117
Iwata T, Sakamoto KQ, Takahashi A, Edwards EWJ, Staniland IJ, Trathan PN, Naito N (2012) Using a mandible accelerometer to study fine-scale foraging behavior of free-ranging Antarctic fur seals. Mar Mammal Sci 28:345–357
Jarman SN, McInnes JC, Faux C, Polanowski AM, Marthick J, Deagle BE, Southwell C, Emmerson L (2013) Adélie penguin population diet monitoring by analysis of food DNA in scats. PLoS One 8:e82227
Jurasz CM, Jurasz VP (1979) Feeding modes of the humpback whale *Megaptera novaeangliae* in southeastern Alaska USA. Sci Rep Whales Res Inst Tokyo 31:69–84
Kear AJ (1992) The diet of Antarctic squid: a comparison of conventional and serological gut contents analyses. J Exp Mar Biol Ecol 156:161–178
Kock K-H (1992) Antarctic fish and fisheries. Cambridge University Press, Cambridge
Kock K-H, Barrera-Oro E, Belchier M, Collins MA, Duhamel G, Hanchet S, Pshenichnov L, Welsford D, Williams R (2012) The role of fish as predators of krill (*Euphausia superba*) and other pelagic resources in the Southern Ocean. CCAMLR Sci 19:115–169
LaRue MA, Lynch HJ, Lyver P, Barton K, Ainley DG, Pollard AM, Ballard G (2014) Establishing a method to estimate Adélie Penguin populations using remotely sensed imagery. Polar Biol 37:507–517
Laws RM (1977) Seals and whales of the Southern Ocean. Philos Trans R Soc B 279:81–96
Leaper R, Cooke J, Trathan PN, Reid K, Rowntree V, Payne R (2006) Global climate drives Southern right whale (*Eubalaena australis*) population dynamics. Biol Lett 2:289–292
Lockyer C (1972) The age at sexual maturity of the southern fin whale (*Balaenoptera physalus*) using annual layer counts in the ear plug. ICES J Mar Sci 34:276–294
Lockyer C (1974) Investigation of the ear plug of the southern sei whale, *Balaenoptera borealis*, as a valid means of determining age. ICES J Mar Sci 36:71–81
Lockyer C (1981a) Growth and energy budgets of large baleen whales from the Southern Hemisphere. In: Mammals in the seas III, general papers and large cetaceans, FAO fishery series 5. FAO, Rome, pp 379–487
Lockyer C (1981b) Estimation of the energy costs of growth, maintenance and reproduction in the female minke whale (*Balaenoptera acutorostrata*), from the Southern Hemisphere, south of 40S. Rep Int Whaling Comm 31:337–343
Lubimova TG, Shust KV (1980) Estimation of the Antarctic krill consumption by the main consumer groups. In: Biologicheskie Resursy Antarkticheskogo Krilya, ONTI VNIRO, Moscow, Nauka, pp 320–337
Mackintosh NA (1974) Sizes of krill eaten by whales in the Antarctic. Discov Rep 36:157–178
Marr J (1962) The natural history and geography of the Antarctic krill *Euphausia superba* Dana. Discov Rep 32:33–464

Miller DGM, Hampton I (1989) Antarctic krill (*Euphausia superba* Dana): a review, BIOMASS scientific series 9. Scientific Committee on Antarctic Research, Cambridge
Mori M, Butterworth DS (2006) A first step towards modelling the krill – predator dynamics of the Antarctic ecosystem. CCAMLR Sci 13:217–277
Murphy EJ (1995) Spatial structure of the Southern Ocean ecosystem: predator – prey linkages in Southern Ocean food webs. J Anim Ecol 64:333–347
Murphy EJ, Trathan PN, Everson I, Parks G, Daunt FHJ (1997) Krill fishing in the Scotia Sea in relation to bathymetry, including the detailed distribution around South Georgia. CCAMLR Sci 4:1–17
Murphy EJ, Watkins JL, Reid K, Trathan PN, Everson I, Croxall JP, Priddle J, Brandon MA, Brierley AS, Hofmann EH (1998) Interannual variability of the South Georgia marine ecosystem: biological and physical sources of variation in the abundance of krill. Fish Oceanogr 7:381–390
Murphy EJ, Trathan PN, Watkins JL, Reid K, Meredith MP, Forcada J, Thorpe SE, Johnston NM, Rothery P (2007a) Climatically driven fluctuations in Southern Ocean ecosystems. Proc R Soc Lond B Biol 274:3057–3067
Murphy EJ, Watkins JL, Trathan PN, Reid K, Meredith MP, Thorpe SE, Johnston NM, Clarke A, Tarling GA, Collins MA, Forcada J, Shreeve RS, Atkinson A, Korb R, Whitehouse MJ, Ward P, Rodhouse PG, Enderlein P, Hirst AG, Martin AR, Hill SL, Staniland IJ, Pond DW, Briggs DR, Cunningham NJ, Fleming AH (2007b) Spatial and temporal operation of the Scotia Sea ecosystem: a review of large-scale links in a krill centred food web. Philos Trans R Soc B 362:113–148
Myers RA, Worm B (2003) Rapid worldwide depletion of predatory fish communities. Nature 423:280–283
Nemoto T, Okiyama M, Takahashi M (1985) Aspects of the roles of squid in food chains of marine Antarctic ecosystems. In: Siegfried WR, Condy PR, Laws RM (eds) Antarctic nutrient cycles and food webs. Springer, Berlin, pp 415–420
Nemoto T, Okiyama M, Iwasaki N, Kikuchi N (1988) Squid as predators of krill (*Euphausia superba*) and prey for sperm whales in the Southern Ocean. In: Sahrhage D (ed) Antarctic Ocean and resource variability. Springer, Berlin, pp 292–296
Nicol S, Bowie A, Jarman S, Lannuzel D, Meiners KM, van der Merwe P (2010) Southern Ocean iron fertilization by baleen whales and Antarctic krill. Fish Fish 11:203–209
Øresland V (1990) Feeding and predation impact of the chaetognath *Eukrohnia hamata* in Gerlache Strait, Antarctic Peninsula. Mar Ecol Prog Ser 63:201–209
Pakhomov EA, Perissinotto R (1996) Trophodynamics of the hyperiid amphipod *Themisto gaudichaudii* in the South Georgia region during late austral summer. Mar Ecol Prog Ser 134:91–100
Pakhomov EA, Perissinotto R, McQuaid CD (1996) Prey composition and daily rations of myctophid fishes in the Southern Ocean. Mar Ecol Prog Ser 134:1–14
Pakhomov EA, Atkinson A, Meyer B, Oettl B, Bathmann U (2004) Daily rations and growth of larval krill *Euphausia superba* in the Eastern Bellingshausen Sea during austral autumn. Deep-Sea Res Pt II 51:2185–2198
Parkinson CL (2002) Trends in the length of the Southern Ocean sea-ice season, 1979–99. Ann Glaciol 34:435–440
Parkinson CL (2004) Southern Ocean sea ice and its wider linkages: insights revealed from models and observations. Antarctic Sci 16:387–400
Payne MR (1977) Growth of a fur seal population. Philos Trans R Soc B 279:67–79
Pinkerton MH, Bradford-Grieve JM, Hanchet SMA (2010) Balanced model of the food web of the Ross Sea, Antarctica. CCAMLR Sci 17:1–31
Pinõnes A, Hofmann EE, Dinniman MS, Klinck JM (2011) Lagrangian simulation of transport pathways and residence times along the western Antarctic Peninsula. Deep-Sea Res Pt II 58:1524–1539

Pinõnes A, Hofmann EE, Daly KL, Dinniman MS, Klinck JM (2013) Modelling the remote and local connectivity of Antarctic krill populations along the western Antarctic Peninsula. Mar Ecol Prog Ser 481:69–92

Prévost J (1981) Population, biomass and energy requirements of Antarctic birds, vol 2, BIOMASS Scientific Series. Scientific Committee on Antarctic Research, Cambridge, UK, pp 125–137

Pusch C, Hulley PA, Kock K-H (2004) Community structure and feeding ecology of mesopelagic fishes in the slope waters of King George Island (South Shetland Islands, Antarctica). Deep-Sea Res Pt I 51:1685–1708

Reid K, Forcada J (2005) Causes of offspring mortality in the Antarctic fur seal, *Arctocephalus gazella*: the interaction of density dependence and ecosystem variability. Can J Zool 83:604–609

Reid K, Trathan PN, Croxall JP, Hill HJ (1996) Krill caught by predators and nets: differences between species and techniques. Mar Ecol Prog Ser 140:13–20

Reid K, Croxall JP, Briggs DR, Murphy EJ (2005) Antarctic ecosystem monitoring: quantifying the response of ecosystem indicators to variability in Antarctic krill. ICES J Mar Sci 62:366–373

Reilly S, Hedley S, Borberg J, Hewitt R, Thiele D, Watkins JL, Naganobu M (2004) Biomass and energy transfer to baleen whales in the South Atlantic sector of the Southern Ocean. Deep-Sea Res Pt II 51:1397–1409

Ropert-Coudert Y, Hindell M, Phillips R, Charassin JB, Trudelle L, Raymond B (2014) Biogeographic patterns of birds and mammals. In: De Broyer C, Koubbi P, Griffiths HJ, Raymond B, d'Udekem d'Acoz C, Van de Putte AP, Danis B, David B, Grant S, Gutt J, Held C, Hosie G, Huettmann F, Post A, Ropert-Coudert Y (eds) Biogeographic atlas of the Southern Ocean. Scientific Committee on Antarctic Research, Cambridge, pp 363–388

Ross RM, Quetin LB (1982) *Euphausia superba*: fecundity and physiological ecology of eggs and larvae. Antarct J US 17:166–167

Ross RM, Quetin LB (1983) Spawning frequency and fecundity of the Antarctic krill *Euphausia superba*. Mar Biol 77:201–205

Ruck KE, Steinberg DK, Canuel EA (2014) Regional differences in quality of krill and fish as prey along the Western Antarctic Peninsula. Mar Ecol Prog Ser 509:39–55

Saunders RA, Brierley AS, Watkins JL, Reid K, Murphy EJ, Enderlein P, Bone DG (2007) Intra-annual variability in the density of Antarctic krill (*Euphausia superba*) at South Georgia, 2002–2005: within-year variation provides a new framework for interpreting previous 'annual' estimates of krill density. CCAMLR Sci 14:27–41

Saunders RA, Collins MA, Foster E, Shreeve R, Stowasser G, Ward P, Tarling GA (2014) The trophodynamics of Southern Ocean Electrona (Myctophidae) in the Scotia Sea. Polar Biol 37:789–807

SC-CAMLR (1985) Report of the fourth meeting of the Scientific Committee. CCAMLR, Hobart, Australia

SC-CAMLR (2007) Report of the twenty-sixth meeting of the Scientific Committee. CCAMLR, Hobart, Tasmania, Australia

SC-CAMLR (2008) Report of the twenty-seventh meeting of the Scientific Committee. CCAMLR, Hobart, Tasmania, Australia

Scheffer M, Carpenter S, de Young B (2005) Cascading effects of overfishing marine systems. Trends Ecol Evol 20:579–581

Schmidt K, Atkinson A, Steigenberger S, Fielding S, Lindsay MCM, Pond DW, Tarling GA, Klevjer TA, Allen CA, Nicol S, Achterberg EP (2011) Seabed foraging by Antarctic krill: implications for stock assessment, bentho-pelagic coupling, and the vertical transfer of iron. Limnol Oceanogr 56:1411–1428

Schwaller MR, Southwell CJ, Emmerson LM (2013) Continental-scale mapping of Adélie Penguin colonies from Landsat imagery. Remote Sens Environ 139:353–364

Shreeve RS, Collins MA, Tarling GA, Main CE, Ward P, Johnston NM (2009) Feeding ecology of myctophid fishes in the northern Scotia Sea. Mar Ecol Prog Ser 386:221–236

Siegel V (2005) Distribution and population dynamics of *Euphausia superba*: summary of recent findings. Polar Biol 29:1–22
Siegel V, Watkins JL (2016) Distribution, biomass and demography of Antarctic krill, *Euphausia superba*. In: Siegel V (ed) Biology and ecology of Antarctic krill. Springer, Cham, pp 21–100
Siegel V, Reiss CS, Dietrich KS, Haraldsson M, Rohardt G (2013) Distribution and abundance of Antarctic krill (*Euphausia superba*) along the Antarctic Peninsula. Deep-Sea Res Pt I 77:63–74
Sladen WJL (1964) The distribution of the Adélie and chinstrap penguins. In: Carrick R, Holdgate MW, Prevost J (eds) Biologie Antarctique. Hermann, Paris, pp 359–365
Smetacek V (2008) Are declining Antarctic krill stocks a result of global warming or the decimation of the whales? In: Duarte C (ed) The impact of global warming on Polar ecosystems. Fundacion BBVA, Spain, pp 46–83
Southwell D, Emmerson L, Forcada J, Southwell C (2015) A bioenergetics model for estimating prey consumption by an Adélie penguin population in East Antarctica. Mar Ecol Prog Ser 526:183–197
Staniland IJ, Gales N, Warren NL, Robinson SL, Goldsworthy SD, Casper RM (2010) Geographical variation in the behaviour of a central place forager: Antarctic fur seals foraging in contrasting environments. Mar Biol 157:2383–2396
Steinberg DK, Ruck KE, Gleiber MR, Garzio LM, Cope JS, Bernard KS, Stammerjohn SE, Schofield OME, Quetin LB, Ross RM (2015) Long-term (1993–2013) changes in macrozooplankton off the Western Antarctic Peninsula. Deep-Sea Res I 101:54–70
Takahashi A, Dunn MJ, Trathan PN, Croxall JP, Wilson RP, Sato K, Naito Y (2004) Krill-feeding behaviour in a chinstrap penguin *Pygoscelis antarctica* compared with fish-eating in Magellanic penguins *Spheniscus magellanicus*: a pilot study. Mar Ornithol 32:47–54
Tarling GA, Fielding S (2016) Swarming and behaviour in Antarctic krill. In: Siegel V (ed) Biology and ecology of Antarctic krill. Springer, Cham, pp 279–320
Tarling GA, Cuzin-Roudy J, Thorpe SE, Shreeve RS, Ward P, Murphy EJ (2007) Recruitment of Antarctic krill *Euphausia superba* in the South Georgia region: adult fecundity and the fate of larvae. Mar Ecol Prog Ser 331:161–179
Tormosov DD, Mikhaliev YA, Best PB, Zemsky VA, Sekiguchi K, Brownell RL (1998) Soviet catches of southern right whales *Eubalaena australis* 1951–1971. Biological data and conservation implications. Biol Conserv 86:185–197
Trathan PN, Reid K (2009) Exploitation of the marine ecosystem in the sub-Antarctic: historical impacts and current consequences. Pap Proc R Soc Tasmania 143:9–14
Trathan PN, Watkins JL, Murray AWA, Hewitt RP, Naganobu M, Sushin V, Brierley AS, Demer D, Everson I, Goss C, Hedley S, Katsakina S, Kawaguchi S, Kim S, Pauly T, Priddle J, Reid K, Ward P (2001) The CCAMLR-2000 krill synoptic survey; a description of the rationale and design. CCAMLR Sci 8:1–24
Trathan PN, Brierley AS, Brandon MA, Bone DG, Goss C, Grant SA, Murphy EJ, Watkins JL (2003) Oceanographic variability and changes in Antarctic krill (*Euphausia superba*) abundance at South Georgia. Fish Oceanogr 12:569–583
Trathan PN, Murphy EJ, Forcada J, Croxall JP, Reid K, Thorpe SE (2006) Physical forcing in the southwest Atlantic: ecosystem control. In: Boyd IL, Wanless S, Camphuysen CJ (eds) Top predators in marine ecosystems. Cambridge University Press, Cambridge, pp 28–45
Trathan PN, Forcada J, Murphy EJ (2007) Environmental Forcing and Southern Ocean marine predator populations: effects of climate change and variability. Philos Trans R Soc B 362:2351–2365
Trathan PN, Ratcliffe N, Masden EA (2012) Ecological drivers of change at South Georgia: the krill surplus, or climate variability. Ecography 35:983–993
Trathan PN, Collins MA, Grant SM, Belchier M, Barnes DKA, Brown J, Staniland IJ (2014) The South Georgia and the South Sandwich Islands MPA: protecting a biodiverse oceanic island chain situated in the flow of the Antarctic Circumpolar Current. Adv Mar Biol 69:15–78
Trivelpiece WZ, Hinke JT, Miller AK, Reiss CS, Trivelpiece SG, Watters GM (2011) Variability in krill biomass links harvesting and climate warming to penguin population changes in Antarctica. Proc Natl Acad Sci U S A 108:7625–7628

Walters A, Lea M-A, van den Hoff J, Field IC, Virtue P, Sokolov S, Pinkerton MH, Hindell MA (2014) Spatially explicit estimates of prey consumption reveal a new krill predator in the Southern Ocean. PLoS One 9:e86452

Waluda CM, Hill SL, Peat HJ, Trathan PN (2012) Diet variability and reproductive performance of macaroni penguins *Eudyptes chrysolophus* at Bird Island, South Georgia. Mar Ecol Prog Ser 466:261–274

Ward P, Meredith MP, Whitehouse MJ, Rothery P (2008) The summertime plankton community at South Georgia (Southern Ocean): comparing the historical (1926/1927) and modern (post 1995) records. Prog Oceanogr 78:241–256

Whitehouse MJ, Meredith MP, Rothery P, Atkinson A, Ward P, Korb RE (2008) Rapid warming of the ocean around South Georgia, Southern Ocean, during the 20th century: forcings, characteristics and implications for lower trophic levels. Deep Sea Research I 55:1218–1228

Zerbini AN, Andriolo A, da Rocha JM, Simões-Lopes PC, Siciliano S, Pizzorno JL, Waite JM, De Master DP, Van Blaricom GR (2004) Winter distribution and abundance of humpback whales (*Megaptera novaeangliae*) off Northeastern Brazil. J Cetacean Res Manag 6:101–107

Chapter 10
Parasites and Diseases

Jaime Gómez-Gutiérrez and José Raúl Morales-Ávila

Abstract The Antarctic krill *Euphausia superba* is among the most studied species of the Order Euphausiacea in biological and ecological aspects; however, reports of their parasites and diseases are relatively scarce. A worldwide overview of all parasites known for 48 out 86 extant euphausiid species includes 17 distinct types of epibionts, pathogens, parasites, and parasitoids. So far, only seven of them have been reported interacting with *E. superba* [epibionts: exuviotrophic ciliates (Foettingeriidae) and microplanktophagous ciliates (Suctoridae, *Ephelota*), pathogens: chitinoclastic bacteria and fungi; and trophically transmitted endoparasites: Apicomplexans (Gregarinidae, *Cephaloidophora*), nematode infecting krill's eggs (under laboratory conditions), and histophagous parasites: Apostomatida ciliates of the family Pseudocollinidae]. The epibionts have interspecific associations that strongly depend on the krill's moult cycle, discarding them at each moulting event. Their colonization and intensity show a remarkable synchronization with the krill moulting process at individual, school, and population levels. The social and sometimes highly dense swarms and schools of *E. superba*, its keystone trophic function (both as voracious predator and as prey to multiple predators) should make it a critical vector for trophically transmitted parasites in the food web. However, *E. superba* interacts with a relatively low diversity of epibionts, pathogens, and parasites, in comparison with parasite diversity known for relatively well-studied temperate (*Meganyctiphanes norvegica*, *Euphausia pacifica*) and subtropical (*Nyctiphanes simplex*) euphausiid species. The apparently low parasite diversity of *E. superba* is likely associated with its Antarctic zoogeographic pattern; where, parasites have not invaded the Antarctic krill with the same evolutionary success as have occurred with other euphausiid species from tropical, subtropical, temperate, and even Arctic ecosystems.

Keywords Pathogens • Diversity • Prevalence • Intensity • Social behaviour

---

J. Gómez-Gutiérrez (✉) • J.R. Morales-Ávila
Instituto Politécnico Nacional, Centro Interdisciplinario de Ciencias Marinas, La Paz, Mexico
e-mail: jagomezg@ipn.mx; jrmoralesa@gmail.com

V. Siegel (ed.), *Biology and Ecology of Antarctic Krill*, Advances in Polar Ecology,
DOI 10.1007/978-3-319-29279-3_10

10.1 Introduction

The Antarctic and non-Antarctic krill were separated approximately 20 Mya (Patarnello et al. 1996) and the Antarctic krill *Euphausia superba* Dana 1850 diverged from *Euphausia crystallorophias* Holt and Tattershall 1906 about 2.7 Mya (D'Amato et al. 2008). The high latitudes inhabited by Antarctic krill species are characterised by extreme changes in seasonal conditions, with very low food availability, long periods of darkness and a massive expansion of sea ice in autumn and winter. Thus, Antarctic krill evolved in a year-round gelid ecosystem with multiple overwintering behavioural and physiological strategies and complex interspecific consumer interactions with planktonic, benthic, and nektonic preys, predators, epibionts, pathogens, and parasites (Seear et al. 2012). These last three groups of organisms, seem to be more diverse in euphausiid from tropical, subtropical, temperate, and Arctic ecosystems than in Antarctic ecosystems; where some of them seem not to have successfully invaded the Antarctic Ocean (Klimpel et al. 2010). Seear et al. (2012) suggested that pathogen and parasites likely are responsive to latitudinal clines in environmental factors such as temperature. For example, at the Antarctic Peninsula, winter temperatures are typically below 0 °C, whereas they largely remained at or above 0 °C year-round at northward latitudes (i.e. South Georgia). It is likely that many disease agents are unable to exist below freezing point, but they may otherwise survive, infect and lead to disease above such temperatures (Seear et al. 2012).

Parasitism appeared early in biological evolution as an extremely common, diverse, and successful consumer interaction (Lafferty 1999; Lafferty and Kuris 2002). It became a relevant connection node for the diversification of complex life cycles of parasites; that sometimes during their ontogeny interacts with multiple host species at distinct trophic levels. *E. superba* is frequently considered a waist-wasp species in the Antarctic Ocean ecosystem (Atkinson et al. 2004). It has been proposed that this species strongly influences energy flow and species assemblages in the Antarctic pelagic realm that are complementary with hydro-climatic feedback interactions. This keystone species represents a colossal available biomass population for epibiontic, ectoparasitic, and endoparasitic organisms.

A healthy krill, with a metabolism in steady-state (in homeostasis), has a transparent body (an adaptation to decrease the risk of visual predators in the water column), red-brown chromatophores, a translucent digestive gland (hepatopancreas), relatively large hepato-somatic index (proportion of the hepatopancreas and the cephalothorax length), fast and synchronized heartbeat, active peristaltic movements of the intestine, energetic and synchronous swimming movements, and regular growth, moulting, and gonad development rate and functions. The pathognomonic (characteristic of specific diseases) in euphausiid is, so far, not well studied worldwide compared with decapods of commercial value. However, among the multiple symptoms of sickness or physiological response to epibionts, pathogens, or parasites in euphausiids include lack of transparency of the body (usually an opaque or pale whitish coloration, which indicates mechanical damage or lack of homeostasis in the individual), black spots on their exoskeleton, opaque colouration, and/or non-functional chromatophores. Additionally, gross signs include coloured and mostly opaque or/and pulsing movements in the digestive gland, a relative small hepato-somatic index (proportion size of hepatopancreas/cephalothorax length), and a contracted intestine that sometimes lacks peristaltic movements (signs of prolonged fasting). Also, slow or desynchronized heartbeat, sluggish or erratic swimming capabilities (that sometimes can separate them from their conspecifics, lingering behind the krill schools), irregular or slow growth (including shrinking), moulting, and gonad development (including re-absorption or castration), and in case of pathogens and histophagous ciliates (parasitoids), cause death of the krill host. In recent years, ecologists have come to recognize the enormous influence of parasites and disease in regulating animal populations (Gómez-Gutiérrez et al. 2003; Kuris et al. 2008).

The Antarctic krill is a voracious omnivore that requires highly energy intake to fuel continuous swimming in the water column ($\leq$20 cm s^{-1}). Despite this high-energy intake, this species shows relatively slow growth rates (its longevity ranges from 4 to 7 years) that can reach one of the largest body sizes for epipelagic species in the order Euphausiacea (up to 65 mm total length) (Baker et al. 1990). This species forms some of the largest aggregations, swarms, and schools known for any species of the order Euphausiacea, with maximum reported regional biomass of about two million tonnes, distributed over an area of 100–450 km^2 at densities of up to 2000 individuals m^{-3} and annual estimated biomass ranging between 100 and 500 million tonnes (wet mass) (Macaulay et al. 1984; Watkins 2000; Atkinson et al. 2009; Nowacek et al. 2011). Antarctic krill populations represent a colossal amount of biomass that potentially can interact with epibionts, pathogens, and parasites. Hamner (1984) observed natural synchronized moulting in *E. superba* schools and also a predator-induced pattern of moulting that he called "decoy moulting". These synchronized krill moulting events must have a direct effect on survival, feeding, and infection strategies of epibionts, chitinoclastic bacteria, and gregarines (Apicomplexa), which their life cycles are strongly coupled with krill's moult cycles.

Two landmark monographs summarized most of the published information about parasites of euphausiids before the 1980s decade (Mauchline and Fisher 1969; Mauchline 1980). Then, it was apparently unknown the presence of epibiont, pathogens, and parasites of *E. superba*. There have been almost 35 years without any updated monograph to show what is currently known about interspecific associations of krill with other species, except predator-prey interactions (Mauchline and Fisher 1969; Mauchline 1980, see Chap. 9, Trathan and Hill 2016). Although an extensive review of parasites of marine zooplankton mentioned several euphausiid parasites (Théodoridès 1989), this review did not explicitly mention *E. superba* interaction with epibionts, pathogens, or parasites. We performed a meta-analysis review of reports of epibionts and parasitic organism of crustaceans of the order Euphausiacea published between 1885 and 2013 (120 publications including about 360 records including personal observations in *E. superba* reared in the Australian Antarctic Division, Tasmania Australia krill laboratory, Aug 2009–Jul, 2010). This worldwide review of literature provides us a relatively broad perspective about the diversity, prevalence patterns, intensity, parasite-host size ratio, availability of microhabitats for parasites in euphausiids, and the association of parasitism with the host reproductive strategies to better understand emerging patterns of parasite–host co-evolution. Currently, there are 17 different known types of epibionts, pathogens, parasites, and parasitoids infecting krill (107 known taxa reported in 48 of the 86 extant species of the order Euphausiacea) (Fig. 10.1a). The definitions of trophic strategies used in the present review were defined and explained in detail in Lafferty and Kuris (2002). They report eleven trophic strategy categories based in four dichotomies: (1) number of victims that an individual attacks throughout the life-history stage (to distinguish predators vs. parasites), (2) whether a successful attack eliminates the fitness of the host (to define castrators and parasitoids), (3) if the host must die to further parasite development (to define parasitoids), and (4) presence or absence of density-dependent pathology (macroparasites vs microparasites). Combining these four dichotomies defines seven types of parasitism (typical parasite, pathogen, trophically transmitted typical parasite, trophically transmitted pathogen, parasitic castrator, trophically transmitted parasitic castrator, and parasitoid), three forms of predation (micropredator, social predator, and solitary predators) and, when one considers obligate and facultative combinations of these forms, four types of predators (Lafferty and Kuris 2002). Several of these types of interspecific interactions have been observed in *E. superba* (Table 10.1). All euphausiid's epibionts and parasites have different life strategies, ranging from epibionts (epizootic diatoms, suctorida, and exuviotrophic ciliates, and chitinoclastic bacteria), and ectoparasites (Dajidae isopods), mesoparasites [Ellobiopsidae and Rhizocephalan, this last is a highly uncertain report (Mooney and Shirley 2000)], potential pathogens (bacteria and fungi), hyperparasitic ciliates (*Phtorophyra* sp.), trophically transmitted endoparasites (Apicomplexa, Cestoda, Trematoda, Nematoda, and Acanthocephala), and parasitoids (dinoflagellates, and histophagous Apostomatida *Pseudocollinia* ciliates). With the broad-scale perspective in this chapter, we compared the diversity, prevalence, and intensity of the parasites that are associated with *Euphausia superba* to conceptualize how much is currently known about those interspecific associations in this pivotal species of the Antarctic Ocean. Although *E. superba* (and other krill species) has been the focus of multiple observational and experimental studies; so far in *E. superba*, only 7 out of 17 previously known types of epibionts, pathogens, parasites, and parasitoids that interact with euphausiids have been documented (Fig. 10.1b). These include: epibionts: (1) exuviotrophic apostome ciliates (unidentified species) of the family Foettingeriidae (Kittel and Rakusa-Suszczewski 1988; Rakusa-Suszczewski and Nemoto 1989; Stankovic and Rakusa-Suszczewski 1996) likely of the genus *Gymnodinoides*, but sometimes incorrectly identified cysts of ciliates (found attached to their appendages) of the genus *Ephelota* spp. (Stankovic et al. 2002), (2) microplanktophagous ciliates of the family Ephelotidae (genus *Ephelota*) (Stawiszyńska-Janas and Kittel 1982; Stankovic et al. 2002; Tarling and Cuzin-Roudy 2008), pathogens: (3) chitinoclastic bacteria of the genus *Psychrobacter* and *Pseudoalteromonas* (Miwa et al. 2008) and (4) fungus *Metschnikowia australis* (Donachie and Zdanowaski 1998); and trophically transmitted endoparasites: (5) Apicomplexa (three species of the genus *Cephaloidophora*, family Gregarinidae) (Avdeev 1985, 1987; Avdeev and Avdeeva 1989; Kawaguchi et al. 1999; Takahashi et al. 2003, 2004, 2008, 2009, 2011),

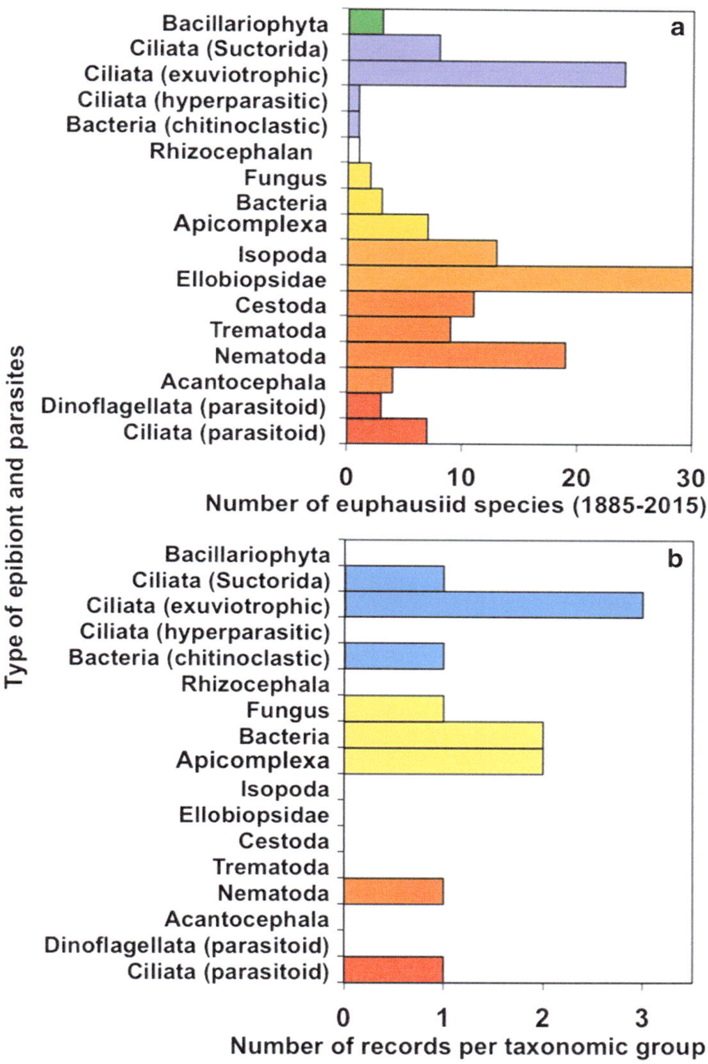

Fig. 10.1 (**a**) Meta-analysis of the 17 types of epibionts (*green* and *blue* bars), pathogens (*light orange bars*), trophically transmitted parasites (*dark orange bars*), and "parasitoids" (*red bars*) that interact with euphausiids (Order Euphausiacea) around the world from a review of 120 published works from 1885 to 2013. (**b**) Types of parasites so far reported for the Antarctic krill *Euphausia superba*

(6) Nematoda (unidentified larvae L1 infecting eggs of *E. superba*) (Robert King and Jaime Gómez-Gutiérrez, pers. observ.), and parasitoids: (7) Apostomatida histophagous ciliates (Stankovic and Rakusa-Suszczewski 1996) probably belonging to the family Pseudocollinidae (Gómez-Gutiérrez et al. 2012; Lynn et al. 2014) (Fig. 10.1b, Table 10.1). Circumstantial evidence suggest potential viral inclusions

Table 10.1 List of epibiont, pathogen, parasite, and parasitoids species assemblage reported interacting with the Antarctic krill *Euphausia superba* in the Antarctic Ocean. Trophic strategies assigned according with criteria of Lafferty and Kuris (2002)

Kuris and Lafferty (2002) classification	Type of parasite	Family	Genus and species	Krill life phase vulnarable	Parasite size range (μm) (μm)	Parasite mean size (mm)	Krill TL length (mm)	Parasite/host total length ratio	Prevalence Range (%)	Prevalence Mean (%)	Intensity	Region of study	References
Epibiont (planktophagous)	Ciliata	Suctoridae	*Ephelotas* spp.	Juvenil and adult	40–420	0.157	65.0	0.002				Southern Ocean, Admiralty Bay	Stawiszyńska-Janas and Kittel (1982)
Epibiont (planktophagous)	Ciliata	Suctoridae	*Ephelotas* spp. (three forms-sizes)	Juvenil and adult	100, 200–250, 450–500	0.250	58.0	0.0043	35–72	55	<95	Southern Ocean, south of Australia	Rakusa-Suszczewski and Nemoto (1989)
Epibiont (exuviotrophic)	Ciliata	Suctoridae	*Ephelotas* spp.	Juvenil and adult			65.0				42	Southern Ocean 63.38 S, 127.10 E	Stankovic and Rakusa-Suszczewski (1996)
Epibiont (planktophagous)	Ciliata	Suctoridae	*Ephelotas* spp.	Juvenil and adult	40–420	0.157	65.0	0.002	<1 %	0.0005	20,000	Southern Ocean, Elephant Island, the South Orkneys, King George Island and Bransfield Strait	Stankovic et al. (2002)
Epibiont (planktophagous)	Ciliata	Suctoridae	*Ephelotas* spp.	Juvenil and adult	40–420	0.157	58.0	0.003	0–35 %	0.35	281	Southern Ocean, South Georgia 54.5S, 37W	Tarling and Cuzin-Roudy (2008)
Epibiont (exuviotrophic)	Ciliata	Foettingeriidae	Unknown spp. (three forms)	Juvenil and adult	27, 38, 46	0.038	54.0	0.0007		84		King George Island, Elephant Island and in the Bransfield Strait	Kittel and Rakusa-Suszczewski (1988)
Epibiont (exuviotrophic)	Ciliata	Foettingeriidae	Unknown spp. (three forms)	Juvenil and adult	35–45, 75–87, 30–37	0.044	55.0	0.0008	100	100	382, 902, 8	Almiratly Bay (an annual cycle) and Wedell Sea	Rakusa-Suszczewski and Filcek (1988)

Epibiont (exuviotrophic)	Ciliata	Foettingeriidae	Unknown spp. (three forms)	Juvenil and adult	35–87	0.061	58.0	0.0011	100	100	<1000	Southern Ocean, south of Australia	Rakusa-Suszczewski and Nemoto (1989)
Epibiont (exuviotrophic)	Ciliata	Foettingeriidae	Unknown spp.	Juvenil and adult			65.0	0.0000	84–100 %	90	450	Southern Ocean, 63.38 S, 127.10 E	Stankovic and Rakusa-Suszczewski (1996)
Epibiont (exuviotrophic)	Ciliata	Foettingeriidae	Unknown spp., erroneously identified as *Ephelota*	Juvenil and adult	70	0.007	65.0	0.0001	0–80 %	0.8	20,000	Elephant Island	Stankovic et al. (2002)
Oportunistic pathogen	Bacteria	g-proteobacteria	*Psychrobacter*	Juvenil and adult	1–2		65 0		2–42 %	0.0350		Southern Ocean, South Georgia	Miwa et al. (2008)
Oportunistic pathogen	Bacteria	g-proteobacteria	*Pseudoalteromonas*	Juvenil and adult	1–2		65.0		2–42 %		0.0350	Southern Ocean, South Georgia	Miwa et al. (2008)
Opportunistic parasite	Fungus	Yeast	*Metschnikowia australis*	Juvenil and adult	45	0.005	65.0	0.0001				Southern Ocean, King George Island	Donachie and Zdanowaski (1998)
Trophically trasmitted parasite	Apicomplexa	Gregarinidae	*Cephaloidophora pacifica*	Juvenil and adult	140–155	0.016	65.0	0.0002	76.4	0.7640	1848	East Pacific and Indian Sector of the Southern Ocean	Avdeev (1985)
Trophically trasmitted parasite	Apicomplexa	Gregarinidae	*Cephaloidophora indica*	Juvenil and adult	102–238	0.024	65.0	0.0004	44.5	0.4450	229	East Pacific and Indian Sector of the Southern Ocean	Avdeev (1985)
Trophically trasmitted parasite	Apicomplexa	Gregarinidae	*Cephaloidophora pacifica*	Juvenil and adult	18–76	0.008	65.0	0.0001	90–100	0.9640		Antarctic Peninsula, Near Syowa station, Pacific and Indian sector the Southern Ocean	Takahashi et al. (2004, 2008)
Trophically trasmitted parasite	Apicomplexa	Gregarinidae	*Cephaloidophora pacifica*	Juvenil and adult	18–76	0.008	65.0	0.0001			195	Southern Ocean, South Georgia	Takahashi et al. (2003, 2008)

(continued)

Table 10.1 (continued)

Kuris and Lafferty (2002) classification	Type of parasite	Family	Genus and species	Krill life phase vulnerable	Parasite size range (μm) (μm)	Parasite mean size (mm)	Krill TL length (mm)	Parasite/host total length ratio	Prevalence		Intensity	Region of study	References
									Range (%)	Mean (%)			
Trophically trasmitted parasite	Apicomplexa	Gregarinidae	*Cephaloidophora pacifica*	Calyptopis to adult	50	0.005	13.0	0.0004	14–79.6	0.4	256	Indian sector of the Southern Ocean	Takahashi et al. (2011)
Unknown	Nematoda	Unknown	Unknown spp.	Eggs								Laboratory Australian Antarctic Division (AAD)	King and Gómez-Gutiérrez (pers observ)
Trophically transmitted castrator	Helminth	Unknown	Unknown spp.	Juvenil and adult	200–1000	0.100	65.0	0.0015	6 %	0.06	50	Southern Ocean, South Georgia	Miwa et al. (2008)
Parasitoid	Cilia	Pseudocollinidae	Unknown spp.	Adults	80–333	0.18	65	0.0028				Southern Ocean, 63.38 S, 127.10 E	Stankovic and Rakusa-Suszczewski (1996)

present within R-cells of the hepatopancreas in <1 % of sampled *E. superba* (Bateman, Hicks, Tarling, Soeffker and Stentiford, WG-EMM-2015/23). However, further studies must confirm such histological observations. However, it is difficult to visualize whether historical research efforts to study epibionts and parasites of *E. superba* (circa 1982 to present) are a precise representation of the apparently low diversity of epibiont and parasitic interactions with this species or whether parasitological studies of *E. superba* are in their infancy. In either case, considerable parasitological research effort remains to be carried out in the Antarctic Ocean in the future to discover the ecological function and consequences of epibiont and parasitic interactions.

Infectious agents—bacteria, fungi, or parasites cause most diseases described in the present chapter. Conditions due to non-infectious causes like cancer (although some cancer can be viral induced), diseases related to prolonged fasting, high levels of persistent organic pollutants (POP), heavy metals, or other toxic substances are not covered in this chapter (Yamamoto et al. 1987; Corsolini et al. 2002; Nash et al. 2008; Poulsen et al. 2012). Aside from the eco-physiological studies of Poulsen et al. (2012) the negative health effects of toxic substances have not been specifically tested and their health consequences are poorly understood. Sub-lethal narcosis (immobility) was observed in non-feeding larval stages of *E. superba* from p,p′-dichlorodiphenyl dichloroethylene (p,p′-DDE) body residues of 0.2 mmol/kg p.w. (Poulsen et al. 2012).

Overall, little is known about euphausiid immune response (i.e. melanization, enhancement of encapsulation, hemocytes, opsonin formation, antibacterial or antifungal activity, production of free radicals, and oxidative stress responses among others), behaviour, influence of epibionts and parasites in the metabolic and reproduction rates, resulting in a poor and fragmented understanding about the effect of epibionts, pathogens, parasites, and parasitoids on euphausiids and presumably zooplankton and nekton krill's predators. However, recently were published two pioneer studies about *E. superba*'s immune system (Seear et al. 2012; Zhao et al. 2013). Seear et al. (2012), using gene expression techniques [cDNA microarrays and quantitative (qPCR)], reported the first *E. superba* study of immune gene expression in any euphausiid species worldwide. *E. superba* showed two major immune gene types: (1) Cathepsins (C and K) localized in the lysosomes and endosomes that degrade intracellular or endocytosed proteins and (2) C-type lectins that contribute to innate immune responses in invertebrates, including prophenoloxidase activation, enhancement of encapsulation, nodule formation of hemocytes, opsonin formation, antibacterial activity, antifungal activity, and injury healing. Seear et al. (2012) mentioned that haemocyanin may also be an additional contributor to the krill immune system given that, in addition to being an oxygen carrier, it is known to have antiviral, antibacterial, and antifungal properties. However, this research did not experimentally challenge krill with any parasite or pathogen (krill was not infected), but detected the expression of genes associated with previously known immune function in shrimps infected with bacteria or virus. Zhao et al. (2013) obtained and purified a preliminary antimicrobial polypeptide (CMCC-1) from *E. superba*. This polypeptide showed cell cytoplasmic membrane destruction and inhibited cell division at the logarithmic phase against the pathogenic bacteria *Staphylococucs aureus*. Despite these research efforts, still being a poorly and fragmented perspective about the effect of parasites upon *E. superba* at the individual, aggregation, population, and species levels. In this chapter, we summarize the parasitological knowledge about the interaction of *E. superba* with each taxonomic group: epibionts, pathogens, parasites, and parasitoids.

10.2 Black Spot Pathogenic Chitinoclastic Bacteria

Lear (1963) was likely the first to mention the occurrence and significance of chitinoclastic bacteria in pelagic waters and zooplankton. Currently is unclear the diversity, density, and relative abundance of bacteria in euphausiid digestive tract and what proportion they have gut-symbiotic or gut-parasitic interactions. Bacteria inhabiting the euphausiid digestive system (stomach, intestine, and digestive gland) are several orders of magnitude more concentrated ($1.6–5.7 \times 10^6$ bacteria mg^{-1}) than in sea water (Rakusa-Suszczewski and Zdanowski 1989). It is unlikely that such high densities are explained by feeding filtering because bacteria are small (<1 μm), suggesting that bacteria reside and multiply in their stomachs (Fevolden and Eidsa 1981; Donachie 1995; Donachie et al. 1995; Donachie and Zdanowski 1998; Denner et al. 2001) (Table 10.2). Bacterial communities collected from the digestive tract of euphausiids (*E. superba*, *E. crystallorophias*, *Thysanoessa macrura* G. O. Sars, 1883) have been studied to understand the role of bacteria in krill spoilage (Kelly et al. 1978) and the digestive function of bacteria and krill health (Donachie and Zdanowski 1998). These studies used culture-dependent techniques, likely resulting in a considerable underestimation of bacterial diversity because only a relatively small fraction of these bacteria can be successfully cultivated from gastrointestinal tracts of invertebrates and vertebrates (10–50 %) (Zoetendal et al. 2004). Overall, Arctic and Antarctic euphausiid species have less diverse bacterial biota than subtropical species (Aguilar-Méndez et al. 2008). The dominant cultured bacteria in *E. superba* are γ-proteobacteria (*Pseudomonas* is ubiquitous in the stomachs of polar krill and *Moraxella*), followed by lower densities of Firmicutes, Actinobacteria, Flavobacteria, and β-proteobacteria (Table 10.2). Stomach bacteria in *E. superba* participate in host digestive processes by producing enzymes and dietary co-factors contributing to proteolytic, lipolytic, and chinitolytic enzyme pools (Rakusa-Suszczewski and Filcek 1988; Dabrowski et al. 1983; Rakusa-Suszczewski and Zdanowski 1989; Donachie et al. 1995; Cieśliński et al. 2005, 2007). It is evident that most bacteria in stomachs of euphausiids participate in digestive processes of the host. Miwa et al. (2008) reported that bacteria may cause potential pathogenic effect when opportunistically increase their numbers when interact with infection inflicted by other krill parasites (Fig. 10.2a, b). Bacteria also have been observed associated with histophagous ciliate infections of northeast Pacific region krill species that, in extreme high intensities, may lead to bacteraemia events (Gómez-Gutiérrez et al. 2012, 2015; Lynn et al. 2014) (Table 10.2). Miwa et al. (2008) is the only published report that specifically proposes that pathogenic bacteria infect *E. superba* (unknown for all other euphausiid species) from South Georgia region causing black spots in different parts of the cephalothorax and trunk (Fig. 10.2a). Their histological observations revealed that the black spots were melanised nodules composed of hemocytes surrounding either bacteria or

Table 10.2 Chronological list of bacterial densities and strain richness associated with Antarctic krill (*Euphausia superba*, *Euphausia crystallorophias*, and *Thysanoessa macrura*)

Euphausiid species	Microbiota	Bacterial densities	References
E. superba	Coryniform like, *Pseudomonas*, *Moraxella* like, *Alcaligenes*, *Acinetobacter*, *Flavobacterium*, *Micrococcus*, *Vibrio*, *Bacillus*	0.875 CFU mg^{-1}	Kelly et al. (1978)
E. superba	*Moraxella* like, *Pseudomonas*, *Alteromonas*	1×10^3 CFU mg^{-1}	Fevolden and Eidsa (1981)
E. crystallorophias	*Moraxella* like, *Alcaligenes*, *Flavobacterium*, Vibrionaceae, *Planococcus*, *Brochothrix thermosphacta Alteromonas*, *Pseudomonas*	0.56 CFU mg^{-1}	Fevolden and Eidsa (1981)
E. superba	*Corynebacterium*, *Micrococcus*, *Pseudomonas*, *Alcaligenes*, *Moraxella*, *Bacillus*, *Flavobacterium*, *Arthrobacter*	3.59×10^6 CFU mL^{-1}	Turkiewicz et al. (1982)
E. superba	*Clostridium bifermentans*, *C. sporogenes*, *C. subterminale*, *Clostridium*	Not estimated	Dabrowski et al. (1983)
E. superba	Cocci	5.7×10^8 cells mg $^{-1}$	Rakusa-Suszczewski and Zdanowski (1989)
E. superba	*Clostridium bifermentans*, *C. sporogenes*, *C. subterminale*, *Clostridium*	Not estimated	Dabrowski et al. (1983)
E. superba	*Flavobacterium*, *Flavobacterium breve*, *Pseudomonas vesicularis*, *Weeksella virosa*, *Moraxella*, *Pasteurella*, *Aeromonas*, *Vibrio*	1.09×10^5 CFU mg^{-1}, 3.28×10^6 mg^{-1} AODC	Donachie (1995), Donachie et al. (1995)
T, macrura	*Pseudomonas*	3.23×10^3 CFU mg^{-1}, 2.22×10^5 AODC mg^{-1}	Donachie (1995), Donachie et al. (1995)

(continued)

Table 10.2 (continued)

Euphausiid species	Microbiota	Bacterial densities	References
E. superba	Gram negative cocci, Gram negative bacilli	1.09×10^5 CFU mg^{-1}, 3.28×10^6 mg^{-1} AODC	Donachie and Zdanowsky (1998)
E. superba	*Psychrobacter proteolyticus*	Not estimated	Denner et al. (2001)
T. macrura	*Pseudoalteromonas*	Not estimated	Cieśliński et al. (2005)
E. superba	*Pseudoalteromonas*	Not estimated	Cieśliński et al. (2007)
E. superba	*Pseudoalteromonas, Psychrobacter*	Not estimated	Miwa et al. (2008)
E. superba	*Non-identified*, SEM Fig. 10.2	Not estimated	Gómez-Gutiérrez unpubl. data

All bacteria were isolated from the digestive tract of krill, except bacteria isolated from black spot melanomas located at the surface of the Antarctic krill exoskeleton. *Pseudoalteromonas*, *Psychrobacter*, *Staphylococcus*, and *Vibrio* are potentially opportunistic pathogenic bacteria (Miwa et al. 2008). Chitinoclastic bacteria may also be opportunistic pathogens (Review information modified from Aguilar-Méndez et al. (2008))

amorphous material (Fig. 10.2b). In 2007, 42 % of the krill had melanised nodules, but prevalences usually range from 2 to 5 %. Most of the nodules had an opening on the body surface of krill (Miwa et al. 2008). Three bacterial strains were isolated from these black spots and classified as *Psychrobacter* or *Pseudoalteromonas*, based on sequences of 16S rRNA gene analysis (Table 10.2).

We have observed chitinoclastic bacterial infection that cause considerable injury, that eventually resulted in death, from live *E. superba* specimens transported from the Antarctic Sea to the Australian Antarctic Division krill laboratory (AAD) located at Kingston, Tasmania, Australia (Robert King and Jaime Gómez-Gutiérrez, pers. observ.) (Fig. 10.2c). Scanning Electron Microscope (SEM) images of *E. superba* black spots and areas with visible black injuries (in cephalothorax, appendages, trunk, and telson) were rapidly colonized by opportunistic rod-shaped bacterial colonies (Fig. 10.2d–f). Specimens developed black spots in the overwhelming high-density tanks where krill are regularly transported in the R/V Aurora Australis. In this case, although bacteremia can eventually cause the death of krill, bacteria cannot be considered, in the strict sense, a parasitoid because bacteria do not require the death of the host to complete their life cycle and they have a density-dependent virulence in the host (Lafferty and Kuris 2002).

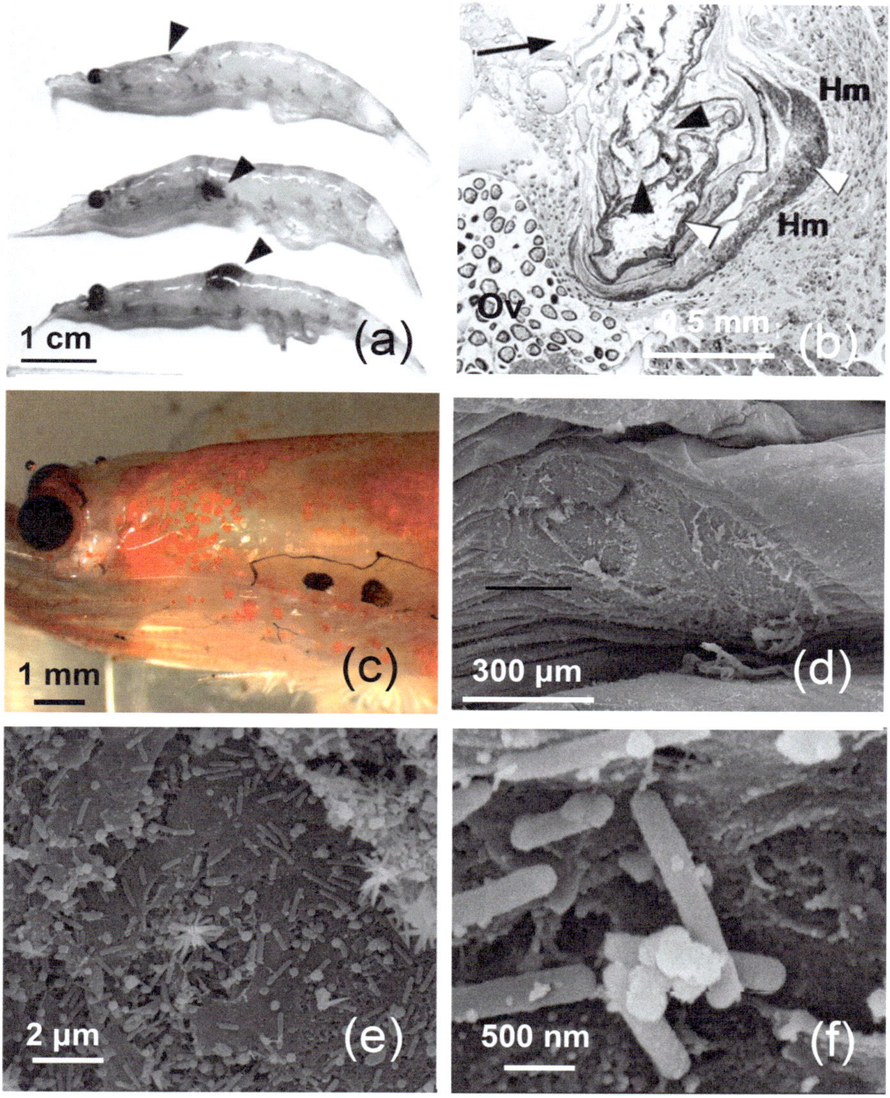

Fig. 10.2 (**a**) Antarctic krill *Euphausia superba* bearing black spots (*arrowheads*). The numbers on the scale are in cm, (**b**) A transverse section of dorsal part of the cephalothorax of a krill with a typical melanised nodule. Bacterial mass (*black arrowheads*) is encapsulated by melanin layers (*white arrowheads*), which is further surrounded by hemocytes (Hm). The *arrow* indicates the opening of the lesion to outside of the body. *Hp* hepatopancreas, *Ov* ovary. HE stain (**a**, **b** reprinted from Miwa, S., Kamaishi, T., Matzuyama, T., Hayashi, T. and Naganobu, M. Histopathology of Antarctic krill, *Euphausia superba*, bearing black spots. Journal of Invertebrate Pathology 98, 280–286, Copyright (2008), with permission from Elsevier). **c** *Black* spot caused by chitinoclasitic bacteria affecting *E. superba* maintained under laboratory conditions in the Australian Antarctic Division, Tasmania, Australia. Scanning electron microscope micrographs show (**c**) zoom of the wound and (**d**–**f**) high population density of rod-shaped chitinoclastic bacteria (**c** Photos and **d**–**f** SEM images taken by JG-G)

10.3 Endoparasitic Apicomplexa (Family Gregarinidae)

Apicomplexan gut-living gregarines, commonly but incorrectly known as sporozoans, infect the digestive tract of annelids and several crustacean taxa (Cirripedia, Amphipoda, Mysidacea, Decapoda, and Euphausiacea) (Dobson 2002; Takahashi et al. 2008). All species are parasites of animals. The apicomplexa infect the stomach, intestine, and midgut gland (hepatopancreas) of euphausiids. Currently, it is known that 7 out the 62 named species of the genus *Cephaloidophora* (Family Cephaloidophoridae), one species of the genus *Lateroprotomeritus*, and several reports of undescribed gregarines are trophically transmitted apicomplexa gregarines of euphausiids from the Mediterranean Sea, Barents Sea, and the Antarctic Ocean (Théodoridès and Desportes 1975; Avdeev 1985; Avdeev and Avdeeva 1989; Théodoridès 1989; Timofeev 2001). So far, gregarines have been reported in only seven euphausiid species (*E. superba*, *M. norvegica*, *Nematoscelis megalops*, *N. atlantica*, *Thysanoessa macrura*, *T. raschii*, and *Stylocheiron abbreviatum*) (Table 10.3). Although most apicomplexan species have monoxenous life cycles involving a single invertebrate host, gregarines are endoparasites relatively difficult to detect (particularly in preserved specimens) due their small cell size (71–144 μm average total length) and the location inside the intestine or the hepatopancreas. Apicomplexan gregarines have a trophic transmission strategy (orofaecal route) and they sometimes attain high prevalences (up to 90 %) in the krill species so far studied. Therefore, it is likely to find new euphausiid hosts with gregarine infections and perhaps new species of gregarines identified with molecular methods.

Avdeev (1985) discovered and described the first two species of apicomplexa gregarine parasites, *Cephaloidophora pacifica* (Fig. 10.3a–c) and *Cephaloidophora indica* (Fig. 10.3d–f), infecting digestive tract of the Antarctic krill *E. superba*, being more prevalent the species *C. pacifica* (75 % of 1848 specimens examined) than *C. indica* (44.5 % of 229 specimens examined) (Table 10.3). Avdeev and Avdeeva (1989) later described two additional species, *Cephaloidophora thysanoessae* infecting *Thysanoessa macrura*, and *Cephaloidophora antarctica* infecting *E. superba*. Following the apicomplexa description by Levine (1988), it appears that the ciliates found reproducing in the gut of *E. superba* (Kawaguchi and Toda 1997) were actually gregarine parasites. Although, they have been studied mostly from adult euphausiid specimens, there are records that Apicomplexa (Eugregarinida) also infect *E. superba* calyptopis and furcilia larval phases (Takahashi et al. 2011). Currently, *C. pacifica* (host *E. superba*) is the best-studied gregarine species infecting euphausiids worldwide, partially because its high prevalence, broad circumpolar distribution, and a significant research effort carried out first by Russian (Avdeev 1985, 1987; Avdeev and Vagin 1987; Avdeev and Avdeeva 1989), followed by Japanese scientists (Takahashi et al. 2003, 2004, 2008, 2009, 2011), and more recently by British scientists (Bateman et al. 2015).

Avdeev (1987) first described the development of gregarines of *E. superba* and Takahashi et al. (2009) conceptualized the known life cycle of gregarines (Order

Table 10.3 Apicomplexa gregarines of the genera *Cephaloidophora* and *Lateroprotomeritus* (the last one of unclear taxonomic affiliation) reported infecting several Antarctic krill *Euphausia superba* and *Thysanoessa macrura*

Gregarine species	Krill host	Size (μm)	Average prevalence (%)	Location (likely distribution range)	Source
C. pacifica	*E. superba*	140–155	76.4 (1848)	East Pacific and Indian Ocean sector of Southern Ocean	Avdeev (1985)
C. indica	*E. superba*	102–238	44.5 (229)	East Pacific and Indian Ocean Sector of Southern Ocean	Avdeev (1985)
C. pacifica	*E. superba*	100	100 (1165)	Kosmonavtov Sea	Avdeev (1987)
C. pacifica	*E. superba*				Avdeev and Vagin (1987)
C. pacifica	*E superba*	No data	No data	Circumpolar	Spiridonov (1996)
C. indica	*E superba*	No data	No data	Antarctic Indian region	Spiridonov (1996)
C. pacifica	*E. superba*	18–76	96.4	Antarctic Peninsula, near Syowa Station, Pacific and Indian sector of the Southern Ocean	Takahashi et al. (2004, 2008)
C. pacifica	*E. superba*	18–76	(195)	South Georgia Region	Takahashi et al. (2003, 2009)
C. pacifica	*E. superba*	40–50	69.7 % (256)	Antarctic Indian region	Takahashi et al. (2011)
C. thysanoessae	*T. macrura*	130–217	29.0 (17)	East Pacific and Indian Sector of the Southern Ocean	Avdeev and Avdeeva (1989)
C. antarctica	*T. macrura*	47–142	2	East Pacific and Indian Ocean Sector of the Southern Ocean	Avdeev and Avdeeva (1989)

The number in parenthesis is the number of krill specimens examined during each study. *Cephaloidophora pacifica* is a species currently considered Antarctic circumpolar

Eugregarinida) (Fig. 10.4). Gregarines typically have high prevalences in the *E. superba* population (up to 87 %). The gregarine parasites have six life stages (Avdeev 1987). Four endoparasitic stages occur inside the digestive tract in krill (sporozoite, cephalin, gamont, and syzygy) and two presumably outside the krill hosts (gametocyst and oocyst), which probably infect an intermediate planktonic host (likely a copepod) (Fig. 10.4). After the gregarine enter the host's intestine, sporozoites excyst from the oocyst and attach to the epithelium (cephalin stage). Cephalins have bodies divided into anterior epimerite, protomerite, and posterior deutomerite and are commonly located in the hind-gut epithelium of their host and liberated into the intestinal lumen. The early developmental stages take place mostly intra-cellularly. The gamont stage, which follows, is mostly found in the

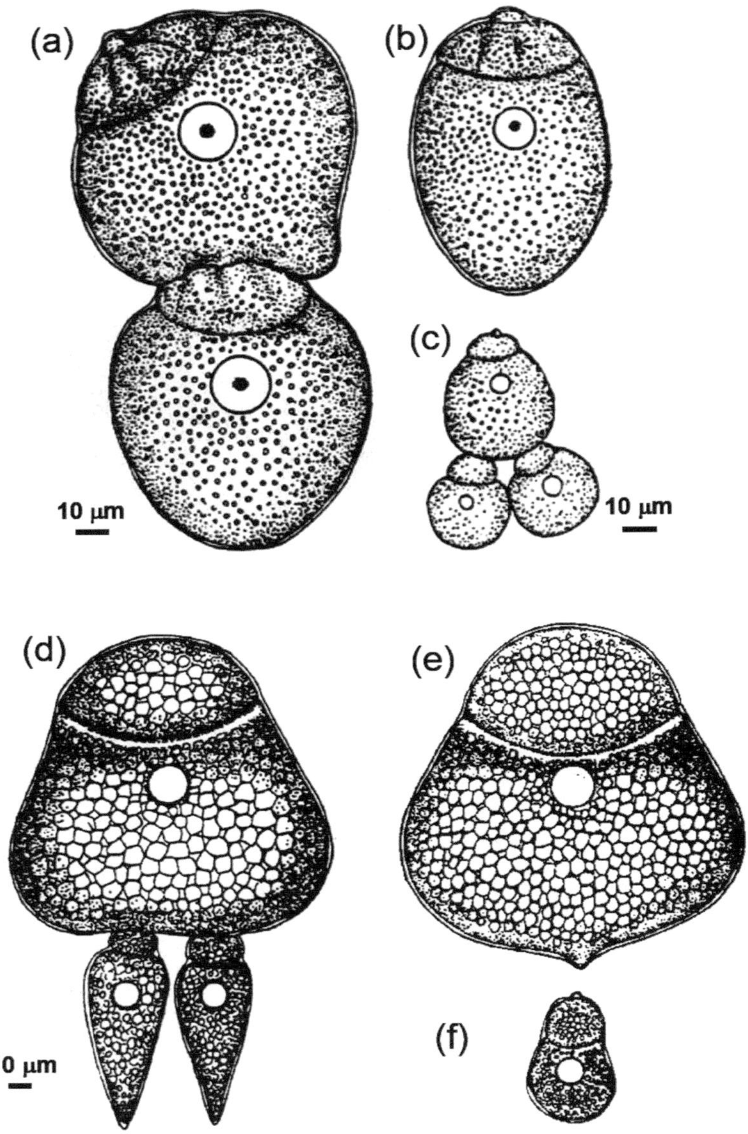

Fig. 10.3 Apicomplexa Gregarinidae that infect *Euphausia superba*. *Cephaloidophora pacifica* (**a**) mature syzygy, (**b**) mature gamont, and (**c**) immature syzygy. *Cephaloidophora indica* (**d**) mature gamont, (**e**) mature syzygy, and (**f**) immature syzygy (Figures are reproduced with permission from Parasitology, Avdeev 1985)

intestinal lumen, as well as the diverticulum of the mid-gut gland. After maturation, they associate head to tail (syzygy) to produce a reproductive gametocyst that will be shed in the host's faeces. Within a few days, mature gametocysts release infective oocysts into the environment to continue the cycle (Fig. 10.4) (Takahashi

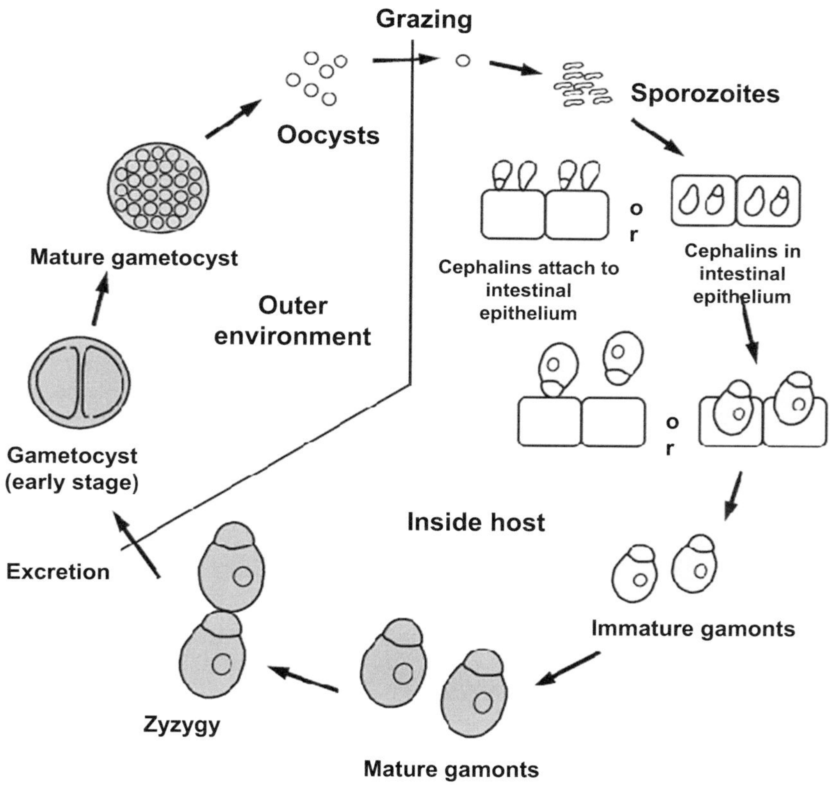

Fig. 10.4 Known general life cycle of gregarines (Order Eugregarinida) (Diagram reproduced from Takahashi et al. (2009) originally published in Polar Biology with kind permission from Springer Science and Business Media)

et al. 2009). The high infection and sometimes high intensity of this parasite prompted Avdeev (1987) to suggest that it must have a pathological effect. Kawaguchi et al. (1999) and Takahashi et al. (2009, 2011) using scanning and transmission electron microscopy concluded that gamonts in the diverticulum appear to damage microvilli, which uptake digested nutrients and secrete various enzymes, and destroy hepatic cells in the mid-gut gland having a significant impact on the nutritional state of the Antarctic krill host. The strategy of gregarines parasitizing the hind-gut epithelium during the cephalin stage may be a positive compromise, not causing a fatal impact on the host, while securing a suitable habitat (Takahashi et al. 2009). *E. superba* frequently attain relatively high gregarine infection rates that may exit in faeces. Because *E. superba* spends most of its life in the epipelagic strata (200 m depth) and produces rapidly sinking faecal pellets, apicomplexans may also sink with faeces, but this process has never been explicitly investigated.

Krill moult every 3–90 days depending on the temperature of their environment (Kawaguchi et al. 2006; Tarling et al. 2006). Their moults include parts of their stomachs and hind-guts, which are covered with cuticle (Ikeda et al. 1984). These biological characteristics of krill do not facilitate the reproductive process of gregarine parasites. If the gametocysts are ejected within the faecal pellets, they may rapidly sink out of the normal vertical range of krill. The sinking velocity of Antarctic krill faecal pellets is estimated to be from 50 to 800 m day^{-1} (Cadee et al. 1992). This process decreases the chance of re-infection with gametocysts in the hosts, and does not support the high prevalences (100 %) sometimes observed in krill. However, Takahashi et al. (2003) observed a possible strategy for avoiding the discharge induced by moulting. The highly motile gamont stage may move to a safety zone where no shedding occurs. Non-motile syzygy and/or gametocysts in early stages of their reproductive development may be discharged during the host's moulting and excretion activities. A possible explanation for this high prevalence rate is the social behaviour of *E. superba* swarms or schools (Hamner 1984; Hamner et al. 1983, 1989), which would increase the chance of krill eating their faecal pellets (coprophagy) and moults before they sink.

Several studies show that *C. pacifica* infects the digestive tract of *E. superba* with average intensities from 87 to 493 cells krill^{-1} (maximum intensity = 8505 cells krill^{-1}) from different locations (Takahashi et al. 2003, 2004, 2008, 2009). Although *C. pacifica* occurs in most *E. superba* populations, its pathologic effect varies greatly. It has been described as low-intensity infection (probably with negligible or minor negative effects) to high intensities in the hepatopancreas in some individuals. In the digestive gland, gamonts reproduce destroying the hepatopancreas tissue that clot the diverticula and loses tissue compactness. With cumulative clots, the hepatopancreas changes its normal coloration (green-yellow to yellow-brown) and becomes dark and opaque (Avdeev and Vagin 1987; Takahashi et al. 2009). These gregarines have a close interspecific association with *E. superba* in all dissected specimens (n = 93) and widely distributed in the Southern Ocean albeit highly aggregated, which is typical of parasites living in marine hosts (Takahashi et al. 2003, 2004, 2008, 2009, 2011).

Unlike other *E. superba*'s parasites that, so far, little is known about their biogeographic distribution, gregarine distribution has been studied in large part of the *E. superba* distribution range in the Antarctic Ocean. Avdeev (1985) described *C. pacifica* that parasitizes *E. superba* throughout most of its range, except the eastern Indian sector where *C. indica* infects *E. superba*. Further studies speculated how these two gregarine species diverged from *E. superba* populations with a biogeographic and paleoceanographic perspective (Dolzhenkov et al. 1987; Spiridonov 1996). However, the biological tag role of the gregarines is still highly controversial and requires future genetic studies. Takahashi et al. (2008) demonstrated that the circumpolar Antarctic distribution of *C. pacifica* infecting *E. superba*, shows little evidence of a supposed geographic and even taxonomic separation between *C. pacifica* and *C. indica*. The current perspective is that *C. pacifica* is present in virtually all of the *E. superba* range, indicating a stable seasonal and parasite-host interaction. Although intensity varies greatly, this is usually >70 % of the infected population (Takahashi et al. 2008, 2009, 2011). Its role on the health of *E. superba* deserves more detailed investigation. On-going systematic histological work carried out by Bateman, Hicks, Tarling, Soeffker and Stentiford (WG-EMM-2015/23, CCAMLR 2015) considered the prevalence of pathogens and diseases in krill collected across the Scotia Sea during the austral summer (Mar–Apr, 2009). Compared to other marine crustaceans, the krill were relatively disease free, with the main parasite being *Cephaloidophora pacifica*.

10.4 Yeasts

Turkiewicz et al. (1982) isolated white budding yeasts from *E. superba*'s alimentary canal. Later, nine psychrophilic yeast strains were isolated from the stomach of *E. superba*, two of them identified as *Leucosporidium antarcticum* and *Metschnikowia bicuspidata australis* (Donachie and Zdanowski 1998). *Leucosporidium antarcticum* is endemic in the Antarctic Sea but not *Metschnikowia*, which has a broader biogeographic distribution range. The yeast *Metschnikowia kamienski* infests the copepod *Eurytemora velox* (Fize et al. 1970). However, the functional biological association between yeast and krill is still unexplored and certainly poorly understood (Donachie and Zdanowski 1998). Based on the free-living habitat of these psychrophilic yeasts and very low abundance (<1 % of the cultured counts), such yeast infections might be opportunistic, presumably with considerably low prevalence rates, although with so far unknown effect on *E. superba* populations. The diversity, pathology, epizootiology, and ecological function of fungi of any euphausiid species in the world are virtually unknown and deserve future research.

10.5 Ciliata

Members of the Class Phyllopharyngea and Oligohymenophorea have evolved in association with Crustacea (Bradbury 1994). The Subclass Suctorida epibionts of euphausiids are ciliates with tentacles that feed on planktonic organisms and reproduce by multiple budding (Stankovic et al. 2002; Tarling and Cuzin-Roudy 2008). The Subclass Apostomatia ciliates that infest (carried out by ectoparasites) or infect (carried out by endoparasites) euphausiids have life cycles that involve at least four distinct, specific feeding strategies, typically reproducing by palintomy or fission: (1) exuviotrophic ciliates (epibionts) that feed exuviotrophically from moult exudates (probably originated as scavengers of the exoskeleton; although the scavenger-feeding mode is now extremely rare), (2) planktotrophic suctorians, (3) histophagous endoparasitic ciliates (parasitoids), and (4) hyperparasitic ciliates (Capriulo and Small 1986; Bradbury 1994; Stankovic and Rakusa-Suszczewski 1996; Landers et al. 2006; Gómez-Gutiérrez et al. 2003, 2006, 2012, 2015; Lynn et al. 2014). The ciliates interacting with *E. superba* are: (1) epizoic sessile predatory suctorian ciliates of the family Ephelotidae that likely cause hydrodynamic drag on krill swimming and may make the host more vulnerable to visual predators (Nicol 1984; Stankovic and Rakusa-Suszczewski 1996; Stankovic et al. 2002; Tarling and Cuzin-Roudy 2008), (2) exuviotrophic ciliates of the family Foettingeriidae that also cause hydrodynamic drag of the swimming host (Lindley 1978; Kittel and Rakusa-Suszczewski 1988; Rakusa-Suszczewski and Nemoto 1989; Stankovic and Rakusa-Suszczewski 1996; Stankovic et al. 2002; Landers et al. 2006, 2007), and (3) histophagous Apostomatida ciliates (Family Pseudocollinidae) that invariably kill the adult hosts being considered as a parasitoid (Capriulo and Small 1986; Capriulo et al. 1991; Stankovic and Rakusa-Suszczewski 1996; Gómez-Gutiérrez et al. 2003, 2006, 2012; Lynn et al. 2014). The last type was originally reported in *E. superba* as unidentified endoparasitic ciliates that "may have a negative (lethal) consequence for the host" (Stankovic and Rakusa-Suszczewski 1996). Because photographs of ciliates from Stankovic and Rakusa-Suszczewski (1996) resemble in size (40 μm) and morphology to the only endoparasitic ciliates known that infect krill (histophagous parasitoid ciliates of the genus *Pseudocollinia*) (Gómez-Gutiérrez et al. 2006, 2012; Lynn et al. 2014) we interpret those ciliates inside *E. superba* must be also histophagous ciliates of the family Pseudocollinidae because their endoparasitic microhabitat (photographed inside the *E. superba* legs). Transmission pathways and identification of these organisms in krill should be further investigated (Gómez-Gutiérrez et al. 2015), particularly since planktonic protozoans are a significant part of the diet of *E. superba* (Schmidt et al. 2006).

10.5.1 Epibiotic Suctorian Ciliates (Subclass Suctorida, Order Exogenida, Family Ephelotidae)

Stawiszyńska-Janas and Kittel (1982) probably provided the first confirmed report of trophic sessile stage suctorian epibionts attached on *E. superba* and *E. crystallorophias* exoskeleton. More detail was provided in further publications (Rakusa-Suszczewski and Filcek 1988; Rakusa-Suszczewski and Nemoto 1989; Stankovic and Rakusa-Suszczewski 1996; Stankovic et al. 2002; Tarling and Cuzin-Roudy 2008).

Ephelotidae epibiont prevalence depends on krill social behaviour and density of the swarms and schools, intermoult period, and size of the host. Several authors suggest that larger, older krill are more likely to be infested than smaller and younger krill (Rakusa-Suszczewski and Filcek 1988; Rakusa-Suszczewski and Nemoto 1989; Stankovic et al. 2002). This hypothesis was explicitly tested and confirmed by Tarling and Cuzin-Roudy (2008), who observed higher prevalences in older krill (Fig. 10.5). Nicol (1984) specifically proposed that, since euphausiids from surface swarms were mature individuals mostly, the high prevalences are the result of senility or delayed ecdysis in reproductive animals. Tarling and Cuzin-Roudy (2008) confirmed that Ephelotidae prevalences were positively correlated

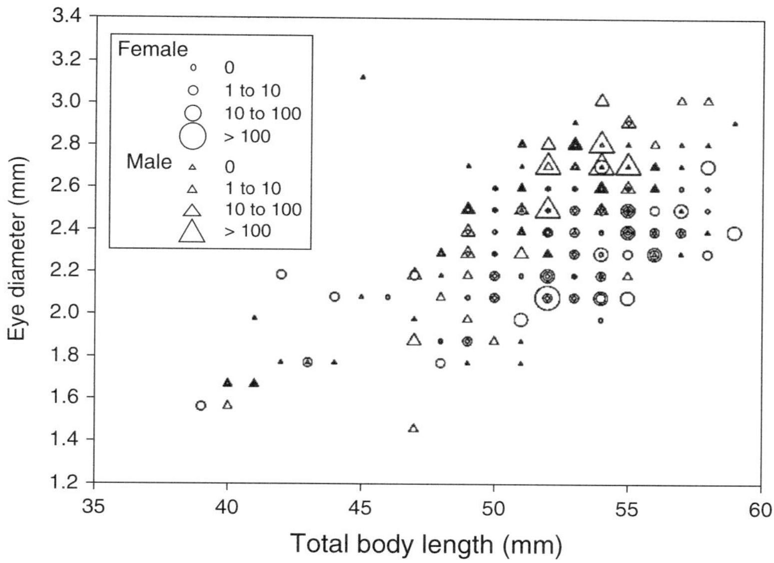

Fig. 10.5 The relationship between body length, eye diameter, and level of suctorian infection in male and female adult krill. Note that symbols are superimposed on each other where body length and eye diameter coincide (Diagram reproduced from Tarling and Cuzin-Roudy (2008) originally published in Polar Biology with kind permission from Springer Science and Business Media)

with eye diameter and the pre-moult stage in *E. superba*. Krill specimens in pre-moult stage were infested as high as 66 % compared to 0 % prevalence in those krill in post-moult stage (Fig. 10.5).

Suctorian ciliates found on juvenile and adult *E. superba* have been invariably identified as *Ephelota* spp. because taxonomy of the genus *Ephelota* has not been firmly established (Nicol 1984; Stankovic et al. 2002; Tarling and Cuzin-Roudy 2008). Rakusa-Suszczewski and Nemoto (1989) proposed that at least three distinct *Ephelota* undescribed species infected *E. superba*, apparently separated per size and type of cyst attachment (body width: small 100 μm, medium 150–175 μm, and large 450–500 μm). Most recently, Stankovic et al. (2002), using small subunit rDNA (SS rDNA), suggested that feeding and budding stage adult suctorians were all members of the same, yet-to-be-named, *Ephelota* species that infected *E. superba* collected from the King George Island region. Further genetic analyses of *COI* from a more extensive range of regions could test whether the *Ephelota* infesting euphausiids is a cosmopolitan species or a multi-species assemblage with distinct biogeographic patterns. Stankovic et al. (2002) also suggested that Antarctic and non-Antarctic ciliate species of *Ephelota* diverged much earlier than Antarctic and non-Antarctic euphausiid species, perhaps implying *Ephelota* species are generalist rather than specialized epibionts of euphausiids.

The epizootiology of suctorian ciliates indicates that they are not frequently detected, but when present, large numbers of krill seem to be infested (Tarling and

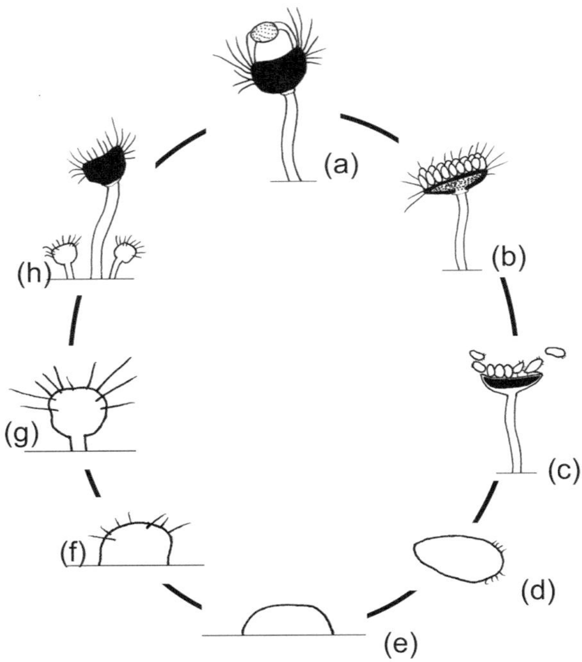

Fig. 10.6 Known general life cycle of suctorian ciliates (Family Suctoridae) in euphausiids. (**a**) Adult (tentacles are used to capture prey and to suck nutrients out of them), (**b**, **c**) adult at budding reproductive stage, (**d**) young swarmer cell with two rings of cilia, (**e**) sessile cell, (**f**) sessile cell developing tentacles, (**g**) young cell with non-contractile stalk, and (**h**) young and adult cells are frequently observed in the same host (*Euphausia pacifica* and *Nematoscelis difficilis*) because life cycle is shorter than the intermoult period (Gómez-Gutiérrez, pers. observ)

Cuzin-Roudy 2008). Intensity varies, with an average of 11 individuals of *Ephelota* spp. per krill, with the suctorian adult phase having a stalk and tentacles (Rakusa-Suszczewski and Nemoto 1989; Stankovic and Rakusa-Suszczewski 1996; Stankovic et al. 2002). Stankovic et al. (2002) interpreted that massive numbers of cysts attached on krill's appendages (ranging from 120 to 900 phoront cysts per host) were also suctorian ciliates. We currently interpret, as earlier studies like Rakusa-Suszczewski and Nemoto (1989), that those ciliate cysts are actually apostome exuviotrophic ciliates (Class Oligohymenophorea, Subclass Apostomatia) that are highly prevalent in krill (Lindley 1978; Landers et al. 2006, 2007), rather than suctorian ciliates (Class Phyllopharingea, Subclass Suctorida) that infest with significantly fewer intensities. The suctorians feed by trapping prey with prehensile and pipe-like suctorial tentacles. The reproductive phase produce budding cells. During the budding process, the tentacles disappear, and a crown of buds emerges in the anterior area of the cell; these swim (transmission stage) and later adhere to the crustacean cuticle, such as *E. superba*. The adhering cell develops protruding tentacles and a short stalk that eventually develop into a solid cylindrical structure. Suctorida ciliates feed and reproduce by budding continuously in the same host resulting in suctorians individuls of different sizes in the same krill host. This overlapping cohort occurs because suctorians complete their life cycle in only few hours (Fig. 10.6). Our direct observations under shipboard laboratory conditions of cyst ciliates swimming inside the *Euphausia pacifica* moult (Oregon coast) show that they are exuviotrophic feeders (growing

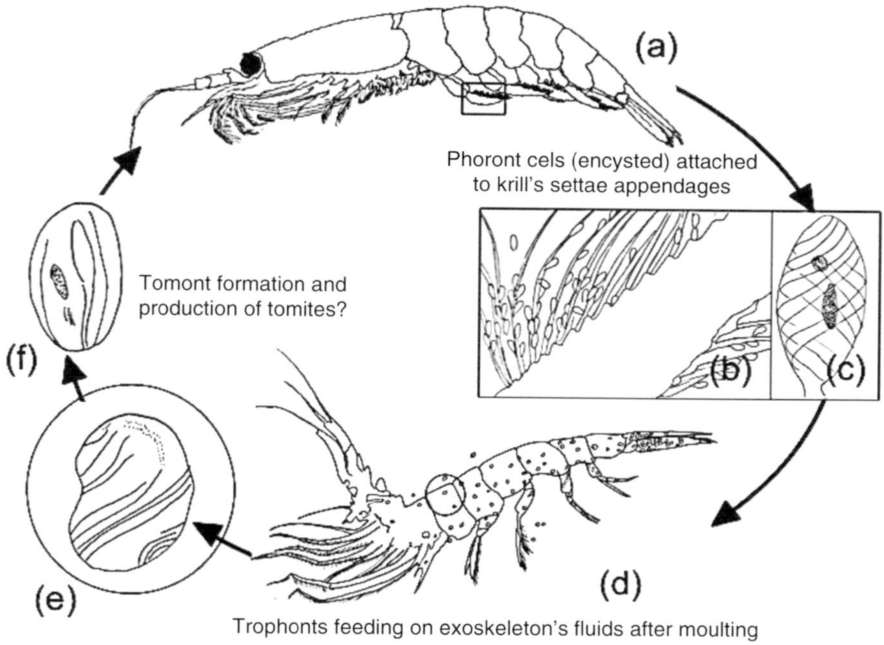

Fig. 10.7 Known general life cycle of apostome exuviotrophic ciliates (Family Foettingeriidae) in *Euphausia superba*. (**a**) *E. superba* infested with phoront cysts of the Apostome ciliates, (**b**) zoom of the krill appendages showing how ciliate cysts attach to setae and (**c**) cyst showing ciliate row pattern (Draw based from photographs of Stankovic et al. 2002), (**d**) trophont cells feeding on exoskeleton's fluids after the krill moulted, (**e**) detail of trophont cell, and (**f**) so far little studied tomont (reproductive) and tomite (transmission) stage inferred from Landers et al. (2006)

and changing coloration as they ingest more moult's fluid) and reproduce by cell division (Landers et al. 2006). Thus, our perception is that ciliate cysts attached to krill's appendages do not feed neither reproduce like suctorian ciliates attached on the cephalothorax and abdomen of *E. superba*.

10.5.2 Epibiotic Exuviotropic Ciliates (Subclass Apostomatia, Order Apostomatida, Family Foettingeriidae)

The Foettingeriidae exuviotrophic ciliates (encysted phoront stage) infest appendages of the euphausiids (thoracic limbs and pleopods, on setae, and between setae) of juveniles and adults of both sexes with intensities up to 900 phoront cysts per host (Lindley 1978; Stankovic and Rakusa-Suszczewski 1996) (Fig. 10.7a–c). When the krill moults, the ciliates excyst into a feeding stage (trophont) that feeds on the fluids of the exuvia (Fig. 10.7d–e), transforming into a reproductive tomont stage that divides by fission to produce multiple tomites (the transmission stage) before infecting another crustacean host as a phoront (Fig. 10.7f). These exuviotrophic ciliates seem to complete their life cycle exclusively infesting krill (Landers et al. 2006, Fig. 10.7a–f). Their prevalence is usually closely related with the proportion of individuals in the population in the intermoult and premoult stages. Postmoult krill do not bear phoronts attached to their swimming appendages (Tarling and Cuzin-Roudy 2008).

Protistan epibionts were first noted as phoronts (resting cysts) on nine krill species in the North Atlantic with prevalences ranging from 3 to 16 % (Lindley 1978). Similarly, at least three different forms of unidentified encysted phoront cysts were later discovered on *E. superba* in the Antarctic Ocean (Kittel and Rakusa-Suszczewski 1988; Rakusa-Suszczewski and Filcek 1988; Rakusa-Suszczewski and Nemoto 1989; Stankovic and Rakusa-Suszczewski 1996; Stankovic et al. 2002). All these epibiont phoront ciliates were not assigned to any genus or species because their life cycle was then unknown due to the virtually exclusive examination of preserved samples. However, Rakusa-Suszczewski and Filcek (1988) suggested their "form 1" was similar to *Spirophrya* ciliates observed from various crustaceans. *Gymnodinioide*s *pacifica* Landers et al. 2006 was the first exuviotrophic ciliate properly named that infest the thoracic and abdominal appendages of six krill species in the Oregon coast, USA (Landers et al. 2006, 2007). Under laboratory conditions, they observed how these phoronts have an exuviotrophic ectocommensal life cycle strategy (Landers et al. 2006). Lindley (1978) identified as Apostomatida ciliates the cysts attached to swimming appendages of several euphausiid species that inhabit in the North Atlantic and Banas (1981), Kittel and Rakusa-Suszczewski (1988), Rakusa-Suszczewski and Filcek (1988) also reported similar taxonomic assignation to cysts attached on *Euphausia superba* from the Antarctic Sea. However, Stankovic et al. (2002), based on SS rDNA evidence, assigned such phoront stages to Suctoridae (*Ephelota* spp.) (see their Figs. 6a–d, 7, and 8), concluding that all cysts that infest *E. superba* belong to one so far non-described species of *Ephelota* and suggested that Lindley's (1978) interpretation of exuviotrophic ciliates was incorrect. The Stankovic et al. (2002) interpretation is not correct, based on evidence obtained from ciliates from *E. superba* and other krill species in the world: (1) A search of GenBank (Jan, 2014) of the 18S rDNA sequences (with a current considerably greater amount of information than was available in 2002) shows that *Ephelota* species (Phyllopharingea, Suctoria, Exogenida, Ephelotidae) are closely related to Suctorida species of the Order Endogenidae (Acinetidae), while the genus *Gymnodinoides* cysts from euphausiid appendages (Oligohymenophorea, Apostomatida; Foettingeriidae) are actually closely associated with histophagous species of the genus *Pseudocollinia* (Olygomenophorea, Apostomatida, Pseudocolliniidae family). The family Ephelotidae and Foettingeriidae actually belong to distinct classes and therefore is unlikely they are the same species, as suggested by Stankovic et al. (2002). Stankovic et al. (2002) possibly analysed a conservative part of the 18S rDNA; more precise species identification can be done using cytochrome oxidase (*COI*) mitochondrial gene (Hebert et al. 2003); (2) Direct observations of live ciliates show distinct feeding strategies: the resting cysts (tomite stage attached to the krill's appendages) excyst and transform into the feeding stage (trophont) that feeds osmotrophycally from the fluids of the moult (exuviotrophic strategy); the suctorian ciliates (Ephelothidae) have a raptorial feeding strategy, using their prehensile tentacles. This means that Foettingeriidae actually obtain energy from the krill's moult fluids, while the Ephelotidae species use krill exclusively as a substrate (basibiont) without obtaining energy from krill; (3) Reproduction of both types of ciliates are distinct. The Foettingeriidae tomont cells reproduce using typical ciliate cell division, while the Ephelothidae reproduce using multiple budding from a crown (relatively unusual in ciliates); (4) Both ciliates that infect *E. superba* (in similar life phases) show distinct morphologic features: (a) In Stankovich et al. (2002) their Fig. 7 shows a phoront cell with nine kineties (typical of apostome Foettingeriidae phoronts cells) (compare with our Fig. 10.7c) rather than 12–18 rows (typical of suctorian phoront cells); (b) The swarmer stage (tomite transmission stage) of suctorian cells have their ciliates restricted to rings around the anterior end and not spiraled around the cell, like Foettingeriidae ciliates, and (c) Stankovich et al. (2002) mention the encysted ciliates have a "stalk of the cyst" (see their Fig. 8, showing a TEM image), which is in fact a typical Foettingeriidae apostome peduncle attachment secreted by the cell during the encysting process that it is distinct from the *Ephelota* stalk that develops as a solid cylindrical structure; and (5) Suctorian cells attached to euphausiids show considerably smaller intensities per host and prevalence in the population than Foettingeriidae ciliates attached to the euphausiid appendages (Tarling and Cuzin-Roudy 2008; Gómez-Gutiérrez pers. observ.). In short, there are several lines of evidences that ciliates of the genus *Ephelota* and *Gymnodinoides* that infest euphausiids are different species from a distinct classess, having distinct consumer, reproductive, and life cycles strategies (Landers et al. 2006, 2007; Fernandez-Leborans 2013). Further *COI* gene analyses could provide the required precision to distinguish species and solve the taxonomic discrepancies about the identification of ciliates in *E. superba* and other krill species worldwide.

10.5.3 Endoparasitic Histophagous Ciliates (Parasitoids)

Stankovic and Rakusa-Suszczewski (1996) is the only report of an unidentified endoparasitic histophagous ciliate infection of *E. superba* [mentioned as "intramuscular *Apostoma* trophonts"] (see their Figs. 10.1, 10.2, 10.3, and 10.4) observed from preserved krill specimens. Although this study was reported after the first publication of endoparasitic ciliates infecting *Thysanoessa inermis* in the Bering Sea (Capriulo and Small 1986; Capriulo et al. 1991), Stankovic and Rakusa-Suszczewski (1996) did not associate them with ciliates of the then known endoparasitic *Collinia beringensis*. Currently it is known that all endoparasitic ciliates that infect krill are obligate histophagous ciliates of the genus *Pseudocollinia*, family Pseudocollinidae (Gómez-Gutiérrez et al. 2003, 2006, 2012; Lynn et al. 2014) observed in at least seven of the most abundant krill species from the northeast Pacific (Bering Sea to Gulf of California) and northwest Atlantic Ocean (Kulka and Corey 1984; Lynn et al. 2014). If future morphological and molecular evidence show that endoparasitic ciliates that infect *E. superba* reported by Stankovic and Rakusa-Suszczewski (1996) actually belong to the genus *Pseudocollinia*, this would be the first published record of histophagous apostome ciliates in the southern hemisphere. They interpreted these endoparasitic ciliates as associated with the phoront cysts (Foettingeriidae) that adhere to the appendages setae and reporting a potential link in a life cycle. However, it is well established that they are actually distinct ciliate species with distinct life cycles, morphology, and feeding strategies (although phylogenetically closely related) (Landers et al. 2006, 2007; Lynn et al. 2014). However, Stankovic and Rakusa-Suszczewski (1996) correctly interpreted that those endoparasitic ciliates "may have a negative (lethal) consequence for the host", being so far the only known parasitoid reported infecting *E. superba*. We use the term "parasitoid" with caution, because, although dinoflagellates and histophagous ciliates infecting krill match with the typical characteristic of the parasitoid definition (i.e. which must kill its host to continue their life cycle), the fact that they occur with remarkable high intensity per krill-host, is quite anomalous when compared with for the typical terrestrial parasitoid definition (Gómez-Gutiérrez et al. 2015).

10.6 Trophically Transmitted Helminths (Nematoda)

Helminths (Trematoda, Cestoda, Acanthocephala and Nematoda) include generalist trophic transmitted endoparasites that infect plankton and nekton in different life phases. In the Australian Antarctic Division krill state-of-art laboratory located at Kingston, Tasmania (Australia), we observed small, unidentified nematodes (<400 μm length) inside just-hatched eggs (with intensity of one or rarely two nematodes) of *E. superba* reared under laboratory conditions (Robert King and Jaime Gómez-Gutiérrez pers. observ., Fig. 10.8a,b). As far as we know, this is the first record of a nematode infecting eggs of any krill species worldwide and, the only report of occurrence of any helminth parasitizing *E. superba*. Although there

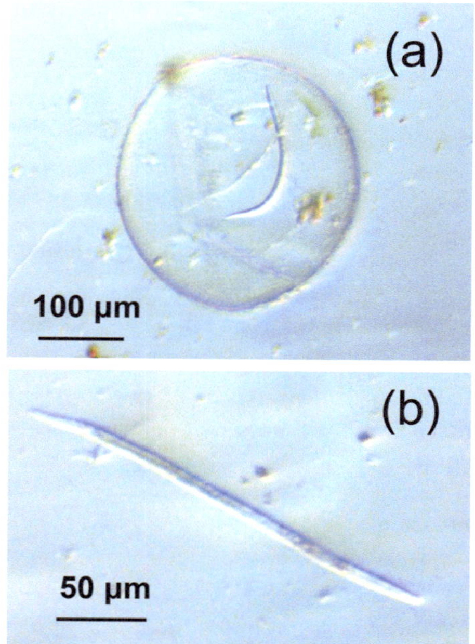

Fig. 10.8 Parasitic nematode found in eggs of *Euphausia superba* reared under laboratory conditions. (**a**) Unidentified larvae of nematode occurring inside just-hatched eggs. (**b**) Nematode (<400 µm length) freed from *E. superba* egg (Robert King and Gómez-Gutiérrez, pers. observ. at Australian Antarctic Division, Kingston, Tasmania, Australia)

are numerous records of helminth infections in Antarctic fish, seabirds, and marine mammals, so far, despite considerably research effort to find such helminthic infections, there are no published records of helminth infecting *E. superba* in the field (Kagei 1969, 1974, 1979; Kagei et al. 1978). Earlier studies reported that the Antarctic krill *E. superba* is free of *Anisakis* spp. infection [34,879 specimens analysed (Kagei 1974; Kagei et al. 1978) and 91,771 specimens analysed (Kagei 1979)]. The same was proposed for Antarctic marine mammals (Kagei and Kureha 1970). However, this perspective is changing because recent research efforts show life cycle biology, specificity, and geographical distribution of Tramatoda, Cestoda, Acanthocephala and Nematoda of Antarctic fishes (Rocka 2006). *Anisakis simplex* and *Anisakis pegreffii* infect migratory myctophids (*Gymnoscopelus nicholsi* and *Electrona carlsbergi*, intermediate hosts that feed on Antarctic krill), and other krill's predators like minke whales (definitive host), and elephant seal *Mirounga leonine* (accidental host) in the Antarctic (Klimpel et al. 2010). So far, the main invertebrate host vectors of such nematode infections are unknown. The endemic myctophid *Electrona antarctica* did not have nematode infections. The occurrence in migrating myctophids coupled with rare findings from other teleosts and regular introduction events through migrating whales lead them to conclude that *A. simplex* and *A. pegreffii* were introduced from northern latitudes outside the Antarctic. Seal worms of *Contracaecum* and *Pseudoerranova* genera clearly dominate the

Antarctic anisakid nematode fauna infecting fish, seals, and cetaceans, but so far, evidence suggest those nematodes species do not infect *E. superba* (Kagei 1974, 1979; Kagei et al. 1978). The icefish *Chaenocephalus aceratus* (especially specimens >30 cm total length) are heavily infected with the nematode *Contracaecum* spp., but no nematodes are found in fish <22 cm in males or females. Larvae of nematodes are often long living, resulting in an accumulation in the fish and a positive correlation between length and age and infection intensity (infection rates increases rapidly for *C. aceratus* >22 cm and attains a mean level >90 %). The reason for infection with nematodes can be deduced from the food items of the Channichthyidae. *Ch. aceratus* feeds on fish, krill, mysids, amphipods, and tunicates and *Champsocephalus gunnari* feeds mainly on krill (Permitin and Traverdiyeva 1972). Krill, however, is not a intermediate host for nematodes in the Southern Ocean (Kagei 1974, 1979; Kagei et al. 1978), so it is not surprising that *Ch. gunnari* is free of nematodes as well as the krill-eating small *Chionodraco* sp. and juveniles (<22 cm) of *Ch. aceratus*. As they grow larger (>22 cm), the latter two Channichthyidae change their main diet to potential intermediate hosts of nematodes; and the *Contracaecum* sp. infection prevalences increase (Siegel 1980a, b). Thus, observational evidence, so far available, indicates that *E. superba* really seems to be "clean" of helminths as concluded Siegel (1980a, b). Evidently, scientists must focus on investigating helminthic infections in the Antarctic ecosystems because they have been detected in several krill predators that migrate seasonally to this gelid ocean during the high-production austral spring and summer (Klimpel et al. 2010).

10.7 The Role of Swarming Behaviour in the Transmission of Parasites and Pathogens

Kuris et al. (1980) proposed based on biogeography island theory "individual host organisms are unequivocal islands where infection is equivalent to immigration of the parasite population and extinction represent the loss of a parasite population either from natural death of the parasites with short life spans, competition from other parasite populations, and/or host defensive responses". Thus, *E. superba* may be regarded as islands for parasites at several levels of organization: (1) individuals, (2) aggregations, swarms, or schools, and/or (3) populations. The euphausiids have an interspecific, and likely size-dependant intraspecific variability of social behaviour (patchiness) (Décima et al. 2010). It ranges from species where individuals are solitary swimmers to social interactions that result in the formation of aggregations, swarms, or even schools at different spatial and time scales (Hamner 1984; Ritz 1994; Ritz et al. 2011; Watkins 2000; Nowacek et al. 2011). Krill social behaviour is closely associated with multiple, significant ecological and physiological processes like reproduction, food searches, predator avoidance strategies, moulting,

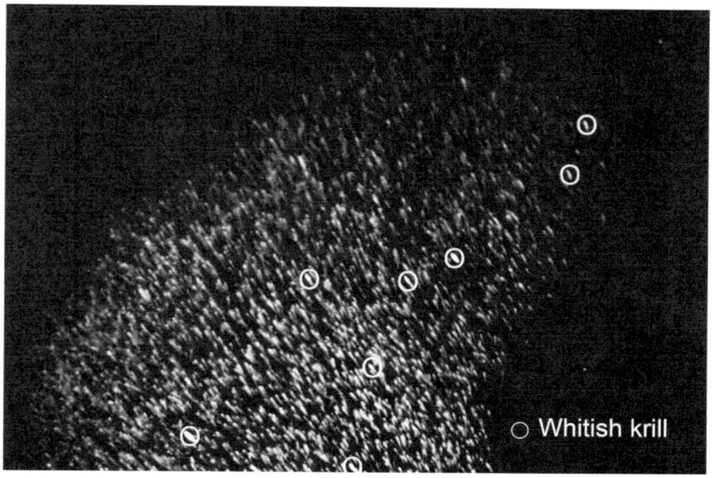

Fig. 10.9 School of *Euphausia superba* in Croker Passage off Antarctic Peninsula. Swimming direction in photo is obliquely downward from *left* to *right*. No individuals occur outside the school. Within the school, krill are closely packed at extremely high density. Unhealthy, whitish animals (*circles*) are easily distinguished (Photo reproduced with permission Koninklijke Brill NV (1984) from Hamner W. M., Aspects of schooling of *Euphausia superba*, Journal of Crustacean Biology)

and parasitic transmission, among others. Hamner (1984) specifically discussed the possible parasite transmission within swarms and among swarms and schools of *E. superba*. Euphausiids species that form dense schools/aggregations seem to interact with a more diverse parasitic assemblage than those that form low-density aggregations because parasite transmission is facilitated by social interaction (Gómez-Gutiérrez et al. 2010). Krill surface swarming behaviour with densities ranging between 100 individuals m^{-3} to 1.5×10^6 individuals m^{-3} may decrease the nearest neighbour distance that facilitates parasite transmission (Fig. 10.9) (Hamner 1984; Hamner et al. 1983; Nicol 1984). This suggests that several parasites require host species with dense and high abundance and compact swarming/schooling behaviour to complete their life cycles during the long-term evolutionary process of speciation. Aggregating behaviour of *E. superba* can develop in early larval stages, as young as furcilia IV (Hamner et al. 1989), possibly enhancing parasitic transmission after the ontogenetic formation of social aggregations and swarms. Hamner (1984) observed that *E. superba* swarms sometimes have opaque ("whitish") individuals (presumably necrotic) positioned behind the schools that are unable to swim as fast as healthy individuals and indicating that schooling may have zoonotic disadvantages (Fig. 10.9). The causes of opaque appearance in *E. superba* are still unknown. It is well known that apostome histophagous ciliates change the colour of the krill host (Gómez-Gutiérrez et al. 2006, 2012), but in *E. superba* it is unknown because histophagous endoparasitic ciliates were observed only in frozen kill specimens (Stankovic and Rakusa-Suszczewski 1996).

Based on the biogeography island theory (Kuris et al. 1980) and assuming similar parasite transmission rates of relatively less virulent parasites, it would be expected that long-lived euphausiid hosts with larger individual biomass would offer longer and more potential sites (or microhabitats) for parasites than short-lived with small individual biomass euphausiid hosts. This is a paramount inference because it would predict that eggs and larvae (with development times within days-week and small individual biomass) are comparatively less likely to be parasitized (or smaller number of parasitic types), than juveniles and adults (intraspecific ontogenetic vulnerability). Additionally, smaller euphausiid species like *Stylocheiron microphthalma* Hansen, 1910 or *Stylocheiron suhmi* G. O. Sars, 1883 (<7 mm and longevity likely <1 year) should have relatively less diverse parasitic fauna than larger species, such as *E. superba* (6.5 cm, longevity of 5–7 years) or *Thysanopoda* species (<15 cm) (interspecific vulnerability). However, an overview of all epibionts and parasites known for 48 of 86 current extant euphausiid species include at least 17 distinct types (epibionts, parasites, pathogens, and parasitoids) (Gómez-Gutiérrez et al. 2010). Only seven of them have been reported in *E. superba* [epibionts: exuviotrophic ciliates (family Foettingeriidae) and microplanktophagous ciliates (family Suctoridae genus *Ephelota*), pathogens: chitinoclastic bacteria and fungus; and trophically transmitted endoparasites: Apicomplexa (family Gregarinidae genus *Cephaloidophora*), Nematoda, and endoparasitic histophagous ciliates (family Apostomatidae)]. The massive and dense aggregations, swarms, and schools of *E. superba*, their keystone function as voracious predators (phytoplankton, benthic microalgae, marine snow, and mesozooplankton), and prey for multiple predator species (macrozooplankton, fish, squids, sea birds, and marine mammal) should make it a critical vector for trophically transmitted parasites in the Antarctic food web. However, comparing parasite diversity of *E. superba* with those for other well studied krill species of the world (*Meganyctiphanes norvegica*, *Euphausia pacifica*, and *Nyctiphanes simplex*), *E. superba* apparently interacts with a relatively low diversity of parasitic taxa (Fig. 10.10). However, future studies must confirm this ontogenetic and interspecific parasite diversity pattern because, so far, relatively few scientists have been studied parasites of krill worldwide. *Euphausia superba*'s diversity, prevalence, and intensity of parasites is less than expected from the theory of island biogeography predicted from the relatively large body size and the colossal *E. superba* population biomass, but consistent with the hypothesis that low temperatures prevailing in the Antarctic Sea are not favourable for parasites and pathogens (Seear et al. 2012). Multiple parasitic taxa diversity with wide global zoogeographic patterns could overlap with *E. superba* range in the circumpolar Antarctic Ocean. Current knowledge indicates that multiple parasites apparently have not invaded the Antarctic ecosystem with the same evolutionary success (Klimpel et al. 2010) as has occurred with euphausiid species in tropical, subtropical, temperate, and even Arctic ecosystems.

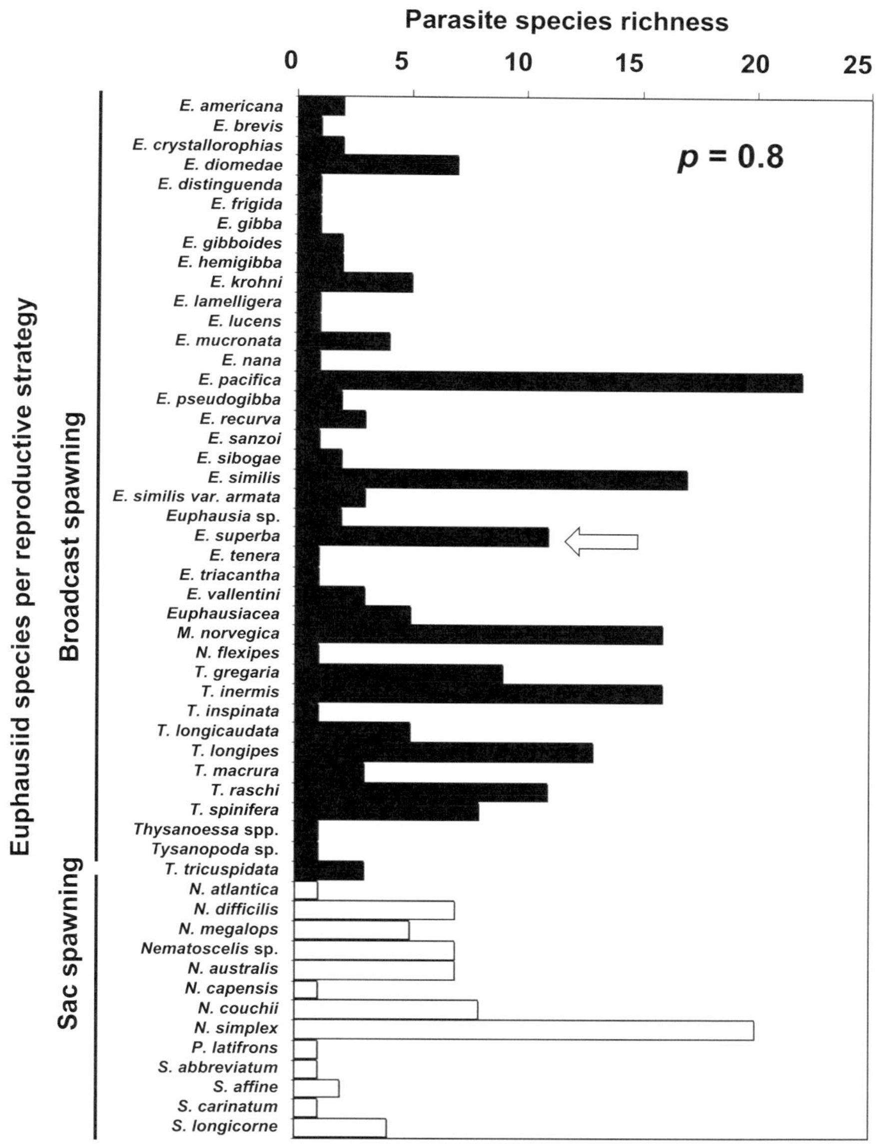

Fig. 10.10 Meta-analysis of epibiont, mesoparasite, parasite, and parasitoid species richness known for each of the 48 out of 86 current extant euphausiid species around the world, comparing krill species of relatively better known symbiotic relationship with the Antarctic krill *Euphausia superba*, comparing species with distinct reproductive strategies (broadcast versus sac-spawning species), from 1885 to 2013 (120 publications) plus personal observations (Gómez-Gutiérrez)

Acknowledgements This research was partially supported by the Instituto Politécnico Nacional, Centro Interdisciplinario de Ciencias Marinas (IPN-SIP 2012–2016) and CONACyT 2012–C01–178615. J.G.G. was supported by an SNI fellowship, COFAA–IPN, and EDI–IPN grants and J.R.M.A. was supported by a CONACyT PhD and a BEIFI–IPN grants. We thank Ira Fogel (CIBNOR) for the English editing of the manuscript and we deeply thank Guest Editor Volker Siegel for inviting us to write this chapter and his valuable information that provide us about nematodes in the Antarctic Ocean. We deeply thank to Mario J Aguilar Méndez for help to gather information about bacteria of krill worldwide.

References

Aguilar-Méndez MJ, López-Cortés A, Gómez-Gutiérrez J (2008) Bacterial diversity associated with parasitoidism of the ciliate *Collinia* spp. in euphausiids from the Gulf of California. Paper presented at the 12th International Society on Microbial Ecology (ISME). Microbial diversity: sustaining the blue planet, Cairns, Australia, August 17–22, 2008

Atkinson A, Siegel V, Pakhomov E et al. (2004) Long-term decline in krill stock and increase in salps within the Southern Ocean. Nature 432(7013):100–103

Atkinson A, Siegel V, Pakhomov EA et al (2009) A re-appraisal of the total biomass and annual production of Antarctic krill. Deep-Sea Res I 56:727–740

Avdeev VV (1985) New species of gregarines of genus *Cephaloidophora* parasites of *Euphausia superba*. Parasitology 1:1–6

Avdeev VV (1987) Certain specific characteristics of the development of gregarine *Cephaloidophora*, a parasite of *Euphausia superba*. Parasitology 21(4):580–582

Avdeev VV, Avdeeva NV (1989) On gregarine fauna from planktonic crustaceans from Antarctica. In: Lebedev BI (ed) Parasites of animals and plants: collected papers. Far East Division of Russian Acad. Sci, Vladivostok, pp 40–44

Avdeev VV, Vagin AV (1987) Pathogenic effect of gregarine *Cephaloidophora pacifica* Avdeev on the organism *Euphausia superba* Dana. Parasitology 21(6):741–743

Banas PT (1981) Apostomatous ciliate association with the antarctic krill, *Euphausia superba* (Master's thesis, Texas A and M University)

Bateman K, Hicks R, Tarling G et al (2015) Chiller killers – first steps towards identifying krill pathogens. Paper presented at the Commission for the Conservation of Antarctic Marine Living Resources (CCAMLR), Working Group on Ecosystem Monitoring and Management (WG-EMM-2015/23). Warsaw, Poland, 6–17 July 2015

Bradbury PC (1994) Parasitic protozoa of molluscs and crustacea. In: Kreier JP (ed) Parasitic protozoa, vol 8, 2nd edn. Academic, San Diego, pp 139–263

Cadee GC, Gonzalez H, Schnack-Schiel SB (1992) Krill diet affects fecal string setting. Polar Biol 12:75–80

Capriulo GM, Small EB (1986) Discovery of an apostome ciliate (*Collinia beringensis* n. sp.) endoparasitic in the Bering Sea euphausiid *Thysanoessa inermis*. Dis Aquat Organ 1:141–146

Capriulo GM, Pedone MJ, Small EB (1991) High apostome ciliate endoparasite infection rates found in the Bering Sea euphausiid *Thysanoessa inermis*. Mar Ecol Prog Ser 72:203–204

Cieśliński H, Kur J, Bialkowska A et al (2005) Cloning, expression, and purification of a recombinant cold-adapted beta-galactosidase from Antarctic bacterium *Pseudoalteromonas* sp. 22b. Protein Expr Purif 39:27–34

Cieśliński H, Bialkowska A, Dlugolecka A et al (2007) A cold-adapted esterase from psychrotrophic *Pseudoalteromas* sp. strain 643A. Arch Microbiol 188:27–36

Corsolini S, Romeo T, Ademollo N et al (2002) POPs in key species of marine Antarctic ecosystem. Microchem J 73(1):187–193

D'Amato ME, Harkins GW, de Oliveira T et al (2008) Molecular dating and biogeography of the neritic krill *Nyctiphanes*. Mar Biol 155:243–247

Dabrowski J, Mierzejewski J, Skoczek A (1983) Isolation of mesophilic anaerobes from Antarctic krill (*Euphausia superba*). Acta Microbiol Pol 32:95–98

de C Baker A, Boden BP, Brinton E (1990) A practical guide to the euphausiids of the world. Natural History Museum Publication, London

Décima M, Ohman MD, Robertis AD (2010) Body size dependence of euphausiid spatial patchiness. Limnol Oceanogr 55(2):777–788

Denner EB, Mark B, Busse HJ et al (2001) *Psychrobacter proteolyticus* sp. nov., a psychrotrophic, halotolerant bacterium isolated from the Antarctic krill *Euphausia superba* Dana, excreting a cold-adapted metalloprotease. Syst Appl Microbiol 24:44–53

Dobson EE (2002) Phylum Apicomplexa Levine, 1970: order Eugregarinorida Léger, 1900. In: Lee JJ, Leedale G, Patterson D, Bradbury PC (eds) Illustrated guide to the Protozoa, 2nd edn. Society of Protozoologists, Lawrence, pp 205–288

Dolzhenkov VN, Avdeev VV, Tirnonin VP (1987) To the problem of the population structure of the Antarctic krill *Euphausia superba* Dana. In: Makarov RR (ed) Biological-oceanographical investigations of the Pacific sector of Antarctic. VNIRO Publishers, Moscow, pp 176–186 (in Russian)

Donachie SP (1995) Ecophysiological description of marine bacteria from Admiralty Bay (Antarctica), and the digestive tracts of selected euphausiidae. In: Rakusa-Suszczewski S, Donachie SP (eds) Microbiology of Antarctic marine environments and krill intestine, its decomposition and digestive enzymes. Department of Antarctic Biology, Polish Academy of Sciences, Warsaw, pp 101–196

Donachie SP, Zdanowski MK (1998) Potential digestive function of bacteria in krill *Euphausia superba* stomach. Aquat Microb Ecol 14:129–136

Donachie SP, Saborowski R, Gerrit P et al (1995) Bacterial digestive enzyme activity in the stomach and hepatopancreas of *Meganyctiphanes norvegica* (M. Sars, 1857). J Exp Mar Biol Ecol 188:151–165

Fernandez-Leborans G (2013) Epibionts on the krill (*Euphausia pacifica*) from the E coast of Japan. Acta Zool (Stockholm) 94:167–176

Fevolden SE, Eidsa G (1981) Bacteriological characteristics of Antarctic krill (Crustacea; Euphausiacea). Sarsia 66:77–82

Fize A, Manier JF, Maurand J (1970) Sur un cas d'infestation du copepoda *Eurytemora velox* (Lillj) par une levure du genre *Metschnikowia kamienslu*. Ann Parasitol Hum Comp 45:357–363

Gómez-Gutiérrez J, Peterson WT, Robertis AD et al (2003) Mass mortality of krill caused by parasitoid ciliates. Science 301:339

Gómez-Gutiérrez J, Peterson WT, Morado FJ (2006) Discovery of a ciliate parasitoid of euphausiids off Oregon, USA. *Collinia oregonensis* n. sp. (*Apostomatida*: *Colliniidae*). Dis Aquat Organ 71:33–49

Gómez-Gutiérrez J, Robinson CJ, Kawaguchi S et al (2010) Parasite diversity of *Nyctiphanes simplex* and *Nematoscelis difficilis* along the Northwestern coast of Mexico. Dis Aquat Organ 88:249–266

Gómez-Gutiérrez J, Strüder-Kypke M, Lynn D et al (2012) *Pseudocollinia brintoni* gen. nov. sp. nov. (Apostomatida: Colliniidae) a parasitoid ciliate infecting the euphausiid *Nyctiphanes simplex*. Dis Aquat Organ 99:57–78

Gómez-Gutiérrez J, López-Cortez, A, Aguilar-Méndez MJ et al (2015) Histophagous ciliate *Pseudocollinia brintoni* and bacterial assemblage interaction with krill *Nyctiphanes simplex*: I. Transmission process. Dis Aquat Organ. 116:213–226. doi:10.3354/dao02922

Hamner WM (1984) Aspects of schooling of *Euphausia. superba*. J Crust Biol 4:67–74

Hamner WM, Hamner PP, Stran SW et al (1983) Behavior of Antarctic krill, *Euphausia superba*: chemoreception, feeding, schooling and molting. Science 220:433–435

Hamner WM, Hamner PP, Obst BS (1989) Field observations on the ontogeny of schooling of *Euphausia superba* furciliae and its relationship to ice in Antarctic waters. Limnol Oceanogr 34:451–456

Hebert PD, Cywinska A, Ball SL (2003) Biological identifications through DNA barcodes. Proc R Soc B 270(1512):313–321

Ikeda T, Nash GV, Thomas PG (1984) An observation of discarded stomach with exoskeleton moult from Antarctic krill *Euphausia superba* Dana. Polar Biol 3:241–244

Kagei N (1969) Life history of nematodes of the genus *Anisakis*. Saishin Igaku 24:389–400

Kagei N (1974) Studies on *Anisakid* Nematoda (Anisakinae) (IV). Survey of *Anisakis* larvae in the marine crustacea (Japanese). Bull Inst Publ Health 23(2):65–71

Kagei N (1979) Euphausiids and their parasites (I). Geiken Tsushin 328:53–62

Kagei N, Kureha K (1970) Studies on anisakid Nematoda (Anisakinae) I. Survey of *Anisakis* sp. on marine mammals collected in the Antarctic Ocean. Bull Inst Publ Health Tokyo 19(3):193–196

Kagei N, Asano K, Kihata M (1978) On the examination against the parasites of Antarctic krill, *Euphausia superba*. Sci Rep Whales Res Inst 30:311–313

Kawaguchi S, Toda T (1997) Discovery of ciliates reproducing in the gut of Antarctic krill. Polar Biol 18:158–160

Kawaguchi S, Hosie G, Nicol S et al (1999) Do gregarines cause damage to midgut gland and intestinal epithelium of Antarctic krill? Paper presented at the Second International Symposium of krill, Santa Barbara, CA August 22–26, 1999

Kawaguchi S, Candy S, King R et al (2006) Modelling growth of Antarctic krill. I. Growth trends with sex, length season, and region. Mar Ecol Prog Ser 306:1–15

Kelly MD, Lukaschewsky S, Anderson CG (1978) Bacterial flora of Antarctic krill (*Euphausia superba*) and some of their enzimatic properties. J Food Sci 43:1196–1197

Kittel W, Rakusa-Suszczewski S (1988) Biological characteristics of *Euphausia superba* Dana (BIOMASS III, November 1986–January 1987). Pol Polar Res 9:315–325

Klimpel S, Busch MW, Kuhn T et al (2010) The *Anisakis simplex* complex off the South Shetland Islands (Antarctica): endemic populations versus introduction through migratory hosts. Mar Ecol Prog Ser 403:1–11

Kulka DW, Corey S (1984) Incidence of parasitism and irregular development of gonads in *Thysanoessa inermis* (Kroeyer) in the Bay of Fundy (Euphausiacea). Crustaceana 46(1):87–94

Kuris AM, Blaustein AR, Alió JJ (1980) Hosts as islands. Am Nat 116:570–586

Kuris AM, Hechinger RF, Shaw JC et al (2008) Ecosystem energetic implications of parasite and free-living biomass in three estuaries. Nature 454:515–518

Lafferty KD (1999) The evolution of trophic transmission. Parasitol Today 15(3):111–115

Lafferty KD, Kuris AM (2002) Trophic strategies, animal diversity and body size. Trends Ecol Evol 17(11):507–513. doi:10.1016/S0169-5347(02)02615-0

Landers SC, Gómez-Gutiérrez J, Peterson WT (2006) *Gymnodinioides pacifica* n. sp., an exuviotrophic ciliated protozoan (Ciliophora, Apostomatida) from euphausiids of the Northeastern Pacific. Eur J Protistol 42:97–106

Landers SC, Gómez-Gutiérrez J, Peterson WT (2007) The fine structure of the phoront of *Gymnodinioides pacifica*, a ciliated protozoan (Ciliophora, Apostomatida) from euphausiids of the Northeastern Pacific. Eur J Protistol 43:239–249

Lear DW (1963) Occurrence and significance of chitinoclastic bacteria in pelagic waters and zooplankton. In: Oppenheimer CH (ed) Symposium on marine microbiology. CC Thomas, Publisher, Springfield, pp 594–610

Levine ND (1988) Progress in taxonomy of the Apicomplexan protozoa. J Protozool 35(4):518–520

Lindley JA (1978) Continuous plankton records: the occurrence of apostome ciliates (Protozoa) on Euphausiacea in the North Atlantic Ocean and North Sea. Mar Biol 46:131–136

Lynn DH, Gómez-Gutiérrez J, Strüder-Kypke MC et al (2014) Ciliate species diversity and host-parasitoid codiversification in the apostome genus *Pseudocollinia* (Ciliophora, Apostomatia,

Pseudocolliniidae) that infect krill, with description of *Pseudocollinia similis* n. sp., a parasitoid of the krill *Thysanoessa spinifera*. Dis Aquat Organ 112(2):89–102
Macaulay MC, English ST, Mathisen OA (1984) Acoustic characterization of swarms of Antarctic krill (*Euphausia superba*) from Elephant Islands and Bransfield Strait. J Crust Biol 4(spec. No. 1):16–44
Mauchline J (1980) The biology of mysids and euphausiids. In: Blaxter JHS, Russell FS, Younge M (eds) Adv Mar Biol 18:1–681
Mauchline J, Fisher LR (1969) The biology of the euphausiids. In: Russell FS, Younge M (eds) Adv Mar Biol 7:1–454
Miwa S, Kamaishi T, Matzuyama T et al (2008) Histopathology of Antarctic krill, *Euphausia superba*, bearing black spots. J Invert Pathol 98:280–286
Mooney JR, Shirley TC (2000) New hosts, prevalence, and density of the ellobiopsid parasite *Thalassomyces fagei* on euphausiids in Prince William Sound, Alaska. J Crust Biol 20:320–325
Nash SMB, Poulsen AH, Kawaguchi S et al (2008) Persistent organohalogen contaminant burdens in Antarctic krill (*Euphausia superba*) from the eastern Antarctic sector: a baseline study. Sci Total Environ 407(1):304–314
Nicol S (1984) *Ephelota* sp. a suctorian found on the euphausiids *Meganyctiphanes norvegica*. Can J Zool 62:744–746
Nowacek DP, Friedlaender AS, Halpin PN et al (2011) Super-aggregations of krill and humpback whales in Wilhelmina Bay, Antarctic Peninsula. PLoS ONE 6(4):e19173. doi:10.1371/journal.pone.0019173
Patarnello T, Bargelloni L, Varotto V et al (1996) Krill evolution and the Antarctic Ocean currents: evidence of speciation as inferred by molecular data. Mar Biol 126:603–608
Permitin YY, Traverdiyeva MJ (1972) The food of some Antarctic fishes in the South Georgia area. J Ichthyol 12:104–114
Poulsen AH, Kawaguchi S, Kukkonen JV et al (2012) Aqueous uptake and sublethal toxicity of p, p′-DDE in non-feeding larval stages of Antarctic krill (*Euphausia superba*). Environ Pollut 160:185–191
Rakusa-Suszczewski S, Filcek K (1988) Protozoa on the body of *Euphausia superba* Dana from Admiralty Bay (the South Shetland Island). Acta Protozool 27:21–30
Rakusa-Suszczewski S, Nemoto T (1989) Ciliates associations on the body of krill (*Euphausia superba* Dana). Acta Protozool 28:77–86
Rakusa-Suszczewski S, Zdanowski MK (1989) Bacteria in krill (*Euphausia superba* Dana) stomach. Acta Protozool 28:87–90
Ritz DA (1994) Social aggregation in pelagic invertebrates. Adv Mar Biol 30:156–156
Ritz DA, Hobday AJ, Montgomery JC et al (2011) Social aggregation in the pelagic zone with special reference to fish and invertebrates. Adv Mar Biol 60:1–161
Rocka A (2006) Helminths of Antarctic fishes: life cycle biology, specificity and geographical distribution. Acta Parasitol 51:26–35
Schmidt K, Atkinson A, Petzke K-J et al (2006) Protozoans as a food source for Antarctic krill, *Euphausia superba*: complementary insights from stomach content, fatty acids, and stable isotopes. Limnol Oceanogr 51:2409–2427
Seear PJ, Goodall-Copestake WP, Fleming AH et al (2012) Seasonal and spatial influences on gene expression in Antarctic krill *Euphausia superba*. Mar Ecol Prog Ser 467:61–75
Siegel V (1980a) Parasite tags for some Antarctic channichthyid fish. Arch Fischereiwissenschaft 31(2):97–103
Siegel V (1980b) Quantitative investigations on parasites of Antarctic channichthyid and nototheniid fishes. Meeresforschung 28(2/3):146–156
Spiridonov VA (1996) A scenario of the late-pleistocene-holocene changes in the distributional range of Antarctic krill (*Euphausia superba*). Mar Ecol 17:519–541
Stankovic A, Rakusa-Suszczewski S (1996) Parasitic protozoa on appendages and inside the body of *Euphausia superba* Dana. Polish Polar Res 17(3–4):169–171

Stankovic A, Borsuk P, Koper M et al (2002) Studies on *Ephelota* sp., an epizoic suctorian found on Antarctic krill. Polar Biol 21:827–832

Stawiszyńska-Janas M, Kittel W (1982) Epizoiczne sysydlaczki (Protozoa, Suctorida) na *Euphausia superba* Dana i *Euphausia crystallorophias* Holt et Tattersal. Polskie Badania Polarne 1970–1982 Toruń, pp 239–248

Takahashi KT, Kawaguchi S, Kobayashi M et al (2003) Parasitic eugregarines change their spatial distribution within the host digestive tract of Antarctic krill, *Euphausia superba*. Polar Biol 26:468–473

Takahashi KT, Kawaguchi S, Kobayashi M et al (2004) The variability in abundance of eugregarines living in the Antarctic krill. Polar Biosci 17:16–25

Takahashi KT, Kobayashi M, Kawaguchi S et al (2008) Circumpolar occurrence of eugregarinid protozoan *Cephaloidophora pacifica* associated with the Antarctic krill, *Euphausia superba*. Antarct Sci 20:437–440

Takahashi KT, Kawaguchi S, Toda T (2009) Observation by electron microscopy of a gregarine parasite of Antarctic krill: its histological aspects and ecological explanations. Polar Biol 32(4):637–644

Takahashi KT, Kawaguchi S, Kobayashi M et al (2011) Eugregarinid infection within the digestive tract of larval Antarctic krill, *Euphausia superba*. Polar Biol 34:1167–1174

Tarling GA, Cuzin-Roudy J (2008) External parasite infestation depends on moult-frequency and age in Antarctic krill (*Euphausia superba*). Polar Biol 31:121–130

Tarling GA, Shreeve RS, Hirst AG et al (2006) Natural growth rates in Antarctic krill (*Euphausia superba*): I. Improving methodology and predicting intermolt period. Limnol Oceanogr 51:959–972

Théodoridès J (1989) Parasitology of marine zooplankton. Adv Mar Biol 25:117–177

Théodoridès J, Desportes O (1975) Sporozoaires d'Invertébrés pélagiques de Villefrenche-sur-mer (étude descriptive et faunistique). Protistologica 11:205–220

Timofeev SF (2001) Quantitative indices of the infection with gregarinids (Sporozoa: Gregarinina) in the euphausiid *Thysanoessa raschii* (Crustacea: Euphausiacea) from the Barents Sea. Parazitol 35(3):235–240 (in Russian)

Trathan PN, Hill SL (2016) Chapter 9, The importance of krill predation in the Southern Ocean. In: Siegel V (ed) The biology of Antarctic krill: *Euphausia superba* Dana 1850, Advances in Polar Ecology, Springer, Cham, pp 321–350

Turkiewicz M, Galas E, Kalinowska H (1982) Microflora of Antarctic krill (*Euphausia superba*). Acta Microbiol Polonic 31:175–184

Watkins J (2000) Aggregation and vertical migration, Chapter 4. In: Everson I (ed) Krill biology, ecology and fisheries, Fish and aquatic resources, series 6. Blackwell Science, Oxford, pp 80–102

Yamamoto Y, Honda K, Tatsukawa R (1987) Heavy metal accumulation in Antarctic krill *Euphausia superba*. Paper presented at the Proceedings of the NIPR Symposium on Polar Biology, Tokyo, pp 198–204

Zhao L, Yin B, Liu Q et al (2013) Purification of antimicrobial peptide from Antarctic Krill (*Euphausia superba*) and its function mechanism. J Ocean Univ China 12(3):484–490

Zoetendal EG, Chen B, Koike S et al (2004) Molecular microbial ecology of the gastrointestinal tract: from phylogeny to function. Curr Iss Intest Microbiol 5:31–48

Chapter 11
The Fishery for Antarctic Krill: Its Current Status and Management Regime

Stephen Nicol and Jacqueline Foster

Abstract Antarctic krill has been fished commercially in the Southern Ocean since the 1970s and has been consistently the largest fishery, by tonnage, in the region since then. The fishery has seen changes in the nations involved, with early catches dominated by vessels from the USSR, Japanese vessels in the middle years and, more recently, most of the catch has been taken by vessels from Norway. A variety of products have emerged from the fishery with early efforts aimed at human consumption but latterly, the bulk of the catch has been used as high-end aquaculture feed with a small but valuable fraction being used to produce krill oil. The fishery has been managed by the international Commission for the Conservation of Antarctic Marine Living Resources which recognised the potential threat to the marine ecosystem through krill harvesting and which has implemented a precautionary approach to management of the fishery. Currently the fishery catches approximately 300,000 tonnes annually, all from the South Atlantic, where the precautionary catch limit has been set at 5.6 million tonnes. The fishery and its management regime will face challenges in the future with the emergence of new technologies, increased catches by new entrants and environmental changes.

Keywords Euphausiid • Harvesting • Commercial • CCAMLR • Catch

11.1 Introduction

The fishery for Antarctic krill (*Euphausia superba*) has been operating in the Southern Ocean since the 1970s, and the development of the fishery has been regularly documented (Budzinksi et al. 1985; Nicol and Foster 2003). Following the initiation of the fishery the scientific community recognised the poor level of

S. Nicol (✉)
Institute for Marine and Antarctic Studies, University of Tasmania, Private Bag 129, Hobart, TAS 7001, Australia
e-mail: krill1953@gmail.com

J. Foster
4 Liriope Drive, Kirkwood, QLD 4680, Australia
e-mail: jacquikf@hotmail.com

V. Siegel (ed.), *Biology and Ecology of Antarctic Krill*, Advances in Polar Ecology,
DOI 10.1007/978-3-319-29279-3_11

information concerning krill and the Southern Ocean ecosystem and an ambitious international scientific program (BIOMASS – Biological Investigations on Marine Antarctic Systems and Stocks) was established to provide information that might be used to regulate the fishery. The remit of the BIOMASS Program was to "gain a deeper understanding of the structure and dynamic functioning of the Antarctic marine ecosystem as a basis for the future management of potential living resources" (El-Sayed 1977, 1994). This initiative was an important predecessor to the negotiation in the late 1970s of an international regime tasked with managing the harvest of krill and other Southern Ocean marine resources (CCAMLR – the Convention on the Conservation of Antarctic Marine Living Resources) (Frank 1983; Edwards and Heap 1981). The main concern in the development of CCAMLR was the effect that a dramatic reduction in the size of krill stocks would have on populations of krill predators, most notably, penguins, seals, whales, fish and seabirds (Howard 1989). The Convention on the Conservation of Antarctic Marine Living Resources (CCAMLR) was adopted in 1980 and it entered into force 2 years later.

CCAMLR was initially known as the "krill convention" (Fernholm and Rudback 1989; Nicol 1991) because of its predominant focus on ensuring the ecosystem-based management of Southern Ocean krill stocks. However, it was not until the early 1990s that CCAMLR began the process of regulating the krill fishery (Nicol 1992). Since then there has been a slow accumulation of management measures that have collectively given effect to many aspects of the ecosystem approach to management of the krill fishery that was originally envisaged in the Convention. Because the krill fishery did not grow in size as quickly as was originally predicted there was little pressure to develop a comprehensive management regime, resulting in the Commission allowing the fishery to develop slowly as CCAMLR focussed on other emerging issues.

In this chapter, we will concentrate mainly on recent developments in the fishery and its management but we provide an outline of the early years and references to published works that deal comprehensively with the early history of the fishery. We describe the current state of the fishery: where it operates, who is involved, how much is caught and the seasonal nature of the operation. We then deal with the nature of the fishery itself: how it operates, the products derived from the catch and the trends in technology and product types. Finally we address the management of the fishery and the scientific and environmental issues that will make future management of this large stock a challenge.

11.2 The Development of the Antarctic Krill Fishery

The commercial krill fishery has operated in the Southern Ocean since the mid-1970s and several publications have recorded the history and development of the fishery (Table 11.1). There have been a number of phases in the development of the fishery (Fig. 11.1). The early years of the fishery were characterised by low

Table 11.1 Published reviews of the krill fishery

Reference	Topic
Eddie (1977)	Harvesting of krill
Everson (1977)	Living resources of the Southern Ocean
Grantham (1977)	Utilisation of krill
McElroy 1984	Economics of krill harvesting
Budzinski et al. (1985)	Processing and marketing of krill products
Miller (1991)	History and management of the krill fishery
Nicol and Endo (1997, 1999)	Krill fisheries of the world their development and management
Miller and Agnew (2000)	Management of Antarctic krill fishery
Ichii (2000)	Krill harvesting
Nicol and Foster (2003)	Update on developments in the krill fishery
Nicol et al. (2012)	Update on developments in the krill fishery

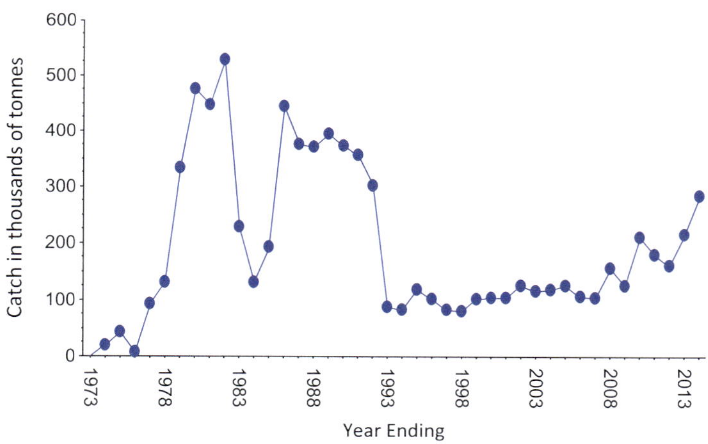

Fig. 11.1 Catches of krill in the Southern Ocean from 1973 to 2014 (CCAMLR 2014)

catches, primarily aimed towards research and product development (Grantham 1977) with vessels from the Soviet Union (from the 1961/1962 season) and Japan (from the 1972/1973 season) conducting exploratory fisheries. In the late 1970s and early 1980s there was a rapid increase in fishing effort, with total catches reaching levels not seen since. This period of high catch rates was primarily a result of fishing by vessels from the USSR, but also included catches by vessels from Japan, Poland, former East Germany and Chile (Table 11.2). This era marked the move of the fishery from an exploratory/experimental phase to a developmental/commercial phase (Budzinski et al. 1985; Nicol and Endo 1997).

There was a sharp decline in catches in the fishery in the mid-1980s, which has been attributed to technical difficulties (Nicol and Endo 1997; Budzinski et al. 1985). Further development of processing techniques was undertaken and

Table 11.2 Reported annual krill catches, in tonnes, by fishing nations at the start of each decade and for the most recent fishing season (Data from CCAMLR (2015) and SC-CCAMLR (2014))

	Period ending				
	1980	1990	2000	2010	2014
Japan	36,275	62,187	67,377	29,910	
Poland	226	1275	13,696	7007	
USSR	440,516	302,376			
Chile		4501			9460
East Germany		396			
Republic of Korea		4040	7525	43,805	52,639
Ukraine				14,023	8928
USA			1561		
Norway				120,429	165,899
Russia				8065	
China				1956	48,102
Total	477,017	374,775	90,159	211,984	285,028

catches quickly recovered before once again declining rapidly in 1991 coinciding with the break-up of the Soviet Union (Nicol and Endo 1997). From the mid-1990s, the krill fishery remained low reaching a catch level of 125,823 tonnes in 2009. The catches during this period were dominated by Japanese and Korean vessels. Since 2009 the catch has increased more rapidly reaching a level of nearly 300,000 tonnes in 2013/2014. This phase has been associated with the entry into the fishery of Norwegian companies and in recent years, from increased catches by vessels from Korea and by the emergence of China as a fishing nation (SC-CCAMLR 2014).

The catch of the early exploratory fishery was aimed at developing harvesting and processing techniques (Eddie 1977; Nicol and Endo 1997). Once the fishery began to develop commercially, catches were used in products for human consumption and meal for animal feed (Eddie 1977). Into the 1980s the scope for use of krill broadened with interest in krill as a source of: enzymes, other proteins, lipids, chitin (Budzinski et al. 1985). Since the 1990s the catch has been used mainly for the production of aquaculture feed, either as meal or as whole frozen krill, and more recently, krill oil as a human health supplement (Nicol et al. 2000, 2012; Nicol and Foster 2003).

11.2.1 Spatial and Temporal Trends

Although the krill fishery was initially circumpolar, it has been concentrated in the South Atlantic (CCAMLR Area 48) since the mid-1990s (Fig. 11.2). The fishery first targeted the South Indian Ocean but over the decades this area became less important and the South Atlantic became the dominant fishing ground. Today the catch is mainly taken from the waters surrounding the Antarctic Peninsula and

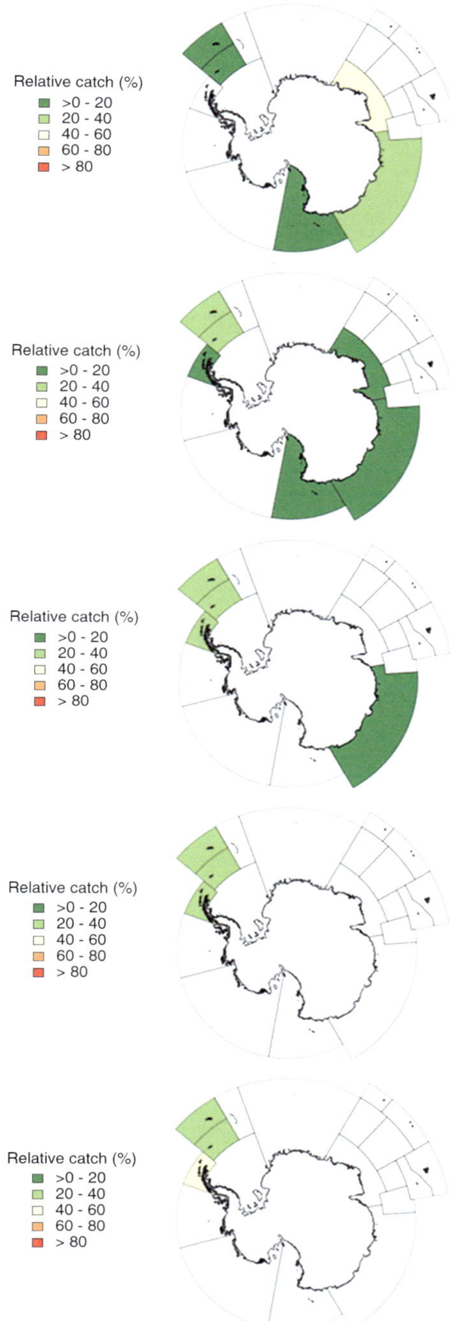

Fig. 11.2 Distribution of krill catches in each decade (1970s, 1980, 1990, 2000, 2010s) using public domain data in the CCAMLR Statistical Bulletin (Figure produced by Lucy Robinson, CCAMLR Secretariat)

South Shetland Islands (subarea 48.1), the South Orkneys (subarea 48.2) and from the South Georgia area (subarea 48.3). This concentration of operations in the South Atlantic is primarily operational as the South Atlantic fishing grounds are closer to port services and are increasingly free of ice for longer periods than the rest of the Southern Ocean (Kawaguchi et al. 2009). Additionally, the Southwest Atlantic is estimated to contain 40 % of the global krill stock (Atkinson et al. 2009). The fishing grounds in the South Atlantic have also changed over the history of the fishery, with the importance of subarea 48.3 declining and that of subarea 48.1 increasing. The krill fishery has also changed its peak season of operation. In the early years of the fishery, most krill were taken in summer and autumn, with lowest catches being taken in spring (Fig. 11.3). In recent years krill catches have shown a reversal in this historic trend with lowest catches occurring over summer, peaking in late autumn, with very little fishing activity in spring (Fig. 11.3).

Some of these temporal and spatial shifts in the fishery over time have been attributed to reductions in winter sea-ice extent in the south of the Scotia Sea which allow winter fishing in subareas 48.1 and 48.2 (Kawaguchi et al. 2009). There has also been a recent tendency of the fishing vessels to avoid catching krill that are actively feeding on algae because this adversely affects the type of product that can be manufactured (Kawaguchi et al. 1999; Ichii 2000; Nicol et al. 2000). The production of krill oil has also led to vessels concentrating on fishing in autumn and winter when krill are richest in lipids (Kim et al. 2014). Additionally, since 1993, vessels fishing in the waters around South Georgia – historically the only ice free krill fishing grounds – have been subject to a license fee imposed by the Government of South Georgia (Agnew 2004) which may act as an incentive for operators to fish in other areas.

11.2.2 Changes in Fishing Technologies

The krill fishery has traditionally used large midwater trawls with a mesh liner towed by large (80+ m) freezer trawlers (Budzinski et al. 1985). These trawls can have a horizontal opening of 60 m and a vertical opening of 50 m. Trawls are generally towed for up to an hour and landed catches are of the order of 10 tonnes. Some of these trawls have been emptied by a pumping system but more usually the catch is landed on deck or into a storage tank. The catch per hour of vessels using conventional trawls has increased recently and this may be a result of changes in fishing behaviour or processing technology.

One of the Norwegian companies that entered the fishery in 2009, Aker Biomarine, has developed a patented "eco-harvesting technology" (patent WO2005004593) that allows much higher catch rates than conventional trawls (800 tonnes/day compared to 400 tonnes per day) (Fig. 11.4). This technology lands krill on the vessel alive and in good condition rendering them suitable for the manufacture of a wider range of products. The eco-harvesting technology uses a hose attached to the cod-end of a small (20×20 m mouth opening) net that is

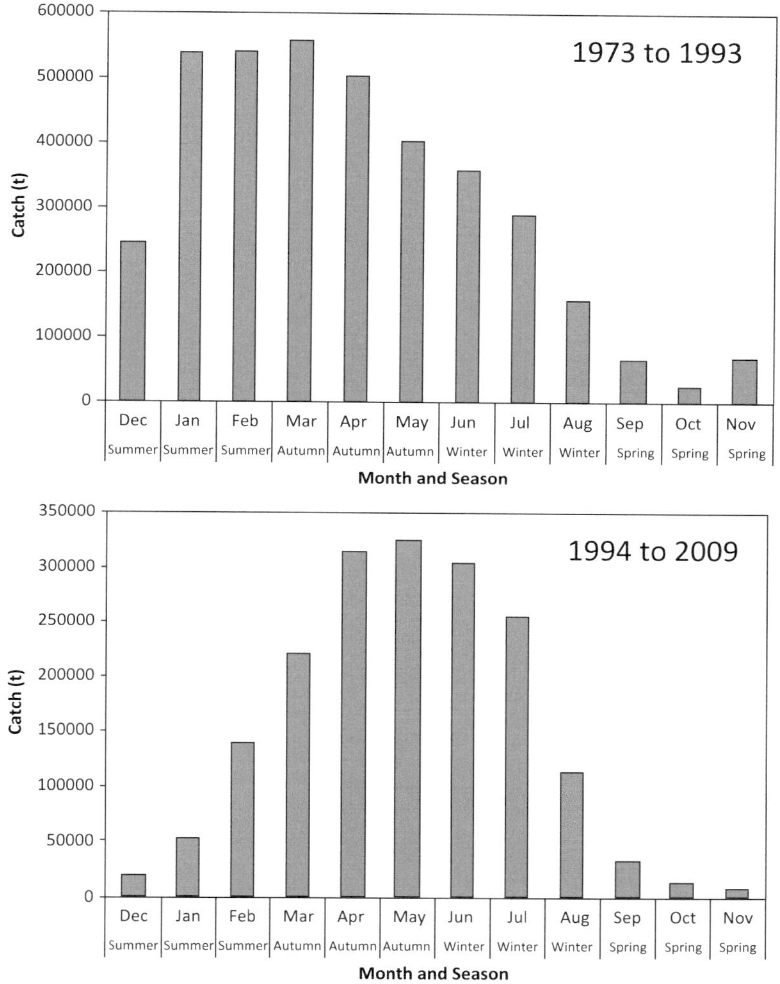

Fig. 11.3 Change in the seasonal distribution of the krill catch in the South Atlantic (Data supplied by the CCAMLR Secretariat)

lowered into a krill swarm. Air is pumped into the cod-end and this lifts water and krill to the surface and onto the ship's conveyor that leads to the factory (Fig. 11.5). The net can be left in the water for weeks at a time and the fishing operation proceeds continuously with the catch rate being determined by the ability of the ship's factory to process the catch. It is also claimed that the continuous fishing system results in less waste of landed catch, allows for accurate estimates of tonnage caught, reduces bycatch and may reduce escape mortality of krill. Escape mortality refers to the krill that are killed by fishing equipment that do not end up in

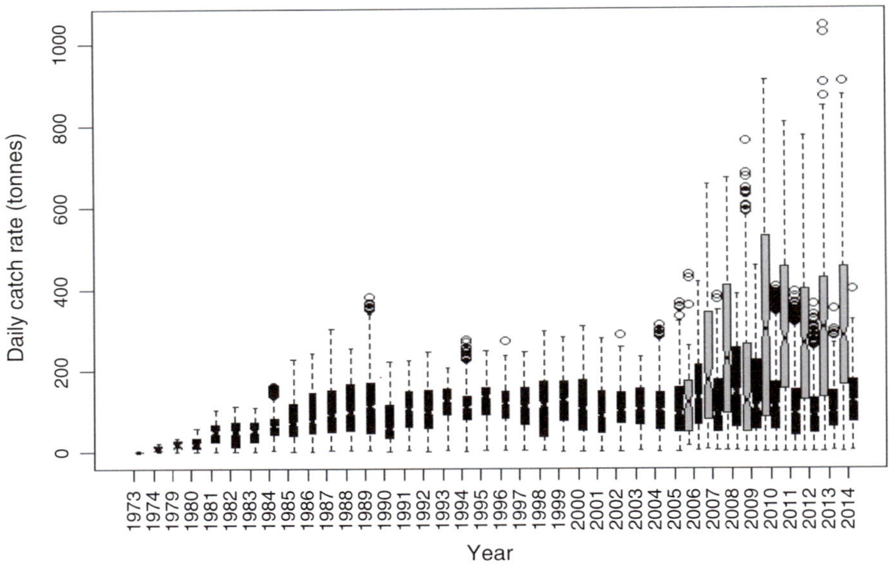

Fig. 11.4 Daily catch (tonnes) of individual fishing vessel in the krill fishery. *Black* is conventional trawls and *Grey* is the continuous fishing system. In the *boxplot* the notch indicates the median with the filled bars over the 25–75 percentile, the max and min (*bars*) and any values that lie outside 1.5 times the IQR (Plot and data supplied by CCAMLR Secretariat)

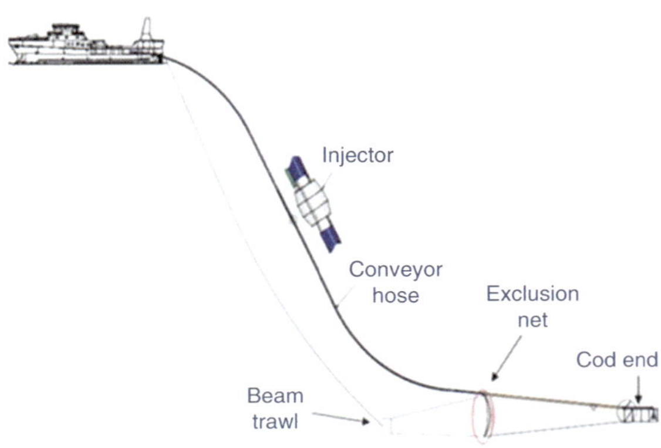

Fig. 11.5 The "Eco-harvesting technology", also known as the continuous pumping system, developed by Aker Biomarine. The injector supplies compressed ait into the cod end and the catch is lifted up the conveyor hose in a packet of water (Illustration courtesy of Aker Biomarine)

the catch. There has been concern within CCAMLR that large trawls may inflict considerable mortality on the krill population that passes through the wider mesh panels on the wings of the net or through being struck by the towing lines (SC-CCAMLR 2011 para 3.12). Modelling studies and some experimental research has failed to clarify the magnitude of this problem (CCAMLR-WG-EMM 2014, paragraphs 2.23–2.24).

11.2.3 Estimating the Total Removal of Krill

A second factor that affects the estimate of mortality of krill attributable to the fishery is the uncertainty surrounding the quantity of krill landed on fishing vessels. The krill catch is generally estimated from the quantity of product from each vessel. It has become apparent that the conversion factors used to relate the weight of product to the tonnage of the catch can vary widely between vessels. Additionally, with the entry into the krill fishery of a different harvesting technology, CCAMLR has focussed on the issue of more accurate assessment of the total catch (SC-CCAMLR 2014, paragraphs 3.4 and 3.5).

11.3 Products from the Krill Fishery

11.3.1 Early Developments

Krill are a difficult raw material; they are easily crushed in nets and need to be rapidly processed once caught because they spoil rapidly once on deck. Krill need to be processed within 1 h of landing if they are to be used for human consumption and within 3 h if they are destined for animal feed (Budzinski et al. 1985; Nicol and Endo 1999). With the initial interest in the commercial fishery in the 1970s the FAO commissioned a series of Fisheries Technical Papers (Budzinski et al. 1985; Eddie 1977; Grantham 1977) addressing the harvesting, utilisation and possibilities of processing and marketing products made from Antarctic krill. In the mid-1970s small catches were being used for product and processing development, with the nutritional value of the products being a high priority (Sidhu et al. 1970). These products included krill meal for animal feed, krill protein paste, frozen cooked whole krill, dried krill, minced muscle and tail meats, all for human consumption (Lyubimova et al. 1973; Siebert et al. 1980; Oehlenschläger and Schreiber 1981; Eddie 1977).

Product development continued throughout the 1980s and early 1990s. However, in the mid-1980s product development research stalled for several years when the industry came to terms with the discovery of high fluoride content of krill (Nicol and Endo 1997; Budzinski et al. 1985). High levels of fluoride that exist in the exoskeleton, particularly in the carapace (Soevik and Braekkan 1979) impeded the use of krill in animal feeds (Yoshitomi et al. 2007) and human products due to health concerns associated with the accumulation of fluoride in the bones and tissues which can be toxic at high levels (Adelung et al. 1987). This problem was eventually overcome with research undertaken into ways to decrease and/or mitigate the quantity of fluorine that would be included in the end product as a result of the inclusion of krill as an ingredient (Tenuto-Filho and Alvarenga 1999;

Tenuto-Filho 1993; Budzinski et al. 1985; Wang et al. 2011). Research continues into methods to reduce fluoride in krill products so that they can be utilised for humans and domestic animals (Xie et al. 2012; Wang and Yin 2012; Zhang et al. 2014). Because many marine vertebrates are tolerant of high fluoride levels in their prey a greater focus on the use of krill as an aquaculture feed developed (Grave 1981; Julshamn et al. 2004) and low fluoride aquaculture feeds have been developed and successfully trialled (Yoshitomi et al. 2007).

The mid-1990s shift into the development of higher value, lower yield products, was predicted to occur in the FAO Technical Papers of the late 1970s and mid-1980s. Grantham (1977) suggested that the initial phase of the krill fishery would see competition for a place in the market, not for the resource, a situation which is highly unusual in commercial fisheries. This prediction was based on the high estimated abundance of krill (biomass estimates at the time ranged from 60 to 5000 million tonnes) (Voronia 1983; Lyubimova et al. 1973) and the low demand for the product in the marketplace at that time (Grantham 1977). The demise of the Soviet Union also meant that the fishery had to move towards a more economic mode of operation. To increase revenues from the fishery, the industry has examined producing a variety of both high value products as well as bulk commodities that are destined for animal feed.

As part of the notification system that CCAMLR has in place, companies provide an indication of the mix of products that they expect to produce. For the 2014/2015 season the frequency of product inclusion in notifications, from the 20 notifications for the 2014/2015 krill fishery was: whole frozen 18, boiled 3, meal 15, oil 3, other product 5 (frozen krill meat, hydrolysate, lipid complex, paste, peeled) (information supplied by the CCAMLR Secretariat). The krill fishery is still dominated by the bulk production of frozen krill and meal; however, the production of smaller quantities of more specialised products probably increases the economic yield from the fishery considerably. The krill fishery operates in waters distant from most ports and in extremely difficult environmental conditions. It is not a fishery that is entered on a whim and because of its relatively low returns and high costs is one that will reward companies with a long term investment in the industry.

Despite the difficulties involved in fishing for krill and of krill as a raw product there is continuing commercial interest in the krill fishery. An index of this interest is the level of patent activity and 1040 krill-related patents have been lodged between 1976 and 2014 (Foster et al. 2011, updated 2014). The most active period has been between 2006 and 2014 when 34 % (a total of 355) of all krill-related patents were lodged (Fig. 11.6). Most recent patents relating to processing (45 % of all patents 2010–2014) were lodged for methods and processes for extracting high-quality krill products including low-fluoride, high quality oils and krill meals that are suited to use in medical applications. Medical uses comprise 29 % of recent patents indicating the industry's focus on products for the high value pharma and nutraceutical sectors.

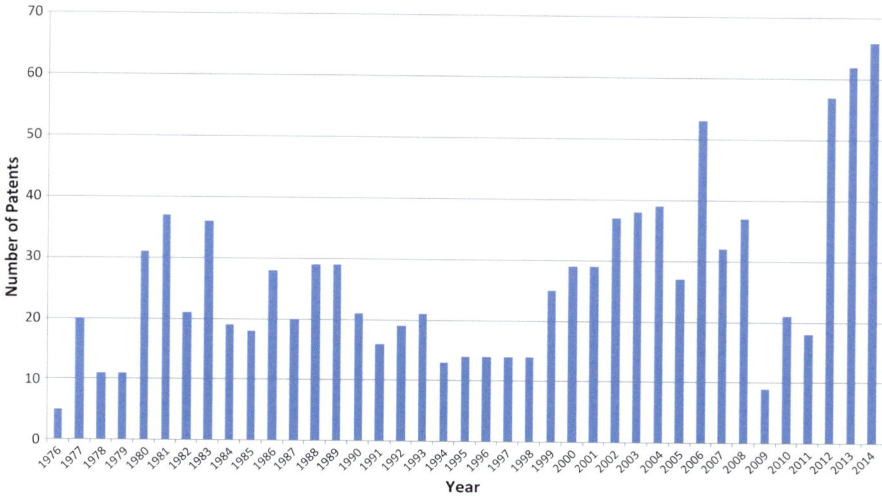

Fig. 11.6 Number of krill-related patents lodged from 1976 to 2014 (Data from the krill patent database presented in Foster et al. (2011) updated in March 2014)

There has also been a shift in the Nations represented in the lodgement of krill-related patents over the past several years. Norwegian interest in the krill fishery has been accompanied by an increase in patent activity. From 1976 to 2008 Norway only represented 2 % of all krill-related-patents lodged but between 2010 and 2014, this increased to 8 %. From 2012 to 2014 China lodged a total of 129 krill-related patents, representing 70 % of all patents lodged. Patents from China have related primarily to new processing technologies for the extraction of high quality products. This level of activity indicates the serious Chinese interest in krill fishing.

11.3.2 Krill for Human Consumption

Since the commencement of the krill fishery there has been an emphasis on producing high quality protein for human consumption, although there have been difficulties in producing a product that is marketable, safe and economically viable (Budzinski et al. 1985; Nicol and Endo 1997; Nicol et al. 2000). Currently very little of the krill caught is destined for use as human food with the bulk going to aquaculture and to produce krill oils which are used as a health supplement. There is still considerable research being carried out into krill for human consumption (Tou et al. 2007; Chi et al. 2013) but the future of krill as a human food will depend largely on finding an economically effective method of removing fluoride (Jung et al. 2013).

11.3.3 Krill as Fishing Bait and Aquarium Feed

A large proportion (~34 %) of the Japanese krill catch in the 1990s and early 2000s was used to attract fish in recreational fishery and freeze dried krill is still used as a commercially available feed for aquarium fish (Nicol et al. 2000). With the Japanese fishing fleet now absent from the krill fishery it is difficult to estimate the current proportion of the catch that is used for these purposes.

11.3.4 Krill Meal for Aquaculture

Production of carnivorous fish through aquaculture has increased dramatically over the last three decades and this has led to an increase in demand for fish meals and marine byproducts. Because of declining fish stocks, the aquaculture industry is investigating the supplementation of currently used proteins and additives with lower cost alternatives that are readily available (Yoshitomi et al. 2007; Davis and Arnold 2000). Some research has focused on determining the properties of krill meal compared to commonly used fish meals (Giogios et al. 2009). Because fishing for krill is an expensive activity krill meal is also expensive and is most commonly used as a high value source of protein, feeding enhancing agents, colouration, growth promoters and sources of Omega-3 oils to aquaculture feeds (Floreto et al. 2001) rather than the primary ingredient. Krill meal has a nutritional value equal to, or surpassing, that of regular fish meals (Yoshitomi et al. 2007; Karlsen et al. 2006; Opstad et al. 2006; Gaber 2005) when used as a substitute in the diets of various farmed species including Atlantic cod, Atlantic salmon and Pacific white shrimp. Due to the increasing price of fish meals, producers are constantly striving to improve the uptake of feeds to minimise feed wastage and provide better economic returns (Smith et al. 2005). Krill oil and chitosan have also been shown to increase the nutritional value and shelf life of various fish fillet products (Duan et al. 2010).

11.3.5 Chitin

Krill contain a number of valuable chemicals that could be extracted as part of the post-capture manufacturing process. The shell of krill contains chitin (up to 40 % dry weight) and this is a chemical for which there is a developing market (Yanase 1981) which could be a potentially plentiful by-product from the fishery. In the traditional processing of rolling krill for a meat product, the exoskeleton has usually been a waste product but is increasingly seen as valuable commodity in its own right, with research into the quality of the waste product being undertaken by Polish researchers in the early 1980s (Naczk et al. 1981) and more recently (Tou et al. 2007).

Chitin is the main source of production of chitosan, which is used in a number of applications, such as a flocculating agent, a wound healing agent and a delivery vehicle for various pharmaceuticals. A number of publications have focussed on the utilisation of krill for these purposes (Bustos et al. 2003). Due to their high biomass and current underutilisation, krill are major producers of chitin in the world's oceans and the fishery for Antarctic krill is potentially a large source of chitin for industry (Nicol and Hosie 1993).

11.3.6 Pigments

The characteristic red colouration of Antarctic krill comes from the pigment astaxanthin, and other carotenoid pigments in the exoskeleton (Nicol et al. 2000). There has been considerable research into the addition of krill-derived pigments to aquaculture feed to improve colouration in farmed species (Savage and Foulds 1987; Kalinowski et al. 2007; Suontama et al. 2007; Moretti et al. 2006; Floreto et al. 2001). Synthetic colourants have generally been used for pigmentation of aquaculture species but consumer demand for more natural products is increasing (Moreira et al. 2006).

11.3.7 Krill Oils

As early as 1970 it was recognised that, as well as being an excellent source of protein, krill could also be a source of oil for consumption by humans and animals (Sidhu et al. 1970). There is an increasing demand for diet supplements from marine oils rich in Omega-3 fatty acids (Arthi 2009; Shahidi and Wanasundara 1998; Hill 2013). Fish oils have been subject to large amounts of R&D in fish processing, feed, functional food and nutra- and pharmaceutical markets over the past 15 years. The Omega-3 and Omega-6 properties of seafood have long been recognised as having health benefits (Fereidoon and Wanasundara 1998) including prevention and treatment of arthritic conditions, cardiovascular disease, PMS and facial appearance.

The high fatty acid properties of krill oil make it effective in the treatment of autoimmune murine lupus (Chandrasekar, et al. 1996), cardiovascular disease (Batetta et al. 2009; Bunea et al. 2004) and arthritis (Deutsch 2007). Other research indicates that properties of krill, including its enzymes, antioxidant astaxanthin and chitin, make it effective in the treatment of liver disease (Tandy et al. 2009) and maintenance of general health and wellbeing (Bridges et al. 2010). It has been suggested that the increased bioavailability of the Omega-3 and Omega-6 fatty acids may make krill oil a superior treatment for cardiovascular disease and other health conditions, compared to regular fish oils (Kidd 2009), however, research continues into whether krill oil is clinically superior to fish oils (Backes and Howard 2014).

In 2004 the European market for Omega-3 fatty acid products was valued at approximately US$194 million with more than three-quarters of these oils being from marine based sources (Nichols and Nelson 2007). However, whilst the market for the use of fish oils in nutraceutical products is high in value, it remains relatively small in volume, estimated to represent only 5–6 % of total world fish oil production (Nichols 2007). Krill oil production, currently at less than 1000 tonnes per year, uses only ~10 % of the tonnage of krill landed. Even if the market of krill oil increases greatly in the near future the extra demand could readily be met from existing catches.

11.3.8 Enzymes

The suite of proteolytic enzymes contained within krill's digestive gland was initially a significant problem for the fishery because they led to rapid spoilage of catch (Nicol et al. 2000). However, these hydrolytic enzymes, in particular the proteases, carbohydrases, nucleases and phospholipases, have wide ranging uses in different medical applications (Nicol et al. 2000). Applications include treatment of spinal injuries (Melrose et al. 1995) and necrotic wounds (Karlstam et al. 1991) and in clinical drug applications (Nicol et al. 2000).

11.4 Management of the Krill Fishery

11.4.1 Background

Antarctic krill is classified by the FAO as an underexploited fish stock (stocks that are currently harvested but could withstand an increased level of harvest) (Maguire et al. 2006). Only 2 % of stocks monitored by the FAO are classed as underexploited (FAO 2008). The current krill catch in Antarctic waters is ~300,000 tonnes a year which is only 3.5 % of estimated precautionary catch level of some 8.6 million tonnes (CCAMLR figures) so the potential for increase is considerable. The careful management of the krill fishery into the future is essential given its key role ecological as the cornerstone of the Antarctic food chain.

Management of the krill fishery is the responsibility of the Commission for the Conservation of Antarctic Marine Living Resources (CCAMLR). CCAMLR first met in 1982 and currently has 25 Members and 9 acceding States. CCAMLR is responsible for the conservation of living resources (with the exception of seals and whales which are managed by other instruments), both targeted and dependent and

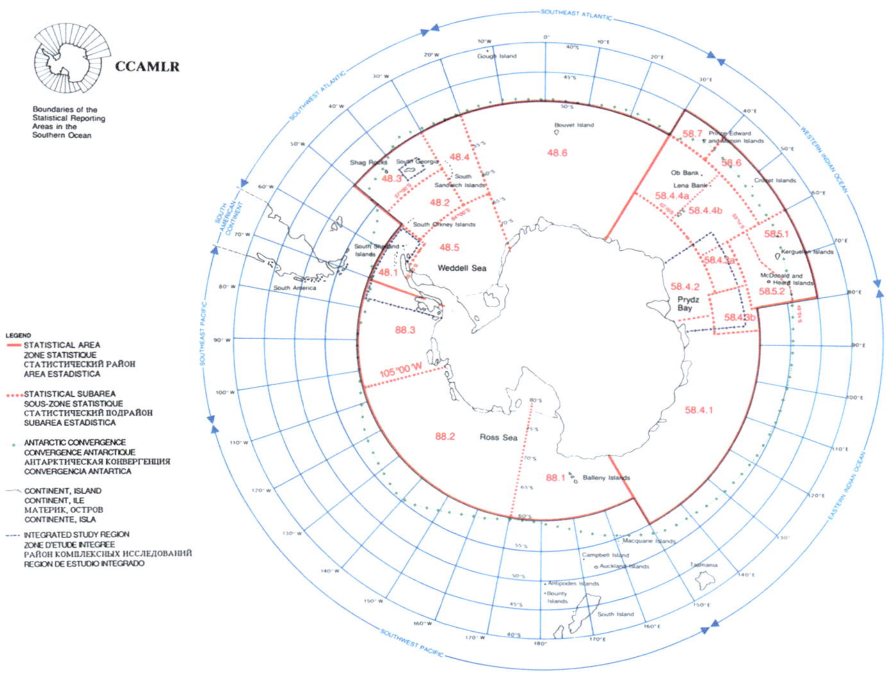

Fig. 11.7 The CCAMLR Convention Area (Figure supplied by CCAMLR)

associated species, in the Southern Ocean (CCAMLR 2009b), (the Southern Ocean is defined for this purpose as the 32.9 million square kilometres of ocean between the Antarctic Polar Front and the continent) (CCAMLR 2009a; Nicol and Endo 1997). The CCAMLR Area is further divided for management purposes into areas based on the FAO Statistical Areas. (Fig. 11.7).

CCAMLR has adopted both a 'precautionary' and 'ecosystem approach' to fisheries management that attempts to combine conservation and "rational use" (CCAMLR 2009b; Miller 2002). CCAMLR's precautionary approach requires that management decisions should have a low risk of long-term adverse effects, an important factor when working with scientific information that has a degree of uncertainty (Hewitt and Linen Low 2000). The ecosystem approach differs from conventional single species fisheries management approaches by managing target species whilst taking into account the needs of dependent and related species. Article II (3) of the Convention set outs the remit for taking an ecosystem approach to the management of Southern Ocean resources:

> 3. Any harvesting and associated activities in the area to which this Convention applies shall be conducted in accordance with the provisions of this Convention and with the following principles of conservation:
>
> (a) prevention of decrease in the size of any harvested population to levels below those which ensure its stable recruitment. For this purpose its size should not be allowed to fall below a level close to that which ensures the greatest net annual increment;

(b) maintenance of the ecological relationships between harvested, dependent and related populations of Antarctic marine living resources and the restoration of depleted populations to the levels defined in subparagraph (a) above; and

(c) prevention of changes or minimisation of the risk of changes in the marine ecosystem which are not potentially reversible over two or three decades, taking into account the state of available knowledge of the direct and indirect impact of harvesting, the effect of the introduction of alien species, the effects of associated activities on the marine ecosystem and of the effects of environmental changes, with the aim of making possible the sustained conservation of Antarctic marine living resources.

CCAMLR has two primary objectives in managing the krill fishery. Firstly, ensuring that overall krill biomass is not compromised and secondly, ensuring that krill biomass is adequate to meet the needs of land-based krill predators (Hewitt and Low 2000). The phrase "dependent and related species" has not been further elaborated by CCAMLR but it refers to those animals such as whales, seals and seabirds that depend on krill for food and those species that coexist in the ecosystem such as other species of krill.

The Commission implements management directives through Conservation Measures (CMs), which are adopted by consensus. Currently, the krill fishery is managed through a series of such Conservation Measures relating to aspects of operation of the fishery including: precautionary catch limits, gear restrictions, data reporting, notification of intent to fish, minimisation of incidental mortality, observer deployment as well as measures regulating exploratory fisheries. Although there has not been a systematic approach to incorporating ecosystem considerations and precaution when managing the krill fishery, the Commission has addressed a range of issues and the overall result of this is to give some practical effect to the principles in Article II of the Convention (Table 11.3). These measures include rules that are specifically aimed at ensuring both ecosystem-based and precautionary management of the krill fishery as well as several measures which are commonly implemented by fisheries management regimes to collect basic fishery data including catch and effort statistics. Some of these measures are specific to the krill fishery whilst others apply across all fisheries operating in the Area.

11.4.2 Precautionary Catch Limits

The first Conservation Measure to specifically regulate the krill fishery (Conservation Measure 51-01 originally numbered 32/X setting precautionary catch limits on krill in Area 48) was adopted in 1991 some 10 years after the Convention was signed (Nicol 1991, 1992). This measure has been updated several times, most notably following the collection of new data from a synoptic survey in 2000 (Hewitt et al. 2004a) (see Chap. 2, Siegel and Watkins 2016). Precautionary catch limits have subsequently been established for Division 58.4.2 in 1992 (revised using new data collected in 2006), and for Division 58.4.1 in 1996 (Table 11.4). These precautionary catch limits are periodically updated using data from new biomass

Table 11.3 Conservation measures applicable to the krill fishery. Specific measures for krill are presented above those that apply to all fisheries (Information from CCAMLR website http://www.ccamlr.org/, accessed March 2014)

CM Number	Adopted	Title	Effect
51-01	2010 (first adopted in 1991)	Precautionary catch limitations on *Euphausia superba* in Subareas 48.1, 48.2, 48.3 and 48.4	Highly precautionary catch level;
			Trigger level;
			Subdivision of catch;
			Adopted prior to expansion of the fishery
51-03	2008 (first adopted in 1992)	Precautionary catch limitation on *Euphausia superba* in Division 58.4.2	Highly precautionary catch level;
			Division of catch;
			Trigger level;
			Observer requirement;
			Adopted prior to expansion of the fishery
51-02	2008 (first adopted 1996)	Precautionary catch limitation on *Euphausia superba* in Division 58.4.1	Highly precautionary;
			Division of catch;
			Gear restrictions to reduce incidental mortality;
			Adopted prior to expansion of the fishery
21-03	2014 (first adopted 2006)	Notifications for krill fisheries	Detailed information on intentions to fish for krill;
			Details of fishing gear types, area to be fished, expected catch, type of vessel, numbers of vessels
51-04	2014 (first adopted 2008)	General measures for exploratory fisheries for *Euphausia superba*	Controls expansion of new fisheries;
			Limits catch;
			Notification requirement;
			Observer requirement
51-06	2014	General measure for scientific observation in fisheries for *Euphausia superba*	Partial Observer coverage mandatory for two seasons;
			Enhances monitoring of fishery;
			Increased biological data collection including by catch
51-07	2014 (first adopted 2009)	Distribution of the trigger limit in the fishery for *Euphausia superba* in Subareas 48.1, 48.2, 48.3 and 48.4	Distributes catch limit across Subareas;
			Prevents taking of large catches from localised areas;
			Protects land breeding predators
23-06	2014	Data Reporting System for *Euphausia superba* Fisheries	Stipulates data requirements from the fishery and reporting intervals
25-03	1991	Minimisation of the incidental mortality of seabirds and marine mammals in the course of trawling	Restricts use of fishing gear;
			Controls deployment of fishing gear;
			Reduces level of incidental mortality.

(continued)

Table 11.3 (continued)

CM Number	Adopted	Title	Effect
33–01	1992	Limitation of the by-catch of *Gobionotothen gibberifrons*, *Chaenocephalus aceratus*, *Pseudochaenichthys georgianus*, *Notothenia rossii* and *Lepidonotothen squamifrons* in Subarea 48.3	Conserves finfish stocks;
			Fishing vessels must act to limit by-catch
21-02	1993	Notification for exploratory fisheries	Controls expansion of new fisheries;
			Notification of fishing intentions
10-04	1998	Automated satellite-linked vessel monitoring systems (VMS)	Applied to krill fisheries in 2007;
			Tracks fishing activities;
			Will allow management of fishing activities within SSMUs
26-01	2006	General environmental protection during fishing	Prohibits use and disposal of environmentally damaging plastics normally used in fishing activities;
			Reduces entanglements and incidental mortality
31-02	2007	General measure for the closure of all fisheries	A fishery can be closed if stock is overexploited
91-03	2009	Protection of the South Orkney Islands southern shelf	In line with Article II;
			Designates an MPA;
			Prohibits fishing to conserve biodiversity

Table 11.4 Precautionary catch limits on the krill fishery (Data from CCAMLR www.CCAMLR.org)

Conservation measure	Year first adopted	Year of survey from which current limit calculated from	Area/division	Total catch limit (millions of tonnes)	Trigger level (millions of tonnes)
51-01	1991	2000	48.1, 48.2, 48.3, 48.4	5.61[a]	0.62
51-02	1996	1996	58.4.1	0.44	n/a
52-03	1992	2006	58.4.2	2.645[a]	0.452

[a]These catch limits are those calculated from more recent surveys conducted in 2000 (Area 48) and 2006 (58.4.1). See text for an explanation of terms

surveys and as a result of changes in the method by which the biomass is calculated from acoustic survey results. The adoption of the first precautionary catch limits for the krill fishery in Area 48, in 1991 (as Conservation Measure 32X) is heralded as a significant achievement in line with the precautionary approach (Fabra and Gascón 2008).

Precautionary catch limits for the krill fishery use data on the krill biomass which have been derived from dedicated acoustic surveys within the relevant CCAMLR statistical area. The original precautionary catch limits in Area 48 and Division 58.4.2 used the results of the BIOMASS surveys of the early 1980s which were the only acoustic surveys at an appropriate scale (Hewitt et al. 2004a). Because the BIOMASS data was not collected using CCAMLR-specific protocols there was a need to obtain more systematic acoustic data to ensure that precautionary catch limits were based on the most reliable information (Hewitt et al. 2004a). There have been three large-scale surveys (see Chap. 2, Siegel and Watkins 2016) that have been conducted for the purposes of setting precautionary catch limits: the BROKE survey of Division 58.4.1 in 1996 (Nicol et al. 2000), the CCAMLR 2000 survey in 2000 (Hewitt et al. 2004a) and the BROKE West survey of Division 58.4.2 in 2006 (Nicol et al. 2010).

Precautionary catch limits are calculated using an estimate of the potential yield of the krill population derived from the regional krill biomass estimates, data on natural mortality, individual growth rates and estimates of inter-annual recruitment variability. These estimates were calculated using the Krill Yield Model and later the Generalised Yield Model (Kock et al. 2007). These models take into account predator requirements through the mortality term and through a three part decision rule. This decision rule was developed for determining the proportion of the krill population (γ) which could be harvested each year with minimal risk, thus addressing Article II, 3 of the Convention (Constable et al. 2000; Miller 2002; Croxall and Nicol 2004).

(iii) Choose $\gamma 1$ such that the probability of the median krill spawning stock biomass declining below 20 % of its pre-exploitation median level over a 20-year harvesting period is 10 %.
(ii) Choose $\gamma 2$ so that the median krill spawning stock biomass after over 20 years of fishing is 75 % of the pre-exploitation median level.
(iii) Select the lower value of $\gamma 1$ and $\gamma 2$ as the level for γ for the calculation of krill yield.

The decision rules take explicit account of the needs of predators by allowing a krill escapement of 75 %. The revised yield model and the three-part decision rule were adopted by the Commission in 1994. These rules provided the first example of specifying the objectives of CCAMLR in scientifically interpretable and measurable terms (Constable et al. 2000).

A further level of precaution was added when CCAMLR introduced trigger levels in 1993. A trigger level is a level of catch (below the precautionary catch limit) which must not be exceeded until a mechanism to subdivide the overall catch limit amongst smaller management units has been agreed. Trigger levels have been employed to ensure that large krill catches are not taken from restricted areas. The trigger level of 620,000 tonnes set in CM 51-01, was derived from the aggregate of the highest annual commercial catches taken from each of the Subareas in Area 48 (SC-CCAMLR 2008, paragraph 3.33). Those in Division 58.4.2 were calculated on a pro rata basis using the proportions from Area 48 (CCAMLR 2008). These trigger levels are in place to ensure that the fishery develops in an orderly manner, with no irreversible impacts on predator species. In 2009 Conservation Measure 51-07 was adopted, providing for the interim distribution of the trigger level catch across Subareas 48.1, 48.2, 48.3 and 48.4. This measure was adopted to avoid the trigger level catch being taken from small areas, alleviating potential pressure on predator populations. The trigger level in Area 48 is set at a level that is only 1 % of the estimated biomass for this area of 60.3 million tonnes, (SC-CCAMLR 2010). The fishery, even if it is operating at the trigger level, would be unlikely to be a significant competitor for krill with the major consumers in the ecosystem (see Chap. 9, Trathan and Hill 2016).

In 2009, the adoption of two Conservation Measures – 51-04 and 51-05 (Table 11.3) served to further control the development of new krill fisheries in the CCAMLR Area. Essentially, the krill fishery now cannot expand into an unsurveyed area without an approved data collection plan and the catch in any unsurveyed subarea or Division is limited to 15,000 tonnes. Further, in these areas no more than 75 % of the catch limit can be taken within 60 n miles of known breeding colonies of land-based krill-dependent predators including seals, birds and penguins. The adoption of Conservation Measures 51-04 and 51-05 mean that there are limits on the krill fishery throughout the Convention Area either as a result of scientifically estimated precautionary catch limits or through regulations in un-surveyed areas. Prior to 1991, the krill fishery could operate in all regions of the CCAMLR area without limitation, by 1996 all the major fishing grounds were covered by precautionary catch limits and by 2009 krill fishing in the entire Convention Area was subject to catch limits or restrictions. CCAMLR has made considerable progress in managing the rational use of its largest fishery whilst allowing for the conservation of the world's largest unexploited marine biomass.

11.4.3 Limiting By-Catch

CCAMLR's ecosystem approach to management of the krill fishery requires that the effects of fishing on dependent and related species are taken into account. This means that the direct effects of fishing on other species – bycatch and incidental mortality – are taken into account. A number of Conservation Measures have been adopted which aim to reduce the level of by-catch taken in krill fisheries, as well as to reduce the incidental mortality of seabirds and marine mammals in krill fishing operations (Table 11.3). There is still uncertainty whether the mortality of young and larval fish is a significant issue in the krill fishery and whether such mortality varies with space or time (Okuda and Kiyota 2012). Incidental mortality of seals, penguins and seabirds has never constituted a serious problem in the krill fishery but a number of Conservation Measures have been adopted to minimise fisheries impacts on birds and seals. Assessing larval fish mortality and recording incidental mortality of seals and seabirds are major tasks for scientific observers on krill fishing vessels (Sabourenkov and Appleyard 2005).

11.4.4 CCAMLR's International Observer Scheme

The CCAMLR Scheme of International Scientific Observation was adopted in 1992 under Article XXIV of the Convention (Sabourenkov and Appleyard 2005). It is an important source of scientific information that is essential for assessing the impact of fishing on the ecosystem, including the status of target populations, as well as those of related and dependent species. The scheme also plays a crucial role in developing approaches to reducing the impact of fishing on the ecosystem by collecting data on the effectiveness of mitigation measures.

All vessels fishing in CCAMLR fisheries are required to carry an observer for some of their fishing operations. At CCAMLR meetings there has been considerable debate over the level of observer coverage in the krill fishery (Agnew et al. 2010) and despite the requirement for 100 % coverage by an international observer in all other CCAMLR fisheries the krill fishery has a target coverage of only 50 % using either international or national observers (CCAMLR 2014).

Scientific observers record information on the gear configuration (including measures to reduce incidental mortality of seabirds and marine mammals), fishing operations (including catch composition), the measurement of green weight and conversion factors and biological measurements of target and by-catch species. The absence of systematically-collected data collected from the krill fishery by recognised scientific observers has meant that it has been difficult to assess the significance of issues such as the bycatch of larval fish in the krill fishery (Agnew et al. 2010).

11.4.5 Krill Fishery Trends

CCAMLR has had persistent difficulties in obtaining information on potential future developments in the krill fishery. The Scientific Committee has repeatedly requested such information (Table 11.5). The absence of publicly available information on krill fishery developments led CCAMLR to formalise the process whereby Members notify their intention to fish for krill in the upcoming season.

Table 11.5 Information requested by the Scientific Committee on the economics of the Antarctic krill fishery

Information requested	Report reference
Past and current krill market information	SC-CCAMLR (2008, paragraph 2.6)
Past and current market prices for krill products	SC-CCAMLR (2009, paragraph 2.7)
Past and current market prices for krill products	SC-CCAMLR (2000, paragraph 2.4)
Updated information on krill processing, market developments, economic analyses	SC-CCAMLR (2001, paragraph 2.4)
Information on economics of the fishery and market developments	SC-CCAMLR (2001, paragraph 3.8)
Information on commercial market prices, economic marketing and technological information and demand for aquaculture feeds	SC-CCAMLR (2002, paragraph 4.11)
Ability to predict trends in the fishery still hampered by a lack of information on technological and economic developments	SC-CCAMLR (2004, paragraph 4.17)
Noted change in pattern of fishery operation in regards to participants, products composition and harvesting technologies	SC-CCAMLR (2005, paragraph 4.11)
Information on how publicly available information in trade press etc. may be used to inform WG deliberations	SC-CCAMLR (2008, paragraph 4.9)

11.4.6 Notifications

In 2006 CCAMLR adopted Conservation Measure 21-03 making it mandatory for Members to notify of their intention to fish for krill in the coming season. Member nations have to provide information on: the expected catch level, the area to be fished, details of the vessel and gear types being used to fish and the expected products to be derived from catch. This Measure was to aid the Commission in tracking development of the krill fishery in light of increasing interest in the fishery and in the absence of other tools for tracking the fishery's development (CCAMLR 2006, paragraph 12.24). From 2003, the Commission had relied on an informal, voluntary system whereby Members would verbally inform the Scientific Committee on their plans for krill fishing in the upcoming season (CCAMLR 2003, paragraphs 4.37–4.39). This voluntary system resulted in little substantial information being submitted to allow the Commission to monitor market interest or fishery trends (Croxall and Nicol 2004).

Since Conservation Measure 21-03 was adopted there has been a marked change in the number and type of notifications being made to the Commission. There has been an increase in the tonnage of krill notified since the scheme was adopted (Fig. 11.8). However, there is a large discrepancy between the level of catch notified by Members and the actual catch that has occurred, with notified catches

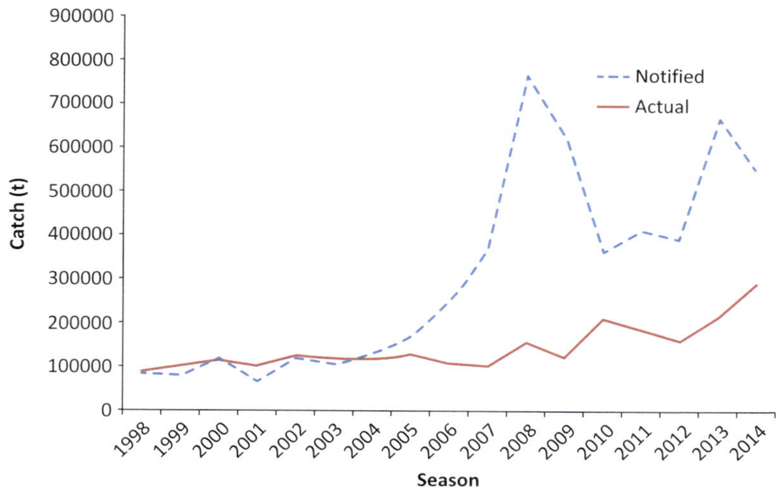

Fig. 11.8 Krill catch and notification data from 1992 to 2014 (CCAMLR 2014). Note that mandatory notifications data was not required until the 2006 season

consistently being considerably higher than actual catches for the duration of this scheme. Although notifications have exceeded the trigger level of 620,000 tonnes for Area 48, actual catches have never reached even half this level in since 1991. In the 14 years prior to mandatory notifications, actual catches were higher than notified catches in 65 % of cases, resulting in an underestimation of the future krill catch. Since 2006, notifications have always been higher than actual catches. Additionally, the notification process has also led to predictions of a larger krill fishing fleet and projections of a larger number of nations being involved in the fishery. It appears that the notification process is being used by Member nations to keep their options open – they cannot fish unless they notify and the number of vessels and the projected catch are over-stated to retain operational flexibility (CCAMLR 2007, paragraph 3.25 and CCAMLR-2008, paragraphs 8.13–8.21). The discrepancy between notified and actual catches could result in an incorrect assumption of if, or when, the trigger level may be reached, as well as an increased workload for the Secretariat in processing inaccurate notifications. To prevent excessive notifications some Members have proposed that a fee be associated with lodging notifications and that a penalty apply for those vessels that notified but don't fish (CCAMLR 2008, paragraph 8.19).

The notification process has not allowed CCAMLR to confidently predict the next season's krill catch, the participants in the fishery nor the size of the fishing fleet. It has however, forced Member and non-Member nations to indicate their interest in the fishery. Further refinement of the process may allow the system to be both predictive as well as indicative.

11.4.7 Economic Data

Investment in products and technology that are specific to the krill fishery is evident in the number and type of krill-related patents being lodged (see Fig. 11.6) (Nicol and Foster 2003; Nicol et al. 2012; Foster et al. 2011). CCAMLR recognised that such information provided an insight into developments in the industry that could lead to future expansion of the krill fishery (SC-CCAMLR 1999, paragraphs 4.9–4.11). Patents and patent databases are increasingly being used for detecting investor interest in emerging technologies (Schiermeier 2010) and for identifying products and technologies relating to particular ingredients (Pilkington 2004). Patented technologies and products represent significant investment by the assignee in the krill industry (to lodge and maintain a patent is a costly process) and therefore can be used as an indicator of likely future activity in the krill marketplace. The CCAMLR Scientific Committee agreed that it would be useful if the Secretariat could maintain a patent database to provide annual updates on these trends (SC-CCAMLR 2009, paragraphs 4.10–4.12).

11.4.8 CCAMLR's Management Structures

The CCAMLR Commission is the decision-making body that ultimately agrees on management measures relating to conservation and fisheries in the Convention Area. The Commission is advised by a Scientific Committee and these two bodies meet annually. The Scientific Committee, in turn, relies on the advice of specialised Working Groups one of which, the Working Group on Ecosystem Monitoring and Management (WG-EMM), is concerned with the krill-based ecosystem and the krill fishery. It is at the annual meeting of WG-EMM that the development of management approaches to the krill fishery are proposed and discussed. CCAMLR and its subsidiary bodies are supported by a Secretariat, based in Hobart, Tasmania, which provides a wide range of services including data management and strategic research. The Secretariat produces an annual Krill Fishery Report which is presented at WG-EMM and provides information that includes: krill catches, notifications, scientific observations and bycatch. Information from these reports becomes available through the Reports of the Scientific Committee and on CCAMLR's website (www.ccamlr.org) and provides a valuable annual update on the status of the fishery.

11.5 Future Management of the Krill Fishery

11.5.1 Small Scale Management Units

There has also been recognition that management of the fishery needs to move from catch limits established for large scale areas to much smaller scale areas (Constable and Nicol 2002). In 1991, when the first Conservation Measure (32X) allocating a precautionary catch limit for Area 48 was adopted by the Commission, the need to consider distribution of precautionary catch limits across smaller spatial areas was first raised (CCAMLR 2008, paragraph 12.63 and Miller 2002). The trigger level in Area 48 has been subdivided (see above) but there is not yet agreement on a subdivision of the overall catch limit. There have been moves to subdivide the precautionary catch limit down to much smaller scales – small scale management units (SSMUs) – to avoid potential effects that concentrated fishing effort might have on the ecosystem (Hewitt et al. 2004b). Spatial boundaries of 15 SSMUs for Area 48 were established in 2002 (CCAMLR 2002, paragraph 4.5), but a mechanism for the subdivision of the precautionary catch limit across the SSMUs is yet to be agreed upon. A number of candidate schemes have been proposed and discussed by within CCAMLR (Hewitt et al. 2004b):

1. the spatial distribution of historical catches by the krill fishery;
2. the spatial distribution of predator demand;
3. the spatial distribution of krill biomass;
4. the spatial distribution of krill biomass minus predator demand;
5. spatially explicit indices of krill availability that may be monitored or estimated on a regular basis; and
6. structured fishing strategies in which catches are rotated within and between SSMUs.

Because there has been no agreement on a method for subdividing the precautionary catch across the SSMUs, in 2009 the Commission adopted an interim measure which divides the precautionary catch amongst four of the most commonly fished Subareas in Area 48 (Conservation Measure 51-07). The trigger level for each subarea was based on the proportion of the krill biomass found in each subarea in the CCAMLR 2000 survey: subarea 48.1: 155,000 tonnes, subarea 48.2: 279,000 tonnes, Subarea 48.3: 297,000 tonnes and Subarea 48.4: 93,000 tonnes (SC-CCAMLR 2009). This subdivision has resulted in the closure of the krill fishery four times when the subdivided trigger level (155,000 tonnes per year) for subarea 48.1 has been reached. CM 51-07 was renewed in 2011 and for a further two years in 2014.

11.5.2 Feedback Management

The current management regime for the krill fishery has always been viewed as an interim scheme that would remain in place until a more elaborate and comprehensive management approach was developed (Nicol and de la Mare 1993). The long-term aim of CCAMLR has been to the development of a feedback management procedure for the krill fishery (adaptive management whereby management measures are regularly adjusted based on ecosystem monitoring indices) (Nicol and de la Mare 1993; Fabra and Gascón 2008). This system would see management measures regularly changed in response to continually updated information from the ecosystem, aimed at detecting impacts from fishing and responding accordingly, before irreversible change occurs (Constable et al. 2000; Constable 2011; Hill and Canon 2013). The development of a feedback management procedure for a remote fishery such as that for Antarctic krill is a complex process and CCAMLR has recognised that its implementation will be difficult, agreeing to a staged approach (CCAMLR 2014). In the interim, a number of other measures are being explored to ensure that the management process uses the maximum amount of information available.

11.5.3 Integrated Assessments

Catch limits on the krill fishery have been established using the results of very large scale acoustic surveys of krill biomass. Such surveys are extremely labour intensive, expensive and take considerable planning and analysis. CCAMLR is considering a move from such basin-wide biomass surveys to an integrated stock assessment (SC-CCAMLR 2014, paragraph 3.52). An integrated krill stock assessment model could be used to provide regular stock assessments based on data from various sources, including: scientific surveys, fishing vessel surveys, observers and the CCAMLR Ecosystem Monitoring Program (CEMP). Robust estimates of the consumption of krill by predators would help to scale biomass estimates. The model would account for differences in the timing of different data sources, and gear-specific selectivity.

11.5.4 Research from Krill Fishing Vessels

Until very recently, most of the research required for the management of the krill fishery was obtained using fisheries-independent surveys (Nicol et al. 2000, 2010, and Hewitt et al. 2004a). There are few research vessels operating in Antarctic waters at any one time and krill research is only one of a number of priority areas for study. Additionally, research vessels are usually limited to studies of 1–2 months in length. Fishing vessels, on the other hand, operate throughout most of the year and sample krill throughout the day (Kawaguchi and Nicol 2007). An International Scientific Observer scheme has been operated by CCAMLR since its inception but scientific observer coverage on krill fishing vessels has been patchy and no information from this scheme has yet been used for management purposes. Recently, CCAMLR has recognised that obtaining information from krill fishing vessels will be essential for future management of the fishery (Godø et al. 2014). Regular, dedicated acoustic surveys using krill fishing vessels have already commenced (Krafft et al. 2011) but there is far more research that can be conducted from the commercial fleet. Suggestions for future research include studies into basic biology of krill, biochemistry of krill and oceanography of the krill fishing grounds. There are even suggestions for the dedicated use of krill trawlers for research purposes and for the construction of hybrid krill fishing-research vessels (Godø et al. 2014). Information from fishing vessels is seen as crucial in the development of feedback management for the krill fishery (SC-CCAMLR 2014 paragraph 3.34).

11.5.5 Industry-Based Approaches to Management

There has been a recent trend in global fisheries for industry to adopt measures over and above those imposed by management bodies such as CCAMLR (Washington and Ababouch 2011). These approaches include third party certification and initiatives that demonstrate the credentials of the companies or industry in pursuing sustainable harvesting. Fisheries throughout the world are confronted with increased market demands to document conformity in relation to the principles of sustainable harvesting. Scientific knowledge and documentation of the ecosystem effects of fisheries is essential for many management bodies, and also within the market place. The achievements of the Marine Stewardship Council (MSC) certification process for marine products indicates how scientifically-based information and documentation directly affects the product price and market access.

The Norwegian company that currently harvests the greatest tonnage of krill, Aker Biomarine, was awarded MSC certification for their "Eco-Harvesting" krill fishing activities (Aker 2010). This certification is awarded to fisheries that undertake their harvesting operations in an environmentally sustainable manner (MSC 2010). Aker uses this certification as a marketing tool that sets its Qrill™ and Superba™ products apart from competitor's products as Aker at the time was the only company with the certification for their krill fishing activities. Jacquet and Pauly (2010) have raised concerns that the certification process overlooked reports from the scientific community that krill stocks are declining and ignored the unsustainable use of krill in fishmeal although these concerns were subsequently addressed by the certifying body (MSC 2010). In 2014 the Aker krill fishery was re-assessed and was re-certified (MSC) with no conditions placed on the company and no adverse comments received on the certification (Schulz 2014). A second Norwegian company, Olympic fisheries, is currently undergoing MSC assessment

(MSC 2015). Although the MSC process addresses krill fishing by company and by vessel their assessments do reflect the overall state of the fishery and its management.

In 2010 several krill fishing companies formed the Association of Responsible Krill Fishing Companies (ARK) with the active involvement of the Worldwide Fund for Nature (WWF). ARK coordinates and cooperates with CCAMLR on the provision of research and information on krill, and the krill fishery and its impact on the ecosystem, with the aim of contributing to CCAMLR's work on managing the krill fishery sustainably (ARK 2015). This initiative arose out of a recognition that the safety of long-term investment in the krill industry would lie in it being well managed and been seen to be well managed. In 2015 ARK had member companies from Norway (2), Chile and Korea and has been an official observer at CCAMLR meetings since 2011. It has begun the process of involving the krill fishing industry in the development of feedback management approach (SC-CCAMLR 2013, paragraphs 10.12–10.15). ARK members currently catch ~80 % of the krill harvested in the Southern Ocean and ARK is in a position to be influential in the future development of the krill fishery.

11.5.6 Environmental Footprint of Krill Fishery

Distant-water fishing operations are frequently criticised for the resources they consume and the emissions they produce. The Aker Biomarine krill fishery was subjected to a Life Cycle Analysis to quantify and characterize life cycle environmental burdens associated with their fishing and processing operations (Parker and Tyedmers 2012). The Fuel Use Intensity (litres per tonne of landed catch) of Antarctic krill is relatively low when compared to high-value species such as cod, tuna or lobster, but much higher than for other fisheries targeting species for reduction into meal and oil e.g. ten times that of the Peruvian anchovy fishery. The fuel consumption by a resupply vessel used to transport meal and oil from the fishing vessel to port accounts for roughly the same scale of energy use and emissions as fuel consumption on the fishing vessel. This suggests that the distance between the fishing vessel and port plays a particularly important role in the life cycle environmental impacts of krill products.

11.5.7 Climate Change and the Krill Fishery

The life history and population dynamics of Antarctic krill are likely to be impacted by a changing climate that results from increasing levels of CO_2 in the atmosphere (Flores et al. 2012), and may result in a reduced habitat range for Antarctic krill (Hill et al. 2013). The reproductive output and recruitment success of krill has been related to the extent, timing and duration of winter sea ice cover (Kawaguchi and

Satake 1994, Quetin and Ross 2003). Krill growth has also been observed to decrease above a temperature optimum of 0.5° C (Atkinson et al. 2006). The early developmental stage of krill will be vulnerable to the increased levels of CO_2 projected to occur within their habitat range to the year 2100 and beyond (Kawaguchi et al. 2013). Overall, the cumulative impact of climate change is most likely to be negative. Changes in the distribution and abundance of krill as a result of climate change are, obviously, likely to affect the fishery. Already the effect of changes in sea ice extent and seasonality have changed the behaviour of the fishing fleets (Kawaguchi et al. 2009). It is uncertain whether a warming surface layer may affect the vertical distribution of krill (Schmidt et al. 2011) but the fishing fleet is already fishing much deeper in some areas (Kawaguchi et al. 2006).

11.6 Conclusions

The krill fishery has been operating for over 40 years in Antarctic waters with nearly 8 million tonnes being harvested during this period. Despite concerns over the effects of the krill fishery on the Southern Ocean ecosystem, there have been no observations of an effect of the harvesting of krill on dependent or related species. This lack of an effect can be traced to the level of fishing remaining low compared to the large biomass and because of the precautionary management regime implemented by CCAMLR. There is currently unprecedented interest in krill as a source of marine proteins and oils and more nations are involved in the fishery than in the past. The future of the fishery, and of the ecosystem, will be played out within CCAMLR and this organisation will either succeed or fail based on its ability to effectively manage its largest fishery.

Acknowledgements We would like to thank the CCAMLR Secretariat, in particular Drew Wright, Keith Reid, Lucy Robinson and David Ramm, for their assistance and patience.

References

Adelung D, Buchholz F, Culik B, Keck A (1987) Fluoride in tissues of krill *Euphausia superba* Dana and *Meganyctiphanes norvegica* M. Sars in relation to the moult cycle. Polar Biol 7:43–50

Agnew DJ (2004) Fishing south: the history and management of South Georgia fisheries. Government of South Georgia

Agnew DJ, Grove P, Peatman T, Burn R, Edwards CTT (2010) Estimating optimal observer coverage in the Antarctic krill fishery. CCAMLR Sci 17:139–154

Aker (2010) www.akerbiomarine.com. Accessed 23rd July 2010

ARK (2015) www.ark-krill.org/. Accessed Feb 2015

Arthi V (2009) Omega 3 concentrates – window of opportunities. Frost & Sullivan Market Insight, 27 Aug 2009. http://www.frost.com/prod/servlet/market-insight-top.pag?docid=178244151. Accessed 2nd Nov 2009

Atkinson A, Shreeve RS, Hirst AG, Rothery P, Tarling GA, Pond DW, Korb RE, Murphy EJ, Watkins JL (2006) Natural growth rates in Antarctic krill (Euphausia superba): II. Predictive models based on food, temperature, body length, sex, and maturity stage. Limnol Oceanogr 51(2):973–987
Atkinson A, Siegel V, Pakhomov EA, Jessopp MJ, Loeb V (2009) A re-appraisal of the total biomass and annual production of Antarctic krill. Deep-Sea Res I Oceanogr Res Pap 56 (5):727–740
Backes JM, Howard PA (2014) Krill oil for cardiovascular risk prevention: is it for real? Hosp Pharm 49(10):907–912
Batetta B, Griinari M, Carta G, Murru E, Ligresti A, Cordeddu L, Giordano E, Sanna F, Bisogno T, Uda S, Collu M, Bruheim I, Di Marzo V, Banni S (2009) Endocannabinoids may mediate the ability of (n-3) fatty acids to reduce ectopic fat and inflammatory mediators in obese zucker rats. J Nutr 139(8):1495–1501
Bridges KM, Gigliotti JC, Altman S, Jaczynski J, Tou JC (2010) Determination of digestibility, tissue deposition and metabolism of the omega-3 fatty acid content of krill protein concentrate in growing rats. J Agric Food Chem 58:2830–2837
Budzinski E, Bykowski P, Dutkiewicz D (1985) Possibilities of processing and marketing of products made from Antarctic krill, FAO fisheries technical paper no. 268. FAO, Rome, p. 46p
Bunea R, Farrah KE, Deutsch L (2004) Evaluation of the effects of Neptune Krill Oil on the clinical course of hyperlipidemia. Altern Med Rev 9(4):420–428
Bustos R, Romo L, Yáñez K, Díaz G, Romo C (2003) Oxidative stability of carotenoid pigments and polyunsaturated fatty acids in microparticulate diets containing krill oil for nutrition of marine fish larvae. J Food Eng 56:289–293
CCAMLR (2002) Report of the twenty-first meeting of the commission. CCAMLR, Hobart
CCAMLR (2003) Report of the twenty-second meeting of the commission. CCAMLR, Hobart
CCAMLR (2006) Report of the twenty-fifth meeting of the commission. CCAMLR, Hobart
CCAMLR (2007) Report of the twenty-sixth meeting of the commission. CCAMLR, Hobart
CCAMLR (2008) Report of the twenty-seventh meeting of the commission. CCAMLR, Hobart
CCAMLR (2009a) www.ccamlr.org. Accessed 1st Oct 2009
CCAMLR (2009b) Report of the twenty-eighth meeting of the commission. CCAMLR, Hobart
CCAMLR (2014) Report of the thirty-third meeting of the commission. CCAMLR, Hobart
CCAMLR (2015) Stat Bull, 26 http://www.ccamlr.org/en/data/statistical-bulletin
CCAMLR-WG-EMM (2014) Report of the working group on ecosystem monitoring and management (Punta Arenas, Chile, 7 to 18 July 2014)
Chandrasekar B, Troyer DA, Venkatraman JT, Fernandes G (1996) Tissue specific regulation of transforming growth factors beta by omega-3 lipid-rich krill oil in autoimmune murine lupus. Nutr Res 16(3):489–503
Chi H, Li X, Yang X (2013) Processing status and utilisation strategies of Antarctic krill (*Euphausia superba*) in China. World J Fish Mar Sci 5(3):275–281
Constable AJ (2011) Lessons from CCAMLR on the implementation of the ecosystem approach to managing to fisheries. Fish Fish 12(2):138–151
Constable AJ, Nicol S (2002) Defining smaller-scale management units to further develop the ecosystem approach in managing large-scale pelagic krill fisheries in Antarctica. CCAMLR Sci 9:117–131
Constable AJ, de la Mare WK, Agnew DJ, Everson I, Miller D (2000) Managing fisheries to conserve the Antarctic marine ecosystem: practical implementation of the Convention on the Conservation of Antarctic Marine Living Resources (CCAMLR). ICES J Mar Sci 57:778–791
Croxall JP, Nicol S (2004) Management of Southern Ocean fisheries: global forces and future sustainability. Antarct Sci 16(4):569–584
Davis DA, Arnold CR (2000) Replacement of fish meal in practical diets for the pacific white shrimp, *Liptopenaeus vannamei*. Aquaculture 185:291–298
Deutsch L (2007) Evaluation of the effect of Neptune Krill Oil on chronic inflammation and arthritic symptoms. J Am Coll Nutr 26(1):39–48

Duan J, Jiang Y, Cherian G, Zhao Y (2010) Effect of combined chitosan-krill oil coating and modified atmosphere packaging on the storability of cold-stored lingcod (*Ophiodon elongates*). Food Chem 122(4):1035–1042

Eddie GC (1977) The harvesting of krill. Southern Ocean Fisheries Survey Programme. FAO, Rome, GLO/SO/77/2, 76p

Edwards DM, Heap JA (1981) Convention on the conservation of Antarctic marine living resources: a commentary. Polar Rec 20(127):353–362

El-Sayed SZ (ed) (1977) Biological investigations of marine Antarctic systems and stocks (BIOMASS), vol 1. Scott Polar Research Institute, Cambridge

El-Sayed SZ (ed) (1994) Southern Ocean ecology: the BIOMASS perspective. Cambridge University Press, Cambridge

Everson I (1977) The living resources of the Southern Ocean. Southern Ocean Fisheries Survey Programme. FAO, Rome, GLO/SO/77/1, 156p

Fabra A, Gascón V (2008) The Convention on the Conservation of Antarctic Marine Living Resources (CCAMLR) and the ecosystem approach. Int J Mar Coast Law 23:567–598

FAO (2008) The state of world fisheries and aquaculture. FAO Fisheries and Aquaculture Department, Rome. 178p

Fereidoon S, Wanasundara UN (1998) Omega-3 fatty acid concentrates: nutritional aspects and production technologies. Trends Food Sci Technol 9:230–240

Fernholm B, Rudback GT (1989) Marine resource management of the Antarctic. Ambio 18(1):68–70

Flores H, Atkinson A, Kawaguchi S, Krafft BA, Milinevsky G, Nicol S, Reiss C, Tarling GA, Werner R, Bravo Rebolledo E, Cirelli V, Cuzin-Roudy J, Fielding S, Groeneveld JJ, Haraldsson M, Lombana A, Marschoff E, Meyer B, Pakhomov EA, Rombolá E, Schmidt K, Siegel V, Teschke M, Tonkes H, Toullec JY, Trathan PN, Tremblay N, Van de Putte AP, van Franeker JA, Werner T (2012) Impact of climate change on Antarctic krill. Mar Ecol Prog Ser 458:1–19

Floreto EAT, Brown PB, Bayer RC (2001) The effects of krill hydrolysate-supplemented soy-bean based diets on the growth, colouration, amino and fatty acids profiles of juvenile American lobster, *Homarus americanus*. Aquac Nutr 7:33–43

Foster J, Nicol S, Kawaguchi S (2011) The use of patent databases to detect trends in the krill fishery. CCAMLR Sci 18:135–144

Frank RF (1983) The convention on the conservation of Antarctic marine living resources. Ocean Dev Int Law J 13(3):291–345

Gaber MMA (2005) The effect of different levels of krill meal supplementation of soybean-based diets on feed intake, digestibility, and chemical composition of juvenile Nile Tilapia *Oreochromis niloticus*, *L*. J World Aquacult Soc 36(3):346–353

Giogios I, Grigorakis K, Nengas I, Papasolomontos S, Papaioannou N, Alexis MN (2009) Fatty acid composition and volatile compounds of selected marine oils and meals. J Sci Food Agric 89:88–100

Godø OR, Reiss C, Siegel V, Watkins JL (2014) Commercial fishing vessels as research vessels in the Antarctic – requirements and solutions exemplified with a new vessel. CCAMLR Sci 21:11–1

Grantham GJ (1977) The utilization of Krill. Southern Ocean Fisheries Survey Programme. FAO, Rome: GLO/SO/77/3 61p

Grave H (1981) Fluoride content of salmonids fed on Antarctic krill (*Euphausia superba*). Aquaculture 24:191–196

Hewitt RP, Low EHL (2000) The fishery on Antarctic Krill: defining an ecosystem approach to management. Rev Fish Sci 8(3):235–298

Hewitt RP, Watkins J, Naganobu M, Sushin V, Brierley AS, Demer D, Kasatkina S, Takao Y, Goss C, Malyshko A, Brandon M, Kawaguchi S, Siegel V, Trathan P, Emery J, Everson I, Miller D (2004a) Biomass of Antarctic krill in the Scotia Sea in January/February 2000 and its

use in revising an estimate of precautionary yield. Deep-Sea Res II Top Stud Oceanogr 51(12):1215–1236
Hewitt RP, Watters G, Trathan PN, Croxall JP, Goebel ME, Ramm D, Reid K, Trivelpiece WZ, Watkins JL (2004b) Options for allocating the precautionary catch limit of krill among small-scale management units in the Scotia Sea. CCAMLR Sci 11:81–97
Hill SL (2013) Prospects for a sustainable increase in the availability of long chain omega 3s: lessons from the Antarctic Krill fishery. In: Omega-6/3 fatty acids. Humana Press, New York, pp 267–296
Hill SL, Cannon M (2013) A potential feedback approach to ecosystem-based management: model predictive control of the Antarctic krill fishery. CCAMLR Sci 20:119–137
Hill SL, Phillips T, Atkinson A (2013) Potential climate change effects on the habitat of Antarctic krill in the Weddell quadrant of the Southern Ocean. PLoS One 8(8):e72246. doi:10.1371/journal.pone.0072246
Howard M (1989) The convention on the conservation of Antarctic marine living resources: a five-year review. Int Comp Law Q 38:104–149
Ichii T (2000) Krill harvesting. In: Everson I (ed) Krill biology, ecology and fisheries. Blackwell Science, Oxford, pp 228–262
Jacquet J, Pauly D (2010) Seafood stewardship in crisis. Nature 467(2):28–29
Julsham K, Malde MK, Bjorvatn K, Krogedal P (2004) Fluoride retention of Atlantic salmon (*Salmo salar*) fed krill meal. Aquac Nutr 10(1):9–13
Jung HR, Kim MA, Seo YS, Lee YB, Chun BS, Kim SB (2013) Decreasing effect of fluoride content in Antarctic krill (*Euphausia superba*) by chemical treatments. Int J Food Sci Technol 46(6):1252–1259
Kalinowski CT, Izquierdo MS, Schuchardt LE (2007) Dietary supplementation time with shrimp shell meal on red porgy (*Pagrus pagrus*) skin colour and carotenoid concentration. Aquaculture 272:451–457
Karlsen Ø, Suontama J, Olsen RE (2006) Effect if Antarctic krillmeal on quality of farmed Atlantic cod (*Gadus morhua* L.). Aquac Res 37(16):1676–1684
Karlstam B, Vincen J, Johansson B, Bryno C (1991) A simple purification method of squeezed krill for obtaining high levels of hydrolytic enzymes. Prep Biochem 21:237–256
Kawaguchi S, Nicol S (2007) Learning about Antarctic krill from the fishery. Antarct Sci 19(2):219–230
Kawaguchi S, Satake M (1994) Relationship between recruitment of the Antarctic Krill and the degree of ice cover near the South Shetland Islands. Fish Sci 60(1):123–124
Kawaguchi S, Ichii T, Naganobu M (1999) Green krill, the indicator of micro-and nano-size phytoplankton availability to krill. Polar Biol 22(2):133–136
Kawaguchi S, Nicol S, Taki K, Naganobu M (2006) Fishing ground selection in the Antarctic krill fishery: trends in patterns across years, seasons and nations. CCAMLR Sci 13:117–141
Kawaguchi S, Nicol S, Press AJ (2009) Direct effects of climate change on the Antarctic krill fishery. Fish Manag Ecol 16:424–427
Kawaguchi S, Ishida A, King R, Raymond B, Waller N, Constable A, Nicol S, Wakita M, Ishimatsu A (2013) Risk maps for Antarctic krill under projected Southern Ocean acidification. Nat Clim Chang 3(9):843–847
Kidd PM (2009) Integrated brain restoration after ischemic stroke – medical management, risk factors, nutrients, and other interventions for managing inflammation and enhancing brain plasticity. Altern Med Rev 14(1):14–35
Kim M-A, Jung H-R, Lee Y-B, Chun B-S, Kim SB (2014) Monthly variations in the nutritional composition of Antarctic krill *Euphausia superba*. Fish Aquat Sci 17(4):409–419
Kock K-H, Reid K, Croxall J, Nicol S (2007) Fisheries in the Southern Ocean: an ecosystem approach. Philos Trans R Soc B 362:2333–2349
Krafft BA, Skaret G, Calise L (2011) Antarctic krill and apex predators in the South Orkney Islands area 2011, surveyed with the commercial fishing vessel Saga Sea. Rapport fra Havforskningen; Nr. 6–2011

Lyubimova TC, Naumov AG, Lagunov LL (1973) Prospects of the utilisation of krill and other unconventional resources of the world ocean. J Fish Res Board Can 30(12/2):2196–2201

Maguire J-J, Sissenwine M, Csirke J, Grainger R, Garcia S (2006) The state of world highly migratory, straddling and other high seas fishery resources and associated species, FAO fisheries technical paper no. 495. FAO, Rome. 84p

McElroy JK (1984) Antarctic fisheries. History and prospects. Mar Policy 8(3):239–259

Melrose J, Hall A, Macpherson C, Bellenger CR, Ghosh P (1995) Evaluation of digestive proteinases from the Antarctic krill *Euphausia superba* as potential chemonucleolytic agents. In vitro and in vivo studies. Arch Orthop Trauma Surg 114:145–152

Miller DGM (1991) Exploitation of Antarctic marine living resources: a brief history and a possible approach to managing the krill fishery. S Afr J Mar Sci 10:321–339

Miller DGM (2002) Antarctic krill and ecosystem management—from Seattle to Siena. CCAMLR Sci 9:175–2002

Miller D, Agnew D (2000) Management of krill fisheries in the Southern Ocean. In: Everson I (ed) Krill biology, ecology and fisheries. Blackwell Science, Oxford, pp 300–337

Moreira AC, Muller ACA, Pereira N, Antunes A (2006) Pharmaceutical patents on plant derived materials in Brazil: policy, law and statistics. World Patent Inf 28:34–42

Moretti VM, Mentasti T, Bellagamba F, Luzzana U, Caprino GM, Giani I, Valfrè F (2006) Determination of astaxanthin stereoisomers and colour attributes in flesh of rainbow trout (*Oncorhynchus mykiss*) as a tool to distinguish the dietary pigmentation source. Food Addit Contam 23(11):1056–1063

MSC (2010) www.msc.org. Marine Stewardship Council. Accessed 3 Aug 2010

MSC (2015) www.msc.org/track-a-fishery/fisheries-in-the-program/in-assessment/southern-ocean/olympic-seafood-antarctic-krill. Accessed Feb 2015

Naczk M, Synowiecki J, Sikorski ZE (1981) The gross chemical composition of Antarctic krill waste. Food Chem 7:175–179

Nichols PD (2007) Fish oil sources. In: Breivik H (ed) Long-chain omega-3 speciality oils. The Oily Press, Cambridge, p 314

Nichols PD, Nelson MM (2007) Marine oils in Australasia. In: O'Connor, J (Chairman) Handbook of Australasian edible oils. Oils and Fats Specialist Group of NZIC. Auckland, p 297

Nicol S (1991) CCAMLR and its approaches to management of the krill fishery. Polar Rec 27(162):229–236

Nicol S (1992) Management of the krill fishery; was CCAMLR slow to act? Polar Rec 28(165):155–157

Nicol S, De la Mare W (1993) Ecosystem management and the Antarctic krill. Am Sci 81(1):36

Nicol S, Endo Y (1997) Krill fisheries of the world, FAO fisheries technical paper no. 367. FAO, Rome, p 81

Nicol S, Endo Y (1999) Krill fisheries – their development, management and ecosystem implications. Aquat Living Resour 12(2):105–120

Nicol S, Foster J (2003) Recent trends in the fishery for Antarctic krill. Aquat Living Resour 16:42–45

Nicol S, Hosie GW (1993) Chitin production by Krill. Biochem Syst Ecol 21(2):181–184

Nicol S, Pauly T, Bindoff NL, Strutton PG (2000) "BROKE" a biological/oceanographic survey off the coast of East Antarctica (80–150°E) carried out in January–March 1996. Deep- Sea Res II 47(12/13):2281–2297

Nicol S, Meiners K, Raymond B (2010) Editorial. BROKE-West, a large ecosystem survey of the South West Indian Ocean sector of the Southern Ocean 30–80°E (CCAMLR Division 58.4.2). Deep-Sea Res II 57(9–10):693–700

Nicol S, Foster J, Kawaguchi S (2012) The fishery for Antarctic krill – recent developments. Fish Fish 13(1):30–40

Oehlenschläger J, Schreiber W (1981) A functional protein concentrate (FKPC) from Antarctic Krill (*Euphausia superba*, Dana 1850). Z Lebensm Unters Forsch 172:393–398

Okuda T, Kiyota M (2012) Analysis of variability of krill size and fish by-catch in the Japanese krill fishery based on scientific observer data. CCAMLR Sci 19:31–47
Opstad I, Suontama J, Langmyhr E, Olsen RE (2006) Growth, survival and development of Atlantic cod (*Gadus morhua* L.) weaned onto diets containing various sources of marine protein. ICES J Mar Sci 63:320–325
Parker RW, Tyedmers PH (2012) Life cycle environmental impacts of three products derived from wild-caught Antarctic krill (*Euphausia superba*). Environ Sci Technol 46(9):4958–4965
Pilkington A (2004) Technology portfolio alignment as an indicator of commercialisation: an investigation of fuel cell patenting. Technovation 24:761–771
Quetin LB, Ross RM (2003) Episodic recruitment in Antarctic krill *Euphausia superba* in the Palmer LTER study region. Mar Ecol Prog Ser 259:185–200
Sabourenkov EN, Appleyard E (2005) Scientific observations in CCAMLR fisheries – past, present and future. CCAMLR Sci 12:81–98
Savage GP, Foulds MJ (1987) Chemical composition and nutritive value of Antarctic krill (*Euphausia superba*) and southern blue whiting (*Micromesistius australis*). N Z J Mar Freshw Res 21:599–604
SC-CCAMLR (1999) Report of the eighteenth meeting of the scientific committee. CCAMLR, Hobart
SC-CCAMLR (2000) Report of the nineteenth meeting of the scientific committee. CCAMLR, Hobart
SC-CCAMLR (2001) Report of the twentieth meeting of the scientific committee. CCAMLR, Hobart
SC-CCAMLR (2002) Report of the twenty-first meeting of the scientific committee. CCAMLR, Hobart
SC-CCAMLR (2004) Report of the twenty-third meeting of the scientific committee. CCAMLR, Hobart
SC-CCAMLR (2005) Report of the twenty-fourth meeting of the scientific committee. CCAMLR, Hobart
SC-CCAMLR (2008) Report of the twenty-seventh meeting of the scientific committee. CCAMLR, Hobart
SC-CCAMLR (2009) Report of the twenty-eighth meeting of the scientific committee. CCAMLR, Hobart
SC-CCAMLR (2010) Report of the twenty-ninth meeting of the scientific committee. CCAMLR, Hobart
SC-CCAMLR (2011) Report of the thirtieth meeting of the scientific committee. CCAMLR, Hobart
SC-CCAMLR (2013) Report of the thirty-second meeting of the scientific committee. CCAMLR, Hobart
SC-CCAMLR (2014) Report of the thirty-third meeting of the scientific committee. CCAMLR, Hobart
Schiermeier Q (2010) Green patents corralled. Nature 465(6):21
Schmidt K, Atkinson A, Steigenberger S, Fielding S, Lindsay MCM, Pond DW, Tarling GA, Klevjer TA, Allen CS, Nicol S, Achterberg EP (2011) Seabed foraging by Antarctic krill: implications for stock assessment, bentho-pelagic coupling, and the vertical transfer of iron. Limnol Oceanogr 56(4):1411–1428
Schulz H (2014) Nutra-Ingrediants USA.com. Aker says MSC recertification eliminates any lingering doubts on sustainability of krill. http://www.nutraingredients-usa.com/Suppliers2/Aker-says-MSC-recertification-eliminates-any-lingering-doubts-on-sustainability-of-krill. Accessed Feb 2015
Shahidi F, Wanasundara UN (1998) Omega-3 fatty acid concentrates: nutritional aspects and production technologies. Trends Food Sci Technol 9:230–240
Sidhu GS, Montgomery WA, Holloway GL, Johnson AR, Walker DM (1970) Biochemical composition and nutritive value of krill (*Euphausia superba* Dana). J Sci Food Agric 21:293–296

Siebert G, Kühl J, Hannover R (1980) Nutritional experiments with krill. In: Connell JJ (ed) Advances in fish science and technology. Blackwell Scientific Publications Ltd, London, pp 332–334

Siegel V, Watkins J (2016) Distribution, biomass and demography of Antarctic krill, *Euphausia superba*. In: Siegel V (ed) Biology and ecology of Antarctic krill. Springer, Cham, pp 21–100

Smith DM, Tabrett SJ, Barclay MC, Irvin SJ (2005) The efficacy of ingredients included in shrimp feeds to stimulate uptake. Aquac Nutr 11:263–272

Soevik T, Braekkan OR (1979) Fluoride in Antarctic Krill (*Euphausia superba*) and Atlantic Krill (*Meganyctiphanes norvegica*). J Fish Res Board Can 36:1414–1416

Suontama J, Kiessling A, Melle W, Waagbø R, Olsen RE (2007) Protein from Northern krill (*Thysanoessa inermis*), Antarctic krill (*Euphausia superba*) and the Arctic amphipod (*Themisto libellula*) can partially replace fish meal in diets to Atlantic salmon (*Salmo salar*) without affecting product quality. Aquac Nutr 13:50–58

Tandy S, Chung RWS, Wat E, Kamili A, Berge K, Griinari M, Cohn JS (2009) Dietary krill oil supplementation reduces hepatic steatosis, glycaemia, and hypercholesterolemia in high-fat-fed mice. J Agric Food Chem 57:9339–9345

Tenuto-Filho A (1993) Fluorine removal during production of krill paste and protein concentrates. Acta Alimaentaria 22:269–281

Tenuto-Filho A, Alvarenga RCC (1999) Reduction of the bioavailability of fluoride from Antarctic krill by calcium. Int J Food Sci Nutr 50:297–302

Tou JC, Jaczynski J, Chen-Chen Y (2007) Krill for human consumption: nutritional value and potential health benefits. Nutr Rev 65(2):63–77

Trathan PN, Hill S (2016) The importance of krill predation in the Southern Ocean. In: Siegel V (ed) Biology and ecology of Antarctic krill. Springer, Cham, pp 321–350

Voronia NM (1983) Biomass and production of Antarctic krill (*Euphausia superba* Dana). Oceanology 23(6):760–762

Wang Z, Yin X (2012) Comparison of the defluoridation efficiency of calcium phosphate and chitin in the exoskeleton of Antarctic krill. Adv Polar Sci 23(3):149–154

Wang L, Xue C, Wang Y, Yang B (2011) Extraction of proteins with low fluoride level from Antarctic krill (*Euphausia superba*) and their composition analysis. J Agric Food Chem 59(11):6108–6112

Washington S, Ababouch L (2011) Private standards and certification in fisheries and aquaculture: current practice and emerging issues. Food and Agriculture Organization of the United Nations, Rome

Xie CL, Kim HS, Shim KB, Kim YK, Yoon NY, Kim PH, Yoon HO (2012) Organic acid extraction of fluoride from Antarctic Krill *Euphausia superba*. Fish Aquat Sci 15(3):203–207

Yanase M (1981) Chemical composition of the exoskeleton of Antarctic krill. Bull Tokai Reg Fish Res Lab 83:1–6

Yoshitomi B, Aoki M, Oshima S (2007) Effect of total replacement of dietary fish meal by low fluoride krill (*Euphausia superba*) meal on growth performance of rainbow trout (*Oncorhynchus mykiss*) in fresh water. Aquaculture 266:219–225

Zhang L, Lu X, Wang Z, Qin L, Lin Z, Yuan L, Zhang W, Yin X (2014) Evaluation of the biological toxicity of fluorine in Antarctic krill. Adv Polar Sci 25(1):38–45

Glossary

ACC	Antarctic Circumpolar Current
ADCP	Acoustic Doppler Current Profiler
AFD	ash-free dry weight
AKES	Antarctic Krill and Ecosystem Studies
APF	Antarctic Polar Front
ARK	Association of Responsible Krill Fishing Companies
ASF	Antarctic Slope Front
AUV	Autonomous Underwater Vehicle
BAS	British Antarctic Survey
BIOMAPER-II	Bio-optical Multifrequency Acoustical and Physical Environmental Recorder II
BIOMASS	Biological Investigations of Marine Antarctic Systems and Stocks
bp	base pair
BROKE	Baseline Research on Oceanography, Krill and the Environment
C	carbon
CC	Coastal Current
CCAMLR	Commission for the Conservation of Antarctic Marine Living Resources
cDNA	complementary DNA
CDW	Circumpolar Deep Water
CEMP	CCAMLR Ecosystem Monitoring Program
Chl-a	Chlorophyll a
CIII	calyptopis-III
CM	CCAMLR Conservation Measure
COI	cytochrome c oxidase, mitochondrial subunit 1 gene
CPR	Continuous Plankton Recorder
CPUE	Catch Per Unit Effort
cry	cryptochrome
CS	citrate synthase

V. Siegel (ed.), *Biology and Ecology of Antarctic Krill*, Advances in Polar Ecology,
DOI 10.1007/978-3-319-29279-3

DD	constant darkness
DG	digestive gland
DGGE	denaturing gradient gel electrophoresis
DHA	Docosahexaenoic acid
DI	Differentiation Index
DM	dry mass
DNA	Deoxyribonucleic acid
DR	daily ratio
DVM	diel vertical migration
DW	dry weight
EEZ	Exclusive Economic Zone
ELISA	Enzyme-linked immuno-sorbent assay
ENSO	El Niño Southern Oscillation
EPA	Eicosapentaenoic acid
EPOS	European Polarstern Studies
EST	expressed sequence tag
FA	Fatty acids
FAO	Food and Agriculture Organisation of the United Nations
FIBEX	First International BIOMASS Experiment
FI-VI	furcilia-I-VI
GAMLR	German Antarctic Marine Living Resources Program
GC%	The percentage of GC base pairs in a DNA molecule
German AMLR	German Antarctic Marine Living Resources Program
GI	growth increment
GYM	General Yield Model
HOAD	3-hydroxyacyl-CoA dehydrogenase
HTS	high throughput sequencing
IGR	instantaneous growth rate
IMP	intermoult period
LAKRIS	Lazarev Krill Study
LD	light dark
LFD	length frequency distribution
LIDAR	Light Detection and Ranging
LSSS	Large Scale Survey System
LTER	Long Term Ecosystem Research
MBES	MultiBeam EchoSounder
MDH	malatdehydrogenase
MI	moul interval
mRNA	messenger RNA
MSC	Marine Stewardship Council
mtDNA	mitochondrial DNA
MYA	million years ago

N	nitrogen
ND1	nicotinamide adenine dinucleotide, mitochondrial subunit 1 gene
NGS	Next Generation Sequencing
NNDs	Nearest Neighbour Distances
NOCC	Number of crystal cones
O	oxygen
OA	Ocean Acidification
OAWRS	Ocean Acoustic Waveguide Remote Sensing
p,p′-DDE	p,p′-dichlorodiphenyl dichloroethylene
P/B	Production Biomass ratio
PAR	Photosynthetically active radiation
PCR	Polymerase Chain Reaction
POC	particulate organic carbon
POP	Persistent organic pollutants
PSA	Pacific South America
PUFA	Polyunsaturated fatty acids
qPCR	quantitative polymerase chain reaction
RACE	rapid amplification of cDNA-ends
RAD	Restriction site associated DNA marker
rDNA	Ribosomal deoxyribonucleic acid
RMT	Rectangular Midwater Trawl
RNA	ribonucleic acid
ROV	Remotely Operated Vehicle
SACCF	Southern Antarctic Circumpolar Current Front
SAM	Southern Annular Mode
SBF	Southern Boundary of Antarctic Circumpolar Current Front
SCAR	Scientific Committee on Antarctic Research
SC-CCAMLR	CCAMLR's Scientific Committee
SCUBA	Self-Contained Underwater Breathing Apparatus
SDS	sexual development stage
SDWBA	Stochastic Distorted Wave Born Approximation
Sea WiFS	Sea-viewing Wide Field-of-view Sensor
SEM	scanning electron microscope
SHAPE	Shoal Analysis and Patch Estimation System
SIBEX	Second International BIOMASS Experiment
SIMCO	sea ice microbial community
SLP	Sea Level Pressure
SO-GLOBEC	Southern Ocean Global Ocean Ecosystem Dynamics
SOI	Southern Oscillation Index
SSMU	Small Scale Management Units
SST	Sea Surface Temperature
ST	stomach
SUIT	Surface and Under Ice Trawl

tRNA	transfer ribonucleic acid
UCDW	Upper Circumpolar Deep Water
US-AMLR	US Antarctic Marine Living Resources Program
WAP	Western Antarctic Peninsula
WF	Weddell Front
WG-EMM	CCAMLR's Working Group on Ecosystem Monitoring and Management
WOCE	World Ocean Circulation Experiment
WSC	Weddell Scotia Confluence
WW	wet weight
WWF	Worldwide Fund for Nature

Subject Index

V. Siegel (ed.), *Biology and Ecology of Antarctic Krill*, Advances in Polar Ecology,
DOI 10.1007/978-3-319-29279-3

Genera and Species Index

V. Siegel (ed.), *Biology and Ecology of Antarctic Krill*, Advances in Polar Ecology,
DOI 10.1007/978-3-319-29279-3

MIX
Papier aus verantwortungsvollen Quellen
Paper from responsible sources
FSC® C105338

If you have any concerns about our products,
you can contact us on
ProductSafety@springernature.com

In case Publisher is established outside the EU,
the EU authorized representative is:
Springer Nature Customer Service Center GmbH
Europaplatz 3, 69115 Heidelberg, Germany

Printed by Libri Plureos GmbH
in Hamburg, Germany